Please return or renew this item by the last date shown.
You may renew items (unless they have been requested
by another customer) by telephoning, writing to or calling
in at any library.   100% recycled paper   BKS 1 (5/95)

International Series on the
Science of the Solid State

General Editor: BRIAN PAMPLIN

VOLUME 16

CRYSTAL GROWTH

SECOND EDITION

Other Pergamon Titles of Interest

Other Titles in the International Series on the
Science of the Solid State

Editor: BRIAN PAMPLIN

An important new review journal
Progress in Crystal Growth and Characterization
Editor-in-Chief: BRIAN PAMPLIN
Free specimen copy available on request

NOTICE TO READERS

Dear Reader

If your library is not already a standing order customer or
subscriber to this series, may we recommend that you place a
standing or subscription order to receive immediately upon
publication all new issues and volumes published in this
valuable series. Should you find that these volumes no longer
serve your needs your order can be cancelled at any time
without notice.

The Editors and the Publisher will be glad to receive
suggestions or outlines of suitable titles, reviews or symposia
for consideration for rapid publication in this series.

ROBERT MAXWELL
Publisher at Pergamon Press

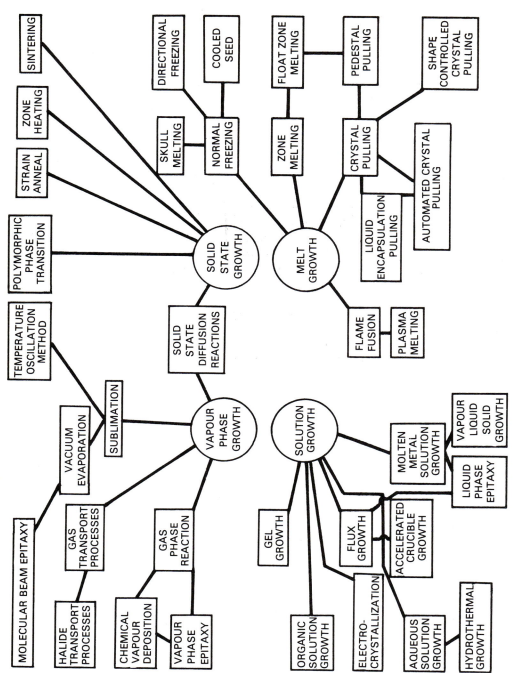

Crystal growth techniques.

CRYSTAL GROWTH

Edited by

BRIAN R. PAMPLIN
School of Physics, University of Bath

SECOND EDITION

PERGAMON PRESS
OXFORD · NEW YORK · TORONTO · SYDNEY · PARIS · FRANKFURT

U.K.	Pergamon Press Ltd., Headington Hill Hall, Oxford OX3 0BW, England
U.S.A.	Pergamon Press Inc., Maxwell House, Fairview Park, Elmsford, New York 10523, U.S.A.
CANADA	Pergamon of Canada, Suite 104, 150 Consumers Road, Willowdale, Ontario M2J 1P9, Canada
AUSTRALIA	Pergamon Press (Aust.) Pty. Ltd., P.O. Box 544, Potts Point, N.S.W. 2011, Australia
FRANCE	Pergamon Press SARL, 24 rue des Ecoles, 75240 Paris, Cedex 05, France
FEDERAL REPUBLIC OF GERMANY	Pergamon Press GmbH, 6242 Kronberg/Taunus, Hammerweg 6, Federal Republic of Germany

First edition 1975

Second edition 1980

British Library Cataloguing in Publication Data

Crystal growth. — 2nd ed. — (International series on the science of the solid state; vol. 16).
1. Crystals — Growth
I. Pamplin, Brian Randall   II. Series
548'.5   QD921   79-42662
ISBN 0-08-025043-2

Printed and bound in Great Britain by
William Clowes (Beccles) Limited, Beccles and London

00 171 153 001

God slumbers in the rock.
He sparkles in the crystals
He breathes in the plant
He dreams in the animal
He awakens in man.

This Indian proverb is inscribed on a plaque outside the Almeda Plaza Hotel in Kansas City. The second line has been added by the Editor because we are now coming to realize the central role ordained for crystals in the evolution of planet Earth.

Preface to the Second Edition

THE changes to this new edition are so extensive that it is almost a completely new book. Yet I reproduce my preface to the first edition because, strangely, I hardly wish to alter a word of that. Indeed, the book itself is still not out of date—it makes a good historical introduction to this 1980 edition. Very little of it has proved inaccurate, but, on the other hand, our subject has marched on to pastures and triumphs new.

For instance, the emphasis on GaAs has moved on in development laboratories to III V alloys by techniques like chemical vapour deposition (CVD) and molecular beam epitaxy (MBE)—both new chapters with new authors. Rather too many pages were devoted to flux growth last time; so now we have two chapters on solution growth—high and low temperatures respectively.

Another completely new feature is hydrodynamics. This is an area which has come to the fore in the ten years since the first edition was conceived. Together with CVD, crystal pulling, and MBE, this is the area I find the most active and promising in crystal growth today. This topic tends to be a high cost area of crystal growth activity, but it is of interest and importance to humbler growers of more mundane crystals. There are, of course, many other examples of progress, and these are reflected in updating of other chapters, particularly Chapter 1.

Chapter 1 includes a massive bibliography of books and source material, and has been completely rewritten with Fig. 1.1, "Classification of Crystal Growth Techniques", as a central key. No scheme of classification can hope to receive universal acclaim as a guide to a diverse subject like crystal growth, but this at least provides a "spider diagram" in which every technique finds its slot. I hope it will summarize for the newcomer the available categories of crystal growth method and indicate the relationships between techniques. I shall be pleased to receive constructive criticism so that it may be improved for the third edition.

One change of emphasis will be noted. Previously authors attempted to list all crystals grown by a given technique. Over the past ten years such lists have become too long for inclusion in a single introductory volume. The number of adamantine semiconducting materials which have been grown by a variety of techniques has increased from a few hundred to several thousand and there are over ten thousand references in this one subfield alone—it is beyond the capacity of one man to review fully.

My grateful thanks go to all authors new and old for their co-operation, patience, and hard work—and particularly for heeding the instruction not to be verbose so that we can hold down the cost of this volume. I would also like to thank various Californian friends, particularly Denis Elwell, for helpful advice while I was planning this volume and the companion review journal during my sabbatical leave at Stanford University, and to Bob Draper of Bath University who compiled the index.

I must also thank the editors, publishers, and printers for their co-operation and good work which has ensured that the final product is worthy of the contributions and published not too long after the writing.

BRIAN PAMPLIN

Scientific Advisers & Co.,
15 *Park Lane,*
Bath BA1 2XH
February 1979

Preface to the First Edition

THIS book was conceived when I was giving an annual optional course of lectures on Crystal Growth for final-year students in physics at Bath University. I felt the need for a single comprehensive source book. During the planning of it some extra topics were included and some possible topics—like a chapter on phase diagrams—had to be omitted. No doubt other crystal growers will feel that it could be better and I hope subsequent volumes will be able to complement this one. It is certainly not possible to include all that a crystal grower must know in one volume.

Much crystal growth is still art and technique rather than science. The conditions for successful growth of good single crystals of a new material cannot be predicted, but a large body of theory, experience and data has now been built up. This volume attempts to review the present position in the art and science of crystal growth. The eighteen authors are experts in their chosen areas and they give the state of the art in each topic.

Good single crystals are essential for a variety of scientific and commercial purposes. They are needed for scientific appraisal of the crystallography, topography and tensor properties of all crystalline material—organic, inorganic and metallic. Silicon and germanium were listed as metals until pure crystals became available just before the Second World War. Modern solid-state electronics is based on a crystal growth revolution which has now made possible the commercial scale growth of large dislocation-free crystals of silicon and GaAs, the production of rods of laser materials like ruby and sapphire, epitaxial growth, and integrated circuit technology and the production of magnetic materials and piezoelectrics like quartz and TGS. Synthetic gem-stones are an important by-product of these advances—but the jewellery trade still prefer the far less perfect natural crystals! Highly perfect crystals of materials like sodium chloride, lithium fluoride and some organic compounds are needed for optical and X-ray spectrometry and use in equipment like electron-probe microanalysis which are in turn used to help study the growth of yet more crystal materials.

The techniques of zone refining, crystal pulling, and epitaxial growth studied intensively for electronic materials are now being applied to new materials. Dendritic growth and the use of whiskers and fibres in composite materials have achieved commercial importance and the future holds great promise. A researcher from a motor manufacturer attending a recent international conference was heard to comment in the bar that one day it might be possible to grow a motor-car—perhaps it will have to be fuelled by hydrogen!

Crystal growth is centuries old. Metals, salt and sugar industries—to name but a few—all use crystal growth and can have their products improved by attention to our subject. Few indeed are the members of the public who realize how great is the debt of our modern technological world to those who studied crystal structure and crystal growth as a pure research activity. Indeed I, for one, feel that we would have a happier world if some of the money that has been so lavishly spent on nuclear physics had been spent on crystal physics.

I would like to thank all my co-authors for their hard work and forbearance and to hope that our readers will accept as inevitable the repetitions and changes in terminology that are bound to occur when a team of co-authors working in different environments pool their contributions into one volume, without time for collaboration. My thanks go also to our publishers and printers for their co-operation and help in this difficult enterprise.

BRIAN PAMPLIN

Scientific Advisers & Co.,
15 *Park Lane, Bath, England*

Contents

3. Hydrodynamics of Crystal Growth Processes   65

L. F. DONAGHEY

CHAPTER 1

Introduction to Crystal Growth Methods

BRIAN R. PAMPLIN
School of Physics, University of Bath, UK

1.1. Main Categories of Crystal Growth Methods

Crystal growth needs the careful control of a phase change. Thus we may define three main categories of crystal growth methods:

Growth from the solid: S → S processes involving solid–solid phase transitions.
Growth from the melt: L → S processes involving liquid–solid phase transitions.
Growth from the vapour: V → S processes involving vapour–solid phase transitions.

We shall introduce a fourth main category which is strictly already included in the above definitions:

Solution growth: growth of solute from an impure melt.

This is done primarily because growth from the melt is such a large and important category and because solution growth methods differ from methods used for pure melt growth. So we have four main categories of crystal growth techniques:

Solid growth, *Vapour growth,* *Melt growth,* and *Solution growth.*

Figure 1.1 shows how these main categories break down into subfamilies of related growth techniques. It is just one of several possible ways of classifying the main growth methods. Since the conception of the first edition of this book, molecular beam epitaxy (MBE) has grown out of vacuum evaporation into a new and exciting method of controlled thin film growth and it may develop into a main category, since it is significantly different from vapour growth in that molecular beam not molecular diffusion is the feed mechanism.

Some familiar proper names have been deliberately omitted from this chart. No disrespect is intended to Czochralski, Stockbarger, Bridgman, and many others. It is felt that descriptive names are more helpful to the newcomer. Some of the many workers to whom credit is due are mentioned in the later text and following chapters as each main method and modification is described. Another reason for avoiding proper names is that there is often much dispute as to who should be commemorated.

1

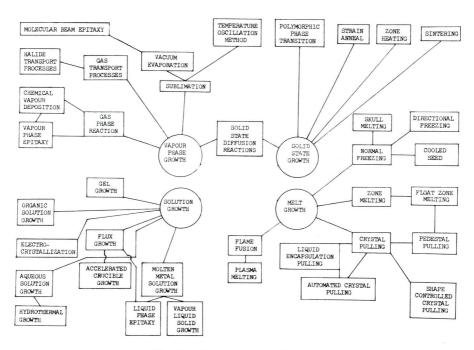

FIG. 1.1. Crystal growth techniques.

1.2. The Chemical Physics of Crystal Growth

If the crystal is in dynamic equilibrium with its parent phase the free energy is at a minimum and no growth will occur. For growth to occur this equilibrium must be disturbed by a change of the correct sign, in temperature, pressure, chemical potential (e.g. saturation), electrochemical potential (e.g. electrolysis), or strain (solid state growth). The system may then release energy to its surroundings to compensate for the decrease in entropy occasioned by the ordering of atoms in the crystal and the evolution of heat of crystallization. In a well-designed growth process just one of these parameters is held minimally away from its equilibrium value to provide a driving force for growth.

Crystal growth then is a non-equilibrium process and thought must be given to the temperature and concentration and other gradients and the fact that heat of crystallization is evolved and must be removed to the surroundings. At the same time the crystal growth process must be kept as near equilibrium and as near to a steady state process as possible. This is why control of the crystal growth environment and a consideration of growth kinetics both at the macroscopic and the atomic levels are of vital importance to the success of a crystal growth experiment. It is particularly important to avoid constitutional supercooling and the breakdown of the crystal–liquor interface that this can cause.

In some growth techniques there is no crystal initially present. Here we meet the nucleation problem which in essence is due to the fact that the surface-to-volume ratio of a small particle is much higher than for a large crystal. Surfaces cost energy because of discontinuities in atomic bonding ("dangling bonds"). Thus the nucleation of a new phase (or a new layer of atoms on an atomically flat surface) is a discontinuous not a quasi-equilibrium process. This is the reason why pure melts supercool (often by many tens of degrees centigrade) and solutions become supersaturated. Thus the growth system departs considerably from equilibrium before a crystal nucleates, and when it comes the new born crystal grows very rapidly at first and is full of defects, some of which propagate on into the later stages of near equilibrium growth. Crystal growers thus seek to use methods where a seed crystal can be introduced into the system to avoid the nucleation problem.

In the last two decades great strides have been made toward achieving crystal perfection motivated by the needs of the electronics and optics industries. While thermodynamics excludes the possibility of growing a perfect crystal, gross defects like grain boundaries, voids, and even dislocations can be eliminated with care and point defects, like impurities, vacancies, interstitials, and antistructure disorder can be minimized by attention to the growth environment and purity of reagents and apparatus. It is for this reason that high vacuum and crucibleless techniques have received much attention.

1.3. Solid Growth Techniques

1.3.1. INTRODUCTION

Solid state growth requires atomic diffusion except in the case of martensitic transformations. At normal temperatures such diffusion is usually very slow except in the case of superionic materials where the small cation is quite mobile. Thus solid state growth techniques are seldom employed when other methods can be used. Figure 1.2, taken from my review article,[1] shows that most of these techniques were known and used before 1900 and, compared with the other growth methods, have not been changed very substantially this century. Solid state growth techniques for the production of single crystals are of very small significance; however, annealing, heat treatment, sintering, and quenching are, of course, metallurgical processes of great importance in tailoring the properties of materials.

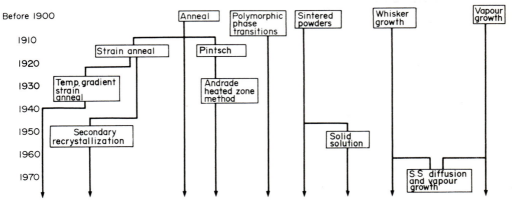

FIG. 1.2. Evolution of solid growth techniques.

I shall devote some space to this subject since no other chapter in this book mentions them.

1.3.2. ANNEALING TECHNIQUES

If a polycrystalline metal rod is held at an elevated temperature below the melting point for many hours some grains may grow at the expense of their neighbours. Since grain boundaries contain more free energy than bulk crystal this process can be seen to lower the free energy of the rod. However, such growth is unreliable and incomplete. Two main techniques were introduced early this century to improve it.

(a) *Strain annealing.* It was found that cold-worked metals showed more grain growth than unstrained samples. Some favoured grains grow at the expense of others— occasionally the whole sample may become a single crystal. Thirteen "laws of grain growth" were enunciated in a book published in 1924;[2] these are quoted in ref. 1. It became clear that a critical strain can induce nucleation and growth of new grains— *secondary recrystallization*. Below the critical strain only normal coarsening or primary crystallization occurs. Above it many new grains nucleate and grow. At the critical strain ideally one grain nucleates and grows to encompass the whole specimen before there is time for any substantial statistical chance of a second nucleation. By this method single crystal aluminium bolts and other shaped crystals have been produced. Temperature gradient strain annealing using a two-zone furnace was successful in causing just one grain to grow down specimens of aluminium, lead, copper, and tin. This has been called the grain boundary migration technique.[3]

Rutter and Aust[4] welded a single crystal of lead to a polycrystalline specimen and used a strain anneal technique to cause this seed crystal to grow down the bar. This neatly avoids the nucleation problem but cannot easily be used for high melting point metals.

(b) *Zone heating.* In the early days of tungsten filament lamps it was found that wires which were single crystals or of a large grain size held up better and suffered less from "filament sag". Bottger[5] describes how finely divided tungsten plus a binder was extruded into a wire and passed through a temperature in excess of 2000°C to produce the

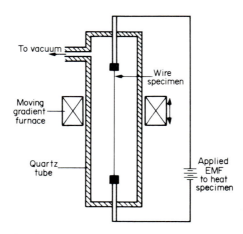

FIG. 1.3. Zone heating method for the production of single crystal wires.

commercial wire sold by the Pintsch Company in Germany early this century.[6] In 1937 Andrade[7] modified this *Pintsch process* to produce single crystal wires of molybdenum and tungsten. This is illustrated in Fig. 1.3. The wire is electrically heated *in vacuo* and a small travelling heater passes a hot zone up and down the wire. Provided the wire is fine enough to have a single grain across its cross-section, the whole length can be grown into a single crystal.

The zone heating method is a forerunner of the techniques of zone melting, zone refining, zone levelling, and float zone refining.

1.3.3. SINTERING AND HOT PRESSING

The annealing of a precompressed powder—*sintering*—and the annealing of a powder under pressure—*hot pressing*—can lead to the homogenization of alloys and to grain growth. Both have been used for many years in metallurgy as tools for studying phase diagrams and as production techniques for producing ingots or shaped components of difficult alloys. Neither offers much scope for the growth of good crystals.

They are complicated processes often encompassing a variety of interrelated phenomena such as recrystallization, solid state diffusion, and crystal growth from solid, liquid, and gas phases as well as changes in porosity, density, and mechanical strength. In the case of mixtures, chemical reactions may also occur. By sintering, intermediate phases—ordered compositions existing some 500° below the solidus—were found in copper–gold alloys Cu_3Au, $CuAu_3$, $CuAu$, and $CuAuII$ (which is a super-superlattice). These phases are formed by careful annealing which permits atomic diffusion to the ordered structure. Such ordering is generally achieved quicker in small particles and grains than in a single crystal.

Sintering was used intensively in the field of pnictide and chalcogenide alloy semiconductors as a method of obtaining homogeneous equilibrium samples of difficult solid solutions of mixed III V and II VI compounds and $III V - III_2 \square VI_3$ and $II VI - III_2 \square VI_3$

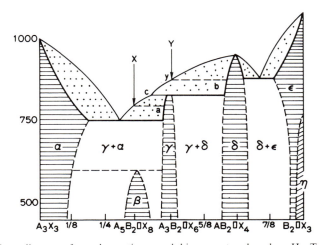

FIG. 1.4. Phase diagram of an adamantine pseudobinary system based on Hg_3Te_3–$In_2\square Te_3$.

alloys in the initial pioneering studies of the 1950s by the groups led by Goryunova[8] and Woolley.[9] Pamplin[10] has worked out the compositions at which ordering is most likely in these adamantine alloys. No ordering in the Ge–Si, III V – III'V', or II VI – II'VI' systems has yet been observed, but it is common when vacancies are plentiful on the cationic sublattice. This is the case in II VI – III$_2 \square$ VI$_3$ alloys illustrated in Fig. 1.4 (HgTe–In$_2 \square$Te$_3$ is an example).

In this hypothetical diagram the phases $\propto$, δ, and ε melt congruently and so crystals of these may be melt grown. The γ phase is a peritectic which may perhaps be grown by a suitable vapour phase or molten metal solution method. The third intermediate phase β can only really be prepared by a solid state annealing or sintering method since it only exists below 600°. The ordered end component phase η is also solid state grown, and it can be grown by a polymorphic phase transition from the zinc blende structure ε phase. Notice that initial freezing of melts of composition X and Y results in solid material of compositions a and b respectively.

Crystals of YIG, BeO, Al$_2$O$_3$, and ZnO grown from sintered powders are mentioned by Aust[11] and Laudise[12] and recently crystals of mm dimensions of a new superionic spinel AgInSnS$_4$ were grown on top of a sintered powder of composition Ag$_3$In$_3$SnS$_8$.[13]

Cycling of the temperature during sintering sometimes called *mineralization* has been used. This is an example of the *Pendelöfen* technique. This is a German word meaning oscillating oven.

1.4. Melt Growth Techniques

1.4.1. INTRODUCTION

Melt growth is undoubtedly the best method for growing large single crystals of high perfection relatively rapidly. It has been used for a great many metals, semiconductors, ionic crystals, and a few organic compounds. Often, especially with semiconductors and laser host crystals, impurities (dopants) can be deliberately added and homogeneously dispersed in a large percentage of the grown crystals. So these techniques have been developed largely in the electronics, optics, and synthetic gemstone industries, and a vast body of detailed expertise has been built up.

However, melt growth normally requires that the material melt congruently (that is it does not decompose below or near its melting point) and has a manageable vapour pressure at its melting point. Thus a great many materials cannot be grown from the melt. These include many hydrated and anhydrous salts, most organic crystals, and virtually all biological materials. Because good ideas and techniques developed in one technique are often carried over and applied to other methods, it is not possible to classify melt growth techniques in an unambiguous way. However, it is convenient here to divide melt growth into four main groups of techniques:

(1) *Normal freezing:* ingot gradually frozen from one end.
(2) *Crystal pulling:* crystal grows on a seed withdrawn from the melt.
(3) *Zone melting:* a molten zone is passed through an ingot.
(4) *Flame fusion or pedestal growth:* crystal grows below a melt which is fed from above.

The supercooling and nucleation problem is eliminated where possible by using a seed crystal. This is usually possible in the latter three groups but is often difficult to achieve in the normal freezing of a boule of melt.

1.4.2. NORMAL FREEZING, DIRECTIONAL FREEZING, OR BRIDGMAN–STOCKBARGER METHOD

The most straightforward and inexpensive melt growth technique is normal freezing—a molten ingot is gradually frozen from one end to the other (Fig. 1.5). When this is achieved by the use of a two-zone furnace, it is called the Bridgman–Stockbarger method. The usual configuration is vertical with the melt in an ampoule being lowered slowly from the hot zone to the cooler zone which is below the melting point. This was the method used in the United States by Bridgman[14] and others[15–18] for the growth of large metal single crystals and later by Stockbarger[19,20] for the growth of optical quality alkali halide crystals for prisms and lenses. It was also developed independently in Europe by Tammann[21] and Obreimov and Schubnikov.[22]

The nucleation problem can be overcome by a variety of devices (Fig. 1.6):

(1) Use of a conical bottom to the crucible so that the coolest region is essentially a point. A cooled gas stream can be directed at this point or a conducting rod (of quartz or copper) attached thereto.

(2) Use of a capillary. The initial rapid growth after nucleation before near-equilibrium conditions are reached takes place in the capillary and (hopefully!) a single seed grows out into the bulk of the melt.

(3) Necking. One or more bulbous extensions are provided below the main ampoule to contain the initial rapid growth and to allow only one seed of fast-growing orientation to pass into the melt.

(4) Melt back. Sometimes the damaging effects of rapid initial growth can be alleviated by raising the temperature (or reversing the pulling) to melt some of the nucleated crystallites.

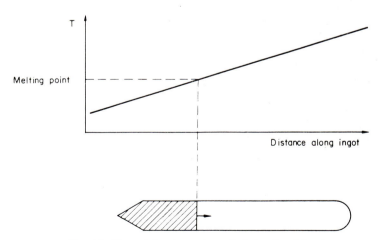

FIG. 1.5. Normal freezing—also called directional freezing.

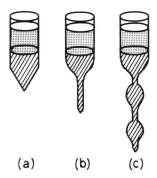

(a)   (b)   (c)

FIG. 1.6. Techniques for nucleating single crystals from the melt: (a) conical bottom, (b) capillary, (c) necking.

If the material expands on freezing and would crack the vertical ampoule or if it is desirable to use a graphite or vitreous carbon boat, horizontal normal freezing is used. This is also the more convenient configuration when the vapour pressure of a component is to be controlled by either liquid encapsulation or having a connection to a source at a fixed temperature (see Fig. 1.9).

There are three ways of moving the freezing interface:

(1) *Moving ampoule:* the melt is drawn through the temperature gradient.
(2) *Moving furnace:* the melt is stationary and the furnace moves.
(3) *Static freeze:* when the temperature gradient is approximately linear as in Fig. 1.5 the temperature of the furnace may be gradually reduced causing the freezing isotherm to run down the ingot. The term directional freezing is sometimes restricted to this technique, which was originated by Stöber.[23]

If, in a normal freeze situation with constant cross-section as in Fig. 1.5, there is present an impurity with segregation constant k and initial concentration C_0, then the composition C of the material freezing when a fraction g of the melt has solidified is given by

$$C = kC_0(1 - g)^{k-1}.$$

This is called the normal freeze equation and applies if there is complete mixing in the liquid phase but no diffusion in the solid phase and k is a constant. Since for most impurities k is appreciably different from unity, the middle region of the crystal is purified. Impurities, like P in GaAs, which have a preference for the solid phase, freeze out early while those, including most dopants in Si, which prefer the liquid phase, are swept to the end. This is the basic principle of zone refining (see below), but it also means that crystals have a graded impurity content especially in the last part to freeze.

Sometimes directional freezing is used to obtain graded solid solution material in a system were complete solid solution occurs. This is illustrated in Fig. 1.7, which applies to a great many pairs of adamantine compounds like InAs–GaAs, to Ge–Si, and to many metal systems like Cu–Au. When a boule of composition C is frozen, the first to freeze has composition D—richer in component B than the liquid. Thus the liquid becomes richer in component A and the composition of the liquid at the freezing interface moves in the direction shown by the arrow on the liquidus at C. The composition of freezing solid is

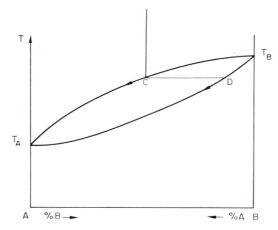

FIG. 1.7. Production of a graded composition ingot. The diagram illustrates complete solid solution phase diagram of A and B.

given by the arrow on the solidus at D. Provided growth is slow enough to avoid constitutional supercooling, the resulting ingot will be graded in composition.

1.4.3. COOLED SEED METHOD

Nacken[24] used a seed crystal attached to copper rod which was cooled by a stream of air in a jacket at the other end to initiate growth in a melt. Kyropoulos[25,26] some 10 years later used a similar cooled seed method for the growth of alkali halides. He dipped an air-cooled platinum tube into the melt and then withdrew it so that just the bottommost seed crystal remained in contact with the melt to nucleate growth.

Adams and Lewis[27] used the cooled seed method to grow large ice crystals and the method has been used commercially for alkali halides.[28] It can be used to grow much fatter crystals than the pulling method and uses simpler apparatus; however, this latter method has achieved far greater popularity and success.

1.4.4. CRYSTAL PULLING

Crystal pulling dates from Czochralski's[29] work on the speed of crystallization of metals published in 1918. Its real importance, however, starts in the early 1950s when Teal and Little[30] at Bell Labs. developed it for the production of pure and doped crystals of germanium and silicon and even grew pn junctions. It has subsequently been used for some III V semiconductors using, when there is a high vapour pressure of arsenic or phosphorus, the liquid encapsulation technique ("LEC pulling").[31] In the early 1960s pulling in air was pioneered for many laser host materials like $CaWO_4$.

Dislocation free crystals have been grown by necking (Fig. 1.8)—a technique originally introduced to produce a single crystal from a polycrystalline seed. Dash[32-34] grew dislocation free Si in the late fifties and since then dislocation free pulled crystals of Ge,[35] GaAs,[36,37] InP,[38] GaP,[39] Al,[40] Cu,[41,42] Ag,[43] and Ni[44] have been reported.

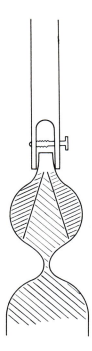

FIG. 1.8. Necking. The illustration shows how a polycrystalline seed crystal may be necked down to yield a single crystal ingot, which will also have a lower dislocation density than the seed. The neck can be repeated more than once if desired.

Because the crystal grows strain free it is difficult to control its shape. Automated automatic diameter control methods have been introduced. Recently[45,46] the use of a die to aid shape-controlled crystal pulling has proved promising. The method has been given the clumsy name "edge-defined film-fed crystal growth" or "EFG pulling" for short.

Since Chapter 7 is devoted to crystal pulling, I shall cut short the discussion of this technique although it is currently the most important method available for the growth of near perfect crystals of large size. Many hundreds of tons of these single crystals are now produced annually.

1.4.5. ZONE MELTING

We have seen, section 1.3.2, that zone heating was a forerunner of zone melting or zone refining. It was developed by Pfann in 1952[47] for germanium and has been well discussed in several books. Because of its importance both as a crystal growth and a purification technique it is the subject of a special chapter (Chapter 8) in this book. It may be classified as an S–L–S process.

Figure 1.9 shows a zone melting situation where there is a volatile component such as As in InAs or GaAs. The tube containing the boat of material for zone refining is kept at a

Fig. 1.9. Zone refining (also called zone melting). The molten zone, heated here by r.f. induction, is moved along the ingot. The separate boat at a lower temperature contains a volatile component and the temperatures are adjusted to prevent any net transport to or from the molten zone.

suitable ambient temperature and a movable heater melts a molten zone which is passed down the ingot. A seed crystal can be introduced at the starting end if single crystal growth is desired. A second boat containing the volatile component is kept at just the right temperature to ensure no net loss of that component from the molten zone.

Float zone refining[48−56] is used with a silicon rod vertical and the zone free from contact with any crucible material to produce the purest single crystal known to man. It is a vital stage in the production of substrates for integrated circuits.

Sometimes the growing crystal above the molten zone and the feed ingot below it are rotated and pushed or pulled independently. If a fat source ingot is pushed slowly into the molten zone and a slimmer crystal is pulled out, the process is sometimes called "pedestal pulling" or "crystal pushing" or "differential pulling". Float zone refining can be thought of as a crucibleless crystal pulling method.

1.4.6. FLAME FUSION TECHNIQUES

Like float zone refining, just mentioned, the Verneuil method has the great advantage of being crucibleless. It was originated before the turn of the century.[57,58] A modern version is pictured in Fig. 4.8. The essential features are a seed crystal the top of which is molten and is fed with molten drops of source material coming usually as a powder through a flame or plasma.

The original interest was centred on the synthesis of sapphires and rubies—gemstones based on alumina, corundum, Al_2O_3—and often the work was clandestine because jewellers much preferred natural stones though these were less perfect. Michel's book[59] gives details of improvements to Verneuil's technique and the dopants needed. For example, chromic oxide must be added to the alumina powder for rubies plus a little iron to imitate the rubies from Thailand. Iron and titanium oxides are both required for sapphires. Many other natural gemstones can be imitated.

Verneuil and his successors used a flame of hydrogen burning in oxygen to melt the falling powder—the technique is still often called "flame fusion". Burner design was improved by Merker[60] who introduced the tricone in which hydrogen passes down the inner tube and two outer tubes carry oxygen. Essentially the same technique is still used to produce rubies and sapphires for watch bearings and similar applications, although crystal pulling is now the preferred method for oxide crystals of high quality. Other methods of heating include arc imaging, plasma heating, and electron beam heating.

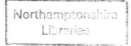

1.4.7. ARC FUSION TECHNIQUES

Drabble and Palmer[61] have described an arc method which can be classified as a variant of the flame fusion technique. The arc is struck between a nickel cathode and a NiO anode the top of which is kept molten by the heat of the arc. Nickel is transported from cathode to anode oxidizing as it comes. The method is restricted to a few oxides (Co, Fe, Ti, U) and ferrites.

High melting point oxides such as MgO, CaO, SrO, ZrO_2, and BaO may be grown as single crystals inside a charge of their own powders. The arc electrodes are buried in the charge which melts near their tips. When this melt is allowed to freeze slowly, large single crystal volumes are frequently found.[62−64] This is a crucibleless normal freezing method.

1.5. Solution Growth Methods

Melt growth often merges in practice almost imperceptibly into solution growth. If one considers the segregation (rejection of impurity atoms), which commonly occurs in melt growth, it is evident that the growing crystal is in equilibrium not with pure melt but with a solution or liquid alloy. However, the distinction between the two groups of techniques is usually clear enough: in melt growth the solvent (major component) freezes whereas in solution growth it is the solute which crystallizes usually well below its melting point (assuming it even has one!).

In this book we have a chapter on Low Temperature Solution Growth (Chapter 10) and another on High Temperature Solution Growth (Chapter 12). This conveniently divides the topic into two roughly equal if arbitrary parts, but see also Bulk Crystallization (Chapter 14). In Fig. 1.1 I have divided it up basically by the nature of the solvent. This is justified by the fact that generally speaking crystals are only grown from one type of solvent—aqueous, organic, flux (usually an ionic compound), or metal. Then electrocrystallization and gel growth seem to need separate classifications, but it is all rather arbitrary.

There is yet another useful classification. Apart from the two last-named categories, solution growth can be classified by the method of obtaining supersaturation into three groups of methods:

  (a) *Temperature change:* cooling (or in rare cases, heating) the solution.
  (b) *Solvent extraction:* usually by evaporation (in rare cases solvent addition is required, e.g. ref 65).
  (c) *Circulation:* a two-temperature system in which the solvent passes nutrient from source to seed. Unlike the other two this method may often be made a more or less continuous process as opposed to a batch process, and is thus favoured in industry.

Solution growth methods are at least an order of magnitude slower than melt growth generally speaking and, of course, generally less pure. However, for many commercial materials like salt, sugar, some fertilizers, hydrated materials, and some compounds that decompose before melting, it is the only viable process. Also many high purity, high perfection materials for the electronics and optical industries are produced from solution. KDP and ADP are grown from aqueous solutions, quartz hydrothermally, magnetic bubble domain materials from fluxes, and III V alloys from molten metals.

Aqueous solution growth has produced the largest crystals known to man. Buckley[66] mentions that Bounds has produced the world's biggest artificial crystals, e.g. an octahedron of alum weighing 240 lb. This crystal, I have been told, took over three years to grow and was turned over in its bath daily to ensure reasonable symmetry of its growth habit.

With other solvents, fluxes and metals particularly, liquid phase epitaxy (LPE) is a technique of major importance. Here, by contrast, a layer only a few microns thick is grown on to the parent crystal slice. Epitaxial growth of III V compounds and alloys is an important production tool for devices like light-emitting diodes (LEDs).[67] It is discussed in Chapter 11.

1.6. Vapour Phase Growth

Sublimation, e.g. of sulphur, was practised by the alchemists and subsequent chemists more as a purification process than a crystal growth method. This is the simplest and the only pure vapour growth method. It has been used in recent decades for the production of high quality bulk crystals of materials like CdS and HgI_2 in which all elements present are volatile at readily attainable temperatures. It has also led (as Fig. 1.1 indicates) to vacuum evaporation and the exciting new technique of MBE (molecular beam epitaxy) which, because of its growing importance for integrated optics and special semiconductor devices, now deserves a chapter to itself. MBE is also the subject of a book.[77]

Piper and Polich[68] and subsequently others (e.g. ref. 69) have used elaborate sublimation methods for CdS. Figure 1.10 illustrates the principles. ZnS, CdI_2, and other volatile materials may be grown as large single crystals by careful use of this method of which there are several variants, but, as indicated above, it is not of wide application. It also suffers from the disadvantage of being a closed tube method.

Also developed during sublimating was the temperature oscillation method (TOM) pioneered by Schöltz[70] in Germany; this *Pendelöfen* technique has been applied

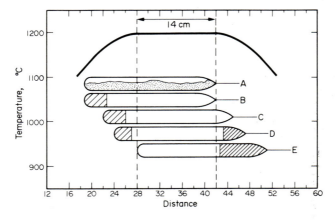

FIG. 1.10. Growth by sublimation of CdS. Stages A: tube filled with purified cadmium and sulphur which are reacted and driven, B, to the cold end. In C, D, E a CdS crystal is nucleated and grown at the pointed end of the tube by motion of the tube through the hot zone (rather like zone melting).

particularly to HgI_2. It has two variants, either periodic oscillation of the source temperature (POST), or periodic oscillation of the crystal temperature (POCT). The periodic temperature oscillations are designed ideally to cause and keep a single nucleation. Minor secondary nucleations evaporate during the heating cycle. Schöltz has tried linear and radial temperature gradients. There is currently considerable discussion of the merits of temperature oscillation and "melt back" in crystal growth generally, although current practice in most growth methods favours as steady a temperature as possible.   Contrasting with these physical vapour deposition methods are various chemical methods (see Stringfellow's discussion (Chapter 5) and Fig. 5.1, p. 182). This, which has been called "impure vapour growth", is to sublimation as solution growth is to melt growth. Other chemical elements are present besides those in the wanted crystal. In Fig. 1.1 these are divided (rather arbitrarily) into halogen transport methods and gas phase reactions. Another useful division is into open tube and closed tube methods, usually sublimation and halogen transport are done in closed tubes and CVD in open tubes, which has an obvious commercial advantage.

Chemical transport methods are widely used in the study of new semiconducting, insulating, and magnetic crystals. They were pioneered by Schäfer[71] and Nitsche[72] and are widely used in closed tube arrangements, and are discussed by Kaldis.[73] For example, small crystals of most adamantine compounds may be grown in sealed quartz ampoules when about 5 mg cc^{-1} of I_2 are added to a powdered charge of the compound in vacuum. Crystals grow at the cold end of the tube (Fig. 1.11). Crystals of $CuGaSnSe_4$ and Cu_2GeSe_3 have recently been grown in this apparatus with *Pendelöfen* of the crystals, showing that it is possible to have simultaneous transport of three non-volatile metals.[74] The function of the iodine is to transport the non-volatile metal atoms by formation of a volatile iodide at the low temperature end and its decomposition in the higher temperature growth zone of a directional or a two-zone furnace. The tube should be tilted by some $10°$ to enhance convection, which is the principal feed mechanism. Most III V, II VI, I III VI$_2$ and II IV V$_2$ compounds can be grown this way—among many other examples.[75]

Solid solution material may also be grown as in our work and that of Yamamoto. Iodine is not the only transporting agent used for non-volatile metals. These elements

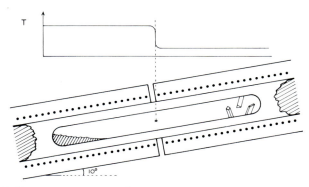

FIG. 1.11. Iodine vapour transport of compounds containing non-volatile metals. The source material at the higher temperature is transported to the cold zone where *Pendelöfen* temperature oscillations can limit the number of crystals that nucleate.

usually have volatile chlorides, fluorides, and bromides too. HCl and Cl_2 were used recently to prepare single crystal samples of $U_{1-x}Th_xO_2$ solid solution.[76]

These closed tube methods are very useful in research but too costly for production where gas phase reactions, open tube methods, are preferred. Several, like the silane process for epitaxial silicon or the gas phase reaction between $GaCl_3$ and As_4 to give epitaxial GaAs, are of considerable commercial importance and are discussed in Chapter 5. A recent book[78] describes the production of quaternary IIIV alloys for advanced electronic and optoelectronic applications.

1.7. Choosing a Crystal Growth Method

For bulk growth of high quality single crystal material seeded melt growth (e.g. crystal pulling or float zone melting) is undoubtedly the best method available today for congruently melting materials. As in the cases of Si, GaAs, and GGG it is fast, efficient, and can be automated, and it produces the most perfect crystals possible. The use of microgravity conditions of space processing are exciting for the future, since the elimination of gravity induced problems could compensate for the increased expense—especially if the products are to be used in space.

When a material is wanted in thin film form with accurately controlled doping, composition, and quality on available substrate material as in the semiconductor and photonics industries (e.g. III V alloys[74]), vapour deposition, especially CVD or MBE, is an appropriate method, although sometimes solution growth is preferred.

In other cases solution growth is the next best choice—using a seed crystal if possible. Failing this chemical transport is often a good method, particularly if *Pendelöfen* is used to limit nucleations. This is a particularly useful method of obtaining small crystals of new materials for scientific research. Sometimes, as with biological crystal growth,[79] and gel growth,[80] the conditions cannot readily be chosen—they are fixed by the system.

It is the aim of this book to introduce and provide reference material for the major crystal growth techniques and industries of the 1980s. We also try and look ahead from the present emphasis in melt, solution, and vapour growth to new areas like space processing and MBE. Our subject is evolving in an exciting way and is at the very basis of modern technological advance.

1.8. The Literature of Crystal Growth

There are a very large number of published articles, books, and proceedings of schools and conferences on the various aspects of crystal growth theory and practice. There are two journals devoted to the subject: *The Journal of Crystal Growth* and the review article journal *Progress in Crystal Growth and Characterization*. Several other journals, like *Materials Research Bulletin*, carry a large percentage of articles on the subject, and many journals in physics, chemistry, materials science, and electrical engineering carry occasional articles on crystal growth or which contain some relevant material.

This literature is scattered and difficult to collate, so books like this are valuable introductions to the subject. At the end of this chapter I have listed by year of publication most of the key works on crystal growth and related topics. There are five headings: A,

Crystal Growth; B, Phase Diagrams; C, Crystal Structure and Defects; D, Conference Proceedings; E, Other Sources of Information.

It is difficult to choose between these books, but the following are specially recommended to the beginner starting a serious study of crystal growth. General books to read or dip into: Buckley,[A1] Gilman,[A16] Laudise,[A40] Hartman.[A46] For melt growth Brice[A48] and for zone refining Pfann's[A28] classic work should be consulted. Schäfer's[A22] book is still a good guide to vapour transport, and Henisch[A39] is the only one on gel growth. As an introduction to phase diagrams, which are so important to crystal growers, Prince's[B15] book is specially helpful. Kröger's[C35] *Chemistry of Imperfect Crystals* should be on the reading list, and for X-ray topography Tanner's[C36] recent book is recommended. Parthé's book[C34] is recommended for those interested in tetrahedral compounds, and Shay and Wernick[A57] for the specialized chalcopyrite compounds. Elwell and Scheel[A52] is a first-rate introduction to flux growth.

Conference proceedings give a picture of the changing emphasis in crystal growth activity. The first serious conference was held in Bristol in 1949[D1] and metal crystals featured large. Dislocations and whiskers were the main themes of the Cooperstown Conference in 1958.[D2] By the start of the series of International Conferences on Crystal Growth, ICCG-1, Boston, 1966[D5] to ICCG-5, Boston again,[D15] semiconductors had taken the limelight and electronic materials generally predominated though not to the exclusion of such diverse topics as snow flakes, eutectics, biological crystals, and polymers. The next conference in this series, ICCG-6, is being held in Moscow in 1980.

In section E I have listed the information services which exist to help those who wish to grow or obtain crystals. Several diligent crystal growers have compiled lists of available crystals. I list the addresses of the originators of the five such lists which I know to exist.

Acknowledgements

This chapter was rewritten while the author was on sabbatical leave at the Center for Materials Research, Stanford University, and he would like to thank NATO for a senior research fellowship and his hosts and Bath University for an exciting year in "Silicon Valley".

References

1. PAMPLIN, B. R. (1977) *Progress in Crystal Growth and Characterization* **1**, 5.
2. JEFFRIES, Z. and ARCHER, R. S. (1924) *The Science of Metals*, McGraw-Hill, New York.
3. AUST, K. T. (1963) in A16 below.
4. RUTTER, J. W. and AUST, K. T. (1961) *Trans. Met. Soc. AIME* **221**, 641.
5. BOTTGER, W. (1917) *Z. Electrochem.* **23**, 121.
6. ALTERTHUM, H. (1924) *Z. Phy. Chem.* **110**, 1.
7. ANDRADE, E. dA C. (1937) *Proc. Roy. Soc.* A **163**, 16.
8. GORYUNOVA, N. A. (1961) *The Chemistry of Diamond-like Semiconductors*, Chapman & Hall, London.
9. WOOLLEY, J. C. in A15 below.
10. PAMPLIN, B. R. (1960) *Nature (London)* **188**, 136.
11. AUST, K. T. (1963) in A16 below.
12. LAUDISE, R. A. (1970) in A41 below.
13. PAMPLIN, B. R., OHACHI, T., MAEDA, S., NEGRETE, P., ELWORTHY, T. P., SANDERSON, R., and WHITLOW, H. J. (1977) *Proc. Third Conf. on Ternary Semiconductors*, Institute of Physics, London Conference Series No. 35, pp. 21 and 35.

14. BRIDGMAN, P. W. (1925) *Proc. Am. Acad. Arts Scs* **60**, 303.
15. DAVEY, W. P. (1925) *Phys. Rev.* **25**, 248.
16. ELAM, C. F. (1926) *Proc. Roy. Soc.* A **112**, 289.
17. DILLON, J. H. (1930) *Rev. Sci. Instr.* **1**, 36.
18. QUIMBY, S. L. (1932) *Phys. Rev.* **39**, 345.
19. STOCKBARGER, D. C. (1936) *Rev. Sci. Instr.* **7**, 133.
20. STOCKBARGER, D. C. (1949) *Discuss. Faraday Soc.* **5**, 294, 299.
21. TAMMANN, G. (1925) *Metallography*, translated Dean and Swenson, Chemical Catalogue Co., New York, p. 26.
22. OBREIMOV, J., and SCHUBNIKOV, L. (1924) *Z. Physik* **25**, 31.
23. STÖBER, F. (1925) *Z. Krist.* **61**, 299.
24. NACKEN, R. (1915) *Neuer Jahrb. Mineral Geol.* **2**, 133.
25. KYROPOULOS, S. (1926) *Z. Anorg. Chem.* **154**, 308.
26. KYROPOULOS, S. (1930) *Z. Physik* **63**, 849.
27. ADAMS, J. M. and LEWIS, W. (1934) *Rev. Sci. Instr.* **5**, 400.
28. MENZIES, A. C. and SKINNER, J. (1949) *Discuss. Faraday Soc.* **5**, 306.
29. CZOCHRALSKI, J. (1918) *Z. Phys. Chem.* **92**, 219.
30. TEAL, G. K. and LITTLE, J. B. (1950) *Phys. Rev.* **78**, 647.
31. METZ, E. P. A., MILLER, R. C., and MAZELSKI, J. (1962) *J. Appl. Phys.* **33**, 2016.
32. DASH, W. C. (1958) in *Growth and Perfection of Crystals* (ed. R. H. Doremus, B. W. Roberts and D. Turnbull), Wiley, New York.
33. DASH, W. C. (1959) *J. Appl. Phys.* **30**, 459.
34. DASH, W. C. (1960) *J. Appl. Phys.* **31**, 736 and 2275.
35. OKKERSE, B. (1959) *Philips Tech. Rev.* **21**, 340.
36. STEINEMANN, A. and ZIMMERLI, V. (1967) *J. Phy. Chem. Solids* **28**, Supp. 1, 81.
37. GRABMAIER, B. C. and GRABMAIER, J. G. (1972) *J. Cryst. Growth* **13, 14**, 635.
38. SEKI, Y., MATSUI, I. and WATANABE, H. (1976) *J. Appl. Phys.* **47**, 3374.
39. ROKSNOER, P. J., HUITBREGTS, J. M. P. L., VAN DE WIJGERT, W. M., and DE KOCK, A. J. R. (1977) *J. Cryst. Growth* **40**, 6.
40. HOWE, S. and ELBAUM, C. (1961) *Phil. Mag.* **6**, 1227.
41. FEHMER, H. and WELHOFF, W. (1972) *J. Cryst. Growth* **13, 14**, 257.
42. BUCKLEY-GOLDER, I. (1977) *J. Cryst. Growth* **40**, 189.
43. TANNER, B. K. (1973) *Z. Naturf.* **28A**, 676.
44. NARAMOTO, H. and KAMADA, K. (1974) *J. Cryst. Growth* **24, 25**, 531 and **30**, 145.
45. CHALMERS, B., LA BELLE, H. E., and MLAVSKY, A. I. (1972) *J. Cryst. Growth* **13, 14**, 84.
46. STEPANOV, A. V. (1969) *Bull. Acad. Sci. USSR* **33**, 1775.
47. PFANN, W. G. (1952) *Trans. AIME* **194**, 747.
48. THEURER, H. C. (1952) US patent 3060123.
49. KECK, P. H. and GOLAY, M. J. E. (1953) *Phys. Rev.* **89**, 1297.
50. KECK, P. H. and VAN HORN, W. (1953) *Phys. Rev.* **91**, 512.
51. KECK, P. H., GREEN, M., and POLK, M. L. (1953) *J. Appl. Phys.* **24**, 1479.
52. KECK, P. H., LEVIN, S. B., BRODER, J., and LIEBERMAN, R. (1954) *Rev. Sci. Instr.* **25**, 298.
53. KECK, P. H., VAN HORN, W., SOLED, J., and MACDONALD, A. (1954) *Rev. Sci. Instr.* **25**, 331.
54. KECK, P. H. (1954) *Physica* **20**, 1059.
55. EMEIS, R. (1954) *Z. Naturf.* **9A**, 67.
56. MULLER, S. (1954) *Z. Naturf.* **9B**, 504.
57. VERNEUIL, A. (1902) *Compt. Rendue* **135**, 791.
58. VERNEUIL, A. (1904) *Am. Chim. Phys.* **8**, ser. 3, 20.
59. MICHEL, H. (1926) *Die Kunstliche Edelsteine*, 2nd edn., W. Diebener, Leipzig.
60. MERKER, L. (1947) *Fiat Tept.* No. 1001.
61. DRABBLE, J. R. and PALMER, A. W. (1966) *J. Appl. Phys.* **37**, 1978.
62. RABENAU, A. (1964) *Chemie–Ingenieur–Tech.* **36**, 542.
63. SCHUPP, L. J. (1968) *Electrochem. Tech.* **6**, 219.
64. BUTLER, C. T., STURM, B. J., and QUINCY, R. A. (1971) *J. Cryst. Growth* **8**, 197.
65. HILLS, M. E. (1970) *J. Cryst. Growth* **7**, 257.
66. BUCKLEY, H. (1951) see A1 below, p. 64.
67. WILLIAMS, E. W. and HALL, R. (1978) *Luminescence and the Light Emitting Diode*, Pergamon Press, Oxford.
68. PIPER, W. W. and POLICH, S. J. (1961) *J. Appl. Phys.* **32**, 1278.
69. SHARMA, S. O. and MALHOTRA, L. K. (1971) *J. Cryst. Growth* **10**, 199.
70. SCHÖLTZ, H. (1967) in *Crystal Growth* (ed. H. S. Peiser), Pergamon Press, Oxford.
71. SCHÄFER, H. (1964) see A22 below.
72. NITSCHE, R. (1967) *Fortschr. Mineral.* **44**, 231.

73. KALDIS, E. (1974) in A51 below.
74. PAMPLIN, B. R. and LOPEZ, A. S. (1980) *J. Cryst. Growth*, to be published.
75. PAORICI, C., ZANNOTTI, L., and ZUCCALLI, G. (1978) *J. Cryst. Growth* **43**, 705.
76. KAMEGASHIRA, N., OHTA, K., and NAITO, K. (1978) *J. Cryst. Growth* **44**, 1.
77. PAMPLIN, B. R. (ed.) (1980) Special Issue of *Progress in Crystal Growth and Characterization* on *Molecular Beam Epitaxy*.
78. PAMPLIN, B. R. (ed.) (1979) Special Issue on Quaternary III V Alloys, ibid.
79. PAMPLIN, B. R. (ed.) (1980) Special Issue on Biological Crystal Growth, ibid.
80. HENISCH, H. K. (1970) see A39 below.

TOPICS RELATED TO CRYSTAL GROWTH

A. *Crystal Growth*

1. BUCKLEY, H. (1951) *Crystal Growth*, Wiley, New York.
2. VERMA, A. R. (1953) *Crystal Growth and Dislocations*, Butterworths, London.
3. HOLLAND, L. (1956) *Vacuum Deposition of Thin Films*, Chapman & Hall, London.
4. RUDDLE, R. W. (1957) *Solidification of Castings*, Inst. of Metals.
5. JACKSON, K. A. (1958) *Liquid Metals and Solidification*, Am. Soc. for Metals.
6. LAWSON, W. D. and NIELSON, S. (1958) *Preparation of Single Crystals*, Butterworths, London.
7. HOLDEN, A. and SINGER, P. (1960) *Crystals and Crystal Growing*, Anchor-Doubleday, New York.
8. PARR, N. L. (1960) *Zone Refining and Allied Techniques*, Newnes.
9. GRUBEL, R. O. (ed.) (1961) *Metallurgy of Elemental and Compound Semiconductors*, Interscience, New York.
10. MULLIN, J. W. (1961) *Crystallization*, Butterworths (2nd edition just published).
11. ROLSTEN, R. F. (1961) *Iodide Metals and Metals Iodides*, Wiley, New York.
12. VAN HOOK, A. (1961) *Crystallization, Theory and Practice*, Reinhold, New York.
13. HURLE, D. T. J. (1962) *Mechanisms of Growth of Metal Single Crystals from the Melt*, Prog. in Mat. Sci., Vol. 10, Pergamon Press, Oxford.
14. SMAKULA, A. (1962) *Einkristalle*, Springer-Verlag.
15. WILLARDSON, R. K. and GOERING, H. L. (1962) *Compound Semiconductors*, Vol. I, *Preparation of III–V Compounds*, Reinhold, New York.
16. GILMAN, J. J. (ed.) (1963) *The Art and Science of Growing Crystals*, Wiley, New York.
17. HIRTH, J. P., POUND, G. M. (1963) *Condensation and Evaporation*, Macmillan, London.
18. HIRTH, J. P. and POUND, G. M. (1963) *Nucleation in Condensation and Evaporation*, Prog. in Mat. Sci., Vol. 2, Pergamon Press, Oxford.
19. CHALMERS, B. (1964) *Principles of Solidification*, Wiley, New York.
20. HOLDEN, A. and THOMPSON, R. H. (1964) *Growing Crystals with a Rotary Crystallizer*, Bell Laboratories, New York.
21. RUTTER, E., GOLDFINGER, P., and HIRTH, J. P. (eds.) (1964) *Condensation and Evaporation of Solids*, Gordon & Breach, New York.
22. SCHÄFER, H. (1964) *Chemical Transport Reactions*, Academic Press, New York.
23. WINEGARD, W. C. (1964) *An Introduction to the Solidification of Metals*, Inst. of Metals, London.
24. BYRNE, J. G. (1965) *Recovery, Recrystallization and Grain Growth*, Macmillan, London.
25. BAMWORTH, A. W. (1965) *Industrial Crystallization*, Leonard Hill, London.
26. HIRTH, J. P. and POUND, G. M. (1965) *Crystal Growth*, Prog. in Mat. Sci., Vol. 2, Pergamon Press, Oxford.
27. BRICE, J. C. (1965) *The Growth of Crystals from the Melt*, North-Holland, Amsterdam.
28. PFANN, W. G. (1966) *Zone Melting*, 2nd edn., Wiley, New York.
29. SCHILDKNECHT, H. (1966) *Zone Melting*, Academic Press, New York.
30. BORKIS, J. O. and RAZUMNEY, G. A. (1967) *Fundamental Aspects of Electro-Crystallization*, Plenum Press, New York.
31. KNIGHT, C. A. (1967) *The Freezing of Supercooled Liquias*, Van Nostrand.
32. KOZLOVA, O. G. (1967) *Rost Kristallov* (in Russian), Moscow 12d, Moscow University.
33. POWEL, C. F., OXLEY, J. H., and BLOCKER, J. M. (1967) *Vapour Deposition*, Wiley, New York.
34. ZETTLEMOYER, A. C. (ed.) (1967) *Nucleation*, Dekker, New York.
35. OUSIENKO, D. E. (ed.) (1968) *Growth and Imperfections of Metallic Crystals*, Consultants Bureau, New York.
36. STRICKLAND-CONSTABLE, R. F. (1968) *Kinetics and Mechanism of Crystallization*, Academic Press, New York.
37. PETROV, T. G., TREWUS, E. B., and KASATKIN, A. P. (1969) *Growing Crystals from Solution*, Consultants Bureau, New York.
38. SITTIG, M. (1969) *Semiconductor Crystal Manufacture*, Noyes Development.
39. HENISCH, H. K. (1970) *Crystal Growth in Gels*, Penn. State Univ. Press.

40. LAUDISE, R. A. (1970) *The Growth of Single Crystals*, Prentice-Hall, Englewood Cliffs, N.J.
41. LEFEVER, R. A. (ed.) (1971) *Aspects of Crystal Growth*, Dekker, New York.
42. CONNOLLY, T. F. (ed.) (1972) *Semiconductors: Preparation, Crystal Growth and Properties*, Plenum Press, New York.
43. MULLIN, J. W. (1972) *Crystallization*, 2nd edn., Butterworths, London.
44. EVANS, C. C. (1972) *Whiskers*, Mills & Boon, London.
45. TARJÁN, I. and MÁTRAI, M. (eds.) (1972) *Laboratory Manual on Crystal Growth*, Akadémiai Kiadó, Budapest.
46. HARTMAN, P. (1973) *Crystal Growth: An Introduction*, Elsevier, New York.
47. LOBACHEV, A. N. (ed.) (1973) *Hydrothermal Synthesis of Crystals*, Plenum Press, New York.
48. BRICE, J. C. (1973) *The Growth of Crystals from Liquids*, North-Holland–Elsevier, New York.
49. WILKE, K. TH. (1973) *Kristallzüchtung*, VEB Deutscher Verlag der Wissenschaften, East Berlin.
50. FLEMINGS, M. C. (1974) *Solidification Processing*, McGraw-Hill, New York.
51. GOODMAN, C. H. L. (ed.) (1972) *Crystal Growth—Theory and Techniques*, Vol. I, Plenum Press, London.
52. ELWELL, D. and SCHEEL, H. J. (1975) *Growth of Crystals from High-temperature Solutions*, Academic Press, New York.
53. HANNAY, N. B. (ed.) (1975) *Treatise on Solid State Chemistry*, Vol. 5, *Changes of State*, Plenum Press, New York.
54. MATHEWS, J. W. (ed.) (1975) *Epitaxial Growth*, Academic Press, New York.
55. UEDA, R. and MULLIN, J. B. (1975) *Proc. 2nd Int. School on Crystal Growth*, North-Holland, Amsterdam.
56. PAMPLIN, B. R. (ed.) (1975) *Crystal Growth*, Pergamon Press, Oxford.
57. SHAY, J. L. and WERNICK, J. H. (1975) *Ternary Chalcopyrite Semiconductors—Growth, Electronic Properties and Applications*, Pergamon Press, Oxford.
58. LEFEVER, R. A. and WILCOX, W. R. (eds.) (1976) *Preparation and Properties of Solid State Materials*.
59. GOODMAN, C. H. L. (ed.) (1978) *Crystal Growth Theory and Techniques*, Vol. 2, Plenum Press, New York.
60. PAMPLIN, B. R. (ed.) (1980) *Molecular Beam Epitaxy*, Pergamon Press, Oxford.

B. *Phase Diagrams*

1. GIBBS, J. W. (1928) *The Collected Works of J. Willard Gibbs*, Longmans, Green.
2. VOLMER, M. (1939) *Kinetik der Phasenbildung*, Steinkopff, Leipzig.
3. MARSH, J. S. (1944) *Principles of Phase Diagrams*, translated by B. A. Rogers, McGraw-Hill, New York.
4. MASING, G. (1944) *Ternary Systems*, Reinhold, New York.
5. FRENKEL, J. (1946) *Kinetic Theory of Liquids*, OUP, Oxford.
6. RICCI, J. E. (1951) *The Phase Rule and Heterogeneous Equilibrium*, Van Nostrand, Amsterdam.
7. —— (1948) *Metals Handbook*, Am. Soc. for Metals.
8. HUME-ROTHERY, W., CHRISTIAN, J. W., and PEARSON, W. B. (1952) *Metallurgical Equilibrium Diagram*, IPPS.
9. COTTRELL, A. H. (1955) *Theoretical Structural Metallurgy*, Arnold, 2nd edn.
10. LEVIN, E., MCMURDIE, H. F., and HALL, F. P., *Phase Diagrams for Ceramists*, Am. Ceram. Soc., Vol. 1, 1964; Vol. 2, 1969.
11. RHINES, F. N. (1956) *Phase Diagrams in Metallurgy*, McGraw-Hill, New York.
12. TEMPERLEY, H. N. V. (1956) *Changes of State*, Cleaver-Hume, London.
13. HANSEN, M. and ANDERKO, K. (1958) *Constitution of Binary Alloys*, McGraw-Hill, New York.
14. VOGEL, R. (1959) *Die heterogenen Gleichgewichte*, Akad. Verlag Geest Portig, Leipzig.
15. PRINCE, A. (1965) *Alloy Phase Equilibria*, Elsevier, Holland; Cleaver-Hume, London.
16. UBBELOHDE, A. R. (1965) *Melting and Crystal Structure*, OUP, London.
17. WEST, D. R. F. (1965) *Ternary Equilibrium Diagrams*, Macmillan, New York.
18. FERGUSON, F. D. and JONES, T. K. (1966) *The Phase Rule*, Butterworths, London.
19. GORDON, P. (1968) *Principles of Phase Diagrams in Material Systems*, McGraw-Hill, New York.
20. ALPER, A. M. (ed.) (1970) *Phase Diagrams: Materials Science and Technology*, 3 Vols., Academic Press, New York.
21. KAUFMAN, L. and BERNSTEIN, H. (1970) *Computer Calculation of Phase Diagrams, with special reference to refractory metals*, London, Academic Press. (Refractory Materials Series, Vol. 4.)
22. TAMÁS, F. and PÁL, I. (1970) *Phase Equilibria Spatial Diagrams: phase diagrams, their interpretation and anaglyph representation*, Iliffe, New York.
23. EHLERS, E. G. (1972) *The Interpretation of Geological Phase Diagrams*, W. H. Freeman, San Francisco.
24. LEVIN, E. M., ROBBINS, C. R., and MCMURDIE, H. F. (1964 and 1969) Supplementary volumes to *Phase Diagrams for Ceramists*, Am. Ceram. Soc.
25. MOFFATT, W. G. (1976) *Binary Phase Diagrams Handbook*, General Electric Co., Schenectady.

C. *Crystal Structure and Defects*

1. BRAGG, W. H. and W. L. (1933) *The Crystalline State*, Bell, London.
2. BARBER, R. M. (1941) *Diffusion in and through Solids*, C.W.P.
3. SCHMID, E. and BOAS, W. (1950) *Plasticity of Crystals*, Hughes.
4. HONESS, A. P., *Nature, Origin and Interpretation of Etch Pits in Crystals*, Wiley, New York.
5. BARRETT, C. S. (1952) *Structure of Metals*, McGraw-Hill, New York.
6. JOST, W. (1952) *Diffusion in Solids, Liquids and Gases*, Academic Press, New York.
7. SCHOCKLEY, W. (ed.) (1952) *Imperfections in Nearly Perfect Crystals*, Wiley, New York.
8. COTTRELL, A. H. (1953) *Dislocations and Plastic Flow in Crystals*, O.U.P.; (1958) *Vacancies and other Point Defects in Metals and Alloys*, Inst. of Metals, London.
9. DARKEN, L. S. and GURRY, R. W. (1953) *Physical Chemistry of Metals*, McGraw-Hill, New York.
10. READ, W. T. (1953) *Dislocations in Crystals*, McGraw-Hill, New York.
11. DIENES, G. J. and VINEYARD, G. H. (1957) *Radiation Effects in Solids*, Interscience, New York.
12. MCLEAN, D. (1957) *Grain Boundaries in Metals*, OUP, Oxford.
13. NYE, J. F. (1957) *Physical Properties of Crystals*, OUP, Oxford.
14. VAN BURREN, H. G. (1961) *Imperfections in Crystals*, North-Holland, Amsterdam.
15. DAMASK, A. C. and DIENES, G. J. (1963) *Point Defects in Metals*, Gordon & Breach, New York.
16. KNOX, R. S. (1963) *Theory of Exitons*, Academic Press, New York.
17. SHEWMON, P. G. (1963) *Diffusion in Solids*, McGraw-Hill, New York.
18. WYCKOFF, R. W. G. (1963) *Crystal Structures*, 2nd edn., Interscience, New York.
19. COTTRELL, A. H. (1964) *Theory of Crystal Dislocations*, Blackie.
20. FRIEDEL, J. (1964) *Dislocations*, Pergamon Press, Oxford.
21. KRÖGER, F. A. (1964) *Chemistry of Imperfect Crystals*, North-Holland, Amsterdam.
22. REED-HILL, R. E. (1964) *Physical Metallurgy Principles*, Van Nostrand.
23. AMELINCKX, S. (1965) *Direct Observation of Dislocations*, Academic Press, New York.
24. BRAGG, L. and CLARINGBULL, G. F. (1965) *Crystal Structure of Minerals*, Bell, New York.
25. CAHN, R. W. (1965) *Physical Metallurgy*, North-Holland, Amsterdam.
26. HIRSCH, P. B., HOWIE, A., NICHOLSON, R. B., PASHLEY, D. W., and WHELAN, M. J. (1965) *Electron Microscopy of Thin Crystals*, Butterworths, London.
27. PEARSON, W. B. (1967) *Handbook of Lattice Spacings and Structures of Metals and Alloys*, Pergamon Press, Oxford.
28. BELL, J. F. (1968) *The Physics of Large Deformation of Crystalline Solids*, Springer-Verlag.
29. HASIGUTI, R. R. (ed.) (1968) *Lattice Defects in Semiconductors*, Penn. State Univ. Press.
30. BOLLMANN, W. A. (1970) *Crystal Defects and Crystalline Interfaces*, Springer-Verlag.
31. KELLY, A. and GROVES, G. W. (1970) *Crystallography and Crystal Defects*, Longmans.
32. FLYNN, C. P. (1972) *Point Defects and Diffusion*, Clarendon Press, Oxford.
33. HENDERSON, B. (1972) *Defects in Crystalline Solids*, E. Arnold, London.
34. PARTHÉ, E. (1972) *Cristallochemie des Structures Tetraédrique*, Gordon & Breach, New York.
35. KRÖGER, F. A. (1974) *The Chemistry of Imperfect Crystals*, 2nd edn., 3 vols., North-Holland, Amsterdam.
36. TANNER, B. K. (1976) *X-ray Diffraction Topography*, Pergamon Press, Oxford.

D. *Conference Proceedings*

1. Bristol (1949) Faraday Soc., *Crystal Growth*, Butterworths, London, reprinted 1957.
2. Cooperstown (1958) DOREMUS, R. H., ROBERTS, B. W., and TURNBELL, D. (eds.) *Growth and Perfection of Crystals*, Wiley, New York.
3. Epitaxy Conference, FRANCOMBE, M. H. and SATO, H., *Single Crystal Films*, Pergamon Press, Oxford, 1964.
4. U.S.S.R. (various), IVANTSOV, G. P., SHEFTAL, N. N., and SHUBNIKOV, A. V. (eds.) (1958) *Growth of Crystals*, Vols. 1–6, Inst. of Crystallography, Acad. Sci. U.S.S.R. (translated by Consultants Bureau).
5. Boston (1966) *IC Crystal Growth I*, PEISER, H. S. (ed.), Pergamon Press, Oxford, 1967.
6. Brighton (1967) *Solidification of Metals*, Institute of Metals, London.
7. Birmingham (1968), *IC Crystal Growth II* (FRANK, F. C., MULLIN, J. B. and PEISER, H. S., eds.), *J. Cryst. Growth*, Vols. 3, 4.
8. *Growth of Crystals, USSR* (1969) (SHEFTAL', N. N., ed.), Vols. 7 and 8, Plenum Press, New York (D4 continued).
9. Zürich (1970) *First IC on Vapour Growth and Epitaxy* (KALDIS, E. and SCHRIEBER, M., eds.), *J. Cryst. Growth*, Vol. 9.
10. Marseille (1971) *IC Crystal Growth III* (LAUDISE, R. A., MULLIN, J. B., and MUTAFTSCHIEV, B., eds.), *J. Cryst. Growth*, Vols. 13 and 14.
11. Jerusalem (1972) *IC on Vapour Growth and Epitaxy II* (CULLEN, G. W., KALDIS, E., PARKER, R. L., and SCHRIEBER, M., eds.), *J. Cryst. Growth*, Vol. 17.

12. Tokyo (1974) *IC Crystal Growth IV* (Jackson, K. A., Kato, N., and Mullin, J. B., eds.), *J. Cryst. Growth*, Vols. 24/25.
13. Amsterdam (1975) *IC on Vapour Growth and Epitaxy III* (Cullen, G. W., Kaldis, E., Parker, R. L., and Rooymans, C. J. M., eds.), *J. Crystal Growth*, Vol. 40.
14. Zürich (1976) *1st European Conference on Crystal Growth* (Kaldis, E. and Scheel, H. J., eds.), North-Holland, Amsterdam.
15. Boston (1977) *IC on Crystal Growth V, J. Crystal Growth*, Vol. 42.
16. Nagoya (1978) *IC on Vapour Growth and Epitaxy IV*, to be published.
17. Moscow (1980) Venue for ICCG VI.

E. *Other Sources of Information*

Information on crystal growth, properties of materials and their availability is provided by the following organizations:

1. The Research Materials Information Center (RMIC)
   Dr. T. F. Connolly
   Oak Ridge National Laboratory
   PO Box X
   Oak Ridge, Tennessee 37830, USA
2. The Electronics Materials Unit
   Royal Signals and Radar Establishment (RSRE)
   St. Andrews Road
   Great Malvern, Worcs., WR14 3PS
   England
3. Centre de Documentation sur les Synthèses Cristallines
   Laboratoire de Physique Moléculaire et Cristalline
   Faculté des Sciences
   Place Eugène-Bataillon
   F-34 Montpellier, France
4. Scientific Advisers & Co.
   15 Park Lane
   Bath BA1 2XH
   England

There are several compilations of available crystals in addition to the sources mentioned above:

5. International List of Available Electronic Materials, 1976 Defense Research Group, Panel on Physics and Electronics
   North Atlantic Council, OTAN/NATO
   1110 Brussels, Belgium
6. Sources of Single Crystals in the UK and Scandinavia 1978
   Mrs. B. M. R. Wanklyn
   Clarendon Labs.
   Parks Road
   Oxford OX1 3PU
   England
7. Information über Kristallzüchtung, 1976
   Professors Dr. R. Nitsche and A. Raüber
   Kristallographisches Institut der Universität Freiburg
   7800 Freiburg i. Br.
   Hebelstrasse 25, West Germany
8. Documentation sur les Synthèses Cristallines 1976
   Professor A. M. Vergnoux
   42 Rue St. Claire
   87000 Limoges, France
9. An Inventory on Crystal Growth in the Netherlands II
   Dr. B. G. Wienk
   Akzo Zout Chemie Nederland bv.,
   18 Boortorenweg, Hengelo
   Postbus 25, Holland

CHAPTER 2

Nucleation and Growth Theory

B. LEWIS

Plessey Research (Caswell) Ltd., Allen Clark Research Centre, Caswell,
Towcester, Northants, UK

The following view of the molecular state of a crystal when in equilibrium with respect to growth or dissolution appears as probable as any. Since the molecules at the corners and edges of a perfect crystal would be less firmly held in their places than those in the middle of a side, we may suppose that when the condition of theoretical equilibrium is satisfied several of the outermost layers of molecules on each side of the crystal are incomplete toward the edges. The boundaries of these imperfect layers probably fluctuate, as individual molecules attach themselves to the crystal or detach themselves, but not so that a layer is entirely removed (on any side of considerable size), to be restored again simply by the irregularities of the motions of the individual molecules. Single molecules or small groups of molecules may indeed attach themselves to the side of the crystal but they will speedily be dislodged, and if any molecules are thrown out from the middle of a surface, these deficiencies will also soon be made good; nor will the frequency of these occurrences be such as greatly to affect the general smoothness of the surfaces, except near the edges where the surfaces fall off somewhat, as before described. Now a continued growth on any side of a crystal is impossible unless new layers can be formed. ... Since the difficulty in the formation of a new layer is at or near the commencement of the formation, the necessary value of potential may be independent of the area of the side, except, when the side is very small. The value of potential which is necessary for the growth of the crystal will however be different for different kinds of surfaces, and probably will generally be greatest for the surfaces for which σ is least.

On the whole, it seems not improbable that the form of very minute crystals in equilibrium with solvents is principally determined ... by the conditions that σ shall be a minimum for the volume of the crystal except so far as the case is modified by gravity or the contact of other bodies, but as they grow larger (in a solvent no more supersaturated than is necessary to make them grow at all), the deposition of new matter on the different surfaces will be determined more by the nature (orientation) of the surfaces and less by their size and relations to the surrounding surfaces. As a final result, a large crystal, thus formed, will generally be bounded by these surfaces alone on which the deposit of new matter takes place least readily, with small, perhaps insensible truncations. If one kind of surfaces satisfying this condition cannot form a closed figure, the crystal will be bounded by two or three kinds of surfaces determined by the same condition. The kinds of surface thus determined will probably generally be those for which σ has the least values. But the relative development of the different kinds of sides, even if unmodified by gravity or the contact of other bodies, will not be such as to make $\Sigma\sigma s$ a minimum. The growth of the crystal will finally be confined to sides of a single kind.

(J. W. GIBBS, 1878)

2.1. Introduction

These words of Gibbs[1] establish the theme and scope of this chapter. It should be added that the treatment will be broad and applicable to all growth systems in so far as that is possible. Particular emphasis will be placed on the use of known material parameters and on the estimation of those which are not known, so that quantitative results are obtained. For greater rigour and more detailed treatments of individual steps the reader is referred to the cited papers or to reviews of the same area by Strickland-Constable[2] and Parker.[3]

2.2. Crystal Models

2.2.1. ATOMIC BONDING

The atoms of a crystalline solid are arranged in a geometrical pattern in which each unit cell is identically repeated and each atom has a specific structural environment. The processes of nucleation and growth of crystals are strongly influenced by the crystal structure. For the purpose of illustration and discussion we shall use the simple cubic structure illustrated in Fig. 2.1 in which each cube represents one atom and also one unit cell. We shall also suppose that any two atoms with faces in contact have mutual binding energy ϕ_1, i.e. ϕ_1 is the strength of a single nearest neighbour bond. Two atoms which share a cube edge are second nearest neighbours with bond strength ϕ_2. Following Kossel[4] and Stranski[5,6] the various atomic sites on an incomplete {001} face are characterized by differing numbers of bonds of strength ϕ_1 and ϕ_2 as follows:

(1) completed edge, $4\phi_1 + 4\phi_2$
(2) completed corner, $3\phi_1 + 3\phi_2$
(3) completed surface, $5\phi_1 + 8\phi_2$
(4) incompleted edge, $\phi_1 + 3\phi_2$
(5) incompleted corner, $\phi_1 + 2\phi_2$
(6) step edge, $2\phi_1 + 6\phi_2$
(7) completed step, $4\phi_1 + 8\phi_2$
(8) adsorbed atom, $\phi_1 + 4\phi_2$
(9) edge vacancy, $-4\phi_1 - 6\phi_2$
(10) face vacancy, $-5\phi_1 - 8\phi_2$
$\frac{1}{2}$ crystal position, $3\phi_1 + 6\phi_2$

Finally, any interior atom has 6 nearest and 12 next nearest neighbours, i.e. it has binding energy $6\phi_1 + 12\phi_2$.

The positions marked $\frac{1}{2}$, making up a {111} plane, or at a kink on a $\langle 100 \rangle$ step, are the important "half-crystal" positions, which have exactly half the total binding energy of an interior atom. The addition of an atom to a half-crystal position creates another half-crystal position to which an atom can be added. Thus on a {111} plane, atoms can continually be added to the half-crystal positions on the plane and create new half-crystal positions as the crystal is built up plane by plane. On a kinked $\langle 100 \rangle$ step, atoms can be added at the kink site until the row is complete, and, except for the first atom of each row, a complete {001} plane can be built by successive additions at half-crystal positions. Apart from the bonding deficiency associated with the edge rows of atoms, the total binding energy of the complete crystal is equal to the energy of the half-crystal position times the number of atoms in the crystal.

The crystal energy can also be obtained as the energy which must be supplied to separate it into individual atoms. This is $\frac{1}{2}(6\phi_1 + 12\phi_2) = 3\phi_1 + 6\phi_2$ per atom, since in bulk each bond is shared between two atoms. Thus we confirm that the binding energy at

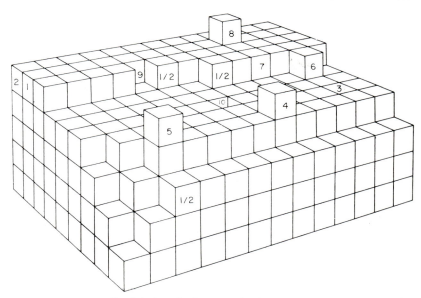

FIG. 2.1. Atomic sites on a simple cubic crystal.

the half-crystal position is equal to the mean binding energy per atom. We can also evaluate $3\phi_1 + 6\phi_2$ as the latent heat or enthalpy λ for the formation of the crystal from the vapour or liquid phase, or from solution.

2.2.2. FORMATION ENERGY OF CLUSTERS ON A CRYSTAL PLANE

In considering addition of atoms to a crystal, or similarly desorption from the crystal, the energy change can be evaluated in any of several ways: Method (a), which has already been used to find the binding energies of the labelled atoms in Fig. 2.1, is the summation of bonds. Method (b) is to consider the change of surface and edge energy of the added atoms and of the crystal, where the surface and edge energy is equal to the bond deficiency $\phi_1/2$ for each missing nearest neighbour bond and $\phi_2/2$ for each missing next nearest neighbour bond. Method (c) is to assign $\lambda = 3\phi_1 + 6\phi_2$ as the bulk bond energy of each atom and subtract the surface and edge energy change between the initial and final states of the crystal. As an example, the adsorption of an atom from the vapour to position 7 can be considered either (a) as contributing bonds $\phi_1 + 4\phi_2$, or (b) as eliminating the surface energy $\frac{1}{2}\phi_1 + 2\phi_2$ associated with one face of the atom, and also with the substrate, total $\phi_1 + 4\phi_2$, or (c) as representing the bulk bond energy of the atom $(3\phi_1 + 6\phi_2)$, minus $(5\phi_1/2 + 4\phi_2)$ surface and edge energy of the adsorbed atom, plus $(\frac{1}{2}\phi_1 + 2\phi_2)$ eliminated surface energy of the substrate, total $\phi_1 + 4\phi_2$.

Method (a) is the simplest for single atoms or small groups of atoms but for large assemblies method (c) is preferable. The surface energy is then the macroscopic parameter, σ per atom or per unit area, which can be determined experimentally for bulk crystalline material, and the bulk bond energy is λ per atom or unit volume.

The surface energy of each crystal plane is different. For the simple cubic structure, the surface energies per atom of a plane surface are $\sigma_{100} = \frac{1}{2}\phi_1 + 2\phi_2$, $\sigma_{110} = \phi_1 + 4\phi_2$, and $\sigma_{111} = 3\phi_1/2 + 6\phi_2$, which is a ratio $1:2:3$. The packing densities n_{100}, n_{110}, and n_{111} are in the ratios $1:\sqrt{2}:\sqrt{3}$ and hence the surface energies per unit area are in the ratios $(\sigma n)_{100}:(\sigma n)_{110}:(\sigma n)_{111}::1:\sqrt{2}:\sqrt{3}$. For the face-centred-cubic structure, which has $\lambda = 6\phi_1$ (neglecting second nearest neighbour bonding), $\sigma_{111} = 3\phi_1/2 = \lambda/4$ and $\sigma_{100} = 2\phi_1 = \lambda/3$ per surface atom, which is a ratio $3:4$. The packing densities are $n_{111}:n_{100}::2:\sqrt{3}$, so the surface energies per unit area are in the ratio $(\sigma n)_{111}:(\sigma n)_{100}::\sqrt{3}:2$. In general the closed packed planes have the lowest surface energies.

2.2.3. SURFACE DIFFUSION

One further parameter will be discussed. It is evident from Fig. 2.1 that an atom absorbed on a low energy face has low binding energy and a high probability of desorption. However, if it can migrate to a step or a kink site it will be more strongly bound. The diffusion distance before desorption is $x_s = (D\tau_a)^{1/2}$ where $\tau_a = (1/v_0)\exp(e_a/kT)$ is the mean desorption time and $D = \frac{1}{4}v_1\exp(-e_d/kT) = v_d\exp(-e_d/kT)$ is the diffusion coefficient; e_a and e_d are the desorption and surface diffusion energies; v_0, v_1, and v_d are, respectively, the attempt frequencies for desorption, a hop in any direction, and a hop in a particular direction, and are all of order $10^{13}\,\text{s}^{-1}$. We arbitrarily assume $v_0 = v_1 = 4v_d$ which gives $x_s = \frac{1}{2}\exp(e_a - e_d)/kT$. For a simple cubic $\{001\}$ plane, $e_a = \phi_1 + 4\phi_2$, $e_d = \phi_2$, and $e_a - e_d = \phi_1 + 3\phi_2 = \lambda/3$. Other crystal structures have been considered by Volmer,[7] who finds that although e_a and e_d vary, the difference $e_a - e_d$ is roughly constant at 0.45λ for any plane. Thus $x_s \sim \frac{1}{2}\exp(0.22\lambda/kT)$.

As an example, for mercury $\lambda_{SV} = 15\,\text{kcal/mol}$, so that for growth from the vapour at $200°\text{K}$ $\lambda/kT = 37$ and $x_s \sim 5000$ sites $\sim 1.5\,\mu\text{m}$. Direct evidence of surface migration has been obtained by Volmer and Esterman[8] and by Sears[9] in experiments in which the edge growth of mercury platelets was about 1000 times faster than the direct impingement rate. It was clear that atoms were migrating to the edges from the major faces. In Sear's experiments the platelet thickness was estimated as 10^{-7} cm, so the diffusion distance was $\sim 500 \times 10^{-7}$ cm $\sim 0.5\,\mu\text{m}$, in reasonable agreement with the estimate above. Some typical values of λ, λ/kT and x_s are given in Table 2.1. It will be noted that the range of values is extremely large and that λ/kT and x_s for mercury are exceptionally high.

TABLE 2.1. *Material Parameters for Some Growth Systems*

System	Material	eV	T (K)	λ/kT	δx_s
Vapour growth	Mercury	0.65	200	37	5000
	Cadmium	1.2	573	23	200
	Ice	0.53	273	22	150
Melt growth	Salol	0.94	314	35	3000
	Silicon	0.4	1074	3.3	2
	Tin	0.07	505	1.6	1.4
Solution growth	Alum	0.29	320	11	12
	ADP	0.09	310	3.5	2
	Sucrose	0.03	273	1.1	1

The fast growth of the edges of mercury platelets occurs because the faces have no capture sites and are not growing. However, if there were steps $<x_s$ apart on these faces and these steps had kinks $<x_s$ apart, then all atoms incident on the surface would have a high probability of reaching a kink. The edges would no longer receive atoms from the faces, and both faces and edges would grow at the direct impingement rate. In other words a capture-site density of $>1/x_s^2$ confers a capture efficiency almost as high as that of a high energy plane.

2.3. Supersaturation, Supercooling, and Volume Energy

2.3.1. GROWTH FROM THE VAPOUR

In the case of growth from the vapour phase, the incidence flux R atoms or molecules per unit area per unit time varies with the vapour pressure P according to the gas kinetic relation

$$R = (2\pi mkT)^{-1/2}P. \tag{2.1}$$

In SI units this gives

$$R\,(\text{m}^{-2}\,\text{s}^{-1}) = 2.6 \times 10^{24} P\,(\text{N m}^{-2})/(MT)^{1/2}, \tag{2.1a}$$

and in c.g.s. units, with pressure in torr,

$$R\,(\text{cm}^{-2}\,\text{s}^{-1}) = 3.5 \times 10^{22} P\,(\text{torr})/(MT)^{1/2}, \tag{2.1b}$$

where m is the atomic or molecular mass, M is the atomic or molecular weight, k is the gas constant per molecule, and T is the absolute temperature. The pressure P_∞ in equilibrium with bulk solid or liquid phase can be written

$$P_\infty = A \exp(-\lambda_{SV}/kT) \tag{2.2a}$$

or

$$P_\infty = A' \exp(-\lambda_{LV}/kT), \tag{2.2b}$$

where for any limited temperature range A, A' are constants and λ_{SV}, λ_{LV} are the latent heats of condensation from the solid and liquid phases, respectively. These relations are plotted in Fig. 2.2.

A solid or liquid phase exposed to the pressure P_∞ will neither grow nor evaporate; these are equilibrium conditions in which the vapour phase is saturated with respect to the condensed phase. A pressure P_a, from an external source, such as is represented by the point a in Fig. 2.2, is supersaturated with respect to the solid phase at temperature T_b, which has an equilibrium pressure P_b. The ratio $P_a/P_b = \alpha$ is the *saturation ratio* and $\alpha - 1$ is the *supersaturation.*

At T_b the solid is in equilibrium with vapour with pressure P_b and has the same chemical potential. The driving force for condensation along the path ab is the free energy difference per atom G_v, between vapour at pressures P_a and P_b,

$$G_v = kT_b \int_{P_a}^{P_b} \frac{dP}{P} = -kT_b \ln \frac{P_a}{P_b} = -kT_b \ln \alpha. \tag{2.3}$$

Elsewhere we use the terminology $\alpha = R/R_\infty = P/P_\infty$. It should be noted that we have defined G_v, which is usually called the volume energy, as the free energy difference per

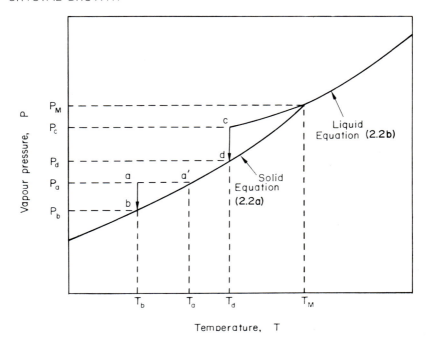

FIG. 2.2. Vapour pressure–temperature relation illustrating the condensation of a solid from the liquid and vapour.

atom. It is more usual to define the volume energy as ΔG_v per unit volume; the relation between these quantities is $G_v = \Omega \Delta G_v$, where Ω is the atomic volume.

The incidence pressure P_a must derive from solid or liquid at a higher temperature than T_b. If the source area is very large compared with the condensation area, then P_a may be in equilibrium with solid at temperature T_a, as represented by the point a' in Fig. 2.2. Then using eq. (2.2b),

$$\alpha = \frac{P_a}{P_b} = \exp\left[\lambda_{SV}\left(\frac{1}{kT_b} - \frac{1}{kT_a}\right)\right], \tag{2.4}$$

$$G_v = -\frac{\lambda_{SV}(T_a - T_b)}{T_a}. \tag{2.5}$$

2.3.2. GROWTH FROM THE MELT

At the melting point T_M both solid and liquid phases are in equilibrium with the saturated vapour P_M. Hence

$$P_M = A\exp(-\lambda_{SV}/kT_M) = A'\exp(-\lambda_{LV}/kT_M). \tag{2.6}$$

If the liquid is supercooled to a temperature T_d, below T_M, then as shown by the points c and d in Fig. 2.2, the vapour pressures P_c and P_d in equilibrium with liquid and solid at the temperature T_d are

$$P_c = A' \exp(-\lambda_{LV}/kT_d), \tag{2.7a}$$

$$P_d = A \exp(-\lambda_{SV}/kT_d). \tag{2.7b}$$

We then divide eqn. (2.7a) by eqn. (2.7b) to obtain the saturation ratio, substitute for A'/A from eqn. (2.6), and put $\lambda_{SL} - \lambda_{LV} = \lambda_{SL}$, the latent heat of fusion, giving

$$\alpha = \frac{P_c}{P_d} = \exp\left[\lambda_{SL}\left(\frac{1}{kT_d} - \frac{1}{kT_m}\right)\right], \tag{2.8}$$

and

$$G_v = -\frac{\lambda_{SL}(T_M - T_d)}{T_M}. \tag{2.9}$$

2.3.3. GROWTH FROM SOLUTION

In solution growth the saturation ratio α is the ratio of the actual concentration C_a to the equilibrium concentration C_b at the interface temperature T_b. Supersaturation can be produced by addition of solute, by evaporation of solvent, by a temperature gradient (particularly in hydrothermal growth), or by supercooling. If the dependence of the equilibrium concentration C_e on temperature is

$$C_e = C_0 \exp(-\lambda_{SS}/kT_e), \tag{2.10}$$

where C_0 is a constant, λ_{SS} is the enthalpy of solution per molecule, and n is the number of ions formed from one molecule of solute, then for a solution supercooled to a temperature T_b from equilibrium at a temperature T_a

$$\alpha = \frac{C_a}{C_b} = \exp\left[\frac{\lambda_{SS}}{n}\left(\frac{1}{kT_b} - \frac{1}{kT_a}\right)\right], \tag{2.11}$$

and

$$G_v = nkT_b \ln \alpha = -\frac{\lambda_{SS}(T_a - T_b)}{T_a}. \tag{2.12}$$

In a fluxed melt, the melting point of the solid is depressed by the presence of flux, and the saturation ratio is again the ratio of the actual and equilibrium concentrations at the growth interface. Supersaturation can be obtained either by reduction in the concentration of flux or by supercooling.

2.4. Basic Nucleation Theory

The equilibrium relations and the definitions of saturation ratio, supersaturation, and volume energy given above are all with reference to the bulk condensed phase. The stability of aggregates decreases with increasing size. In classical terms, the Gibbs–Thomson effect causes the equilibrium vapour pressure to increase with the curvature of the surface, or the surface to volume ratio, which increases as the size decreases. The equivalent atomistic statement is that small clusters are less stable because the mean number of near neighbours and bonds per atom decreases with decreasing size. It follows that small aggregates are less stable than bulk condensate and that in the presence of a mother phase, which is supersaturated with respect to bulk condensate only, those larger than a certain critical size $i*$ are more likely to grow than decay.

Taking account of the size-dependent stability, the concentrates N_i of aggregates of i atoms in equilibrium with the mother phase have a minimum at the critical size $i*$; the formation rate J of supercritical clusters is limited by the density of critical-sized nuclei, such that

$$J \simeq \gamma_{i*} N_i^*, \tag{2.13}$$

where γ_{i*} is the capture rate of monomer by critical nuclei.

Equation (2.13) is only a first approximation to the nucleation rate because when J is finite critical nuclei are not in equilibrium, due to the flux through the system, and their density is smaller than N_i^*. It is also necessary to allow for the reverse flux of decaying supercritical clusters. We therefore commence with a kinetic treatment of nucleation as given by Becker and Döring[10] and Frank.[11]

We define n_i as the kinetically determined density of aggregates of size i which may change size by one atom at a time. The flux J_i between sizes i and $i + 1$ can then be written as

$$J_i = \gamma_i n_i - \delta_{i+1} n_{i+1} \quad (i \gtrless 1), \tag{2.14}$$

where γ_i is the capture rate of mother phase atoms by i-aggregates and δ_{i+1} is the decay rate of $(i + 1)$-aggregates to i-aggregates. The γ_i are all proportional to the density of the mother phase. The δ_i decrease with increasing i and aggregates with $\delta_i > \gamma_{i-1}$ are supercritical.

We now define

$$R_i = \frac{1}{\gamma_i} \prod_{j=2}^{i} (\delta_j/\gamma_{j-1}), \tag{2.15a}$$

$$\psi_i = \gamma_i n_i R_i = n_i \prod_{j=2}^{i} (\delta_j/\gamma_{j-1}). \tag{2.15b}$$

Multiplying through by R_i, eqns. (2.14) then reduce to

$$J_i R_i = \psi_i - \psi_{i+1} \quad (i \gtrless 1), \tag{2.15c}$$

which has the form of Ohm's law for a current J_i flowing between points with potentials ψ_i and ψ_{i+1} connected by a resistance R_i.

In the steady state all the J_i's are equal and we may drop the subscript i. Adding eqns. (2.15c) for $i = 1$ to a size h which is larger than $i*$, the intermediate ψ_i cancel giving

$$J \sum_{i=1}^{h} R_i = \psi_1 - \psi_h. \tag{2.16a}$$

For $i = 1$, $R = 1/\gamma_1$ and $\psi_1 = n_1$. Now monomer in the condensed phase is indistinguishable from monomer in the mother phase. Hence n_1, which is the driving potential in eqn. (2.15a), is the atom concentration in the mother phase. R_i increases with i while $\delta_i > \gamma_{i-1}$ and decreases when $\delta_i < \gamma_{i-1}$. R_i therefore has a maximum (or maxima if the variation of δ_i with i is irregular) where $\delta_i \sim \gamma_{i-1}$. Thus, while i^* does not occur explicitly in eqn. (2.16a), aggregates near the critical size provide the largest R_i in the sum. Beyond i^*, where R_i is decreasing, $\gamma_i n_i$ is approximately constant and ψ_i also decreases; we may therefore choose the size h such that ψ_{h+1} and the R_i for $i > h$ are negligibly small. Equation (2.16a) then reduces to

$$J = n_1 \bigg/ \sum_{i=1}^{h} R_i,\tag{2.16b}$$

which may also be written

$$J = \gamma_1 n_1 \left[1 + \sum_{i=2}^{h} \frac{\gamma_1}{\gamma_i} \prod_{j=2}^{i} (\delta_j/\gamma_{j-1}) \right]^{-1}.\tag{2.16c}$$

The next step required is to evaluate or eliminate the size-dependent δ_i.

Zeldovich[12] showed that the δ_i can be eliminated in favour of hypothetical equilibrium densities N_i, which may be determined thermodynamically. He considered eqn. (2.14) for a hypothetical equilibrium state maintained by a demon which removes clusters beyond size h and redistributes them as monomer. With no flux through the system, a set of equilibrium densities N_i is established. It is postulated that n_1 and the γ_i remain unchanged. It is also assumed that the δ_i remain the same under equilibrium as under steady state conditions. Then from eqn. (2.14) with $J_i = 0$, we find

$$\delta_i/\gamma_{i-1} = N_{i-1}/N_i,\tag{2.17a}$$

and by iteration from $N_1 = n_1$ for $i = 1$,

$$R_i = \frac{1}{\gamma_i} \prod_{j=2}^{i} (\delta_j/\gamma_{j-1}) = n_1/\gamma_i N_i,\tag{2.17b}$$

$$J = \left[\sum_{i=1}^{h} (\gamma_i N_i)^{-1} \right]^{-1}.\tag{2.17c}$$

Again the critical size does not appear, but as in eqns. (2.15) provides the largest term in the sum. It is therefore appropriate to re-write eqn. (2.17c) as

$$J = Z\gamma^* N^*,\tag{2.18a}$$

where * denotes the critical size i^* and

$$Z = \left[\sum_{i=1}^{h} (\gamma^* N^*/\gamma_i N_i) \right]^{-1}.\tag{2.18b}$$

Since the N_i vary with i much more strongly than the γ_i, eqn. (2.18b) may be simplified to

$$Z = \left[\sum_{i=1}^{h} (N^*/N_i) \right]^{-1}.\tag{2.18c}$$

Equations (2.18) represent the nucleation rate as the product of the gross forward rate γ^* and the equilibrium density N^* of critical nuclei, modified by the Zeldovich factor, which allows for depletion of n^* below the equilibrium N^* and for the reverse flux of decaying supercritical clusters. It should be appreciated that the equilibrium densities N_i do not and cannot actually occur.

Recently Katz and co-workers[13-15] have shown that recourse to a hypothetical equilibrium state is not necessary. They consider instead the true equilibrium state obtained by decreasing the incidence flux isothermally from R to R_∞, which is in equilibrium with bulk condensate. This has the advantage, compared with the treatment above, that the state considered is physically realizable and the disadvantage that n_1 and the γ_i have changed. They assume that the δ_i are unchanged and that the γ_i are reduced in the ratio R_∞/R. The nucleation rate for an incidence flux R is then found in terms of the cluster densities $N_i(R_\infty)$ in equilibrium with R_∞. The result is identical to that above if the hypothetical densities $N_i(R)$ in equilibrium with R are related to the $N_i(R_\infty)$ by

$$N_i(R) = \alpha^i N_i(R_\infty), \tag{2.19}$$

where $\alpha = R/R_\infty$ is the saturation ratio.

In the classical model, the $N_i(R)$ are given by the Van't Hoff reaction isotherm for equilibrium with monomer of density n_1 as

$$N_i = n_1 \exp(-G_i/kT), \tag{2.20}$$

where G_i is the free energy of formation of i-aggregates from monomer. Conformity with eqn. (2.19) depends on the evaluation of G_i which is discussed below.

Using eqn. (2.20), Zeldovich[12] wrote eqn. (2.18) in integral form, with i treated as a continuous variable, as

$$Z^{-1} = \int_1^h \exp\left[-\frac{G^* - G_i}{kT} di\right], \tag{2.21a}$$

where G^* is the largest G_i, corresponding to N^* which is the smallest N_i. In the Taylor expansion for G_i in terms of G^* and derivatives at $i = i^*$, the first derivative is zero, and

$$G^* - G_i = -\frac{(i - i^*)^2}{2} \frac{d^2 G_i}{di^2}\bigg|_{i^*}, \tag{2.21b}$$

to second order. Since the integral in eqn. (2.21a) is negligible for $i < 1$ and $i > h$, the limits may be extended to $\pm\infty$ as

$$Z^{-1} = \int_{-\infty}^{\infty} \exp\left[-\frac{(i - i^*)^2}{2kT} \frac{d^2 G}{di^2}\bigg|_{i^*}\right] di. \tag{2.21c}$$

The integral is now of standard form and gives

$$Z = \left[-\frac{1}{2\pi kT} \frac{d^2 G_i}{di^2}\bigg|_{i^*}\right]^{1/2} \tag{2.21d}$$

The equations above apply for any case of nucleation-controlled condensation. However, G_i and the other quantities depend on the system, as we now discuss.

2.5. Three-dimensional Nucleation

2.5.1. NUCLEUS FORMATION ENERGY

The treatment above reduces the kinetic nucleation problem to the thermodynamic one of finding the cluster densities N_i in equilibrium with supersaturated monomer. Classically, as established by Volmer and Weber[16] and Becker and Döring,[10] the formation energy G_i of an aggregate of i atoms from monomer is expressed in terms of volume and surface energies as

$$G_i = G_v(i) + G_s(i) + G_c. \tag{2.22a}$$

$G_v(i)$ is the free energy difference between i atoms of monomer and i atoms of bulk condensate and is equal to $iG_v = -ikT\ln\alpha$, where $\alpha = P/P_\infty = n_i/n_{1\infty}$, as discussed in section 2.3. $G_s(i)$ is the energy expended in creating the surface of the cluster, and for three-dimensional nuclei of equilibrium shape may be evaluated as $i^{2/3}B$, in which B is the product of a shape factor and the surface energy density, and is assumed to be independent of cluster size. With these substitutions

$$G_i = iG_v + i^{2/3}B + G_c. \tag{2.22b}$$

The need for the term G_c was recognized by Becker and others (see ref. 17), and was first evaluated by Lothe and Pound.[18] It is still sometimes (but mistakenly) omitted. All authorities agree that it contains $-G_v$, which cancels the pre-exponential n_1 of eqn. (2.20), so that $N_i \propto \alpha^i$ (rather than α^{i+1}) as is required thermodynamically and for conformity with eqn. (2.19). Otherwise the evaluation remains controversial. The Lothe–Pound treatment concerns the replacement of internal degrees of freedom of bulk condensate by translational and rotational degrees of freedom of isolated clusters. The principal term is the translational one, which contains a factor which cancels the pre-exponential n_1. Inclusion of this term causes substantial disagreement with experimental data for nucleation of water droplets from the vapour. Reiss et al.[19] dispute the need for inclusion of translational and rotational contributions. Nishioka and Pound[20] defend the Lothe–Pound treatment but do not discuss the important n_1-dependent factor. Blander and Katz,[21] in a lucid treatment, bring in the G_v term for thermodynamic consistency with eqn. (2.20), and the entropy contributions, if present, are then the difference between those for i-clusters and for monomer, but uncertainty and difficulty remain in the proper evaluation. Following their argument further, the criterion we now adopt to find G_c is that eqn. (2.20) should be correct for $i = 1$, i.e. that $G_1 = 0$. Putting $i = 1$, $G_1 = 0$ in eqn. (2.22b) we find

$$G_c = -G_v - B. \tag{2.22c}$$

This evaluation is similar to a treatment given by Strickland-Constable[2] in which G_c appears as an integration constant. Substituting back, eqn. (2.22b) can now be written

$$G_i = (i - 1)G_v + (i^{2/3} - 1)B, \tag{2.22d}$$

which is the form given by Hillig.[22]

2.5.2. THE FORMATION ENERGY OF LIQUID NUCLEI

For liquid nuclei with isotropic surface energy σ per surface atom, or $\sigma/\Omega^{2/3}$ per unit area, the equilibrium cluster shape is spherical and eqn. (2.22d) can be written in terms of i or the cluster radius r as

$$G_i = (i - 1)G_v + (i^{2/3} - 1)(36\pi)^{1/3}\sigma, \tag{2.23a}$$

$$= \left(\frac{4\pi r^3}{3} - 1\right)G_v + \frac{4\pi r^2\sigma}{\Omega^{2/3}} - (36\pi)^{1/3}\sigma, \tag{2.23b}$$

where Ω is the atomic volume. Maximization then gives:

$$i^* = -\frac{32\pi\sigma^3}{3G_v^3}, \tag{2.24a}$$

$$r^* = -\frac{2\sigma\Omega^{1/3}}{G_v}, \tag{2.24b}$$

$$G^* = \frac{16\pi\sigma^3}{3G_v^2} - G_v - (36\pi)^{1/3}\sigma, \tag{2.24c}$$

$$N^* = n_1 \exp(-G^*/kT), \tag{2.24d}$$

$$Z = \left(\frac{2}{9\pi}\right)^{1/3} \frac{(\sigma/kT)^{1/2}}{i^{*2/3}}. \tag{2.24e}$$

Substituting $G_v = -kT\ln(P/P_\infty)$ from eqn. (2.3) into eqn. (2.24b) we find

$$kT\ln(P/P_\infty) = 2\sigma\Omega^{1/3}/r^*, \tag{2.24f}$$

which is the Gibbs–Thomson relation for the vapour pressure P in equilibrium with a droplet of radius r^* when the bulk liquid has vapour pressure P_∞.

2.5.3. THE FORMATION ENERGY OF CRYSTALLINE NUCLEI

For crystalline nuclei, the equilibrium nucleus shape, which minimizes the surface energy, depends on the crystal structure. In a spherical, liquid nucleus each surface atom has the same probability of loss because each has an identical environment characterized by the radius r. The binding energy of surface atoms of a crystalline nucleus depends on their crystalline environment. Hence the stability of nuclei depends on the atomic configuration as well as on size.

As a specific example let us consider a simple cubic structure with nearest neighbour bond strength ϕ and latent heat $\lambda = 3\phi$, with respect to the parent phase which may be liquid or vapour. The surface energy $G_s(i)$ can be evaluated atomistically as $\phi/2$ for each unsatisfied bond and is also equal to $i\lambda - E_i$, where E_i is the binding energy. Then, writing $G_v(i)$ as iG_v and G_c as $-G_v - \lambda$, we have

$$G_i = (i - 1)(G_v + \lambda) - E_i. \tag{2.25a}$$

Alternatively, we may use a classical evaluation of G_i similar to that used for liquid nuclei. We treat each cluster as an equilibrium shaped cube of side $i^{1/3}$ and surface area $6i^{2/3}$ atomic units. Then $\sigma = \phi/2 = \lambda/6$, $G_s(i) = 6i^{2/3}\sigma = i^{2/3}\lambda$, and

$$G_i = (i - 1)G_v + (i^{2/3} - 1)\lambda. \tag{2.25b}$$

Since $E_1 = 0$, both equations give $G_1 = 0$ when $i = 1$.

For numerical examples we now put $G_v = -0.7\lambda$, -0.4λ, and -0.2λ corresponding to high, moderate, and low supersaturation, respectively. G_i in units of λ is plotted against i in Fig. 2.3. For each value of G_v the classical G_i has a single maximum and the atomistic G_i has an absolute maximum at almost the same value of i, which is the critical size i^*. When $G_v = 0.7\lambda$ $i^* = 1$, when $G_v = 0.4\lambda$ $i^* = 5$, and when $G_v = 0.2\lambda$ $i^* = 39$.

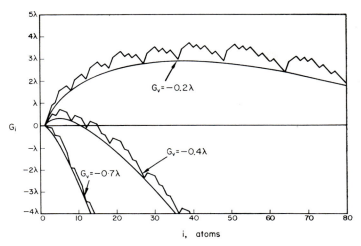

FIG. 2.3. Free energy of formation G_i of a three-dimensional nucleus of i atoms as dependent on i and on the volume energy G_v. G_i and G_v are both in units of the latent heat λ. The jagged curve is a simple atomistic evaluation by eqn. (2.25a) and the smooth curve the classical evaluation by eqn. (2.25b).

The atomistic G_i also has minor minima corresponding to completion of each successive row or plane of atoms and minor maxima for the first atom of each new row or plane. However, eqn. (2.25a) should contain additional entropy terms. In the simple bond model, configurational entropy decreases G_i by $kT\ln w_i$, where w_i is the number of distinguishable configurations. Bonissent and Mutaftschiev[23] have made computer calculations of the formation energy of atomic clusters containing 2–14 atoms. Central forces and internal partition functions were considered. They found that the drastic variations (due to symmetry changes) of the internal energy between neighbouring sites are mostly compensated by variation of the entropies of rotation and vibration. In our case these entropy terms reduce G_i for the symmetrical configurations but have no effect on the high symmetry clusters with $i = n^3$ (n an integer), and their inclusion thus tends to suppress the minor maxima and minima in Fig. 2.3.

Quantitatively suppression of the atomistic maxima depends on λ/kT and i and on the crystal structure. When i is very large, i.e. for crystals which have grown well beyond the three-dimensional nucleation stage, the atomistic maxima are not suppressed when $\lambda/kT \gtrsim 3$, and at low supersaturation they represent a significant barrier to the nucleation of new atomic layers on completed crystal planes. When i is small, as in the initial nucleation case we are now considering, the two evaluations of G_i are in substantial agreement for all reasonable values of λ/kT. This leads us to the important conclusion that classical formulations of G_i, with the surface energy $G_s(i)$ expressed in terms of $i^{2/3}\lambda$ or $i^{2/3}\sigma$, can be used for small values of i, for which, without this supporting evidence, macroscopic surface energy concepts could not be expected to apply.

The advantage of eqn. (2.25b) over eqn. (2.25a) is that it can be differentiated to give G^* and i^* as

$$i^* = (2\lambda/3G_v)^3, \tag{2.26a}$$

$$G^* = 4\lambda^3/27G_v^2 - G_v - \lambda. \tag{2.26b}$$

TABLE 2.2. *Saturation Ratio, Nucleation Barrier and Critical Nucleus Size for Selected Values of the Parameters*
λ/kT *and* G_v

λ/kT	30	10	3	30	10	3
G_v		Saturation α			$\exp(-G^*/kT)$	i^*
-0.7	10^9	10^5	400	10^{-4}	0.4	1	1
-0.4	10^3	50	7	10^{-13}	10^{-2}	1	5
-0.2	8	3	2	10^{-38}	10^{-4}	1	39

Values of the saturation ratio α, the nucleation barrier $\exp(-G^*/kT)$, and the critical size i^* for the three curves in Fig. 2.3 are given in Table 2.2 for $\lambda/kT = 30$, 10, and 3.

The relation $G_s(i) = 6i^{2/3}\sigma = i^{2/3}\lambda$, which was the basis of eqn. (2.25b), was derived for cubic nuclei with the surface energy $\sigma = \lambda/6$ given by the nearest neighbour bond model. For the general case, since we do not usually know the surface energies of each crystal plane, we treat the equilibrium nucleus shape as quasi-spherical and write $G_s(i)$ in terms of the effective surface energy σ as $(36\sigma)^{1/3}i^{2/3}$. Then, eqns. (2.24) apply.

2.5.4. NUCLEATION RATES

Having obtained expressions for i^*, G^*, and Z we now require γ^* and n_1 to find the nucleation rate. We need to distinguish between cases in which the mother phase is homogeneous, i.e. not mixed with any other phase, and those in which monomer is supplied to growing nuclei by diffusion through a carrier gas, a solvent or flux or a solid matrix, or over a solid surface.

When the monomer supply is not diffusion controlled,

$$\gamma^* = RS^*\Omega^{2/3}, \tag{2.27a}$$

$$J = n_1 RZS^*\Omega^{2/3} \exp(-G^*/kT), \tag{2.27b}$$

where $S^*\Omega^{2/3}$ is the nucleus surface area and R is the flux per unit area. For spherical nuclei, $S^* = (36\pi)^{1/3}i^{*2/3}$, Z is given by eqn. (2.24e) and $ZS^* = 2(\sigma/kT)^{1/2}$. The evaluation of n_1 and R depends on the system.

For condensation from the vapour phase, R is given by eqn. (2.1) as $P/(2\pi mkT)^{1/2}$ and $n_1 = P/kT$, so that

$$J = \frac{P^2 ZS^*\Omega^{2/3}}{(2\pi m)^{1/2}(kT)^{3/2}} \exp\left[-\frac{G^*}{kT}\right]. \tag{2.28}$$

For crystallization from the melt, Turnbull and Fisher[24] have evaluated R by transition state theory. If the melt and crystal have similar densities and transition across the interface requires activation energy G_d, $R\Omega^{2/3} = (kT/h)\exp(-G_d/kT)$, where kT/h is an attempt frequency. Then

$$J = \frac{n_1 kTZS^*}{h} \exp\left[-\frac{G^* + G_d}{kT}\right], \tag{2.29}$$

where n_1 is the monomer density in the melt. Desolvation and capture in solution growth is discussed by Bennema[25] and involves a sequence of activated steps. If one of these steps is predominantly rate controlling, a similar relation to melt growth will apply.

In other cases, including dilute solutions, growth from fluxed melts, and the solid state, γ^* is diffusion-controlled. In all cases n_1 is the monomer concentration.

We now wish to examine the range of nucleation rates and critical sizes which may occur and their dependence on α and $G_v = -kT\ln\alpha$. Strictly α occurs in the pre-exponential as well as in G^*, but its effect there is so trivial in comparison that we write

$$J^* = J_\infty \exp(-G^*/kT) \tag{2.30}$$

for the general case and treat J_∞ as a constant.

The three relevant material parameters for crystallization are the latent heat λ, the growth temperature T_g which must be below the melting point T_M, and the surface energy σ. These are most conveniently combined as the dimensionless ratios λ/kT_g and σ/λ.

For growth from the vapour λ_{SV} is known and we shall assume that T_g is just below T_M. The surface energy σ_{SV} can be obtained from creep measurements and values for metals just below the melting point have been collected by Jones.[26] Dufour and Defay[27] give a value for ice. σ_{SV} is usually about 20% higher than σ_{LV}, which can be obtained from surface tension measurements and is known for many materials. λ_{SV}/kT_M is plotted against σ_{SV}/λ_{SV} in Fig. 2.4 for some metals and for ice, and the reason for choosing these ratios as parameters is now evident. Although λ_{SV}, T_M, and σ_{SV} vary widely, most of the points are grouped around $\lambda_{SV}/kT_M = 30$ and $\sigma_{SV}/\lambda_{SV} = 0.15$. There are, of course, exceptions such

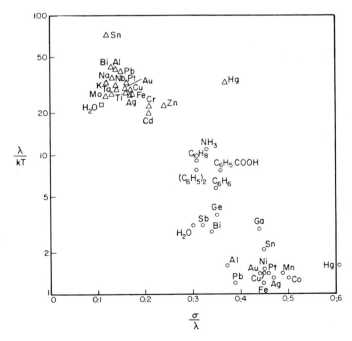

FIG. 2.4. The material parameters λ/kT and σ/λ relevant to the solidification of metals from the vapour △, ice from the vapour □, and metals and compounds from the melt ○.

as mercury, which has unusually high surface energy, tin, which has a low melting point, and ice, which has low surface energy and latent heat. Organics often have rather higher values of λ_{SV}/kT_M. For growth from the vapour G_v can be expressed either as $-kT\ln\alpha$, or as $-\lambda_{SV}(T_s - T_g)/T_s$, where T_s is the source temperature with which the vapour is in equilibrium.

For growth from the melt σ_{SL} is not known by independent measurements, and the growth temperature T_g depends directly on the supercooling which is related to the volume energy by $G_v = -\lambda_{SL}(T_m - T_g)/T_M$. The values of λ_{SL}/kT_g and σ_{SL}/λ_{SL} plotted in Fig. 2.4 are derived from T_g and σ_{SL} obtained from measurements by Turnbull and Cech[28] for the metals and by Thomas and Staveley[29] for the compounds.

J^*/J_∞ from eqns. (2.24) and (2.30) is shown in Fig. 2.5 for the parameter values $\sigma/\lambda = 0.15$, 0.22, and 0.44, as $\log J^*/J_\infty$ also in units of λ/kT_g, and $(T_M - T_g)/T_M$ or $(T_s - T_g)/T_s$. Values of i^* are shown against the curves. Scales for J^*/J_∞ and α are given for $\lambda/kT_g = 30$, which is typical for metals condensing from the vapour, and for J^*/J_∞ for $\lambda/kT = 1$, which is typical for metals from the melt.

We now wish to determine the critical saturation ratio α_{crit} below which the nucleation rate is negligible. For metals from the vapour with $\lambda/kT \sim 30$, $R\Omega^{2/3} \sim 1$ monolayer s^{-1}, $n_1 \sim 10^{13}$ ml^{-1}, and $J^* \sim 1$ ml^{-1} s^{-1} is a reasonable detection limit. This gives $J^*/J \sim 10^{-13}$. The plotted triangles in Fig. 2.5 represent $J^*/J = 10^{-13}$ and α_{crit} for crystallization of metals from the vapour. Typically, $\alpha_{crit} \simeq 10^2$ with $i^* \simeq 10$. Extreme values are $\alpha_{crit} \simeq 4 \times 10^5$ with $i^* \simeq 8$ for Sn, and $\alpha_{crit} \simeq 30$ with $i^* \simeq 20$ for Mo. The square is for condensation of ice from the vapour, for which the vapour pressure at the melting point is particularly high, giving $R\Omega^{2/3} \sim 10^6$ monolayers s^{-1}, $n_1 \sim 10^{17}$ ml^{-1}, and $J^*/J_\infty \sim 10^{-22}$ for $J^* = 1$ ml^{-1} s^{-1}. As a consequence, $\alpha_{crit} \simeq 5$ is exceptionally low and $i^* \simeq 50$ is unusually high.

Bulk liquids normally solidify a few degrees below the melting point, but this is believed to be due to nucleation on solid foreign particles. When liquids are divided into small drops, most drops are free of nucleating particles and experimentally the supercooling required for solidification becomes much greater. The circles in Fig. 2.5 are the experimental values of $(T_M - T_g)/T_M$ plotted against the estimated values of J^*/J_∞, which were 10^{-35} in Turnbull and Cech's[28] experiments. These results were used by Turnbull[30] and by Thomas and Staveley[29] to calculate the values of σ_{SL} and the ratios σ_{SL}/λ_{SP}, which are plotted in Fig. 2.4. They used a nucleation equation similar to (2.24) and (2.30) but omitting G_c, which is small compared with G^*.

For the solid–vapour interface, the density of the vapour is low. σ_{SV} thus represents the unsatisfied bonding per surface atom of a plane surface, for which the experimental value of about a sixth of the total binding energy per atom λ_{SV} is very reasonable. The atom densities of solid and liquid, however, are similar, and Zell and Mutaftschiev[31] have found by examination of a ball model that the liquid in contact with solid is disorganized and imperfectly wets the crystal face. It is, therefore, reasonable that in this case the interface energy σ_{SL} should be a third or a half of λ_{SL}, the binding energy difference per atom between solid and liquid. Nucleation from solution has not been systematically studied so that σ_{SS} and σ_{SS}/λ_{SS} are not known.

Variation of the ratio σ/λ between growth systems and between individual materials in each system is responsible for the differences in nucleation behaviour for similar values of λ/kT. However, the most important parameter is λ/kT. For nucleation from the vapour, λ/kT is large, so that high supersaturation is required and the critical size is small. The

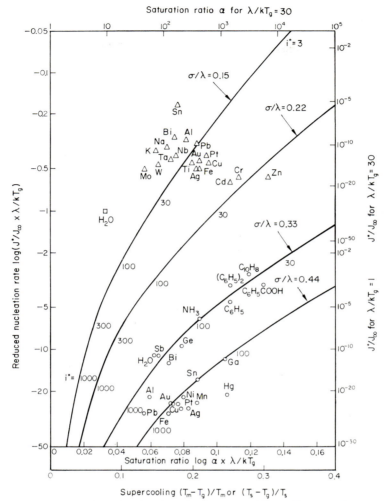

FIG. 2.5. The variation of reduced nucleation rate J^*/J_∞ with saturation α and supercooling $(T_M - T_g)/T_M$ below the melting point T_M, or $(T_s - T_g)/T_s$ below the vapour source temperature T_s. The lines are calculated from eqns. (2.26) and (2.30) for the values of σ/λ shown. The numbers against the curve are the critical nucleus size i^*. The plotted points show the critical condensation conditions for the substances and material parameters shown in Fig. 2.4. Scales are given in units of λ/kT_g and also for $\lambda/kT_g = 30$ and 1, which are typical for growth from the vapour and from the melt, respectively.

nucleation of ice is a notable exception. For nucleation from the melt and from solution, λ/kT, α_{crit}, and i^* cover a wide range of values.

2.6. The Growth of Crystal Surfaces

2.6.1. INTRODUCTION

In section 2.5.3 we found that minor maxima in the variation of the atomistic G_1 are suppressed by configurational entropy terms when i is small. For large crystals, although

G_i given by eqns. (2.25) is negative, the atomistic maxima following completion of a smooth plane can now represent a significant barrier to growth at low supersaturation. If growth of smooth planes is impeded, capture at kinks and steps causes all higher energy surfaces to grow out, so that the growth rate of the crystal is the impeded growth rate on the low energy smooth planes.

The kinetic theory of crystal growth on low energy planes was established by Volmer[7,16] and Becker and Döring[10] with important contributions by Kossel[4] and Stranski.[5] The theory predicted a steep dependence on the saturation ratio of the formation rate of growth of two-dimensional clusters on a low energy plane.

Volmer and Esterman[8] tested the growth of mercury platelets from the vapour with $\alpha \sim 1400$. The edge growth rate was about 10^3 times the direct impingement rate on the edge and the major faces grew slowly until the crystals were about 0.3 mm diameter when thickness growth became detectable and the lateral growth rate decreased. Volmer and Esterman concluded that the nucleation and growth rate on the major faces was negligible while they were small, so that monomer impinging on them was able to migrate to growth sites at the edges. They did not examine the dependence of growth rate on saturation.

Volmer and Schultze[32] tested two-dimensional nucleation growth theory by examining the growth of iodine, phosphorus, and naphthalene as a function of supersaturation. They found that the growth rate was proportional to $\alpha - 1$ down to $\alpha = 1.01$ for iodine and to 1.001 for the other materials, which was well below the saturation at which they expected the growth rate to be negligible. However, they could not positively conclude that nucleation growth did not apply because of uncertainty of their estimate of the cluster edge energy, which has no macroscopic counterpart.

In 1949 the Faraday Society arranged a discussion on crystal growth in Bristol at which Becker[17] and Stranski[33] contributed reviews of nucleation kinetics and the equilibrium forms of crystals. At this meeting Burton and Cabrera[34] presented the results of a fundamental re-examination of both these topics. Previously neglected statistical considerations enlarged and refined the understanding of equilibrium structures. Solutions of the diffusion problem for migrating monomer gave attachment rates at cluster or step edges. Both modifications increased the predicted two-dimensional nucleation rate and decreased the expected critical supersaturation for nucleation growth. However, evaluation of the nucleus formation energy in terms of bond energy showed that the nucleation rate was negligibly small in growth from the vapour at $\alpha = 1.01$. It was thereby firmly established that Volmer and Schultze's[32] experimental results required an alternative source of growth centres. The explanation provided by Frank[35] was screw dislocation growth. Subsequently these developments were published by Burton, Cabrera and Frank[36] (hereafter BCF) in what is deservedly the best known and most frequently cited paper in the field of crystal growth.

Screw dislocation growth was quickly accepted as a major growth mechanism and the BCF formulations have been widely used. However, the material relevant to two-dimensional nucleation has been largely neglected. Although examples of impeded crystal growth at moderate supersaturation, such as Volmer and Esterman's[8] platelet growth with $\alpha \sim 1400$, are not hard to find, it was many years before the BCF results were applied to two-dimensional nucleation growth by Lewis.[37] This disparity can be partly attributed to the different styles of Frank's part of the paper concerned with screw dislocations and growth spirals and Burton and Cabrera's treatment of the more fundamental topics. The physical mechanism of screw dislocation growth is easily understood. Useful end equations are developed and experimental examples are considered and discussed. By contrast, the treatment of the basic groundwork (which occupies about three-quarters of the paper, including six appendices) is so complex and condensed that the physical argument is often difficult to follow. An expression for the nucleus formation energy is obtained but is evaluated only for growth from the vapour at low supersaturation. Application to nucleation and growth rates is not considered.

To redress this balance, the BCF treatment of surface and cluster structure and energy is now presented as simply as possible. Mathematical derivations and proofs which cannot be simplified are cited as results, since the substantiating detail is available in BCF and in a paper on the critical nucleus by van Leeuwen and Bennema.[38] Application to two-dimensional nucleation growth and to screw dislocation growth is then described.

2.6.2. THE EQUILIBRIUM STRUCTURE OF
SURFACES AND STEPS

The equilibrium structure of crystal surfaces has been discussed by BCF,[36] Jackson,[39] Leamy and Jackson,[40] and Leamy et al.[41] Relevant parameters are the unsatisfied bonding and energy e and ε per edge atom along steps, and the unsatisfied bonding and energy s and σ, per surface atom. We denote the values for straight steps and smooth planes by subscript 0. For a Kossel crystal the smooth planes are $\{100\}$ with $s_0 = \sigma_0 = \frac{1}{2}\phi_1 + 2\phi_2 \simeq \lambda/6$ and the straight steps are $\langle 100 \rangle$ with $e_0 = \varepsilon_0 = \frac{1}{2}\phi_1 + \phi_2 = \lambda/6$.

At $0°K$ straight steps and smooth planes have the lowest energy. At $T > 0$ defects such as those illustrated in Fig. 2.1 appear which increase the unsatisfied bonding but decrease the surface energy due to the entropy associated with multiple configurations of the defects. The equilibrium structure minimizes the surface energy.

When $\lambda/kT \lesssim 10$ the only surface defects are adatoms and vacancies. With smaller λ/kT, clusters become more numerous and larger and begin to interact. The problem then becomes a co-operative one because tracing any path from atom to atom over the surface and back to the starting point the total change of level must be zero. A rigorous solution has been found only for a two-level system, i.e. substrate and one level of adatoms and clusters. A solution for the equilibrium surface roughness $(s - s_0)/s_0$, based on Onsager's[42] treatment of ferromagnetism, was obtained by Burton and Cabrera[34] and is plotted as the dotted line in Fig. 2.6. However, the restriction to two levels is unrealistic and seriously distorts the result when λ/kT is small.

Another approach employs a generalization of the Bragg–Williams theory of ordering in which it is assumed that the relevant level probabilities are independent and each unit in a given layer has the same environment. This zero-order approximation rules out clustering. It can be improved to a first-order correction using a method by Bethe[43] which introduces correlation factors which take account of the geometrical constraints and effectively allows clustering. Evaluations for five levels by BCF[36] and for an unlimited number of levels by Leamy and Jackson[40] are plotted as broken and full lines, respectively, in Fig. 2.6. The points in Fig. 2.6 are "experimental" values obtained by computer simulations, as described in section 2.7. It is a remarkable tribute to BCF that their result agrees so well with the experimental simulation. Figure 2.6 is for a Kossel

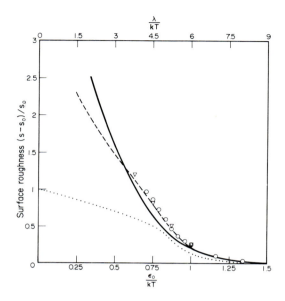

FIG. 2.6. The variation of surface roughness $(s - s_0)/s$ of the $\{100\}$ surface of a Kossel crystal with latent heat λ/kT and the edge energy ε_0/kT of a straight step as obtained by Burton and Cabrera[34] (dotted line, Onsager method), BCF[36] (broken line, five-level Bethe method), and by Leamy and Jackson[40] (full line, multi-level treatment). The points are values obtained in simulation experiments by Gilmer and Bennema[53] $\triangledown$, and by Leamy and Gilmer[54] $\bigcirc$.

crystal but a similar variation of roughness with λ/kT is expected for other structures.

As the roughness of a plane increases, the impediment to growth decreases progressively and when it disappears the growth interface is no longer crystallographic. Experimentally melt growth interfaces become smooth when $\lambda/kT \gtrsim 3$. From solution faceted growth appears to occur for all values of λ/kT. Thus the surface roughness has only qualitative significance in determining growth behaviour. One reason is that it includes many defects, e.g. depressions below the mean level, which do not assist growth.

As discussed in section 2.1, growth depends strongly on the availability of kink sites, which exist only at the edges of steps. We now postulate a low energy surface which contains monatomic steps. These steps may be produced by thermal roughening or in other ways which we consider below. The steps will be either closed loops or extend to the crystal edges, or they may have one or both ends where a screw dislocation meets the surface. Closed steps can only be in equilibrium at under- or supersaturation, i.e. with $R/R_\infty = \alpha \neq 1$. We now consider a crystal in equilibrium, with $\alpha = 1$, for which a step in equilibrium with the surface has a constant mean direction.

At finite temperature all steps have thermally induced kinks. The problem is easier than the surface one because the kinks are non-interacting even at high density, and the topography of a step can be completely described in terms of edge displacements, apart from overhangs, which will be rare. We are interested in the equilibrium spacing of kinks $x_k(\theta)$ and the effective edge energy $\varepsilon(\theta)$ for steps of any mean orientation θ, where $\theta = 0$ is the orientation of the lowest energy step, e.g. a $\langle 100 \rangle$ step on a Kossel crystal.

The kink formation energy for a Kossel crystal, normalized to kT, is

$$w_k = (\phi_1 + \phi_2)/2kT \simeq \lambda/6kT. \tag{2.31}$$

BCF showed that for a $\langle 100 \rangle$ step

$$x_k(0) = \tfrac{1}{2}\exp(w_k) + 1 \tag{2.32a}$$

and that the edge energy is reduced by configurational entropy from $\varepsilon_0(0) \simeq \lambda/6$ for a perfectly straight step to

$$\varepsilon(0) = \varepsilon_0(0) - 2kT\exp(-w_k), \tag{2.32b}$$

$$\simeq \lambda/6 - 2kT\exp(-\lambda/6kT). \tag{2.32c}$$

The configurational term is negligible when $\lambda/kT \lesssim 20$. For a $\langle 110 \rangle$ step with $\theta = 45°$,

$$x_k(45) = 1, \tag{2.33a}$$

$$\varepsilon(45) = \varepsilon_0(45) - 2kT\ln[1 + \exp(-\phi_2/2kT)], \tag{2.33b}$$

$$\simeq \lambda/3 - 2kT\ln 2. \tag{2.33c}$$

For f.c.c. crystals the lowest energy plane is $\{111\}$ and the lowest energy step on a $\{111\}$ plane is $\langle 110 \rangle$. Each atom in bulk has 12 nearest neighbours with bond strength $\phi = \lambda/6$, considering only nearest neighbour bonds. The edge energy per atom of a straight $\langle 110 \rangle$ step is $\varepsilon_0(0) = \lambda/6$ and the kink formation energy is $w_k = \tfrac{1}{2}\phi = \lambda/12$. Thus $\varepsilon_0(0)$ is similar to that for a Kossel crystal, but w_k is lower and hence x_k and $\varepsilon(0)$ are smaller, for a given λ/kT.

In growth from the vapour, λ/kT ranges from about 18 to 40, and in growth from solution or from the melt from about 1 to 30. Values of $x_k(0)$ and $\varepsilon(0)$ for f.c.c. and simple cubic crystals and $\varepsilon(45)$ for simple cubic are shown in Table 2.3 for three values of λ/kT.

Now kinks are the "half-crystal" positions at which atoms have the crystal binding energy λ. Atoms which reach a step edge may migrate to a kink either directly along the edge or by a combination of edge and surface migration. Comparing x_s and x_k in Tables 2.1 and 2.3 we see that $x_s > x_k$ for all λ/kT. Hence any atom which reaches a step edge would have a high probability of reaching a kink on the edge if migrating by surface diffusion alone, and the probability will be increased by edge adsorption and migration. We conclude that there is always a high probability that an atom which reaches a step will

TABLE 2.3. *Kink Spacing and Edge Energy as Dependent on λ/kT for Simple Cubic and Face-centred Cubic Crystals*

$\dfrac{\lambda}{kT}$	Simple Cubic Crystals			Face-centred Cubic	
	$x_k(0)$	$\dfrac{\varepsilon(0)}{kT}$	$\dfrac{\varepsilon(45)}{kT}$	$x_k(0)$	$\dfrac{\varepsilon(0)}{kT}$
30	75	5.0	8.6	7	4.8
18	11	2.9	4.6	3	2.6
9	3	1.1	1.6	2	0.5

reach a kink on the step, i.e. we may treat all sites on the step edge as in equilibrium with the crystal, with attachment probability α.

From Table 2.3 we also see that $\varepsilon(0) < \varepsilon(45)$ for all λ/kT and that for both simple cubic and f.c.c. crystals $\varepsilon(0) \to 0$ for $\lambda/kT \gtrsim 5$. However, from Fig. 2.6 we note that for $\lambda/kT \gtrsim 5$ the two-level and multi-level predictions diverge indicating that the model of steps on an otherwise flat surface is no longer valid, so that eqn. (2.32b) may not apply. Indeed, occurrence of faceted growth with $\lambda/kT < 5$ indicates that $\varepsilon(0)$ does not go to zero.

2.6.3. THE EQUILIBRIUM STRUCTURE AND FORMATION ENERGY OF TWO-DIMENSIONAL NUCLEI

The surface energy change associated with the formation of a two-dimensional nucleus is just the edge energy, which we write as $g_e(i)$, since the surface energy $i\sigma$ of the major surface is balanced by eliminated surface energy $-i\sigma$ of the crystal plane. The formation energy from monomer g_i and the equilibrium density of i-clusters N_i are then

$$g_i = ikT\ln\alpha + g_e(i) + g_c, \tag{2.34a}$$

$$N_i = n_1 \exp(-g_i/kT). \tag{2.34b}$$

In two-dimensional nucleation, monomer and clusters may be regarded as localized on adsorption sites. Defining the site density as unity (the units of N_i are site^{-1}) and using the approximation $\Sigma n_i = n_1$, Lothe and Pound[18] showed that $g_c = kT\ln n_1$. With this substitution eqns. (2.34) may be written

$$\hat{g}_i = -ikT\ln\alpha + g_e(i), \tag{2.35a}$$

$$N_i = \exp(-\hat{g}_i/kT). \tag{2.35b}$$

This formulation has the advantage of directly demonstrating that N_i is proportional to α^i.

We shall now check the validity of eqns. (2.35) for $i = 1$. The adsorption energy of monomer on the surface is $2\sigma_0$, and with an incidence rate R the monomer density is

$$N_1 = \frac{R}{v_0}\exp\left(\frac{2\sigma_0}{kT}\right). \tag{2.36a}$$

In eqns. (2.35) we now put $i = 1$, $\alpha = R/R_\infty$, and $R_\infty = v_\infty\exp(-\lambda/kT)$, which gives

$$N_1 = \frac{R}{v_\infty}\exp\left[\frac{\lambda - g_e(1)}{kT}\right], \tag{2.36b}$$

which agrees with eqn. (2.36a) if $v_0 = v_\infty$ and

$$g_e(1) = \lambda - 2\sigma_0. \tag{2.36c}$$

If $g_e(1)$ and σ_0 are evaluated from the unsatisfied bonding of an adatom, e.g. $g_e(1) = 2\phi$, $\sigma_0 = \frac{1}{2}\phi$, $3\phi = \lambda$ for a Kossel crystal, this relation is satisfied.

The edge energy $g_e(i)$ may be evaluated atomistically as

$$g_e(i) = p_i\varepsilon_0 - kT\ln[k_i + k_i'\exp(-2w_k)], \tag{2.37}$$

where k_i is the number of ground state configurations with p_i unbonded edge sites and k_i' is the number of excited state configurations with excess energy $2kTw_k$. However, this formulation is inconvenient because it is tedious to evaluate for $i \lesssim 6$ and we cannot differentiate to find i^* and g_i^*. The classical assumption is that nuclei have an equilibrium shape independent of i and an edge energy proportional to $i^{1/2}$. Then

$$g_e(i) = \beta i^{1/2}, \tag{2.38a}$$

$$\hat{g}_i = -ikT\ln \alpha + \beta i^{1/2}, \tag{2.38b}$$

and maximizing gives

$$i^* = (\beta/2kT\ln \alpha)^2, \tag{2.38c}$$

$$g_e(i^*) = \beta_i^{*1/2} = 2i^*kT\ln \alpha, \tag{2.38d}$$

$$\hat{g}^* = \beta^2/4kT\ln \alpha. \tag{2.38e}$$

In order that eqns. (2.38c, d, e) should be correct, it is not necessary for eqn. (2.38a) to be valid for all i, but it should be correct at and near i^* where critical clusters are in metastable equilibrium.

If the nuclei have isotropic edge energy ε per atom (i.e. independent of the orientation of the edge) as Volmer[16] assumed, the equilibrium cluster shape is circular with

$$i^*/d^2 = \pi/4 = 0.78, \quad \beta = 2\pi^{1/2}\varepsilon, \tag{2.39}$$

where d is the nucleus diameter. If the nuclei have unkinked $\langle 100 \rangle$ edges with edge energy $\varepsilon_0(0)$ per atom then the equilibrium clusters are square with edge length d

$$i^*/d^2 = 1, \quad \beta = 4\varepsilon_0(0), \tag{2.40}$$

as given by Becker and Döring.[10]

BCF[36] derived the equilibrium cluster shape on a $\langle 100 \rangle$ Kossel crystal surface from the statistics of kinks along a step edge in (metastable) equilibrium which, for $\alpha > 1$, has convex curvature which depends on α and the local orientation θ, and forms a closed loop. By symmetry there is a point with $\theta = 0$ at the middle of each side of an equilibrium nucleus, and such a point is taken as the origin. They then obtain the locus of the equilibrium cluster edge from $x = y = \theta = 0$ at the origin to $x = p(0)$ at $\theta = 90°$ as

$$\frac{y}{d'} = \frac{1}{2w_k}\ln\left[1 - \frac{\sinh^2(w_k x/d')}{\sinh^2(w_k/2)}\right], \tag{2.41a}$$

$$d' = 2w_k/\ln \alpha, \quad \alpha > 1. \tag{2.41b}$$

Equation (2.41a) is not quite correct since the figure is not closed, as it should be, because of neglect of "overhangs" in the kink statistics. However, it is a good approximation for $45° > \theta > -45°$, and the complete nucleus shape can be obtained by repetition over the other three quadrants. Equations (2.41) show that the equilibrium shape depends only on w_k while the equilibrium (critical) size depends on w_k and α.

The equilibrium shape is a classical concept and should not be interpreted as representing the actual configuration of a critical nucleus. Rather it represents the mean of all possible configurations with appropriate statistical weighting. When w_k is very large the equilibrium shape is a square, as in eqn. (2.40). For the moderate value $w_k = 3$ the corners become rounded, and $i^*/d^2 = 0.86$ where $d = d(0)$ is the diameter along or perpendicular to $\theta = 0$. A derivation of eqn. (2.41) and other examples of equilibrium shapes are given by van Leeuwen and Bennema.[38] When $w_k = 1.33$ the shape is circular and for $w_k < 1.33$ $d(45)$ is slightly smaller than $d(0)$. We now introduce a shape factor

$$p = 4(i^*/d^2)^{1/2}, \tag{2.42}$$

and note that for a Kossel crystal for all w_k

$$4 > p > 3.5.$$

Kink statistics may also be used to find d in terms of α and w_k. However, when w_k is small this method is inaccurate due to neglect of overhangs, and it is preferable to obtain the relation from Wulff's theorem. Wulff's theorem states that in a crystal at equilibrium the distances of the faces from the centre of the crystal are proportional to their surface free energies per unit area. BCF derived the corresponding relation for two-dimensional nuclei, which is

$$\tfrac{1}{2}d(\theta) = L\varepsilon(\theta), \tag{2.43a}$$

where $d(\theta)$ is the diameter perpendicular to an edge with orientation θ and edge energy $\varepsilon(\theta)$, and L is a constant characteristic of the size of the critical nucleus. The derivation also gave the total edge energy as

$$g_e(i^*) = 2i^*/L. \tag{2.43b}$$

Comparison with eqn. (2.38d) then shows that $L^{-1} = kT\ln \alpha$ and hence

$$2\varepsilon(\theta)/d(\theta) = kT\ln \alpha. \tag{2.43c}$$

Now eqn. (2.43) applies for all θ and we may therefore now put $\theta = 0$, $d = d(0)$, and use eqn. (2.43c) to obtain

$$d = 2\varepsilon(0)/kT\ln \alpha. \tag{2.44a}$$

Then introducing the shape factor p defined by eqn. (2.42), comparison with eqns. (2.38) shows that

$$\beta = p\varepsilon(0) \tag{2.44b}$$

in which $\varepsilon(0)$ is given by eqn. (2.32b).

Values of p, $\varepsilon(0)/kT$, β/kT, and $\hat{g}^*/kT$ as dependent on λ/kT and α are shown in Table 2.4. The variation of p is relatively unimportant. The important variable is $\varepsilon(0)/kT$, which is

TABLE 2.4. *Shape Factor, Edge Energy, and Critical Nucleus Formation Energy as Dependent on λ/kT and α for Simple Cubic Crystals*

				$\alpha = 1.01$	1.1	2	10
$\dfrac{\lambda}{kT}$	p	$\dfrac{\varepsilon(0)}{kT}$	$\dfrac{\beta}{kT}$	$\dfrac{\hat{g}^*}{kT}$	$\dfrac{\hat{g}^*}{kT}$	$\dfrac{\hat{g}^*}{kT}$	$\dfrac{\hat{g}^*}{kT}$
30	3.9	5.0	19.2	9000	960	132	40
18	3.7	2.9	10.8	2880	304	40	12
9	3.5	1.1	3.7	342	36	5	1.5

proportional to λ/kT when λ/kT is large and tends to zero when λ/kT becomes small. The nucleation rate is proportional to $\exp(-\hat{g}^*/kT)$. Table 2.4 shows that in growth from the vapour with $\lambda/kT > 18$, $\hat{g}^*$ is large and the two-dimensional nucleation rate is very small for $\alpha \lesssim 10$.

BCF[36] used these results only to demonstrate that two-dimensional nucleation growth could not account for Volmer and Schultze's[32] experiments with materials with $\lambda/kT \sim 30$ at $\alpha \sim 1.01$. However, they have much wider significance. It has been shown that the assumption of classical nucleation theory that the cluster edge energy varies as $\beta i^{1/2}$, where β is constant, is justified at constant temperature, but that β is temperature dependent. Furthermore, β and $\hat{g}^*$ have been evaluated for simple cubic crystals in terms of λ and λ/kT. However, these relations will not apply when the nucleus size is too small for the kink statistics to be valid. It is therefore advisable in any calculations to check the value of i^* and to use eqn. (2.37) when i^* is small.

The derivations above are for a Kossel crystal. For real crystals we should base estimates on known properties as we did for three-dimensional nucleation in section 2.5.4, making use of the experimental values of λ/kT and σ/λ given in Fig. 2.4, since there are no experimental measurements of edge energy.

To obtain ε_0 from σ and λ we consider a simple layer of i atoms with the equilibrium shape of a two-dimensional nucleus. Treating the surface energy $2i\sigma$ and the edge energy $pi^{1/2}\varepsilon_0$ as unsatisfied bonding, and λ as the total bonding per atom,

$$i\lambda = 2i\sigma + pi^{1/2}\varepsilon_0 + M(i), \tag{2.45}$$

where $M(i)$ is the mutual bonding between i atoms of the cluster. This equation contains two unknowns, $p\varepsilon_0$ and $M(i)$, both related to the co-ordination within the layer. To proceed further we divide the layer into i single atoms for each of which $M(1) = 0$ and

$$\lambda = 2\sigma + p\varepsilon_0. \tag{2.46a}$$

The possible values of shape factor cover so small a range that the assumption $p = 2\pi^{1/2} = 3.5$, as for circular nuclei, is always acceptable. Thus

$$\varepsilon_0 = (\lambda - 2\sigma)/2\pi^{1/2}. \tag{2.46b}$$

This relation depends on the assumption that σ and $p\varepsilon_0$ are constant down to $i = 1$, which may not be accurate, but nevertheless gives the best available estimate of ε_0 for the general case. From Fig. 2.4, σ/λ lies between 0.12 and 0.22 for solid vapour interfaces, which gives ε_0/λ ranging from 0.22 to 0.16, respectively. For solid–liquid interfaces, σ/λ is larger and ε_0/λ is very small.

To obtain ε from ε_0 we now assume that for the general case a relation similar to eqn. (2.32b) applies, with $w_k = \varepsilon_0/kT$, i.e. that

$$\varepsilon = \varepsilon_0 - 2kT\exp(-\varepsilon_0/kT). \tag{2.47}$$

This relation is plotted in Fig. 2.7 and predicts that ε goes to zero when $\varepsilon_0/kT = 0.85$. Finally, putting $\beta = 2\pi^{1/2}$, eqns. (2.38) give

$$i^* = \pi\varepsilon^2/(kT\ln\alpha)^2, \tag{2.48a}$$

$$r^* = \varepsilon/kT\ln\alpha, \tag{2.48b}$$

$$\hat{g}^* = \pi\varepsilon^2/kT\ln\alpha. \tag{2.48c}$$

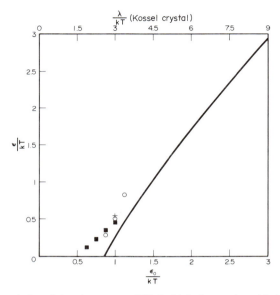

FIG. 2.7. The variation of the edge energy ε/kT of a kinked step edge with λ/kT and with ε_0/kT for a straight step, as given by eqn. (2.47). The points are values derived from simulation experiments by Gilmer and Bennema[53] ■, van Leeuwen and van der Eerden[55] +, and van der Eerden et al.[56] ○.

2.6.4. TWO-DIMENSIONAL NUCLEATION AND GROWTH

Using eqns. (2.18a) and (2.35b) the formation rate of supercritical nuclei is

$$J = Z\gamma^* \exp(-\hat{g}^*/kT), \tag{2.49a}$$

in which the Zeldovich factor for two-dimensional nucleation is

$$Z = \left(\frac{(\ln \alpha)^{1/2}}{4\pi i^*}\right) = \frac{kT}{2\pi\varepsilon}(\ln \alpha)^{3/2}. \tag{2.49b}$$

Following BCF,[36] the supply rate of adatoms to the edge of an isolated circular nucleus of radius r^* by surface diffusion is

$$\gamma^* = \frac{2\pi R x_s^2}{K_0(r^*/x_s)I_0(r^*/x_s)}, \tag{2.49c}$$

where $R = \alpha R_\infty$ is the supply rate of monomer to the surface and K_0, I_0 are Bessel functions.

The net growth rate of circular supercritical clusters is similar to eqn. (2.49c) with $(\alpha - 1)R_\infty$ replacing R, and r replacing r^*. However, the equilibrium shape will not be maintained during growth. The edges will tend to straighten with the orientation of low order steps. It is then more convenient and accurate to consider growth as the advance rate of the edges. For a step separated from neighbouring parallel steps by a distance y, BCF showed that the advance rate is

$$\Gamma_e = 2(\alpha - 1)R_\infty x_s \tanh(y/2x_s). \tag{2.50a}$$

When $y \gg x_s$, as applies for growth of isolated supercritical clusters with diameter $\gg x_s$,

$$\Gamma_e = 2(\alpha - 1)R_\infty x_s, \tag{2.50b}$$

which corresponds to collection of atoms incident within x_s of either side of the edge.

When the crystal size and supersaturation are sufficiently small that $(l^2 J)^{-1}$ is larger than the growth time $l/2\Gamma_e$ to cover a face of edge length l, the growth rate in monolayers s^{-1} is $l^2 J$. This condition requires that the growth rate is less than $F_\infty x_s/l$, where $F_\infty = (\alpha - 1)R_\infty$ is the unimpeded growth rate at saturation α. In some cases this condition could be satisfied with $l \sim 1\,\mu$m, but more often $l \ll 1\,\mu$m would be required. It follows that in most cases the cluster growth rate must be taken into account.

For the large crystal case we consider a square nucleus which grows at rate Γ_e from an edge length $z = 2x_s$ to $z = 2x_s + \Gamma_e t$ over time t. The effective nucleation area on top of this growing cluster is zero at $t = 0$ because of monomer depletion due to its own growth and approximately $4\Gamma_e^2 t^2$ at time t, which gives a nucleation rate $4\Gamma_e^2 t^2 J$. We now integrate $4\Gamma_e^2 t^2 J\,dt$ from $t = 0$ to τ_0 and equate to unity to find $\tau_0 = (4\Gamma_e^2 J/3)^{-1/3}$ as the mean time at which a new nucleus forms on the growing one. Except at high supersaturation, τ_0 is much larger than the growth time from i^* to $2 = 2x_s$ and the growth rate F_0 of new planes is τ_0^{-1}. Thus

$$F_0 = (4\Gamma_e^2 J/3)^{1/3}. \tag{2.51a}$$

The nucleus spacing is such that each nucleus coalesces with its neighbours at about the same time as a new nucleus forms on top. Growth relations of this form have been given by Hillig,[44] Brice,[45] and Bertocci,[46] and have been used for interpretation of simulation studies as we describe in section 2.6.

To establish the dependence on α, ε, and x_s we now substitute for Γ_e and J to find

$$F_0 \simeq (\alpha - 1)^{2/3}\alpha^{1/3}R_\infty x_s^{4/3} \exp - \frac{\pi(\varepsilon/kT)^2}{3\ln\alpha}, \tag{2.51b}$$

in which a quasi-constant term of order unity is omitted. The subscript 0 denotes two-dimensional nucleation growth.

F_0/R_∞ is plotted against supersaturation in Fig. 2.8 for the parameter values $\lambda/kT = 30$ and 6 and $\varepsilon_0/\lambda = 0.2$ and 0.17, which are the values given by eqn. (2.47) for low energy planes with $\sigma/\lambda = 0.14$ and 0.2, respectively. Scales are given for α and for F for $R_\infty = 10^{14}\exp(-\lambda/kT)$ monolayers s^{-1}, which is appropriate for growth from the vapour.

At high supersaturation eqn. (2.51b) does not apply due to the omission of the tanh term in eqn. (2.50a) and an analogous term in eqn. (2.49c) and depletion of the monomer density by capture. These effects cause F_0 to approach the rough plane growth rate F_∞ rather gradually above a saturation ratio α_0 given by

$$\ln\alpha_0 = \pi(\varepsilon/kT)^2/4\ln x_s. \tag{2.52}$$

Below α_0, F_0 drops sharply, as shown in Fig. 2.8. Values of i^* are shown against the curves. The curves are steep and the small difference between $\varepsilon_0/\lambda = 0.2$ and 0.17 changes F_0 by a factor ~ 100, when $\alpha < \alpha_0$. α_0 and $(\alpha_0 - 1)$ are plotted against λ/kT in Fig. 2.9 and range from $\alpha_0 = 1$ for $\varepsilon_0/\lambda = 0.17$ and $\lambda/kT \lesssim 3.3$, or $\varepsilon_0/\lambda = 0.2$ and $\lambda/kT \lesssim 2.8$, which gives $\varepsilon = 0$ and no nucleation barrier, to $\alpha_0 = 10^6$ for $\varepsilon_0/\lambda = 0.2$ and $\lambda/kT = 100$.

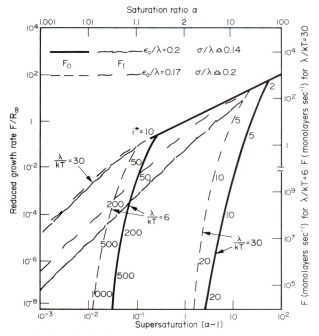

FIG. 2.8. The variation of reduced growth rate F/R_∞ with supersaturation $(\alpha - 1)$ for two values of λ/kT. Scales for these values of λ/kT are also given. The full lines and the continuous screw lines give F_0 (two-dimensional nucleation) and F_1 (screw dislocation growth) for $\varepsilon/\lambda = 0.2$. The broken lines give F_0 and F_1 for $\varepsilon/\lambda = 0.17$. The numbers against the curves are the critical nucleus size i^*.

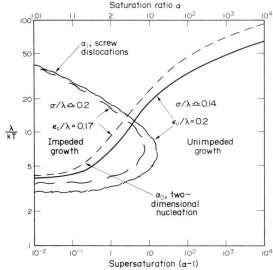

FIG. 2.9. The variation with λ/kT of the transitional saturation ratios α_0 for two-dimensional nucleation and α_1 for screw dislocation growth, plotted as $(\alpha_0 - 1)$ and $(\alpha_1 - 1)$. Below these saturation ratios the growth rate is impeded, and above them it approaches F_∞.

At $\alpha = \alpha_0$ the mean spacing between the steps generated by two-dimensional nucleation is $y_0 = 2x_s$. Below α_0, $y_0 = 2x_s F_\infty / F_0$, i.e. the step spacing increases as the growth rate falls and the surface becomes smoother. Above α_0 the step spacing decreases below y_0 and the surface roughens.

Kashchiev[48] has examined the variation of surface roughness with saturation theoretically for a growth model similar to that described above. Lateral growth and coalescence is treated more rigorously by methods introduced by Kolmogoroff[49] and Avrami.[50] At low supersaturation the growth rate is as in eqns. (2.51) with a slightly different numerical factor. Kashchiev predicts that at low supersaturation the mean height h_{99} above a layer which has just reached 99% coverage is 1.86 layers and the total thickness of the growth interface is 2.7 layers. The surface is thus quite smooth and a significant proportion of incident monomer is desorbed, i.e. the condensation coefficient is much smaller than the rough-plane value of $(\alpha - 1)/\alpha$. As the supersaturation is increased, the growth rate rises towards F_∞ and the surface roughness increases towards an asymptotic limit with $h_{99} = 4.61$ layers and an interface thickness of 8.2 layers.

2.6.5. SCREW DISLOCATION GROWTH

Table 2.4 shows that in all cases of growth from the vapour the two-dimensional nucleation rate at $\alpha = 1.01$ is negligibly small. Thus the observed growth of crystals from the vapour at low supersaturation by Volmer and Schultze[32] requires some other explanation. Frank[35] suggested that screw dislocations having a displacement vector parallel to the crystal face would provide a source of steps, which would persist, and not grow out.

We first consider growth due to a single emergent dislocation on a low energy plane. At equilibrium the dislocation will create a straight step between the point of emergence, which we take as the origin, and the crystal edge. When $\alpha > 1$, the step will advance at rate Γ_e, and since it is pinned at the origin it will rotate with angular velocity $\omega(r) = \Gamma_e/r$ and wind up into a spiral.

However, Γ_e is a function of the curvature $\rho(r)$ and decreases to zero at the critical curvature $\rho^* = r^*$ of eqn. (2.48b). Hence the spiral will wind itself up to a steady state form with $\rho(0) = \rho^*$ and then continue to rotate at an angular velocity ω which is independent of r. The crystal growth rate is then $\omega/2\pi$ monolayers s^{-1}.

The solution to the problem must satisfy the geometric relations for ρ in terms of r and θ, ω in terms of Γ_e/r and θ, and Γ_e in terms of r, ρ, and ρ^*. The Archimedean spiral

$$r = 2\rho^*\theta \qquad (2.53a)$$

has the proper central curvature $\rho(0) = \rho^*$. The spacing y_∞ between turns of this spiral when r is large is

$$y_\infty = 2\pi/\theta'(r) = 4\pi\rho^*, \qquad (2.53b)$$

and the crystal growth rate, which we denote by subscript 1 for a single dislocation, is

$$F_1 = \Gamma_e(\infty)/y_\infty = \Gamma_e(\infty)/4\pi\rho^*, \qquad (2.53c)$$

where $\Gamma_e(\infty)$ is given by eqn. (2.50b) with the validity condition $y_\infty \gg x_s$. When r is large,

$$\omega(\infty) = \Gamma_e(\infty)/2\rho^*, \qquad (2.53d)$$

and $F_1 = \omega/2\pi$ agrees with eqn. (2.53c). Hence this solution is self-consistent for large r. However, $\omega(r)$ decreases for small r, tending to zero at $r = 0$, i.e. the spiral is too tight when r is small. A more accurate solution was obtained by Cabrera and Levine[51] and gives

$$y_\infty = 19\rho^*,\tag{2.54a}$$

$$F_1 = \frac{\Gamma_e(\infty)}{19\rho^*} = \frac{kT\ln\alpha\,\Gamma_e(\infty)}{19\varepsilon} \simeq \frac{(\alpha - 1)\Gamma_e(\infty)}{19(\varepsilon/kT)}\tag{2.54b}$$

in which we have substituted for ρ^* from eqn. (2.48b) and approximated $\ln\alpha$ as $\alpha - 1$ for $\alpha < 1.1$.

As α increases, ρ^* and y_∞ decrease and the validity condition $y_\infty \gg x_s$ is infringed. Two modifications to eqns. (2.54) are required. The first modification is to allow for the diffusion field of the first turn of the spiral on ρ^* at the centre. An approximate treatment of this "back-stress" effect has been given by Cabrera and Coleman.[52] They obtain two asymptotic solutions for the modified turn spacing, which we denote as y_1, as dependent on y_∞ and λ. These solutions are

$$y_1 = y_\infty \quad\text{when}\quad y_\infty \gg 2x_s,\tag{2.55a}$$

$$y_1/2x_s = (y_\infty/2x_s)^{1/3} \quad\text{when}\quad y_\infty < x_s.\tag{2.55b}$$

The second modification, which is needed when $y_1 \lesssim 4x_s$, is to include the tanh term of eqn. (2.50a) in the expression for Γ_e to allow for competition between the diffusion fields of adjacent steps. Then

$$F_1 = \frac{2x_s}{y_1}R_\infty(\alpha - 1)\tanh\frac{y_1}{2x_s},\tag{2.55c}$$

which is valid for all y_1. BCF gave this relation with $y_1 = 4\pi\rho^*$.

To obtain explicit relations for the α-dependence of F_1 we now define a saturation ratio, which we designate α_1, for which $y_1 = 2x_s$. From eqns. (2.55a), (2.54a), and (2.48b)

$$\ln\alpha_1 = 19\varepsilon/2kTx_s.\tag{2.56}$$

Within the validity of eqn. (2.55a) F_1 may now be written as

$$F_1 = \frac{\ln\alpha}{\ln\alpha_1}R_\infty(\alpha - 1)\tanh\frac{\ln\alpha_1}{\ln\alpha}, \quad \alpha \lesssim \alpha_1.\tag{2.57a}$$

The growth rate at $\alpha = \alpha_1$ is $0.76R_\infty(\alpha - 1) = 0.76F_\infty$, which is quite close to the maximum growth rate. When $\ln\alpha < \frac{1}{2}\ln\alpha_1$, $\tanh(\ln\alpha_1/\ln\alpha) \to 1$, and if $\alpha < 1.1$ we may also use the approximation $\ln\alpha = \alpha - 1$. Then

$$F_1 = R_\infty(\alpha - 1)^2/\ln\alpha_1, \quad \ln\alpha < \tfrac{1}{2}\ln\alpha_1 < 0.1,\tag{2.57b}$$

which is quadratic in the supersaturation. When $\alpha > \alpha_1$, y_1 is given by eqn. (2.55b) and the approximation $\tanh z = z - z^3/3$ for $z < 1$ may also be used and eqn. (2.55c) then becomes

$$F_1 = R_\infty(\alpha - 1)\left[1 - \frac{1}{3}\left(\frac{\ln\alpha_1}{\ln\alpha}\right)^{2/3}\right], \quad \alpha > \alpha_1.\tag{2.57c}$$

The term in square brackets varies slowly with α, so F_1 is approximately linear in the supersaturation, and the approach to the maximum growth rate $F_\infty = R_\infty(\alpha - 1)$ is slow. It also follows from eqn. (2.55b) that even when $\alpha > \alpha_1$ $y_1 \sim x_s$, and the growth spiral remains a very shallow cone.

F_1 and α_1 are plotted in Figs. 2.8 and 2.9 for comparison with F_0 and α_0. F_1 is high and α_1 is low for both high λ/kT (because x_s is large) and low λ/kT (because ε is small). When $\lambda/kT > 14$, $F_1 > F_0$ for all α. When $\lambda/kT = 5$ the two growth rate curves cross, two-dimensional nucleation being dominant above $\alpha \sim 1.05$ and screw dislocation growth below.

Experimentally, the square low growth rate predicted below α_1 is fairly rare, but this is because single dislocations are rare. A typical dislocation density is $10^4\,\mathrm{cm}^{-2}$. BCF also considered multiple dislocations. Two dislocations of opposite sign closer than $2\rho*$ have zero activity. If they are further apart than $2\rho*$ they combine to send out closed loops with a step spacing of approximately $4\pi\rho*$ and for any number of such centres more than $4\pi\rho*$ apart the growth rate is the same as that of a single spiral. Two or more dislocations of the same sign each separated by less than $2\pi\rho*$ from its neighbour send out an interleaved system of spirals with separation $y_2 = S/2L$, where S is the excess of dislocations of one sign in a length L. Thus

$$F_2 = \frac{x_s S}{L} R_\infty(\alpha - 1)\tanh\frac{L}{x_s S}, \quad L > 2\pi\rho* > L/S, \tag{2.58}$$

where subscript 2 denotes growth by multiple dislocations. When $L/S = 2\pi\rho*$, $F_2 = F_1$ as for a single dislocation. The growth rate of a face is determined by the group of dislocations with the highest S/L. F_2 varies with $\rho*$ only indirectly via S/L. In general we expect S/L to decrease with increasing $r*$ as $2\pi\rho* = L/S$ increases and more widely spaced dislocations are brought into the most active group. Thus F_2 falls somewhat faster than $(\alpha - 1)$ but slower than $(\alpha - 1)^2$. For an array of length L separated by at least L from other dislocations, F_2 falls as $(\alpha - 1)$. This is the "second linear law" discussed by Bennema[25] which holds until, with increasing $\rho*$, the validity condition $L > 2\pi\rho*$ is violated and the growth rate then falls as $(\alpha - 1)^2$. Thus with multiple dislocations a linear growth rate dependence on $(\alpha - 1)$ does not necessarily imply a growth rate close to F_∞, and a change to the square law $(\alpha - 1)^2$ can occur at $\alpha < \alpha_1$.

When the steps are widely spaced compared with x_s, two-dimensional nucleation and dislocations produce steps independently so that $F = F_0 + F_2$. At higher supersaturation, such that the step spacing is of order x_s, the processes become competitive and both are reduced by the tanh term and depletion of monomer by growth. Thus, as for either separately, the approach to F_∞ with increasing α is gradual.

2.6.6. APPLICATION TO VAPOUR, MELT, AND SOLUTION GROWTH

In growth from the vapour, λ/kT is large and growth at low supersaturation can only occur if there are dislocations. When the supersaturation is high, dislocations are not needed.

The growth of metals and semiconductors from the melt is on the borderline of unimpeded growth even at the lowest supersaturation because of the low values of λ/kT

and ε/kT. Thus in some cases faceted growth occurs at low supersaturation, and in others the growth interface is determined by heat flow rather than by crystal anisotropy.

We expect growth from solution to be similar to growth from the melt. However, experimentally there are many cases in which solution growth produces well-defined crystal planes although λ/kT is very low. We conclude that in solution-growth perfect crystal planes form because of some additional impediment to their growth, related to desolvation or to the adsorption of solvent.

2.7. Simulated Crystal Growth

2.7.1. THE SCOPE AND OBJECTIVES OF SIMULATION STUDIES

Over the last decade theoretical studies of crystal growth have been supplemented by computer simulation of growth processes.

Computer simulation represents the crystal surface by a matrix, typically 50×50 atoms square. Atoms are incident on this surface and are allowed to form bonds, to migrate, and to desorb. The physical processes of incidence, migration, cohesion, and decay are represented by rules written into the computer program. Each statistically determined event, e.g. selection of an incidence site, or the lifetime of an atom in any given configuration, is randomly chosen with the probability appropriate to the specified physical parameters.

A severe limitation of simulation techniques is that the parameters must be within the capability of the computer. The real size scale ranges from 1 atom to 10^4 or more atoms in a cluster. A representative surface area may cover 10^6 or more sites. The real time scale ranges from 10^{-11} s for a migration jump to seconds for a representative growth time. In computer simulation these ranges must be enormously compressed, and direct investigation of real growth kinetics is therefore restricted.

In section 2.6 a model, which represents the essential features of nucleation and growth, has been reduced to analytical form. There is no limitation to the numbers which can be inserted. However, detailed checks between experiment and theory are hampered by measurement difficulties and lack of knowledge of material parameters.

In simulation all the parameters are known, the model is completely defined, and any desired experimental result can be printed out. This situation is ideal for comparison between theory and experiment. Consequently the most useful role for simulation is to test the completeness and accuracy of theory.

2.7.2. EQUILIBRIUM SURFACE STRUCTURE

Gilmer and Bennema[53] and Leamy and Gilmer[54] examined the equilibrium surface structure of a Kossel crystal as dependent on ε_0/kT. The surface configuration was recorded as the height above each lattice point. The triangles and circles plotted in Fig. 2.6 are experimental values of surface roughness $(s - s_0)/s_0$, which is equal to the number of unsatisfied lateral bonds per site. Agreement between the two simulations and the five-level BCF theoretical curve is excellent.

Leamy and Gilmer[54] have also examined the equilibrium structure of a surface containing steps. One short edge of a 40×20 matrix was maintained several layers higher than the other so as to generate the required number of steps. The simulation was run to equilibrium with successively smaller values of ε_0/kT. A parameter E_s, representing the surface roughness associated with the steps, was obtained from the variation of s with step density. E_s rose to a maximum at $\varepsilon_0/kT = 1.2$ and then fell to zero at $\varepsilon_0/kT = 0.8$. At this stage the imposed steps were masked by the steps at the edges of thermally induced clusters. The effect is closely related to the surface roughening in Fig. 2.7 and to the variation of ε in Fig. 2.8.

2.7.3. NUCLEATION AND GROWTH

Bertocci[46] examined the dependence of the simulated crystal growth rate on Γ_e, J, and the array size l^2. At high nucleation rates the growth rate was independent of array size and followed eqn. (2.51a) with a numerical coefficient varying with symmetry from $1.5^{1/3}$ to $2^{1/3}$. At lower nucleation rates, the growth rate fell gradually towards $l^2 J$. Thus the simulation confirms and sets limits to the validity of eqn. (2.51a) and provides a solution for intermediate crystal sizes.

Gilmer and Bennema[53] measured the growth rate of stepped and unstepped surfaces, with and without surface diffusion, as dependent on ε_0/kT and supersaturation. Without steps, eqn. (2.51) is expected to apply. With $\varepsilon_0/kT = 1$ and 0.875, the growth rate varied non-linearly with supersaturation due to the exponential term, and Gilmer and Bennema determined ε from the variation with $\ln \alpha$. For smaller ε_0/kT the growth rate varied almost linearly with $\ln \alpha$ but was still well below the unimpeded growth rate F_∞, and ε can again be calculated, but less accurately. These experimental values of ε are plotted as squares in Fig. 2.7 and lie well above the plot of eqn. (2.47). The range of critical nucleus size in these experiments is between 1 and 60 atoms. We conclude that eqn. (2.47) is inaccurate when i^* is small, but it may still be satisfactory for the large critical nuclei for which it was derived.

With steps maintained in the matrix, the growth rate was consistent with capture at steps alone at high ε_0/kT and low α. With increasing α and decreasing ε/kT, the growth rate was progressively and significantly increased by nucleation growth on the terraces. Surface diffusion with $x_s = 1$ or 2 sites increased the growth rate as predicted by a growth relation similar to eqn. (2.49c). Higher α, small ε_0/kT, larger x_s, and the presence of steps all increased the growth rate towards F_∞. However, even for the most favourable conditions the experimental growth rate remained below F_∞.

van Leeuwen and van der Eerden[55] measured the coverage a by clusters, the area b of the largest cluster, and the monomer concentration c on an initially flat 40×10 matrix as dependent on time and supersaturation for $\varepsilon_0/kT = 1$. After a short induction time, a, b, and c remained approximately constant until a time τ^+ after which a and b increased sharply and c fell. They interpreted τ^+ as the formation time of a growing supercritical cluster, and from its $\ln \alpha$ dependence derived the value of ε plotted as $+$ in Fig. 2.7, which agrees well with Gilmer and Bennema's evaluation from growth rate measurements. Beyond τ^+, b increased irregularly, and in some cases approximately linearly with time. However, the growth rate of a single supercritical cluster is expected to increase gradually beyond i^*. Hence the observed rapid growth rate after τ^+ is probably due to coalescence of two or more clusters, and the linear area increase then corresponds to a decrease of Γ_e as concavities grow out. It is unreasonable to infer that individual clusters have linear area growth, which implies a continuous decrease of Γ_e with increasing cluster size. Indeed, the measurements show an irregular, generally quadratic area growth rate, as we expect, becoming systematically linear only at coverages above 30%.

van der Eerden et al.[56] have measured simulated crystal growth rates on surfaces with isotropic (cubic) and anisotropic (tetragonal) bond energies. Anisotropy increased the growth rate. The isotropic growth rate was numerically similar to that observed by Gilmer and Bennema[53] but was interpreted as being proportional to $(\Gamma_e J)^{1/2}$ (for linear area cluster growth) from which values of ε were obtained which were lower than those found by Gilmer and Bennema. Reinterpretation assuming Γ_e is independent of cluster size gives the values of ε plotted as circles in Fig. 2.7.

van der Eerden et al.[56] also found that the difference between the isotropic and anistropic growth rates decreased as $\ln \alpha$ was increased or ε_0/kT was decreased, becoming equal for conditions giving a rough surface structure. They showed correspondence in this region with a pair approximation model and that the growth rate within the roughening transition region remains below the unimpeded growth rate F_∞ until the supersaturation becomes very large. Thus roughening and the approach to F_∞ occurred gradually over an extended transition region.

Kashchiev et al.[57] examined growth structures at different supersaturations. They found that cluster growth shapes differed noticeably from the equilibrium shape and changed with supersaturation and time, in agreement with theoretical expectations that growth forms are kinetically determined. The interface width at 99% coverage of an initially flat surface varied between limiting values of 2 layers at low supersaturation to 10 at high supersaturation, in fair agreement with Kashchiev's[48] theoretical predictions of 2.7 and 8.2 layers, respectively.

We conclude that simulation studies have generally confirmed the treatment of nucleation and growth described in sections 2.6.3 and 2.6.4 and provide some guidance for cases outside the validity limits of the theoretical formulations.

2.8. Material and Heat Flow in Crystal Growth

In section 2.2 the supersaturation was expressed in terms of concentrations and temperatures. For the nucleation step by which a crystal is first formed, the source concentration, the temperature, and the supersaturation may be uniform and well defined. However, as soon as a crystal is growing the conditions are perturbed by the material and heat flow associated with growth.

2.8.1. GROWTH FROM SOLUTION

Let us consider a crystal growing from a supersaturated solution C_∞ and temperature T_∞. Figure 2.19 illustrates the variation of temperature and concentrations with distance from the face of the crystals. Firstly, the latent heat liberated at the growing faces raises the temperature there to T_b, which is higher than T_∞, and there is a corresponding variation of the equilibrium concentration C_e. In solution growth this variation is normally quite small. Secondly, there is a concentration gradient associated with volume diffusion of solute from solution to the growing face, which depresses the concentration C_a at the face below C_∞. The magnitude of the diffusion drop $\Delta C_d = C_\infty - C_a$ is proportional to the growth rate and also depends on the volume diffusion coefficient and the vigour of stirring.

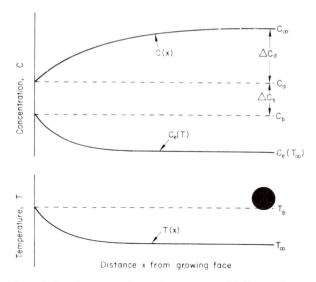

FIG. 2.10. The variation of concentration and temperature with distance from a crystal face in growth from solution.

The effective distance over which this drop occurs is sometimes called the "unstirred layer". The concentration difference $\Delta C_s = C_a - C_b$ is the supersaturation associated with the processes of adsorption, migration, and capture at the crystal surface.

Growth behaviour represents the solution of the continuity equation for the growth flux which relates ΔC_d and ΔC_s, and of the heat balance equations. A treatment of this problem has been given by Gilmer et al.[58] Here we seek only to obtain some qualitative understanding of the effects of material and heat flow. The dependence of growth rate on supersaturation has already been considered and is generally non-linear. Thus at low supersaturations and with vigorous stirring we expect supersaturation to be the controlling parameter, i.e. $\Delta C_s > \Delta C_d$, whereas at high supersaturation and without stirring we may find $\Delta C_d > \Delta C_s$. Furthermore, we must take account of the non-uniformity of ΔC_s associated with the greater ease of volume diffusion to the corners than to the centres of a face.

Although in solution growth we do not, in general, have effective control or knowledge of the supersaturation, we do have excellent control of the growth rate. The mass of crystal which grows during any drop of temperature is simply the mass which must come out of solution to maintain the concentration just above the equilibrium concentration. We are assuming that the change of $\Delta C_s + \Delta C_d$ during growth is small. We can program the temperature drop to give any required growth rate, and under these "forced growth" conditions the supersaturation adjusts itself to provide that growth rate.

2.8.2. GROWTH FROM THE MELT

The controlling parameter in growth from the melt is heat flow associated with the dissipation of the latent heat liberated at the growing face. The direction of heat flow depends on whether the heat is removed through the melt, or the growing crystal, or both. The heat flow provides "forced growth" conditions in which the supersaturation at the interface is again self-regulating to provide the externally imposed growth rate. Fluxed-melt growth, which is described in Chapter 12, is controlled by both material and heat flow and, again, "forced growth" conditions prevail.

Interface stability in growth from the melt and the effects of constitutional supercooling in two-component systems are discussed in Chapter 3 of the first edition.

2.8.3. GROWTH FROM THE VAPOUR

Growth from the vapour and hydrothermal growth are notably different from the cases considered above. In a closed system with a source at temperature T_a and a substrate or seed crystal at temperature T_b, the supersaturation at the crystal surface is known provided that the geometry is suitable and the net growth rate is well within the capacity of the source. In a molecular beam vapour growth system the incidence rate R_a can be determined by pressure measurements or by calibration experiment in which T_b is low, R_b is very small, and complete condensation occurs. The same source conditions can then be used in experiments in which the crystal temperature is raised so that R_b approaches R_a and the supersaturation becomes small.

In a flow system, such as is commonly used in epitaxial growth of semiconductor crystals, vapour is transported by a carrier gas from a hot source zone to a cooler growth zone, and out to waste. The supersaturation is not known, the growth rate is not "forced", and neither can be predetermined, since the concentration profile is not known or controlled.

In section 2.6 we considered nucleation and growth in terms of supersaturation and temperature alone. While the principles which have been established apply to all growth systems, the discussion above indicates the difficulties of quantitative application except in some cases of growth from the vapour phase.

2.9. The Kinetic Generation of Crystal Forms

We have, in section 2.6, distinguished between unimpeded growth at the rate F_x and impeded growth which leads to the production of low energy planes. Experimentally impeded growth produces a variety of crystal forms which we now briefly describe and relate to the generating growth processes.

2.9.1. WHISKERS

Filamentary crystals, or whiskers, are typically $1-50\,\mu m$ in diameter and up to a few centimetres long. The particular features of interest here are fast growth along an axis which is often not a unique crystallographic axis. The inference, which has been experimentally confirmed, is that whiskers contain an axial screw dislocation, which determines the fast growth direction. The side faces of whiskers are low energy planes which grow only slowly, or not at all, and feed the emergent screw at the tip (or at the base) by surface diffusion. For a whisker of radius r, the growth rate may then be x_s/r times the direct impingement rate on the tip, and this factor may be as large as 1000.

For whisker growth to occur, the supersaturation of the side faces must be sufficiently low to prevent two-dimensional nucleation which would allow radial growth. Sears[59] has found that there is a critical supersaturation above which whiskers of Cd, Ag, Zn, and CdS do not grow, which corresponds to the calculated supersaturation for two-dimensional nucleation. In the case of Hg whiskers[60] grown at constant supersaturation, axial growth stops and sideways growth commences when the whiskers are a few millimetres long. This suggests that two-dimensional nucleation is suppressed for a distance x_s which is depleted of adatoms by growth at the tip. When this length is exceeded, nucleation generates steps which propagate along the whiskers, causing general sideways growth, and the axial growth rate then reverts to the direct impingement rate.

2.9.2. NEEDLES AND PLATELETS

In the case of Hg, Cd, and Zn, which all have hexagonal structure, the whisker axis is in the basal plane. At supersaturations too high for whisker growth, platelets occur lying in the basal plane, and thus conforming with the crystal symmetry; it may be inferred that the basal planes have the lowest surface energy, and that in the appropriate range of supersaturation, two-dimensional nucleation and growth occur on the edges but not on the basal planes. In common with whiskers the growth rate of mercury platelets[7] is considerably enhanced by diffusion from the non-growing surfaces, until the major dimension exceeds x_s, when three-dimensional growth occurs. If low energy planes are parallel to a unique crystallographic axis, then needles are produced. The distinction between needles and whiskers is that the fast growing axis of a needle is a symmetry axis.

2.9.3. FLAT FACES

Crystals commonly grow with flat faces although, due to gradients of mass or heat flow, the supersaturation is non-uniform. Discussing this, Frank[61] has stated: "We know how it does that—the growth rate of the whole face is determined by the supersaturation at the point of emergence of the dominant growth-promoting centre in the face—in most cases a screw dislocation." The argument applies equally well if growth originates by two-dimensional nucleation or by both mechanisms. In all cases monomolecular steps migrate across the face at a rate proportional to the local value of $(\alpha - 1)$; as α varies across the face, their spacing changes and compensates for the change of α to produce a uniform growth rate across the whole face. The mechanism only operates if the step spacing is everywhere $\sim x_s$ or larger and the growth rate is less than F_α.

Between one face and another, with substantially uniform α, a smaller surface energy corresponds to a larger ε and r^*, with consequential changes in the growth rate, as illustrated in Fig. 2.8. For a perfect crystal, the lowest energy planes have much lower growth rates than any other planes. We conclude, in agreement with Gibbs, that all other planes will grow out and the crystal will be bounded by the lowest energy planes allowed by the crystal structure. For faces with a single dominant screw dislocation, the growth rate F_1 is proportional to r^*. Hence if the crystal faces all have single dominant screws, the crystal will be bounded by planes with growth rate proportional to r^* and with areas in inverse order of σ. As shown in Fig. 2.8, the growth rate becomes more sensitive to changes of ε_0 when λ/kT is small. This is because ε is now changing more rapidly with ε_0. For multiple dislocations, the growth rate F_2 is dependent on r^*, ε, and α, and also depends on the density and distribution of dislocations.

2.9.4. EQUILIBRIUM AND CHARACTERISTIC HABITS

Experimentally, it is a matter of common observation that crystals often have characteristic habits, with substantially flat faces and well-defined edges and corners. High symmetry crystals may have only one type of face, e.g. {100} faces in the case of NaCl, as expected from the growth kinetics, if these faces have the lowest surface energy. In crystals of lower symmetry, more than one face occurs. For example, the water-soluble piezoelectric crystals ADP and KDP are orthorhombic and have both {100} and {100} faces. The natural habit is an elongated prism with slow-growing {100} prism faces and faster-growing {110} end faces.

The relative growth rates and areas of crystal faces are by no means constant. Buckley,[62] who has made a particular study of crystal habits, gives many examples of drastic modifications produced by impurities. An example is the stimulation of the growth of the {100} prism faces of ADP and KDP by excess cation, which is used in the production of seed crystals. These effects may be explained by disturbance of the growth kinetics by selective adsorption.

In addition to these examples of crystals whose habits conform with the expectations of two-dimensional nucleation growth, there are many others which do not have the simplest crystallographic form. An example is potassium alum, which often has {100} and {110} faces truncating the corners and edges of the basic octahedral {111} form. In many cases these crystals appear to have a consistent characteristic habit, with a tendency for simpler forms to occur at higher supersaturations.

A critical nucleus is in equilibrium with monomer, and has the form which minimizes the surface energy. Curie[63] proposed that the characteristic habits of macrocrystals were similarly determined. For a crystal of i atoms with minimum surface energy there exists a point within it from which the perpendicular distance $(h_{ij})_{eq}$ to each face j is proportional to the surface energy σ_j. This theorem was enunciated by Wulff.[64]

Frank[61] and Strickland-Constable[2] have shown that the Wulff criterion is related to a variant of the Gibbs–Thomson equation for the vapour pressure P_i over each face of a crystal of i atoms, and may be written

$$kT\ln\frac{P_i}{P_\infty} = \frac{2\sigma_j}{(h_{ij})_{eq}},\qquad(2.59)$$

if both σ_j and h_{ij} are in atomic units. For the equilibrium form, all the P_i's are the same. For any non-equilibrium form, h_{ij} and P_{ij} replace P_i and $(h_{ij})_{eq}$. We can also define a saturation ratio of $\alpha_{ij} = P/P_{ij}$ by analogy to $\alpha = P/P_\infty$, with respect to bulk crystal, and find

$$kT\ln\frac{P_{ij}}{P_\infty} = kT\ln\frac{\alpha}{\alpha_{ij}} = \frac{2\sigma_j}{h_{ij}}.\qquad(2.60)$$

For a crystal with non-equilibrium form, the driving energy towards equilibrium is related to the difference between the $\ln\alpha_{ij}$'s, when the h_{ij}'s differ from $(h_{ij})_{eq}$. However, this difference is $\gtrsim \ln \alpha - \ln \alpha_{ij}$ and is inversely proportional to h_{ij}; for the representative value $\sigma_j/kT = 1$ and a micron-sized crystal, it is ~ 0.001, which is much smaller than the lowest supersaturation at which a crystal will grow. We conclude that the driving energy towards the equilibrium form is quite insignificant for macrocrystals, and that crystal habits are determined by the kinetics of growth.

The occurrence and supersaturation dependence of common crystal forms is satisfactorily explained as due to growth assisted by dislocations. All planes with $\alpha < \alpha_1$ appear and the growth rates are roughly proportional to $1/\varepsilon$.

2.9.5. DENDRITES

We have established that a necessary condition for the development of flat faces is that the growth rate is lower than F_∞ everywhere on the face where $F_\infty = (\alpha - 1)R_\infty$ is itself proportional to the supersaturation, and may be non-uniform. Let us use superscripts c for the corners and f for the face centres of a crystal growing from solution. While the mean forced growth rate is less than F_∞^f and F_∞^c, the faces remain substantially flat, with steps spreading from the regions of highest supersaturation or the dominant dislocation centre. Now let the forced growth rate $\bar{F}$ be increased to exceed F_∞^f, which is smaller than F_∞^c due to non-uniform supersaturation. The face growth rate now lags and the corner growth rate increases to maintain the growth rate at $\bar{F}$. Thus $F^f \lesssim F_\infty^f < \bar{F} < F^c < F_\infty^c$. If the growth rate is increased until F^c approaches F_∞^c the system becomes unstable. In order to attain the forced growth rate, the corners must penetrate further into solution of higher concentration, while the faces grow slowly in depleted solution. This is the classic situation of dendritic growth, easily demonstrated as warm ammonium chloride solution cools on a microscopic slide. As the trunk of the dendrite penetrates into the solution, branches grow from its sides, and twigs from the branches, all directed along the fast-growing crystallographic axes.

Snowflakes, which occur in multivarious forms, are examples of dendritic growth of ice from the vapour. They are generated by changes of supersaturation and temperature, and the symmetry of each flake is due to the hexagonal symmetry of the crystal and the identical environment of each branch as the flake falls through the atmosphere.

In growth from the melt, if λ/kT and ε/kT are sufficiently high to present a nucleation barrier to growth on low energy planes, then crystal faces develop at low growth rates and dendritic growth occurs at high rates. It is very common in metal casting, as described in Chapter 13. When λ/kT is small the growth interface is controlled entirely by heat flow considerations. Morphological instability may occur at high growth rates. In both melt and solution growth, stirring makes the supersaturation more uniform over the growth interface and delays the onset of instability as the growth rate is increased.

References

1. J. W. GIBBS, *On the Equilibrium of Heterogeneous Substances, Collected Works*, Longmans Green, New York (1928), p. 325.
2. R. F. STRICKLAND-CONSTABLE, *Kinetics and Mechanism of Crystallisation*, Academic Press, London (1968).
3. R. L. PARKER, in *Solid State Physics* (edited by H. Ehrenreich, F. Seitz, and D. Turnbull), Academic Press, New York, **25** (1970) 151–299.

4. W. KOSSEL, *Nacht. Akad. Wiss. Göttingen, Math.-Physik. Kl.* (1927) 135–43.
5. I. N. STRANSKI, *Z. Phys. Chem.* **136** (1928) 259–78.
6. I. N. STRANSKI and R. KAISCHIEV, *Z. Phys. Chem.* (B) **26** (1934) 31–39.
7. M. VOLMER, *Kinetik der Phasenbildung*, Steinkopf, Dresden (1939).
8. M. VOLMER and I. ESTERMAN, *Z. Phys.* **7** (1921) 1.
9. G. W. SEARS, *J. Chem. Phys.* **25** (1955) 637–42.
10. R. BECKER and W. DÖRING, *Ann. Phys.* **24** (1935) 719–52.
11. F. C. FRANK, *J. Crystal Growth* **13/14** (1972) 154–6.
12. J. ZELDOVICH, *J. Exp. Theor. Phys.* **12** (1942) 525.
13. J. L. KATZ and H. WEIDERSICH, *J. Colloid and Interface Sc.* **61** (1977) 351–5.
14. J. L. KATZ and F. SPAEPEN, *Phil. Mag. B*, **37** (1978) 137–48.
15. J. L. KATZ and M. C. DONOHUE, *Advan. Chem. Phys.* (to be published).
16. M. VOLMER and A. WEBER, *Z. Phys. Chem.* **119** (1926) 277–301.
17. R. BECKER, *Discuss. Faraday Soc.* No. 5 (1949) 55–61.
18. J. LOTHE and G. M. POUND, *J. Chem. Phys.* **36** (1962) 2080–5.
19. H. REISS, J. L. KATZ, and E. R. COHEN, *J. Chem. Phys.* **48** (1968) 5553–60.
20. K. NISHIOKA and G. M. POUND, *J. Cryst. Growth* **24/25** (1974) 571; *Acta Met.* **22** (1974) 1015–21.
21. M. BLANDER and J. L. KATZ, *J. Statistical Phys.* **4** (1972) 55–59.
22. W. B. HILLIG, *Acta Met.* **14** (1966) 1868.
23. A. BONISSENT and B. MUTAFTSCHIEV, *International Conference on Crystal Growth* (1971), Paper A1–4.
24. D. TURNBULL and J. C. FISHER, *J. Chem. Phys.* **17** (1949) 71–73.
25. P. BENNEMA, *J. Cryst. Growth* **1** (1967) 278–86; ibid. **5** (1969) 29–43.
26. H. JONES, *Metal Sci. J.* **5** (1971) 15–18.
27. L. DUFOUR and R. DEFAY, *Thermodynamics of Clouds*, Academic Press, New York (1963).
28. D. TURNBULL and R. E. CECH, *J. Appl. Phys.* **21** (1950) 804–10.
29. D. G. THOMAS and L. A. K. STAVELEY, *J. Chem. Soc.* (1952) 4569–77.
30. D. TURNBULL, *J. Appl. Phys.* **21** (1950) 1022–8.
31. J. ZELL and B. MUTAFTSCHIEV, *J. Cryst. Growth* **13/14** (1972) 231–4.
32. M. VOLMER and W. SCHULTZE, *Z. Physik. Chem.* A, **156** (1931) 1.
33. I. N. STRANSKI, *Discuss. Faraday Soc.* No. 5 (1949) 13–21.
34. W. K. BURTON and N. CABRERA, *Discuss. Faraday Soc.* No. 5 (1949) 33–48.
35. F. C. FRANK, *Discuss. Faraday Soc.* No. 5 (1949) 48–53.
36. W. K. BURTON, N. CABRERA, and F. C. FRANK, *Phil. Trans. Roy. Soc.* A, **243** (1951) 299–358.
37. B. LEWIS, *J. Cryst. Growth* **21** (1974) 29–39.
38. C. VAN LEEUWEN and P. BENNEMA, *Surf. Sci.* **51** (1925) 109–30.
39. K. A. JACKSON, in *Growth and Perfection of Crystals* (edited by R. H. Doremus, B. W. Roberts, and D. Turnbull), Wiley, New York (1958), pp. 339–49.
40. H. J. LEAMY and K. A. JACKSON, *J. Appl. Phys.* **42** (1971) 2121–7; *J. Cryst. Growth* **13/14** (1972) 140–3.
41. H. J. LEAMY, G. H. GILMER, and K. A. JACKSON, in *Surface Physics of Materials* (edited by J. M. Blakely), Academic Press (1975), vol. I, pp. 121–88.
42. L. ONSAGER, *Phys. Rev.* **65** (1944) 117.
43. H. A. BETHE, *Proc. Roy. Soc.* A, **150** (1935) 552.
44. W. B. HILLIG, in *Growth and Perfection of Crystals* (edited by R. H. Doremus, B. W. Roberts, and D. Turnbull), Wiley, New York (1958), pp. 350–60.
45. J. C. BRICE, *J. Cryst. Growth* **1** (1967) 218.
46. U. BERTOCCI, *Surf. Sci.* **15** (1969) 286–302.
47. B. LEWIS and G. J. REES, *Phil. Mag.* **20** (1974) 1253.
48. D. KASHCHIEV, *J. Cryst. Growth* **40** (1977) 29–46.
49. A. N. KOLMOGOROFF, *Bull. Acad. Sci. URSS (CR Sci. Math. Nat.)* **3** (1932) 355.
50. M. AVRAMI, *J. Chem. Phys.* **7** (1939) 1103; ibid. **8** (1940) 212; ibid. **9** (1941) 177.
51. N. CABRERA and M. M. LEVINE, *Phil. Mag.* **1** (1956) 450.
52. N. CABRERA and R. W. COLEMAN, in *The Art and Science of Growing Crystals* (edited by J. J. Gilman), Wiley, New York (1963), p. 3.
53. G. H. GILMER and P. BENNEMA, *J. Appl. Phys.* **43** (1972) 1347–60.
54. H. J. LEAMY and G. H. GILMER, *J. Cryst. Growth* **24/25** (1974) 499.
55. C. VAN LEEUWEN and J. P. VAN DER EERDEN, *Surf. Sci.* (1977) 137–250.
56. J. P. VAN DER EERDEN, C. VAN LEEUWEN, P. BENNEMA, W. L. VAN DER KRUK, and B. P. TH. VELTMAN, *J. Appl. Phys.* **48** (1977) 2124–30.
57. D. KASHCHIEV, J. P. VAN DER EERDEN, and C. VAN LEEUWEN, *J. Cryst. Growth* **40** (1977) 47–58.
58. G. H. GILMER, R GHEZ, and N. CABRERA, *J. Cryst. Growth* **8** (1971) 79–93.
59. G. W. SEARS, *Acta Met.* **3** (1955) 367–9.
60. G. W. SEARS, *Acta Met.* **3** (1955) 361–6.

61. F. C. FRANK, in *Growth and Perfection of Crystals* (edited by R. H. Doremus, B. W. Roberts, and D. Turnbull), Wiley, New York (1958), pp. 1–9.
62. H. E. BUCKLEY, *Crystal Growth*, Wiley, New York (1951).
63. P. CURIE, *Bull. Soc. Franç. Minéral.* **8** (1885) 145–50.
64. G. WULFF, *Z. Krist.* **34** (1901) 449–530.

CHAPTER 3

Hydrodynamics of Crystal Growth Processes

Chevron Research Co., PO Box 1627, Richmond, CA 94802, USA

3.1. Introduction

Hydrodynamics of crystal growth is the study of the flowfields of fluids surrounding crystals during the growth process. The word hydrodynamics literally means water-motion, and brings to mind the solution growth method wherein crystals are formed by crystallization from an aqueous solution. In the more general sense, hydrodynamics applies to the mathematical description of flow in real as well as ideal fluids.

The early contributors to the field of hydrodynamics used the word in a restricted sense to denote the study of flowfields of ideal, nonviscous fluids. The modern field of hydrodynamics now includes problems of real fluids with viscous action. Thus the mathematical description of flowfields in crystal growth processes can be considered to be within the domain of modern hydrodynamics. Indeed, the recent crystal growth literature provides ample evidence that the methods and concepts of modern hydrodynamics are contributing greatly to our understanding of crystal growth phenomena and to quantitative descriptions of crystal growth processes.

The objective of this chapter is to provide a brief review of the fundamentals of classical and modern hydrodynamics as applied to crystal growth processes, and to describe mathematically the flowfields of important liquid phase and vapor phase growth processes. Emphasis is given to boundary layer and shear layer phenomena, to flow in rotating fluids, and to flow in channels. Thermal effects are examined as a significant contribution to flowfields in crystal growth. Finally, mass transfer models are developed for various crystal growth processes, utilizing the flowfield descriptions introduced earlier in the chapter. Citations to fundamental papers are given in preference to a review of the most recent papers in the literature.

3.2. Fundamentals

Description of the flowfields in crystal growth processes requires a knowledge of the physical properties of the fluid phase and an appropriate form of the equations of motion and continuity. Physical properties of the fluid are needed to deduce the transport parameters in the equations of motion. The solution of the transport equations then provides the streamlines of the fluid velocity, the distribution of temperature in the fluid, and, finally, the mass distribution and crystal growth rate.

3.2.1. FLOWFIELDS

The term field refers to a quantity defined as a function of time and position throughout the region of interest. Representation of the field in mathematical terms takes one of two forms—the Lagrangian or the Eulerian. The difference between these two approaches lies in the manner in which the position is identified within the field.

In the Lagrangian representation an element of fluid is followed as it is carried along its velocity trajectory within the field. The coordinates (x, y, z) are those of the fluid element and, because of the motion of the fluid element, the coordinates are functions of time. This approach is used in problems of rigid body dynamics, but is seldom used in fluid mechanics.

In the Eulerian approach, the position (x, y, z) is fixed within the field and thus position and time are independent variables. This approach is preferred in fluid mechanics since the type of information desired is the values of variables at fixed positions within the flowfield.

3.2.2. THE FLOWNET

The flownet is a mathematical description of the flowfield. Pictorially, a flownet is a system of streamlines which follow the fluid velocity and a system of velocity potential lines which lie perpendicular to the streamlines. Taken together, the two groups of lines form a grid. An example of the flownet for creeping flow past a cylinder is shown in Fig. 3.1. The flownet is seen to consist of the expected orthogonal families of curves, with equipotential curves decreasing in magnitude in the direction of flow.

The stream function provides an analytic description of the flowfield by defining the streamlines which are lines drawn tangent to the velocity vectors at each point in the field. If v_x and v_y are the x and y components of the velocity, then the stream function ψ is defined by

$$\psi = -\int v_y \, dx + \int v_x \, dy + \text{const.} \qquad (3.1)$$

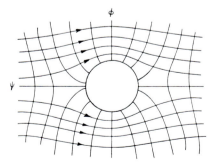

FIG. 3.1. Equipotential lines of the stream function and the velocity potential for rectangular flow past a cylinder.

It can be shown that the stream function satisfies the continuity equation automatically. An important property of the stream function representation of flowfields is that complex flowfields can be represented by the superposition of simpler fields of rectilinear, forced vortex, free vortex, source, and sink flow. The stream function satisfies the Laplace equation, which is

$$\nabla^2 \psi = 0. \tag{3.2}$$

This equation, together with the boundary conditions of velocity on the boundaries of the flowfield, defines the stream function distribution.

Another important function useful for flowfield description is the velocity function ϕ, which is defined by the relation

$$\nabla \phi = \mathbf{v}. \tag{3.3}$$

For all practical flows the Laplace equation in terms of ϕ is satisfied, owing to the continuity equation. It can be shown that the vorticity, defined above, must be zero for the existence of a velocity potential, and for this reason, irrotational flows are also known as potential flows.

The relation between streamlines, which are the constant values of the stream function, and equipotential lines, which are constant values of the velocity potential, can be shown by their relation to the components of the fluid velocity. The streamlines require constant ψ and have a slope $dy/dx = v_y/v_x$, while the velocity equipotentials have a slope of $-v_x/v_y$. Therefore, the equipotential lines are perpendicular to the streamlines. When plotted together, the streamlines and equipotential lines form a flownet. The flownet is of significant importance in estimating the velocity distribution of a flowfield where exact calculations are unobtainable.

3.2.3. NAVIER–STOKES EQUATIONS

The equations of motion of a fluid are derived from Newton's second law. In general, two forces act on a fluid element—gravitational forces which act on the bulk of the fluid G, and body forces which act on the boundary of the fluid B. If we let $D\mathbf{v}/Dt$ represent the time derivative of the velocity of a fluid element which is moving along the streamlines with the stream velocity, then the equation of motion is

$$\rho \frac{D\mathbf{v}}{Dt} \equiv \rho \left(\frac{\partial \mathbf{v}}{\partial t} + \mathbf{v} \cdot \nabla \mathbf{v} \right) = \mathbf{G} + \mathbf{B}, \tag{3.4}$$

where ρ is the fluid density. The body force is external to the fluid and is independent of flow rate, while the surface force depends on the rate at which the fluid is strained by the velocity field present in it. Thus the force acting on the boundary is a tensor quantity arising from force components normal to the surface of the fluid, the hydrostatic pressure, and shear forces acting parallel to the surface of the fluid element.

The normal shear stresses are functions of two parameters λ and μ corresponding to the dynamical and bulk viscosities. Stokes (1845) proposed a simple working relation between these quantities, thereby reducing the number of coefficients needed to describe the normal shear stress by one. A further simplification of the equation of motion was provided by Navier. As described by Rouse and Ince (1959), he assumed that the shear stresses were proportional to a viscosity constant μ, and that the viscosity and fluid density could be assumed to be constant. The combined assumptions resulted in the celebrated Navier–Stokes equation,

$$\rho\left(\frac{\partial \mathbf{v}}{\partial t} + \mathbf{v} \cdot \nabla \mathbf{v}\right) = \mathbf{G} - \nabla p + \mu \nabla^2 \mathbf{v}, \tag{3.5}$$

where $\mathbf{G}$ is the gravitational force vector and p is the hydrostatic pressure. This equation is highly useful in solving for flowfields in crystal growth processes in spite of the fact that the assumptions used do not afford a complete description of the fluid motion, as changes in pressure and density produce temperature variations through the thermodynamic equation of state.

In many flow processes, nonlinearities in the Navier–Stokes equations make the solution of these equations mathematically complex if not impossible. Nevertheless, a number of exact solutions are known to these equations, and these have been reviewed comprehensively by Berker (1963). These include flow to a rotating disk, a flow process occurring in liquid phase epitaxy on a rotating substrate, and flow in a channel—a flow structure encountered in horizontal epitaxial reactors. We shall explore some of the solutions in the subsequent sections of this chapter.

3.2.4. THE VORTICITY TRANSPORT EQUATION

An important hydrodynamic variable is the vorticity which defines the properties of rotational flow. The vorticity is defined in the terms of the fluid velocity,

$$\xi = \tfrac{1}{2}\nabla \times \mathbf{v}. \tag{3.6}$$

For steady state, irrotational flow the vorticity is zero. For frictionless flow in the absence of pressure the Navier–Stokes equations written in terms of the vorticity reduce to the form

$$\frac{\partial \xi}{\partial t} + \mathbf{v} \cdot \nabla \xi = \nu \nabla^2 \xi. \tag{3.7}$$

This equation is known as the vorticity transport equation and indicates that the change in the local and convective vorticity is equal to the rate of dissipation of the vorticity through frictional processes.

Examples of the stream function and vorticity distributions for linear flow past a sphere are shown in Fig. 3.2 for two different values of Reynolds number. As the Reynolds number increases the vortex region is displaced toward the rear of the solid object in the flowfield.

3.2.5. TRANSPORT COEFFICIENTS

The transport properties most important in crystal growth processes are the fluid viscosity, the thermal diffusivity, and the mass diffusivity in solution. The properties of many pure substances have been measured, but the properties of real gases and solutions are often not known, and require estimation or indirect determination through the crystal

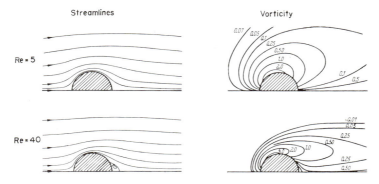

FIG. 3.2. Equipotential lines of the stream function and the reduced vorticity at Reynolds numbers of 5 and 40 for rectangular flow past a sphere.

growth process itself. Estimation methods for real fluids are discussed extensively by Reid and Sherwood (1966).

The dependence of transport properties of gases and liquids on temperature and pressure, as deduced from kinetic theory, are summarized in Table 3.1.

For dilute gases, kinetic theory provides a rigorous basis for prediction, as developed by Chapman and Enskog. Their theory is described in detail by Chapman and Cowling (1951). In this theory, molecules are characterized by a hard sphere diameter σ and a collision integral Ω. For liquids, the collision processes are complex and involve many body interactions. Thus, for liquids the transport properties cannot be deduced from first principles and lattice structures provide a convenient basis for theoretical predictions.

The equations for the transport of momentum and energy are analogous to that for diffusion when put in forms where the transport coefficients have the same dimensions as

TABLE 3.1. *Transport Properties of Gases and Liquids*

Dilute gases	
Viscosity	$\mu = \text{const}\, \dfrac{\sqrt{MT}}{\sigma^2 \Omega_\mu}$
Thermal conductivity	$k = \begin{cases} \text{const}\, \dfrac{\sqrt{T/M}}{\sigma \Omega_\mu}, & \text{monatomic} \\ (\hat{C}_p + 1.25\, R/M)\mu, & \text{polyatomic} \end{cases}$
Diffusivity	$D_{ij} = \text{const}\, \dfrac{\sqrt{T^3(M_i^{-1} + M_i^{-1})}}{\sigma_{ij}^2 \Omega_D \mathscr{P}}$
Liquids	
Viscosity	$\mu = \text{const}\, \rho\, e^{\Delta E/RT}$
Thermal conductivity	$k = \text{const}\, a\, v_s$
Diffusivity	$D_{ij} = \text{const}\, \dfrac{RT}{\mu}$

T, absolute temperature, $^\circ$K; M, molecular weight; $\hat{C}_p$, specific heat per unit mass; R, gas constant; v_s, velocity of sound; a, interatomic spacing; ρ, density

the diffusivity. The coefficients of interest are the kinematic viscosity and the thermal diffusivity defined by

$$v = \frac{\mu}{\rho} \tag{3.8}$$

and

$$\alpha = \frac{k}{\rho C_p}, \tag{3.9}$$

where C_p is the specific heat at constant pressure and ρ is the fluid density.

In addition to the similarity of diffusive processes, the flowfields of different sizes of the same crystal growth process can be dynamically similar, differing mainly in scale. Reduced space and time variables can be formed by dividing these variables by characteristic dimensions of the process. The coefficients in the so reduced equations of motion are dimensionless transport numbers. The most important transport numbers are the Reynolds number, which relates inertial to frictional forces, the Prandtl number, which relates momentum to thermal diffusion, the Schmidt number, which relates momentum to masses diffusion, and the Grashof number, which assesses buoyancy forces in relation to viscous forces. These are defined by:

$$Re = \frac{vL}{v}, \tag{3.10}$$

$$Pr = \frac{v}{\alpha}, \tag{3.11}$$

$$Sc = \frac{v}{D}, \tag{3.12}$$

$$Gr = \beta g L^3 \, \Delta T / v, \tag{3.13}$$

where L is a characteristic length, D the mass diffusivity, β the coefficient of thermal expansion, g the gravitational constant, and ΔT a temperature difference.

A comparison of the magnitudes of physical and transport properties of liquids and gases is shown in Table 3.2. The liquid and gas selected are liquid gallium at the freezing point and hydrogen gas at 600°K. The comparison shows that the transport properties of a low pressure gas are similar in magnitude, whereas for liquids these parameters differ by several orders of magnitude. These examples are useful in assessing the magnitudes of driving forces for flow in crystal growth.

3.3. Flow over Crystals in Solution

An important hydrodynamic problem in solution crystal growth is the description of the flowfields surrounding free or supported crystals in a moving fluid. When crystals of arbitrary shape are displaced through a fluid during crystal growth, the flowfield surrounding the crystal will be distorted from the free stream distribution only in the fluid region adjacent to the crystal and in the flow wake. The actual shape of the flowfield will depend on the shape of the crystal. In addition, the flowfield can exhibit interesting effects such as vortex formation and flow separation at high flow rates.

Table 3.2. *Magnitudes of Transport Properties*

	Liquid Gallium at T_m	Hydrogen Gas at 600°K, 101.3 kPa
Physical properties		
Density, ρ (kg m^{-3})	6095.	4.093
Specific heat, C_p (J kg^{-1}°K^{-1})	399.1	1.455×10^{-3}
Thermal exp. coeff. β (°K^{-1})	1.4×10^{-4}	9.257×10^{-3}
Transport properties		
Kinematic viscosity, v (m^2 s^{-1})	3.36×10^{-7}	3.491×10^{-5}
Thermal diffusivity, α (m^2 s^{-1})	1.4×10^{-5}	5.285×10^{-5}
Diffusion coeff. D (m^2 s^{-1})	$1.66 \times 10^{-7\,\text{(a)}}$	$6.28 \times 10^{-5\,\text{(b)}}$
Transport numbers		
Prandtl number, Pr	2.4×10^{-2}	0.664
Schmidt number, Sc	2.17×10^{2}	0.556
Grashof number, $Gr/\Delta T \cdot L^3$ (°K^{-1} m^{-3})	1.2×10^{10}	7.443×10^{7}

(a) Self-diffusion.
(b) Interdiffusion, SiCl$_4$ in H$_2$.

3.3.1. STOKES FLOW

The problem of determining the flowfield around a sphere falling through a stationary fluid was solved by Stokes (1851). The resulting flowfield is known as Stokes flow or creeping flow, as it requires low flow velocities or Reynolds numbers in the surrounding fluid. The velocity profile far from the sphere suggests that the velocity near the sphere can be represented as the product of radial and angular functions. For spherical symmetry, the only nonvanishing component of the vorticity is the azimuthal component, and this component can also be represented by a similar form. With this functional form substituted into the vorticity transport equation with pressure terms omitted, the resulting differential equation in spherical coordinates is a Euler differential equation with a solution of the form $r^n \sin \theta$. The boundary conditions then give the following equation for the vorticity:

$$\xi_\phi = -\frac{3}{2} v_\infty \frac{R}{r^2} \sin \theta. \tag{3.14}$$

A similar analysis then gives the following equations for the velocity distribution in the flowfield:

$$v_r = v_\infty \left[1 - \frac{3R}{2r} + \frac{1}{2} \left(\frac{R}{r} \right)^3 \right] \cos \theta, \tag{3.15}$$

$$v_\theta = -v_\infty \left[1 - \frac{3R}{4r} - \frac{1}{4} \left(\frac{R}{r} \right)^3 \right] \sin \theta. \tag{3.16}$$

In making this derivation, Stokes assumed that the inertial effects were small compared to pressure and viscous force terms, an approximation which is valid only near the surface of the sphere. At the leading surface of the sphere a boundary layer develops. The reduced velocity distribution for this flowfield is shown in Fig. 3.4 where the reduced normal distance is $\zeta = \sqrt{u_\infty R/\nu}\,(y/R)$.

3.3.2. FLOW AROUND ASYMMETRIC CRYSTALS IN SOLUTIONS

For crystals of arbitrary shape in slowly moving fluid, the shape of the flowfield depends on the asymmetry of the crystal and on the orientation of the crystal with respect to the direction of fluid flow. When the axis of symmetry is not aligned with the flow direction, then the flowfield will not be symmetrical about the crystal, and the difference in pressure forces on the different sides of the crystal will tend to align the crystal with the flow direction.

The calculation of flowfields around unsymmetric crystals is best treated by expansion methods utilizing special functions. For example, Acrivos (1968) has described the method of matched asymptotic expansion.

3.3.3. FLOW SEPARATION

As the velocity of flow increases, or as the pressure of a flowing gas decreases, the smoothness of flow over a crystal surface will be lost at discontinuities in the surface curvature. At such discontinuities, as well as in the region of the trailing surface of the crystal, flow separation is said to occur. Flow separation is exemplified by the high flow rate condition shown in Fig. 3.2. For flow over a sphere, the wake of separated flow will be smaller than for equivalent flow over a faceted crystal. In general, the wake flowfield is extremely difficult to calculate.

3.4. Boundary Layer Phenomena

3.4.1. BOUNDARY LAYERS

When a real fluid moves against a boundary, the fluid sticks to the boundary without slipping. The fluid particles at the wall remain at rest while the fluid particles far from the boundary move with the free stream velocity. Fluid within the vicinity of the boundary experiences a velocity gradient and an associated shear stress which diffuses into the fluid from the wall in the downstream direction of flow. Thus the shear stress is confined to a narrow region near the boundary, and this region increases with the downstream distance

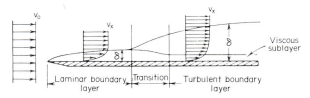

FIG. 3.3. Boundary layer development over a plate showing the laminar, transition, and turbulent regimes of flow.

along the body. The essential description of this flowfield, that the frictional aspects of the flow are confined within a narrow boundary layer and in the wake behind a body within the fluid, is due to Prandtl (1904). When a fluid flows with an initially uniform velocity over a plate, and if the fluid velocity at the plate surface is zero, then a boundary layer will form within the fluid adjacent to the plate. In solution crystal growth, the thickness of the boundary layer determines the rate at which solute can diffuse to the plate. Thus, the growth rate of crystals in solution is dependent upon the shape of the flowfield within the boundary layers.

Figure 3.2 shows a boundary layer developing over a flat plate. The flow within a boundary layer can be laminar or turbulent. Near the leading edge of the plate the flow is laminar, whereas at distances far from the leading edge the flow becomes turbulent with only a thin laminar sublayer adjacent to the plate. Between these regions there is a smooth transition region wherein the turbulent region widens.

In the laminar region the boundary layer increases with downstream distance x. Assuming a linear velocity gradient dv/dy near the boundary, which is of magnitude v_∞/δ, where v is the free stream velocity and δ is the boundary layer thickness, if the frictional force is equal to the inertial force $\rho v_\infty^2/x$, then the boundary layer thickness is given by

$$\delta \sim \sqrt{\frac{vx}{v_\infty}}. \qquad (3.17)$$

Therefore, the boundary layer thickness increases with the square root of the distance from the leading edge of the plate. As the boundary layer increases in thickness, so does the volume of fluid supporting the shear stress. The criterion for the onset of the transition and turbulent regions is the magnitude of the local Reynolds number defined by

$$Re_x \equiv \frac{u_\infty x}{v}. \qquad (3.18)$$

The transition region begins when Re_x equals 200,000, while the turbulent region is encountered for local Reynolds numbers greater than 3,000,000.

In the turbulent region the boundary layer thickness increases with the 0.8 power of the downstream distance. The laminar sublayer thickness is considerably less than that for the laminar region. The experimental studies of Nikuradse (1932) have shown that the velocity distribution above a flat plate is

$$v = (y/\delta)^{1/7} v_\infty. \qquad (3.19)$$

The corresponding boundary layer thickness is

$$\delta \sim x Re_x^{-1/5}. \qquad (3.20)$$

The turbulent regime is seldom encountered in liquid phase crystal growth processes. However, in chemical vapor deposition processes there is a growing trend toward methods which increase the flux of reagents to the surface of growth. Increasing the free stream velocity in chemical vapor deposition reactors not only reduces the boundary layer thickness in the laminar region, but promotes turbulent regions where the growth rate is controlled by diffusion through the laminar sublayer.

The fluid velocity profiles developed on the leading edge of various shapes are surprisingly similar. This similarity is shown in Fig. 3.4 for boundary layer flow past a

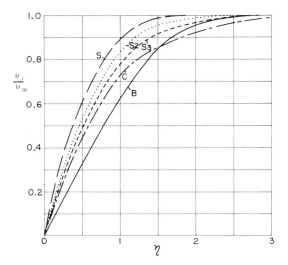

FIG. 3.4. Velocity distribution in boundary layer flow: B, Blasius solution for flow over a plate; C, convergent channel flow; S, boundary layer on a sphere; S2, two-dimensional stagnation flow; and S3, three-dimensional stagnation flow.

plate and sphere, flow on the walls of a convergent channel, and two-dimensional and three-dimensional stagnation flow to a plate. The latter flowfields will be explored subsequently in this chapter.

3.4.2. BOUNDARY LAYER FLOW OVER A FLAT SURFACE

The velocity distribution within the developing boundary layer at the leading edge of a moving plate can be attributed to Blasius (1908). If x is the distance from the plate upstream edge, y is the perpendicular coordinate, and the x-directed pressure gradient is neglected, then the Navier–Stokes equations simplify to the following:

$$v_x\frac{\partial v_x}{\partial x} + v_y\frac{\partial v_y}{\partial y} = v\frac{\partial^2 v_x}{\partial y^2}.$$
(3.21)

Similarly, the continuity equation has the following simple form:

$$\frac{\partial v_x}{\partial x} + \frac{\partial v_y}{\partial y} = 0.$$
(3.22)

In the region immediately behind the boundary layer the shear forces are insufficient to change the magnitude of the viscosity. Therefore we exclude this effect and assume a constant viscosity. We now use the similarity principal that the shape of the boundary layer at a given distance x is similar to that at other distances, and that the boundary layer extends a distance δ, which is a function only of y. Thus, normalizing the x-directed velocity and the y dimension we introduce

$$\eta = \frac{y}{2}\sqrt{\frac{v_\infty}{vx}}.$$
(3.23)

A stream function ψ can be introduced by integrating the equation of continuity. Assume that ψ is related to a dimensionless stream function by

$$f(\eta) = \frac{\psi}{\sqrt{vxv_\infty}}.$$ (3.24)

Then the equations of motion simplify to the following ordinary differential equation:

$$f''' + ff'' = 0,$$ (3.25)

$$f(0) = f'(0) = 0,$$ (3.26)

$$f'(\infty) = 2.$$

Blasius obtained a solution to this equation by the method of series expansion. The results of the calculation are shown in Fig. 3.4. These calculations show that the x-directed velocity increases nearly linearly with distance from the plate, then rather abruptly reaches asymptotically the free stream velocity at a distance δ.

It is not possible to define the boundary layer thickness in other than an arbitrary way owing to the asymptotically smooth velocity profile at the outer edge of the boundary layer. If the linear portion of the velocity profile is extrapolated to the free stream value then the boundary layer thickness is given by

$$\delta \simeq 3.01 \sqrt{\frac{vx}{v_\infty}}.$$ (3.27)

Thus the boundary layer thickness increases with the square root of the distance behind the leading edge of the plate.

In crystal growth as well as in other engineering sciences, there has been a continuing need for rapid approximation methods for describing the velocity profile as a function of the normal distance from a surface. Such a method was developed by von Kármán (1921) and Pohlhausen (1921) for application to incompressible fluids with arbitrary pressure gradients. The essence of the method is to approximate the velocity profile by a polynomial series which approaches the far-field velocity profile in the fluid far from the surface, and which falls to zero at the surface. Although polynomials of up to eleven terms have been reported, only a few terms are needed in engineering applications. The von Kármán–Pohlhausen method has practical utility in describing the flow around crystals of arbitrary shape submerged in moving fluids.

3.5. Flow in Rotating Fluids

3.5.1. FLOW TO A ROTATING DISK SUBSTRATE

Fluid flow to a rotating disk is one of the most celebrated hydrodynamics problems for which the Navier–Stokes equations have an exact solution. The use of a rotating disk substrate in solution crystal growth processes to establish a steady state flowfield also has considerable practical importance, for this type of flowfield allows all points on the surface of the disk to be uniformly accessible for mass transfer.

The forced convective flow to a rotating disk has been studied extensively because of important practical applications in electrochemistry. The fluid motion near the rotating disk is shown schematically in Fig. 3.5a. Because of the nonslip condition for real fluids at the disk surface, fluid adjacent to the disk must rotate with the angular velocity of the disk, while far from the disk the angular velocity of the fluid is zero. The angular velocity imparted to the fluid by the disk causes fluid to be drawn toward it and to be thrown radially outward by centripital force.

The first determination of the steady state flowfield near the rotating disk was given by von Kármán (1921) for a fluid with constant physical properties. von Kármán suggested a separation of variables approach to reduce the time independent Navier–Stokes equations to a set of coupled, nonlinear, ordinary differential equations by defining reduced fluid velocity components, reduced dynamical pressure, and reduced axial distance from the disk as follows:

$$v_x = v\Omega F(\eta), \quad v_y = r\Omega G(\eta), \quad v_z = \sqrt{v\Omega}\, H(\eta), \quad p = \mu\Omega P(\eta), \tag{3.28}$$

where

$$\eta = \sqrt{\Omega/v}\, z, \tag{3.29}$$

and inserting these into the Navier–Stokes equations for constant physical properties, von Kármán obtained the following ordinary differential equations:

$$2F + H' = 0,$$
$$F^2 - G^2 + HF' = F'',$$
$$2FG + HG' = G'', \tag{3.30}$$
$$P' + HH' = H''.$$

The boundary conditions on the disk and far from the disk are given in reduced form by

$$F(0) = 0, \quad G(0) = 1, \quad H(0) = 0,$$
$$F(\infty) = 0, \quad G(\infty) = 0. \tag{3.31}$$

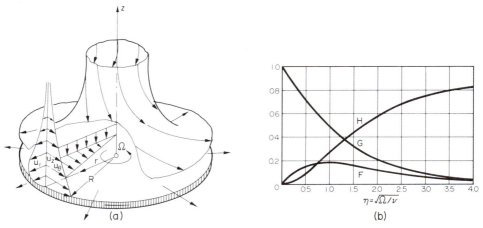

FIG. 3.5. Flow to a rotating disk: (a) velocity components, (b) reduced velocity components.

von Kármán found a solution to these equations by an approximation method. Later, Cochran (1934) calculated more accurate solutions in the form of power series near the disk, and asymptotic series far from the disk by numerical integration. The results are shown in Fig. 3.5b. These calculations show that the region of influence of the rotating disk is limited to a momentum boundary layer of thickness $\delta = 3.6 \sqrt{\dfrac{vx}{u_\infty}}$. More important for crystal growth, the axial velocity distribution is independent of position on the disk surface, and, therefore, the growth rate is uniform over the surface of the disk. An approximate form of the axial velocity profile above the disk surface, obtained by the von Kármán–Pohlhausen method, is the following:

$$v_z = -\sqrt{v\Omega}\,[0.510\,\eta^2 - 0.500\,\eta^3 + 0(\eta^4)]. \tag{3.32}$$

This equation is derived for an infinite disk, but should hold for finite disks to within a boundary layer thickness from the disk edge. For a slowly rotating disk in a rotating fluid, the axial fluid velocity depends on both angular velocities of the disk and the fluid, and, again, the above equation does not hold but is approximate if the angular velocity Ω is replaced by the difference between the angular velocities of the disk and fluid.

Edge effects on fluid flow to rotating surfaces are often considered to be negligible on the assumption that such effects have little influence on flow near the axis of rotation. In most cases of interest in electrochemistry the boundary layer thickness is small compared to the disk radius. However, in crystal growth on slowly rotating crystals, the opposite condition can occur. The radial and aximuthal fluid velocities are proportional to r, and, thus, the radial and azimuthal momenta are most sensitive to the shape of the surface at its outer edge. The effect of edge discontinuities, such as crystal facets, is more pronounced for inward flow than for outward flow since the outward flowfield is stabilized by the flowfield at small radii. For inward flow on irregular, rotating crystal surfaces, stability of the flowfield can be difficult to achieve. In general, edge effects in flowfields are complex and not easily handled analytically.

The interesting transport properties of the rotating disk have been utilized by Suguwara (1972) in a study of the kinetics of chemical vapor deposition of silicon on a rotating silicon disk. Because the diffusional mass transfer rates to the disk are uniform over the disk surface, and because their magnitude can be calculated, it is possible to extract the surface reaction kinetics by the rotating disk method. Owing to the approximations used by Suguwara, the analysis applies to liquid phase epitaxial deposition on a rotating disk substrate from a liquid. However, for deposition from a gas phase, Suguwara equations lead to an error of up to 15%.

3.5.2. FLOW TO A ROTATING FLUID

Crystal growth methods such as Czochralski crystal growth and accelerated crucible crystal growth are associated with a rotating fluid adjacent to a stationary surface or a surface rotating at a different velocity. The flowfield between a stationary wall and a fluid rotating at constant angular velocity was studied by Boedewadt (1940). This flowfield, one of the few exact solutions to the Navier–Stokes equations, is essentially the reverse of the flowfield for flow to a rotating disk. The flow distribution is shown schematically in Fig. 3.6a. Fluid particles rotating in the fluid far from the disk are in equilibrium under the

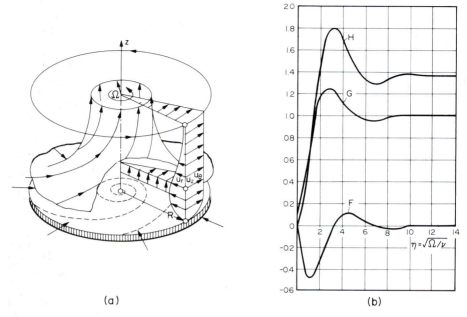

(a)   (b)

FIG. 3.6. Flow to a rotating fluid from a stationary surface: (a) velocity components, and
(b) reduced velocity components.

influence of a centrifugal force which is balanced by a radial pressure gradient given by the
following relation:

$$\frac{\partial p}{\partial r} = \rho r \Omega^2. \tag{3.33}$$

The decreased centrifugal force near the stationary wall causes fluid to flow radially
inward within a boundary layer adjacent to the wall and outward in axial flow in the
rotating fluid. The flow in a boundary layer on the surface, whose velocity distribution
differs from that in the external fluid, is generally referred to as secondary flow.

The Navier–Stokes equations, when reduced velocity functions are introduced through
eqns. (3.28) and when azimuthal derivatives are omitted by reason of symmetry, reduced
to the simple form

$$2F + H' = 0,$$

$$F^2 - G^2 + HF' = F'' + 1,$$

$$2FG + HG = G'',$$

$$P' + HH' = H''.$$

$$\tag{3.34}$$

In this problem the boundary conditions are the following:

$$F(0) = G(0) = H(0) = 0,$$

$$F(\infty) = 0, \quad G(\infty) = 1.$$

$$\tag{3.35}$$

Also, the pressure gradient in the direction normal to the surface can be assumed to be equal to zero. These equations have been solved by Nydahl (1971). The resulting reduced velocity functions are shown in Fig. 3.6b.

For flow to a rotating fluid adjacent to a stationary surface, again the axial velocity profile is independent of position on the surface. Using the von Kármán method, the axial fluid velocity is adequately described by the relation

$$v_z = 0.624 \sqrt{v\Omega}\, \eta. \tag{3.36}$$

3.5.3. FLOW BETWEEN TWO ROTATING PLANE SURFACES

The bounded flow between two rotating plane surfaces differs from that of flow to a rotating disk and flow to a rotating fluid but contains similar properties of both flowfields. The problem of steady fluid flow of a viscous, incompressible fluid between coaxial disks has inspired theoretical attention, again because of the possibility of obtaining an exact solution to the Navier–Stokes equations. Ekman boundary layers can be expected to form on both rotating surfaces containing the fluid. Bachelor (1951) proposed that the fluid outside the surface boundary layers would not approach the von Kármán free disk solution, but would rotate with a constant angular velocity which is intermediate between those of the two surfaces. Numerical solutions by Lance and Rogers (1962), Pearson (1965), and Stephenson (1969) have shown that the qualitative description of Bachelor is essentially correct. Subsequently, fluid flow in an idealized configuration of two coaxial rotating disks in an open fluid has been studied by Schulz-Grunow (1935), Picha and Eckert (1958), and Stewartson (1953).

3.5.4. ACCELERATED CRUCIBLE CRYSTAL GROWTH

An important new technique for improving stirring in high temperature solution crystal growth is accelerated crucible rotation developed by Scheel (1972). In this process the crucible is alternately accelerated and decelerated or accelerated alternately in clockwise and counterclockwise directions. The consequence of this motion is that strong stirring effects are generated within the fluid.

Simulations of the process using potassium permanganate as a coloring agent show that flow in this process can be interpreted in terms of the theoretical flow process adjacent to rotating surfaces. If the crucible is subjected to uniform rotation, mixing is prevented in accordance with the Taylor–Proudman theorem. This effect has been observed also by Carruthers and Nassau (1968).

In accelerated crucible rotation, the acceleration step causes shear layers to grow near the wall and floor of the crucible. Rotation of the crucible floor causes fluid in the center of the crucible to move axially downward to the Ekman layer above the floor, and outward to the Ekman layer adjacent to the cylindrical wall of the crucible. During the deceleration period, the crucible floor rotates more slowly than does the fluid above it, and fluid is pumped radially inward within the Ekman boundary layer adjacent to the floor, and upward along the center axis of the crucible.

An interesting consequence of the method is that during the deceleration period small crystallites and inclusions whose density is greater than that of the fluid are swept to the center of the crucible floor. It is not surprising, therefore, that nucleation should take place preferentially at this site. Also, as crystals at this site increase in size, their surfaces are swept clean of free-floating crystallites by the periodic flowfield induced by crucible acceleration and deceleration. The combined effects of acceleration and thermal convection have been reviewed by Schulz-DuBois (1972).

3.5.5. DETACHED SHEAR LAYERS

A phenomenon encountered in rotating fluids is the detached shear layer. A detached shear layer is similar to a boundary layer in that the fluid velocity varies across the layer but differs in that the shear layer is not in contact with a fluid boundary. This phenomenon arises as a consequence of the Taylor–Proudman theorem, which states that all steady motions in a rotating fluid are two-dimensional with respect to the rotating axis. As developed theoretically by Taylor (1923), this famous theorem is only strictly valid for ideal fluids, but the theorem is approximately valid at large Reynolds numbers.

Detached shear layers can arise in Czochralski crystal growth and in liquid phase epitaxial growth on a rotating disk. The schematic experimental arrangement for Czochralski growth is shown in Fig. 3.7. The arrangement for epitaxial growth on a rotating substrate is similar to that shown in Fig. 3.7 but with the crystal replaced by a rotating disk and with a region of fluid above the disk.

The formation of detached shear layers in a fluid between a disk rotating with angular velocity Ω_1 and a cylindrical crucible rotating with angular velocity Ω_2 has been studied by Hyde and Titman (1967). Their experiments showed that regions of high shear appear not only at the wall and floor of the cylindrical crucible and on the surface of the disk, but also in a cylindrical zone whose axis and radius are the same as those of the surface of crystal growth. This region is the detached shear layer shown by the dashed pairs of lines in Fig. 3.7b. The properties of the flowfield can be represented in terms of the Rossby and Ekman numbers ε and E, defined respectively by

$$\varepsilon = 2(\Omega_2 - \Omega_1)/(\Omega_1 + \Omega_2), \tag{3.37}$$

$$E = 2v/[a^2(\Omega_1 + \Omega_2)]. \tag{3.38}$$

$$\varepsilon \leqslant 16.8\, E^{0.57}. \tag{3.39}$$

Flow velocities within the axisymmetric flow regions are related to the magnitude of the Rossby number, while the thickness of the layer is related to the magnitude of the Ekman number. The boundary layer thickness is approximately given by the relation

$$\delta \sim E^{1/2}. \tag{3.40}$$

The theoretical study of Stewartson (1953) shows that only the Ekman layers adjacent to the cylinder floor and the disk are of order $E^{1/2}$, while the detached shear layer thickness is of order $E^{1/4}$.

The shear stresses in the detached shear layer were sufficient to generate azimuthal variations in the width of the shear layer. The form of the nonaxisymmetric detached shear layer protrusions is a pattern of blunt ellipses when the Rossby number is negative, and is a regular pattern of waves in the planes perpendicular to the axis of rotation when the Rossby number is positive.

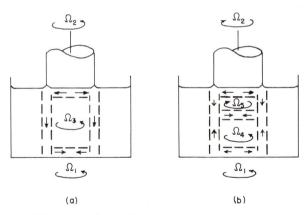

(a)   (b)

FIG. 3.7. Detached shear layers in a fluid between a rotating crystal and a crucible: (a) co-rotation and (b) counter-rotation.

3.5.6. FLOW IN CZOCHRALSKI CRYSTAL GROWTH

The flowfield of Czochralski crystal growth is complex owing to the rotation of both the crystal and crucible, to the effects of the lateral walls of the crucible, and to the often faceted surface of the growing crystal. The complexities of fluid flow descriptions in Czochralski growth have been shown through the experimental studies of Turovsky and Milvidsky (1962), Robertson (1966), Carruthers (1967), Carruthers and Nassau (1968), and Chen (1974). These studies show that upward flow proceeds toward the rotating crystal surface, but that more complex flow patterns occur when the crucible is also rotated. In addition, the thermal cooling of the liquid by the rotating crystal causes a downward flow from the crystal–liquid interface unless the crystal rotation rate exceeds a minimum rate.

Flow visualization of the flowfield in Czochralski crystal growth has been studied extensively by Chen (1974) using dye tracer injection and time-lapse photography. The simulation apparatus consisted of ethylene glycol contained in a cylindrical beaker with an inner diameter of 9.6 cm. A mirror placed at an angle of 45 degrees from the horizontal was placed over the plastic simulation crystal so that both the top and side views of the fluid could be photographed simultaneously. A neutrally buoyant dye dissolved in ethylene glycol was injected into the fluid at the beginning of each experiment.

The results of the flow visualization experiments are shown in Fig. 3.8 for a fluid depth equal to the crystal diameter, which is one-third the crucible diameter. For counter-rotation of the crystal and crucible, with crucible rotation fast enough to control the flow (top row), a detached shear layer forms to separate the convection cell adjacent to the crystal from that adjacent to the crucible. The detached shear layer moves closer to the crystal center when the crystal rotation rate is reduced (second row), while the upper convection cell fills the crucible for crystal rotation only (third row). Co-rotation of the crystal and crucible (rows four and five) produces a modification of the simple flow pattern for crucible rotation only.

A comparison between the experimentally observed flow pattern and those predicted for low viscosity fluids is shown in Fig. 3.9. Because of the finite viscosity of the fluid, the

FIG. 3.8. Dye-traced fluid flow in Czochralski crystal growth:
I: crystal-dominated counter-rotation,
$t(a - e) = 7, 12, 19, 29, 63\,\text{s}$;
II: crucible-dominated counter-rotation,
$t(a - e) = 5, 17, 25, 42, 80\,\text{s}$;
III: crystal-only rotation,
$t(a - e) = 7, 62, 120, 170, 210\,\text{s}$;
IV: crystal-dominated co-rotation,
$t(a - e) = 6, 15, 34, 62, 130\,\text{s}$;
V: crucible-dominated co-rotation,
$t(a - e) = 7, 16, 85, 256, 421\,\text{s}$.

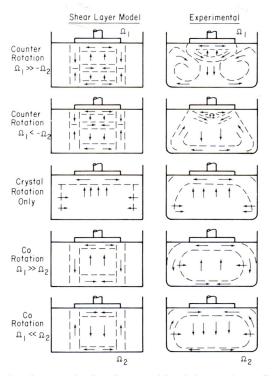

FIG. 3.9. Comparison between the shear layer model and the experimentally determined fluid flow pattern in Czochralski crystal growth.

boundaries between convection cells are distorted. For counter-rotation a central core rotating as a solid body rises toward the vortex ring circulating adjacent to the crystal, while a larger toroidal vortex fills the majority of the crucible. This and earlier studies confirm the complexity of the fluid flow patterns in Czochralski growth.

3.6. Flow in Gas Phase Epitaxial Reactors

Vapor phase epitaxial crystal growth has become increasingly important as a method for preparing planar crystals for electronic and optical devices. In this section, flowfields of several important types of vapor phase reactors are examined.

3.6.1. FLOW IN A STRAIGHT CHANNEL

The flow channel of a horizontal epitaxial reactor is a tube of rectangular cross-section containing a support for single crystal substrate. The channel can be straight or convergent as shown in the cross-sections of Fig. 3.10. In cold wall reactors the support is indirectly heated while the tube wall is forced air cooled. Often the substrate support is inclined at a small angle, so that the channel height decreases with distance in the direction of flow.

The flowfield within a horizontal epitaxial reactor is complicated by gradients in the temperature and concentration profiles within the reactor. In cold wall reactors the heated length is less than the total channel length, and the channel height is sometimes discontinuous at the leading edge of the substrate support. An idealized representation of the channel, shown in Fig. 3.10, is introduced here to demonstrate exact solutions to the Navier–Stokes equations.

If the flow channel is approximated by an isothermal, semi-infinite parallel channel between the wall at $y = b$, $y = -b$ in cartesian coordinates, and concentration profiles are neglected, then the only fluid velocity component is that in the x-direction within the channel. Such a flowfield is called parallel flow since all fluid particles move in only one direction. The equation of continuity then gives $dv_x/dx = 0$, and therefore the velocity v is not a function of x. Also, the Navier–Stokes equation for the y-direction indicates that $dp/dy = 0$, and, therefore, the pressure depends only on x.

The Navier–Stokes equation for the x-direction with a constant viscosity is

$$\frac{\partial v_x}{\partial t} = -\frac{1}{\rho}\frac{dp}{dx} + v\left(\frac{\partial^2 v_x}{\partial y^2} + \frac{\partial^2 v_x}{\partial z^2}\right). \tag{3.41}$$

If the flow is in steady state and there is no variation in the z-direction, only the first two terms on the right hand side of this equation are nonzero. Furthermore, if the pressure gradient is constant, this equation can be integrated with the condition that $v = 0$ on the walls of the channel to give

$$v_x = \frac{-\rho(b^2 - y^2)}{2v}\frac{dp}{dx}. \tag{3.42}$$

The velocity profile is thus a parabolic one with a maximum at the center of the flow channel. The average fluid velocity is two-thirds that of the maximum velocity in the channel. Also, the volume rate of flow Q per unit of channel width is given by

$$Q = -\frac{2b^3\rho}{3v}\frac{dp}{dx}. \tag{3.43}$$

In horizontal, parallel channels of finite length, boundary layers can be expected to form on both walls of the channel near the entrance to the channel. From section 3.4.1, the

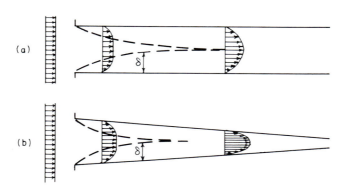

Fig. 3.10. Developing boundary layer flow in a channel: (a) straight channel and (b) convergent channel.

width of the boundary layer increases with the square root of the distance from the entrance of the channel. Without heating effects included, the free stream velocity of the fluid entering the channel is two-thirds of that given by eqn. (3.42) evaluated at $y = 0$. The entrance length is obtained by combining eqn. (3.20) for x with $\delta = b$. The resulting channel entrance length is

$$L = -\frac{\rho^3 b^4}{3v^2}\frac{dp}{dx}.$$

(3.44)

The velocity distribution within a straight channel containing an entrance region with developing boundary layers is shown schematically in Fig. 10a.

Between the inlet region and the fully developed region of the channel is an additional region wherein the velocity profile adjusts from that of the overlapping boundary layers to that of the Poiseuille similar profile. Mohanty and Asthana (1979) examined the inlet flow region in a pipe and found that the length of the intermediate region was three times that of the inlet region. Therefore the influence of the overlapping boundary layers extends an appreciable distance along the flow channel in the horizontal reactor.

Heating of the gas in the channel will raise the maximum velocity within the channel as well as the length of the developing boundary layers by the ratio of the temperature within the channel to that upstream from the channel entrance. An increase in the pressure gradient and fluid velocity within the channel could change the flow into the turbulent regime. In this case the velocity gradients are large at the walls of the channel, and the cross-stream velocity is uniform near the centerline. Whether the transition region is present or not depends on the b/L ratio.

In commercial horizontal reactors for epitaxial silicon growth the channel converges in the direction of gas flow. In addition, heating of the gas entering the reactor increases the cross-stream average velocity. These effects are shown in Fig. 3.10b. If the convergence angle is small, the streamlines are straight lines at different polar angles θ in polar coordinates, and the reactor walls lie at θ_b and $-\theta_b$. Then, flow is in the $-r$ direction, and the radial velocity has the form

$$v_x = \frac{v}{r}f(\theta).$$

(3.45)

If this form is introduced into the Navier–Stokes equations and the pressure is eliminated by combining the equations for the r and θ directions, the following differential equation is obtained:

$$f'''' + 2f'(2 + f') = 0.$$

(3.46)

Hamel (1916) obtained a solution to this differential equation by expressing f as an explicit elliptic function of θ. The velocity profile is shown in Fig. 3.4 for flow in a convergent channel. The consequence is that the velocity profile increases with distance along the channel, but the overall flowfield is otherwise not significantly changed except at high Reynolds numbers, where the velocity gradient at the channel walls is increased. As the flowfield of Fig. 3.10b shows, the entrance region of the reactor contains boundary layers which overlap in a flow transition region, and, finally, a fully developed flow region with a gradually increasing maximum velocity. At sufficiently high flow rates the channel would also support a turbulent region. It is surprising that the above flow models have not been utilized more extensively in predictive and descriptive studies of chemical vapor deposition in horizontal reactors.

3.6.2. FLOW IN VERTICAL CYLINDER REACTORS

The flow channel of the vertical cylinder reactor is the annulus between two cylinders, one of which supports substrates for epitaxial growth from the vapor phase. In reactors supporting a vertically downward flow, the inner surface of the annulus is often tapered so that the channel height decreases in the direction of flow. Characteristics of the flowfield in the vertical cylinder reactor are similar to those of the horizontal reactor, except that lateral wall effects are absent, and the direction of buoyancy lies along the flow channel rather than perpendicular to the flow direction.

Manke and Donaghey (1977) have studied the flowfield in the vertical cylinder reactor with an outward tapering inner surface and downward flow using a modified Hagen–Poiseuille flow model in conjunction with thermal expansion of the gas within the channel. The classical Hagen–Poiseuille relation expresses the volume rate of flow to the pressure gradient in a pipe for a constant fluid viscosity and no slip on the pipe wall, but their derivation of the flowfield is useful in other channel flow problems. If the inner surface is approximated by a right circular cylinder and the channel is assumed to be isothermal, then the z-component of the Navier–Stokes equation is

$$v_z \frac{\partial v_z}{\partial z} = -\frac{1}{\rho} \frac{d\mathscr{P}}{dz} + v\left[\frac{1}{r} \frac{\partial}{\partial r}\left(r \frac{\partial v_z}{\partial r} \right) + \frac{\partial^2 v_z}{\partial z^2} \right], \tag{3.47}$$

where $\mathscr{P}$ represents the combined effect of the static pressure and the gravitational force. If z is the distance from the top of the channel, then the effective pressure is

$$\mathscr{P} = p - \rho g h. \tag{3.48}$$

Since the z-directed velocity is at steady state and independent of z, the equation of continuity and derivative is

$$\frac{\partial v_z}{\partial z} = \frac{\partial^2 v_z}{\partial z^2}. \tag{3.49}$$

When eqn. (3.47) is simplified with eqn. (3.49) and integrated twice with respect to r between the channel inner radius R_1 and outer radius R_2, where the flow is nonslip, one obtains the following velocity distribution:

$$v_z = \frac{\rho}{4v}\left[r^2 - R_2^2 + \frac{(R_2^2 - R_1^2)\ln\left(\dfrac{R_2}{r}\right)}{\ln(R_2/R_1)} \right]\left(\frac{\partial \mathscr{P}}{\partial z} \right). \tag{3.50}$$

This equation describes only the fully developed distribution, which is expected only at large distances from the inlet of the flow channel. The study of Manke and Donaghey (1977) confirms that the entrance region, characterized by boundary layers on inner and outer walls, extends a significant distance along the flow channel, and that thermal expansion of the gas develops throughout the channel.

3.6.3. STAGNATION FLOW REACTORS

A gas phase reactor frequently used for experimental studies of chemical vapor deposition of thin films is one in which a downward flowing gas stream impinges on a horizontal heated substrate. Theurer (1961) has used this crystal growth method for producing epitaxial layers of silicon by the disproportionation of silicon tetrachloride with hydrogen. Maslaiyah and Nguyen (1979) have studied mass transfer in an elongated stagnation flow region. The gas stream meets the substrate surface at right angles and flows away in the perpendicular direction along the surface. The symmetry point on the surface is called the stagnation point. The resulting flowfield is shown in Fig. 3.11.

The flowfield is obtained by solving the Navier–Stokes equations in cylindrical coordinates with the stagnation point as the origin. A solution in the form of a power series was first obtained by Homann (1936), and more extensive tabular results have been reported by Froessling (1940).

If we denote the radial, azimuthal, and cylindrical components of the velocity vector by $v_r(r, z)$, $v_\theta(r, z)$, and $v_z(r, z)$, respectively, and assume rotational symmetry such that $v_\theta(r, z) = 0$ everywhere, then the Navier–Stokes equations become

$$v_z \frac{\partial v_x}{\partial r} + v_z \frac{\partial v_x}{\partial z} = \frac{-1}{\rho} \frac{\partial p}{\partial r} + v\left(\frac{\partial^1 v_x}{\partial r^2} + \frac{1}{r} \frac{\partial v_x}{\partial r} - \frac{v_x}{r^2} + \frac{\partial^2 v_x}{\partial z^2}\right), \tag{3.51}$$

$$v_x \frac{\partial v_z}{\partial r} + v_z \frac{\partial v_z}{\partial z} = -\frac{1}{\rho} \frac{\partial p}{\partial z} + v\left(\frac{\partial^2 v_z}{\partial r^2} + \frac{1}{r} \frac{\partial v_z}{\partial r} + \frac{\partial^2 v_z}{\partial z^2}\right), \tag{3.52}$$

where the continuity equation becomes

$$\frac{\partial v_x}{\partial r} + \frac{v_x}{r} + \frac{\partial v_z}{\partial z} = 0. \tag{3.53}$$

The boundary conditions on the surface and at large axial distance are as follows:

$$v_x(r, 0) = 0, \quad v_z(v, 0) = 0, \quad v_x(r, \infty) = v_0. \tag{3.54}$$

In viscous flow the fluid meets the surface without sliding, whereas in potential flow the fluid is free to slide along the surface. In potential flow the pressure distribution is given in terms of the stagnation pressure by

$$p = p_0 - \tfrac{1}{2}\rho(r^2 + 4z^2)C^2, \tag{3.55}$$

where C is a constant, while in viscous flow the wall friction effect causes a distortion of the z-directed term so that the term $4z^2$ is represented by an arbitrary function of z. The Navier–Stokes and continuity equations can be reduced to a simpler form by a similarity

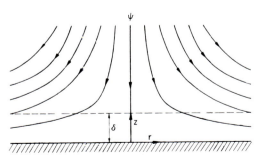

FIG. 3.11. Two-dimensional stagnation flow to a plane.

transformation of variables given by

$$\phi(\zeta) = \frac{-2v_z}{\sqrt{Cv}}, \quad \zeta = \sqrt{C/vz}, \tag{3.56}$$

and because of the continuity equation relating u_x and u_y we obtain a single differential equation for ψ:

$$\phi''' + 2\phi\phi' - (\phi')^2 + 1 = 0,$$
$$\phi(0) = 0, \quad \phi'(0) = 0, \quad \phi'(\infty) = 1. \tag{3.57}$$

This equation can be solved by power series representations. A simple and fairly accurate representation is the following:

$$\phi(\zeta) \cong \sin(1.1\zeta), \quad \zeta < 1.2. \tag{3.58}$$

The calculated reduced velocity profiles for two-dimensional and three-dimensional stagnation flow to a plane are shown in Fig. 3.4, where the reduced surface normal distance is $\eta = \sqrt{a/vz}$ and a is a constant. These distributions are surprisingly similar to the reduced velocity distribution for boundary layer flow to a sphere.

In vapor phase epitaxial reactors the flowfield is distorted from that given above by the temperature gradient at the heated substrate. Nevertheless, the above solution is very useful in calculations of the growth rates by chemical vapor deposition.

Visualization of the flow pattern in the reactor has been reported by Mantle et al. (1975) by reacting CO_2 with $AlCl_3$ in a hydrogen carrier gas to precipitate Al_2O_3 particles, and by reacting $TiCl_4$ reacting $TiCl_4$ with H_2O in an argon carrier gas to cause precipitation of TiO_2 particles in the gas phase. In these experiments, the inlet stream was separated into two axisymmetric, coaxial zones, each of which carried one of the reactive species in the carrier gas. An optical beam in the form of a vertical sheet was passed transversely through an experimental reactor to illuminate the TiO_2 particles in the plane of the sheet of illumination. In experiments where the heated disk was far from the inlet tubes, there was extensive gas phase reaction in the downward flowing gas, and the disk was covered with an apparent boundary layer of precipitate particles resulting from the gas reaction. In other experiments, where the substrate disk was close to the inlet tubes, the downward flowing gas reached the disk in laminar flow with essentially no gas phase reaction. The maximum inlet gas velocity yielding stable, laminar flow was about 40 cm s^{-1}. Decreasing the pressure in the reactor reduced the mixing effect between the two gas streams and helped to suppress recirculation zones, which formed in the reactor below the heated disk.

Flow in the inverted stagnation flow reactor has been studied by Wahl (1975). The experimental and calculated flow patterns for this interesting reactor are shown in Fig. 3.12. In the inverted reactor the chemically reactive gas mixture flows vertically upward and meets a heated surface at normal incidence. The gas then flows radially outward with a stagnation point formed at the center of symmetry on the surface. Numerical solution of the Navier–Stokes equations by finite difference equations has been described for such problems by Gosman et al. (1969) and by Mazille (1973). In this approach the flow velocity profile is specified on the inlet and outlet channels, a nonslip condition is specified on the reactor wall, the Navier–Stokes equations are written in finite difference form on a grid of points within the reactor, and these are solved by successive approximation until a self-consistent solution is reached.

Experimental
flow pattern

Calculated
stream lines

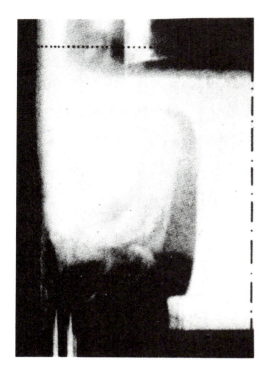

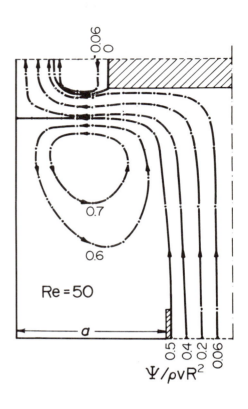

(a)   (b)

FIG. 3.12. Flow in an inverted, three-dimensional stagnation-flow reactor: (a)
experimental flow visualization, and (b) calculated stream lines.

Wahl (1975) solved for the stream function in the inverted, stagnation flow reactor by assuming an incompressible, isothermal gas and a parabolic velocity distribution—Hagen–Poiseuille flow—for gas entering the reactor through a nozzle. The results are shown in Fig. 3.12b for a Reynolds number of 50. Here v is the mean velocity of the gas at the nozzle and R, the nozzle radius, is the characteristic length defining the Reynolds number. Visualization of the flow in this reactor was achieved by the method of Takahashi *et al.* (1970), wherein smoke is produced by the reaction of $TiCl_4$ and H_2O. The resulting flow structure is shown in Fig. 3.12a. This study shows that a recirculating flow toroid is established between the cylinder wall of the reactor and the divergent flow region above the nozzle.

The inverted stagnation flow reactor has the advantage of improved flow stability over the noninverted reactor. Because the gas entering the reactor is usually at room temperature while the surface of deposition is heated, a buoyancy force acts on the gas near the stagnation point and modifies the calculated stream function described in Figs. 3.12b and 3.4. The flow is stable against thermal instabilities in the inverted reactor because the buoyancy force and the inlet gas velocity lie in the same direction. This type of reactor should be highly useful in experimental studies of chemical vapor deposition.

3.7. Thermally Driven Flow

The temperature gradients found in almost every crystal growth method induce convective flow effects which modify the flowfield of forced convection and induce secondary effects on the crystal growth process. In this section some of the concepts underlying thermal convection are reviewed in relation to the overall flow structure.

3.7.1. CONVECTIVE FLOW ON VERTICAL SURFACES

The problem of thermally induced flow is encountered whenever heat is extracted from the surface of growth and when the thermal gradient is not directed opposite to the gravitational vector. Examples of convective flow in crystal growth processes are shown schematically in Fig. 3.13.

Combined forced and thermally driven flowfields are found in liquid phase epitaxial crystal growth, where a substrate is supported vertically in solution, and in solution growth, where seed crystals are displaced horizontally through the fluid. The thermally driven flow of the former example is shown in Fig. 3.13a.

The problem of heat transfer and flow near a vertical surface has been described by Sparrow and Gregg (1960), by Sparrow et al. (1959), and by Acrivos (1968) for vertical flow in the presence of a perpendicular thermal gradient. Their results show that the parameters controlling the relative importance of thermally driven and forced convection are the Grashof number Gr, the Reynolds number Re, and the Prandtl number Pr. The Grashof number measures the force on the fluid due to a gradient in fluid density along the normal distance from the solid surface, which is at a temperature different from that of the fluid.

The Grashof and Reynolds numbers are controlling parameters in problems of forced convective flow, whereas the Prandtl and Reynolds numbers apply to problems of thermally driven convection. The distribution of the fluid velocity adjacent to a cooled, isothermal plate was obtained by Ostrach (1955), who used a similarity transform similar

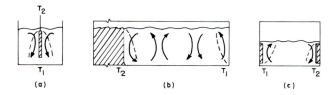

FIG. 3.13. Thermally driven flowfields in crystal growth processes: (a) liquid phase epitaxial growth on a vertical substrate, (b) horizontal normal freezing, and (c) dissolution-driven epitaxial growth.

to that used in the problem of flow to a rotating disk. The reduced fluid velocity, reduced distance from the plate, and the reduced stream function are defined by the following:

$$u = \sqrt{Lg\Delta\rho/\bar{\rho}}, \tag{3.59}$$

$$\eta = y\left(\frac{LGr}{4x}\right)^{1/4} \tag{3.60}$$

$$f(\eta) = \psi\left(\frac{LGr}{4x}\right)^{3/4}, \tag{3.61}$$

where x is the distance from the leading edge of the plate, y is the distance from the plate, L is a characteristic dimension of the plate, and $\Delta\rho$ is the difference in fluid density between that of the plate and that in the fluid, for which $\bar{\rho}$ is the mean value. The dimensionless velocity profiles for different Schmidt numbers are shown in Fig. 3.14 from the results of Ostrach (1955). These calculations show that the tangential velocity increases with distance from the plate and becomes constant over a zone which decreases with increasing Schmidt number.

Forced convection can overcome the effect of natural convection. As the Reynolds number for forced convection increases, one can expect a transition from flow dominated by thermal to forced convection. The study of Sparrow et al. (1959) showed that the flow regime is given by the Gr/Re^2 ratio as follows.

Forced convection: $0 < \dfrac{Gr}{Re^2} < 0.3.$ \hfill (3.62)

Mixed convection: $0.3 < \dfrac{Gr}{Re^2} < 16.$ \hfill (3.63)

Thermally driven convection: $16 < \dfrac{Gr}{Re^2}.$ \hfill (3.64)

The experimental work of Mahajan and Geshart (1979) shows that the velocity and thermal fields are closely coupled. For vertical natural convection, events of flow transition are not completely correlated by the Grashof and Reynolds numbers alone but can depend also on the downstream location and on the local pressure.

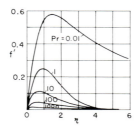

FIG. 3.14. Dimensionless velocity distributions for free convective flow near a vertical plane, at constant Prandtl numbers.

3.7.2. CONVECTIVE FLOW IN FLUIDS HEATED FROM BELOW

When a horizontal fluid layer is heated from below the fluid becomes unstable with respect to density driven convection. The pioneering experiment in thermally driven convection was performed by Bénard who observed spatially periodic convection cells in the thin liquid layer heated from below. The convection cells were the same height as the liquid layer but had cross-sections in the form of hexagons or elongated rectangles. In the hexagonal cells, flow occurred vertically along the axis of the hexagonal cell, then outward, then vertically along the perimeter of the cell. In rectangular cells, flow occurred in rolls about a horizontal axis at mid-height in the fluid.

Bénard's work stimulated Lord Rayleigh to propose and theoretically analyze the buoyancy driven flow systems observed experimentally. Since then a significant amount of knowledge has been accumulated on convective instabilities for the Bénard–Rayleigh problem as well as other geometries. These have been reviewed by Rogers (1976).

The stability of equilibrium states has recently been examined by Kraska and Sani (1979), who showed a hierarchy of stable equilibrium states in the fluid. With increasing thermal gradient, the fluid states change from the undisturbed state to hexagons and undisturbed state, to hexagons, to hexagons and rolls, and, finally, to rolls. The range of the temperature gradient, or Rayleigh number, over which the hexagon state is stable increases with both the Prandtl number and the Nusselt number, and depends also on surface tension forces. Also, the transition value of the thermal gradient, or transition Rayleigh number, for the hexagon to roll transition increases as the stability range of the hexagonal state increases.

The direction of flow in the convection cells depends on whether the density driven buoyancy force dominates over surface tension forces. The flow direction is downward when driven by buoyancy, whereas the flow direction is upward when driven by surface tension variations on the liquid surface.

Density gradients are also known to lead to convective instabilities. Crystal growth in solutions is accompanied by gradients in solute which increase with crystal growth rate. If the gradient is oriented so as to produce a density gradient, which is not vertically upward, then buoyancy driven flow can occur. Indeed, Kishitake (1976) has shown that macrosegregation can occur in Al–Mg alloys due to natural convection. Thus, the diffusion of a second component can influence the onset of convective instabilities. Extending this problem, Griffiths (1979) has shown how a third diffusing component can influence the onset of thermally driven convective instabilities.

3.7.3. HORIZONTAL NORMAL FREEZING

Horizontal normal freezing and zone refining are examples of crystal growth process where forced convection is absent, yet flow occurs by the action of thermal and mass concentration gradients. The flow in the former process has been visualized with transparent models by Utech et al. (1967) and treated theoretically by Chandrasekhar (1961).

A theoretical treatment of the related problem of heat transfer between closely spaced vertical plates is given by Bachelor (1954). The flowfield for this problem is shown schematically in Fig. 3.13b, and consists of rectangular regions of vorticity whose sense of rotation alternates along the fluid. The series of vortices is maintained by the extraction of heat at the freezing interface, where cooling increases the density of the adjacent fluid and causes the fluid in this region to fall. At the opposite end of the liquid region, heating induces a vertical flow in the fluid. Intermediate flow vortices are further stabilized by the presence of a thermal gradient along the liquid region.

A similar flow pattern is to be found in the liquid phase epitaxial growth chamber of Stringfellow and Green (1971). In this arrangement, shown in Fig. 3.13c, a source crystal and a seed crystal are placed on the end walls of a cylinder which is partly filled with solvent. The cylinder is placed within a thermal gradient, where the source dissolves into the solvent. Vortex cells can be expected to form within the liquid zone with widths comparable to the height of liquid in the cylinder. The shape of the vortex cells will be modified by the curved lower surface of the cylindrical container. The convective instabilities can be suppressed, and the liquid phase epitaxial growth rates increased, by oscillatory motion of the cylinder about its axis, thereby inducing forced convective flow.

The magnitude of the flow rates within vortex cells has been estimated by Scheel and Elwell (1973), and found to be less than $0.05 \, \text{cm s}^{-1}$, even in fluids exhibiting extremely large thermal diffusivities (of order 1000).

These authors conclude that thermally driven convection provides flow rates at crystal surfaces which are too low for the prevention of instabilities caused by constitutional supercooling.

Criteria for the onset of thermal instabilities due to convection in crystal growth from the melt have been summarized by Carruthers (1976).

3.7.4. CONVECTIVE INSTABILITIES IN VAPOR PHASE CRYSTAL GROWTH

Convective instabilities can occur in vapor phase crystal growth processes. Unlike the Bénard–Rayleigh problem, vapor phase growth processes are enclosed, and lateral boundaries modify the criteria for flow instabilities.

Convective instabilities in vertical, vapor transport crystal growth have been studied experimentally by Olson and Rosenberger (1979). In their studies, a vertical transport tube with a height to radius ratio of 6 was differentially heated along its length. The results showed that the convective motion in the tube was entirely different from that in the horizontal liquid, and that the critical Rayleigh number for onset of the first mode was 230—much below that for the Bénard–Rayleigh problem.

Asymmetric modes with different numbers of cells within the tube were observed to occur with increasing Rayleigh number. These observations confirm the theoretical predictions of Hales (1937) who predicted multicellular asymmetric rolls in such an enclosure.

Convective instabilities for multi-component vapor transport crystal growth was also studied by Olson and Rosenberger (1979). Experimental studies of a Xe–He system showed stable oscillatory modes similar to those observed in monocomponent gases, but differ in that thermal diffusion and vertical vorticity effects are present. Transitions from one flow regime to another are accompanied by considerable hysteresis, and at a given Rayleigh number there can be two stable states.

Thermally driven flow instabilities in horizontal, chemical vapor deposition reactors have been studied by Takahashi *et al.* (1970), and by Sugawara *et al.* (1970). These studies showed that the flows in horizontal epitaxial reactors for the chemical vapor deposition of silicon follow the predictions of Sparrow *et al.* (1959), except that the mixed flow regime took the form of spirals which formed bilaterally about the mid-plane of the reactor. These spiral flows were stabilized by thermal expansion causing upward flow along the mid-plane of the reactor, and by thermal contraction at the side walls of the flow channel causing downward flow. The theoretical study of Mori (1961) suggested somewhat different criteria for the mixed flow regime, and $Gr/Re^{2.5}$ as the parameter describing the flow regime. Other geometries of vapor phase crystal growth, where thermally induced flows are found, have been reviewed by Curtis and Dismukes (1972).

3.8. Flow-assisted Mass Transfer

Crystal growth rates are strongly influenced by the flowfield in many crystal growth processes. In this section the equations for mass transfer are developed and applied to the crystal growth processes whose hydrodynamic descriptions are presented above.

3.8.1. MASS TRANSFER EQUATIONS

The molar flux of a component in a mixture is expressed by Fick's first law of diffusion, for which there are a number of mathematically equivalent forms. Although this equation is usually written in terms of the molar diffusion flux, a more useful form is Fick's first law written in terms of N_i, the molar flux of species i relative to stationary coordinates:

$$N_i = - x_i \sum_j N_j + cD_{ij}\nabla x_i. \tag{3.65}$$

Here the diffusion flux N_i, defined as directed toward the crystal surface, is the result of two vector quantities, the first of which results from the bulk flow of the fluid, while the second results from diffusion superimposed on the bulk fluid flow.

The time dependence of the mole fraction of species i, X_i, is expressed by the continuity equation. The form of this equation useful for multi-component diffusion is obtained from the flux N_i by its equivalent expression involving the concentration gradient. The resulting equation is

$$\frac{\partial c_i}{\partial t} + \nabla \cdot c_i v = \nabla \cdot D_{ij}\nabla c_i + R_i. \tag{3.66}$$

In this equation, R_i is the molar rate of production of species i per unit volume from chemical reactions, and v is the molar average fluid velocity. This equation is valid for variable total concentration c, or total density ρ, and for variable diffusivity D_{ij} of species i in a mixture j. In the vicinity of surfaces of crystal growth, a useful approximation is that the concentration gradient is one-dimensional along the surface normal coordinate z, with $z = 0$ at the surface. Then, in the absence of chemical reactions, the convective diffusion equation reduces to the following form:

$$v \cdot \nabla c_i = D_{ij}\nabla^2 c_i. \tag{3.67}$$

With concentrations c_0 on the surface of growth and c_∞ in the fluid far from the growth surface, the above equation can be integrated twice to give

$$c_i(z) = c_0 + K \int_0^z \exp\left\{\frac{1}{D_{ij}} \int_0^z v_z dz\right\} dz. \tag{3.68}$$

The constant K is the reduced molar flux to the surface, and is a constant of integration. Thus

$$N_i = D\frac{dc_i}{dz}\bigg|_{z=0} = D_{ij} K. \tag{3.69}$$

Then K has the following magnitude:

$$K^{-1} = (c_\infty - c_0) \int_0^\infty \exp\left\{\frac{1}{D_{ij}} \int_0^z v_z dz\right\} dz. \tag{3.70}$$

This expression shows that the molar flux, or crystal growth rate, is dependent on the fluid velocity profile along the growth surface normal. Thus, accurate descriptions of the flowfields of crystal growth processes are essential to the prediction of crystal growth rates under mass transfer controlled kinetics.

3.8.2. GROWTH RATE OF CRYSTALS IN STOKES FLOW

In Stokes flow, fluids move past a crystal at low Reynolds numbers. For spherical crystals the velocity profile is given by the Stokes equation presented in section 3.3.2. For spherical crystal growth, Acrivos and Taylor (1962) have solved the convective diffusion equation by asymptotic expansion and express the molar flux to the surfaces in terms of the Peclet number $Pe = Re \times Pr$, wherein the crystal radius R is the characteristic dimension. The Acrivos–Taylor expression is

$$N_{ave} = \frac{D_{ij}(c_\infty - c_0)}{R}\left[1 + \frac{1}{4}Pe + \frac{1}{8}Pe^2 \ln Pe + 0(Pe^2)\right]. \tag{3.71}$$

This growth rate expression should apply in solution crystalizers.

Growth rates to crystals in Stokes flow at large Peclet numbers, where the crystal velocity is large or where the diffusivity is small, then convection predominates over diffusion in the fluid everywhere except in a thin boundary layer at the crystal surface. The boundary layer is sufficiently thin that the concentration gradient is large enough to make the diffusion term comparable to the convective term. A solution to the convective diffusion equation was obtained by Levich using a coordinate transformation. Acrivos and Goddard (1965) have extended this solution and express the average molar flux by

$$N_{ave} = \frac{D_{ij}(c_\infty - c_0)}{R}\left[0.461 + 0.496Pe^{1/3} + 0(Pe^{-1/3})\right]. \tag{3.72}$$

The approximations made in obtaining this expression break down in the region of the rear of the crystal.

3.8.3. GROWTH RATE ON A ROTATING SURFACE

Rotating surfaces of crystal growth occur in liquid phase epitaxial growth on a rotating disk substrate and in Czochralski crystal growth. The hydrodynamics of fluid flow presented above are independent of position on the rotating surface, and the fluid velocity toward the disk depends only on the distance from the disk, not on the radial position. Consequently, the convective diffusion equation for constant fluid properties reduces to eqn. (3.67).

Whether convection of diffusion predominates in the growth rate equation depends upon the Schmidt number Sc, and the most accurate solution for the growth rate is due to Newman (1966). For low Schmidt numbers the convected velocity to the disk is small, while the diffusion layer extends to a large distance from the surface of growth. Thus the growth rate depends on the rate of material from fluid far from the disk. The growth rate is then,

$$N \cong 0.8845(c_\infty - c_0)\sqrt{v\Omega}, \quad Sc \ll 1. \tag{3.73}$$

In gases the Schmidt number is near unity, and the growth rate is given by

$$N \cong 0.603(c_\infty - c_0)\sqrt{v\Omega}\, Sc^{-3/7}. \tag{3.74}$$

For large Schmidt numbers, as frequently are found for liquids, the velocity of fluid motion toward the disk is large, and the diffusion boundary layer is comparable to the hydrodynamic boundary layer. Then the growth rate is given by

$$N \cong 0.620(c_\infty - c_0)\sqrt{v\Omega}\, Sc^{-2/3}, \quad Sc \gg 1. \tag{3.75}$$

This expression applies to liquid phase epitaxial growth on a rotating substrate and to Czochralski growth at moderate and high rotation rates.

3.8.4. MASS TRANSFER THROUGH BOUNDARY LAYERS

Boundary layers occur frequently in crystal growth processes, and growth rate equations are needed for both liquids and gases. The rate of mass transfer is again defined by the convective diffusion in two dimensions, and the most useful reduced parameters are the Schmidt number Sc, the Reynolds number Re, and the Peclet number $Pe = Re \times Sc$. Solution of the equations of motion and diffusion is greatly simplified by assuming constant fluid properties.

Calculations of the mass, heat, and momentum flux to the surface beneath the boundary layer have been carried out by Mickley *et al.* (1954), using a convenient set of dimensionless variables:

$$\Pi_v(\eta) = \frac{v_x}{v_\infty}, \quad \Pi_D(\eta) = \frac{c - c_0}{c_\infty - c_0},$$

$$\Lambda_v = 1, \quad \Lambda_D = Sc, \tag{3.76}$$

$$\eta = \frac{y}{2}\sqrt{\frac{v_\infty}{vx}}.$$

The equations of continuity in two dimensions were integrated to eliminate the velocity component v_y by Schlichting and Bussman (1943), who calculated the velocity profile in the boundary layer. Using the reduced variables, the convective diffusion equation, eqn. (3.67), can be integrated twice to give

$$\Pi_i = \frac{\int_0^\eta \exp\left\{-\varLambda_i \int_0^\eta (K + L)\,\mathrm{d}\eta\right\}\mathrm{d}\eta}{\int_0^\infty \exp\left\{-\varLambda_i \int_0^\eta (K + L)\,\mathrm{d}\eta\right\}\mathrm{d}\eta}, \quad i = v, D \tag{3.77}$$

for either the velocity or molar concentration profile. In this equation, K is the dimensionless mass flux from the surface and L is the dimensionless x-directed velocity integral. These terms are defined by

$$K = Sc^{-1}\left(\frac{c_\infty - c_0}{c - c_0}\right)\Pi'_D\Big|_{\eta = 0}, \tag{3.78}$$

$$L = \int_0^\eta 2\Pi_v\,\mathrm{d}\eta. \tag{3.79}$$

For $K = 0$ and $\varLambda = 1$, the profiles for mass, temperature, and velocity are identical within the boundary layer, and the dimensionless gradient at the surface is given by the relation due to Pohlhausen (1921),

$$\Pi'_i\Big|_{\eta = 0} = 0.664\varLambda_i^{1/3}. \tag{3.80}$$

For K decreasing, as well as for $\varLambda$ increasing, the dimensionless slope at the surface increases and the crystal growth rate is larger than that estimated from the isothermal, diffusionless boundary layer flowfield described in section 3.4.1 above.

Because of the similarity of the dimensionless profiles of mass, heat, and momentum when the growth rate is low in a dilute solution, these profiles can be equated using the Pohlhausen approximation, eqn. (3.80). It follows from eqn. (3.80) that the ratio of the boundary layer thicknesses is given by

$$\delta_D/\delta_v = Sc^{1/3}. \tag{3.81}$$

Furthermore, a resulting convenient analogy for low mass transfer rates is

$$\frac{N|_{y=0}}{v_\infty(c_\infty - c_0)}Sc^{2/3} = \frac{0.664}{2}Re^{-1/2}, \tag{3.82}$$

and is the celebrated Chilton–Colburn analogy, as introduced by Colburn (1933) for heat transfer. This relation is valid for Sc greater than about 0.5.

Mass transfer in the boundary layer depends on the Schmidt number. For $Sc = 1$ the thickness of the diffusion layer is about the same as the thickness of the hydrodynamic boundary layer. This condition is representative of gases. When Sc is very large, as it is for liquids, then the diffusion layer is much thinner than the hydrodynamic boundary layer, and only the flowfield very near the surface is important to crystal growth. For large Sc, we expect the growth rate to depend on $S^{1/3}$, while for low Sc the growth rate depends on $Sc^{1/2}$.

3.8.5. GROWTH RATES IN EPITAXIAL REACTORS

Growth rates in epitaxial reactors for silicon deposition from silane or chlorosilanes have been studied by many investigators, and is a subject of intense current research. Eversteyn *et al.* (1970), Eversteyn and Peek (1970), and Rundle (1968, 1971) have developed models for the growth rate of silicon in horizontal epitaxy reactors by considering diffusion through a laminar sublayer adjacent to the growth surface, while the remainder of the channel is in turbulent flow. Fujii *et al.* (1972), Dittman (1974), and Manke and Donaghey (1977) have studied growth rate distributions in vertical cylinder reactors using a Hagen–Poiseuille flow model, a model based on the Chilton–Colburn analogy, and a developing temperature model, respectively.

In all but the latter work, the temperature of the flow channel is assumed to be isothermal. In real reactors, however, the temperature, mass concentration, and velocity profiles change rapidly in the entrance region, even over a constant temperature susceptor. Although several of the above studies recognize the need for incorporating boundary layer theory in the calculation of growth rates, the geometry of the flow channel and the entrance region transients make quantitative calculations difficult. Manke and Donaghey have circumvented this difficulty by solving the transport equations for mass, momentum, and temperature by finite difference equations using variable coefficients in order to properly account for the interaction of mass, velocity, and temperature in the flow channel.

Mass transfer in a channel, wherein deposition occurs on only one side, the side which is a heated graphite susceptor supporting substrates for crystal growth, depends upon the velocity profile and local velocity toward the wall, as described in section 3.8.4 above. If $\dot{M}$ is the total molar flow rate and $\overline{X}$ is the mean molar concentration of reactant in the channel, then the molar growth rate over a length dz of reactor length is equal to the difference between the product $\dot{M}\overline{X}$ entering the channel element and $\dot{M}\overline{X}$ leaving the element. Since for a gas $c = p/(RT)$, the molar flux to the surface of growth over the differential element of length dz and width w is

$$\dot{M}\frac{d\overline{X}}{dz} = \frac{wDp}{d_h RT} Sh(\overline{X} - X_0). \tag{3.83}$$

Here d_h is the hydraulic diameter of the flow channel and also the characteristic length for defining the Sherwood number:

$$Sh = \frac{k_x d_h p}{DRT}, \tag{3.84}$$

where k_x is the mass transfer coefficient defined by $k_x = N_i/\Delta c$. On integration of this differential equation, Manke and Donaghey (1977) obtained

$$\overline{X} = \overline{X}_0 \exp\left\{ -\frac{p}{\dot{M}R} \int_0^z \frac{wD}{d_h T} Sh\,dz \right\}, \tag{3.85}$$

where $\overline{X}_0$ is the mean molar concentration of reactant entering the reactor. For chemical vapor deposition, the growth rate is limited by the thermodynamic efficiency η. The linear growth rate in μm min^{-1} then becomes

$$G_{Si} = 6 \times 10^5 \frac{MDpX_0\eta}{\rho\, d_h RT} Sh \exp\left\{ \frac{p}{\dot{M}R} \int_0^z \frac{wD\,Sh}{d_h T}\,dz \right\}, \tag{3.86}$$

where M and ρ are the molecular weight and density of the deposited silicon.

An important problem in chemical vapor deposition reactors is the determination of the temperature profile, since transport properties are strongly varying with temperature, and since the gas volume in the channel expands linearly with temperature.

The boundary value problem for temperature can be solved by first subtracting the fully developed solution valid at large distances along the channel, then solving the resulting homogeneous differential equation of the Sturm–Liouville type by separation of variables. For a cylindrical channel, Reynolds *et al.* (1963) have solved this problem by an iterative method involving numerical integration.

The stagnation flow epitaxy reactor represents a departure from the channel flow reactor owing to the short radial dimension of the disk of growth, which is the characteristic length of interaction between the gas and solid surface. Kay (1958) has determined the thermal distribution within the flowfield of the stagnation flow from a nozzle. By an analogous method, the molar flux of reagent to the stagnation point can be determined by an integration of the convective diffusion equation along the stagnation contour.

$$N = 0.75 D^{0.6} v^{-0.1} \sqrt{\beta v_\infty / R}. \tag{3.87}$$

Here β is a shape factor and R is the disk radius. A similar equation was derived by Scholtz and Trass (1970) for small diameter jets placed one disk diameter distance from the disk:

$$N = 0.36 D^{0.64} v^{-0.14} \sqrt{v_\infty / R} \, (c_\infty - c_0). \tag{3.88}$$

Both results indicate that the molar flux of reagent to the stagnation point increases with the square root of the inlet gas flow velocity. Furthermore, the temperature dependence of the molar deposition rate is determined by the temperature dependence of the parameters D and v. From the above models and the temperature dependences shown in Table 3.2, the predicted temperature dependence of the growth rate is $T^{0.8}$. This temperature dependence has been confirmed by Lin *et al.* (1977) for the growth of GaAs from trimethyl gallium and arsine in a hydrogen carrier gas.

An application of the finite difference method to the calculation of mass transport limited growth process has been presented by Secrest *et al.* (1971). The method consists of solving the finite difference equations for energy, mass transport, and fluid flow on a grid of points contained within the region defining the container of the crystal growth process. In addition, boundary conditions must be defined on the container walls. Use of this method allows the calculation of mass transfer limited growth rates in vapor phase reactors.

A numerical simulation method utilizing a marching integration of finite difference equations in the direction of flow within a vertical cylinder reactor has been described by Manke and Donaghey (1977, 1977a). In this approach, the transport equations for developing flow within the reactor channel are solved by the method of separation of variables using series expansions in functions satisfying the Sturm–Liouville equation. The coefficients are obtained by defining the boundary conditions at the inlet of the reactor, then integrating in the downstream direction to define the variables at successive rows of mesh points. An example of the calculated distributions of the velocity, temperature, and silicon tetrachloride mole fraction profiles is shown in Fig. 3.15.

A comparison of the results of growth rate models shows the usefulness of the marching integration technique for determining the deposition rate profiles in vapor deposition reactors. A comparison of growth rate models to experimental data is shown in Fig. 3.16,

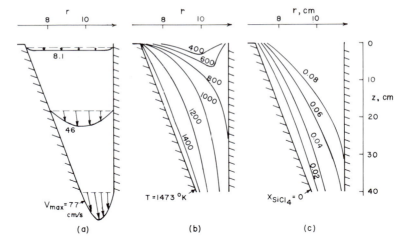

FIG. 3.15. Transport processes in the flow channel of a vertical cylinder reactor; distributions of (a) velocity, (b) temperature, and (c) mole fraction of silicon tetrachloride in hydrogen for the epitaxial deposition of silicon.

based on the calculations of Manke and Donaghey (1977). These results for a cylindrical reactor with a tapered susceptor show that the models of Fujii and Eversteyn predict higher values than the experimental values, especially in the inlet region. The overestimation of the growth rate is based on the erroneous assumption that the temperature is uniform throughout the reactor. On the other hand, the models of Dittman and Chilton–Colburn significantly underestimate the growth rate. These models, developed for flow over a flat plate, assume that the developing boundary layers extend throughout the reactor. Rundle's model, based on a laminar, parabolic velocity

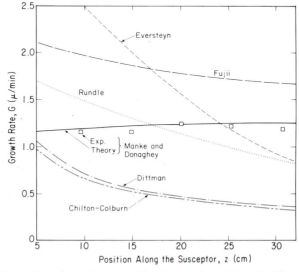

FIG. 3.16. Comparison of growth rate models with experimental data for silicon deposition from 1.15 % silicon tetrachloride in hydrogen within a vertical cylinder reactor.

distribution, gives growth rates closer to the experimental values than do the above models, indicating that the parallel flow model is more representative of the velocity distribution within the flow channel than is the turbulent core, laminar sublayer model used by Eversteyn. It can be seen in Fig. 3.16 that the marching integration method gives an excellent prediction of the growth rates within the reactor.

3.9. Conclusions

There is a continuing need to improve our mathematical descriptions of the flowfields of crystal growth processes. In this chapter, the approaches are useful for calculating the flowfields of many important crystal growth methods. Surprisingly, a number of important methods are amenable, in the isothermal limit, to exact solutions of the Navier–Stokes equations.

It is hoped that the material developed above will be of inspirational and technical value to all those who wish to quantify and understand the fluid flow processes of existing as well as yet to be developed crystal growth processes.

References

ACRIVOS, A. (1968) *Chem. Eng. Educ.* **2**, 62.

ACRIVOS, A. and GODDARD, J. D. (1965) *J. Fluid Mech.* **23**, 273.

ACRIVOS, A. and TAYLOR, T. D. (1962) *The Phys. of Fluids* **5**, 387.

BACHELOR, G. K. (1951) *Q.Jl. Mech. Appl. Math.* **4**, 29.

BACHELOR, G. K. (1954) *Q. Appl. Math.* **12**, 209.

BAN, V. S. (1977) in *6th Int. Conf. on Chemical Vapor Deposition* (Donaghey, L. F., Rai-Choudhury, P., and Tauber, R. N., eds.), The Electrochemical Soc., Princeton, NJ, 66.

BERKER, R. (1963) in *Handbuch der Physik* (Fluegge, S., ed.), VIII/2, 1, Berlin.

BIRD, R. B., STEWART, W. E. and LIGHTFOOT, E. N. (1960) *Transport Phenomena*, Wiley, New York.

BLASIUS, H. (1908) *Z. Math. Phys.* **56**, 1 (English) NACA Report 1121 (1953).

BOEDEWADT, U. T. (1940) *Z. angew. Math. Mech.* **20**, 241.

BRANDLE, C. D. (1977) *J. Cryst. Growth* **42**, 405.

CARRUTHERS, J. R. (1967) *J. Cryst. Growth* **114**, 959.

CARRUTHERS, J. R. (1976) *J. Cryst. Growth* **32**, 13.

CARRUTHERS, J. R. and GRASSO, M. (1972) *J. Appl. Phys.* **43**, 436.

CARRUTHERS, J. R. and NASSAU, K. (1968) *J. Appl. Phys.* **39**, 5205.

CHANDRASEKHAR, S. (1961) *Hydrodynamic and Hydromagnetic Stability*, Clarendon Press, Oxford.

CHAPMAN, S. and COWLING, T. G. (1951) *Mathematical Theory of Non-uniform Gases*, 2nd edn., Cambridge University Press, Cambridge.

CHEN (1974) p.s., Fluid Flow Patterns and Solvent Selection for Liquid Phase Epitaxial Crystal Growth on Rotating Crystals, MS thesis, University of Calif., Berkeley.

COCHRAN, W. G. (1934) *Proc. Cambridge Phil. Soc.* **30**, 365.

COLBURN, A. P. (1933) *Trans. AIChE* **29**, 174.

CURTIS, B. J. and DISMUKES, J. P. (1972) *J. Cryst. Growth* **17**, 128.

CURTIS, B. J. and DISMUKES, J. P. (1973) in *Proc. 4th Int. Conf. on Chemical Vapor Deposition* (eds.), The Electrochemical Soc., Princeton, NJ, 218.

DITTMAN, F. W. (1974) in *Chemical Reaction Engineering* II (Hulburt, H. M., ed.), A. Chem. Soc., Washington, DC, 463.

EVERSTEYN, F. C. and PEEK, H. L. (1970) *Philips Res. Rep.* **25**, 472.

EVERSTEYN, F. C., SEVERIN, P. J. W., BREKEL, C. H. J. v. D., and PEEK, H. L. (1970) *J. Electrochem. Soc.* **117**, 925.

FROESSLING, N. (1940) Verdunstung, Wämeübertragung und Geschwindigkeitsverteilung bei zweidimensionaler und rotationssymmetrischer laminarer Grenzschichtströmung, *Lunds. Univ. Årsskr.* N.F. Avd. **2**, 35, No. 4.

FUJII, E., NAKAMURA, H., HARUNA, K., and KOGA, Y. (1972) *J. Electrochem. Soc.* **119**, 1106.

GOSMAN, A. D., RUN, W. M., RUNCHAL, A. K., SPALDING, D. B., and WOLFSHTEIN, M. (1969) *Heat and Mass Transfer in Recirculating Flows*, Academic Press, London.

GOSS, A. J. and ADLINGTON, R. E. (1959) *Marconi Rev.* **22**, 18.

GREENSPAN, H. P. (1968) *The Theory of Rotating Fluids*, Cambridge University Press, London.

GRIFFITHS, R. W. (1979) *J. Fluid Mech.* **92**, 659.

HALES, A. L. (1937) *Roy. Astro. Soc. Geophys.* Suppl. **4**, 122.
HAMEL, G. (1916) *Jahresb. d. Dt. Math.-Vereinigung* **25**, 34.
HAN, T. and PATEL, V. C. (1979) *J. Fluid Mech.* **92**, 643.
HOMANN, F. (1936) Der Einfluss grosser Zähigkeit bei der Strömung, *Forsh. Ing.-Wes.* **7**, 1.
HURLE, D. T. J. (1973) in *Cryst. Growth: an Introduction* (Hartman, P., ed.), North-Holland, Amsterdam, 210–47.
HYDE, R. and TITMAN, C. W. (1967) *J. Fluid Mech.* **29**, 39.
KAY, W. M. (1958) *Convective Heat and Mass Transfer*, McGraw-Hill, New York.
KISHITAKE, K. (1976) *J. Cryst. Growth* **35**, 98.
KRASKA, J. R. and SANI, R. L. (1979) *Int. J. Heat Mass Transfer* **22**, 535.
LANCE, G. N. and ROGERS, G. N. (1962) *Proc. Roy. Soc. A*, **266**, 109.
LANGLOIS, W. E. (1977) *J. Cryst. Growth* **42**, 386.
LAUDISE, R. A. (1970) *The Growth of Single Crystals*, Prentice-Hall, Englewood Cliffs, New Jersey.
LAUDISE, R. A. (1973) in *Crystal Growth: an Introduction* (Hartman, P., ed.), North-Holland, Amsterdam, 162–97.
LIN, A. L., DAO, V., and DONAGHEY, L. F. (1977) in *Proc. 6th Int. Conf. on Chemical Vapor Deposition* (Donaghey, L. F., Rai-Choudhury, P., and Tauber, R. N., eds.), The Electrochemical Soc., Princeton, NJ, 264.
MAHAJAN, R. L. and GESHART, B. (1979) *J. Fluid Mech.* **91**, 131.
MANKE, C. W. and DONAGHEY, L. F. (1977) *J. Electrochem. Soc.* **124**, 562.
MANKE, C. W. and DONAGHEY, L. F. (1977a) in *6th Int. Conf. on Chemical Vapor Deposition* (Donaghey, L. F., Rai-Choudhury, P., and Tauber, R. N., eds.), The Electrochem. Soc., Princeton, NJ, 151.
MANTLE, H., GASS, H., and HINTERMANN, H. E. (1975) in *Proc. 5th Int. Conf. on Chemical Vapor Deposition* (Blocher, J. M., Jr., Hintermann, H. E., and Hall, L. H., eds.), The Electrochem. Soc. Princeton, NJ, 540.
MASLAIYAH, J. H. and NGUYEN, T. T. (1979) *Int. J. Heat Mass Transfer* **22**, 237.
MASON, P. J. and SYKES, R. I. (1979) **91**, 433.
MAZILLE, J. E. (1973) Etude et résolution numérique d'un problème d'aérothermie, thèse, Grenoble.
MICKLEY, H. S., ROSS, R. C., SQUYERS, A. L., and STEWART, W. E. (1954) NACA Tech. Note 3208.
MOHANTY, A. K. and ASTHANA, S. B. L. (1979) *J. Fluid Mech.* **90**, 433.
MORI, Y. (1961) *J. Heat Transfer* **83**, 479.
NEWMAN, J. S. (1966) *J. Phys. Chem.* **70**, 1327.
NEWMAN, J. S. (1968) *Ind. Eng. Chem. Fundamentals* **7**, 514.
NEWMAN, J. and HSUEH, L. (1967) *Electrochemica Acta* **12**, 417.
NIKURADSE, J. (1932) *Forsch. Arb. Ing.-Wes.* No. 356.
NYDAHL, J. E. (1971) Heat Transfer for the Boedewadt Problem. Dissertation. Colorado State Univ., Fort Collins, Colorado.
OLANDER, D. R. (1967) *Ind. Eng. Chem. Fundamentals*, **6**, 188.
OLSON, J. M. and ROSENBERGER, F. (1979) *J. Fluid Mech.* **92**, 609 and 631.
OSTRACH, S. (1955) NACA Rept. 1111 (Supersedes NACA Tech. Note 2635).
PICHA, K. G. and ECKERT, E. R. G. (1958) *Proc. 3rd US Natl Cong. on Appl. Mech.* 791.
PEARSON, C. E. (1965) *J. Fluid Mech.* **21**, 623.
POHLHAUSEN, K. (1921) *Z. angew. Math. Mech.* **1**, 235.
PRANDTL, L. (1904) in *Proc. 3rd Int. Math. Congr. Heidelberg*, 484.
REID, R. C. and SHERWOOD, T. K. (1966) *The Properties of Gases and Liquids*, 2nd ed., McGraw-Hill, New York.
REYNOLDS, W. C., LUNDBERG, R. E., and MCCUEN, P. A. (1963) *Int. J. Heat Mass Transfer* **6**, 483.
RIEDL, W. J. (1976) in *Advances in Epitaxy and Endotaxy* (Schneider, H. G. and Ruth, V., eds.), Elsevier, Amsterdam, 97-136.
ROBERTSON, D. S. (1966) *Br. J. Appl. Phys.* **17**, 1047.
ROGERS, M. H. and LANCE, G. N. (1960) *J. Fluid Mech.* **7**, 617.
ROGERS, R. H. (1976) *Rep. Prog. Phys.* **39**, 1.
ROUSE, H. and INCE, S. (1959) *History of Hydrodynamics*, Iowa Inst. Hydrodynamics, Iowa City.
RUNDLE, P. C. (1968) *Int. J. Electron.* **24**, 405.
RUNDLE, P. C. (1971) *J. Cryst. Growth* **11**, 6.
RUNYAN, W. R. (1965) *Silicon Semiconductor Technology*, McGraw-Hill, New York.
SCHEEL, H. J. (1972) *J. Cryst. Growth* **13/14**, 560.
SCHEEL, H. J. and ELWELL, D. (1973) *J. Electrochem. Soc.* **120**, 818.
SCHLICHTING, H. (1968) *Boundary-layer Theory*, McGraw-Hill, New York.
SCHLICHTING, H. and BUSSMAN, K. (1943) *Schrift. d. deutsch. Akad. Luftfahrtforschung*, **7B**, Heft 2.
SCHOLTZ, M. T. and TRASS, O. (1970) *AIChE J.* **16**, 82.
SCHULZ-DUBOIS, E. O. (1972) *J. Cryst. Growth* **12**, 81.
SCHULZ-GRUNOW, F. (1935) *Z. angew. Mech.* **15**, 191.
SECREST, B. G., BOYD, W. W., and SHAW, D. W. (1971) *J. Cryst. Growth* **10**, 251.

SHAW, D. W. (1974) in *Crystal Growth Theory and Technique*, Vol. 1 (Goodman, C. H. L., ed.), Plenum, London, 1–48.

SHERWOOD, T. K., PIGFORD, R. L., and WILKE, C. R. (1975) *Mass Transfer*, McGraw-Hill, New York.

SHIROKI, K. (1977) *J. Cryst. Growth* **40**, 129.

SPARROW, E. M. and GREGG, J. L. (1960) *Trans. ASME* **82**, 294.

SPARROW, E. M., EICHHORN, R., and GREGG, J. L. (1959) *Phys. Fluids* **2**, 319.

STEPHENSON, C. J. (1969) *J. Fluid Mech.* **38**, 335.

STEWARTSON, K. (1953) *Proc. Cambridge Phil. Soc.* **49**, 333.

STOKES, G. G. (1845) *Trans. Cambridge Phil. Soc.* **8**, 287.

STRINGFELLOW, G. B. and GREEN, P. E. (1971) *J. Electrochem. Soc.* **118**, 805.

SUGAWARA, K., TAKAHASHI, R., TOCHIKUBO, H., and KOGA, Y. (1970) in *Proc. 2nd Int. Conf. on Chemical Vapor Deposition*, The Electrochemical Soc., Princeton, NJ, 713.

SUGUWARA, K. (1972) *J. Electrochem. Soc.* **119**, 1749.

TAKAHASHI, R., SUGAWARA, K., NAKAZAWA, Y., and KOGA, Y. (1970), in *Proc. 2nd Int. Conf. on Chemical Vapor Deposition*, The Electrochemical Soc., Princeton, NJ, 695.

TAYLOR, G. I. (1923) *Phil. Trans. Roy. Soc. (London)* A, **223**, 289.

THEURER, H. C. (1961) *J. Electrochem. Soc.* **108**, 649.

TUROVSKY, B. M. and MILVIDSKY, M. G. (1962) *Soviet. Phys. (Cryst.)* **6**, 606.

UTECH, H. P., BROWER, W. S., and EARLY, J. G. (1967) in *Crystal Growth* (Peiser, H. S., ed.), Suppl. *J. Phys. Chem. Solids*, 201.

VENNARD, J. K. and STREET, R. L. (1976) *Elementary Fluid Mechanics*, Wiley, New York.

VON KÁRMÁN, TH. (1921) *Z. angew. Math. Mech.* **1**, 233.

WAHL, G. (1975) in *Proc. 5th Int. Conf. on Chemical Vapour Deposition* (Bocher, J. M., Jr., Hintermann, H. E., and Hall, L. H., eds.), The Electrochem. Soc., Princeton, NJ, 391.

WELTY, J. R., WICKS, C. E., and WILSON, R. E. (1976) *Fundamentals of Momentum, Heat, and Mass Transfer*, Wiley, New York.

CHAPTER 4

Environment for Crystal Growth

J. S. SHAH

University of The Andes, Merida, Venezuela†

"And so we see that the poetry fades out of the problem and by the time serious application of exact science begins we are left with only pointer readings."

(EDDINGTON)

4.1. Introduction

Good crystals are grown in good equipment. The design of good equipment needs consideration for creating and controlling the environment in which a crystal with desired properties can grow. Specification of the environment requires specifications of temperature, atmosphere, container materials, constituents of the supply, parameters which control growth velocities, and growth instabilities.

It is therefore necessary to discuss creation, control, and measurement of the above parameters.

4.1.1. GENERAL REMARKS ON INSTRUMENTATION

The term instrument may be used to embrace anything from a simple metre rule to a very complex control signalling equipment. All instruments, however, may be classified or divided into the following categories according to the functions they perform:

1. Detectors.
2. Indicators.
3. Measuring or metering devices.
4. Control systems.

A detector is employed in cases where a *property* or an entity (often called the signal) needs to be detected, but where the precise knowledge of the magnitude of the property is not required. Normally, the design and mode of use of a detector is such that it detects a minimum or a maximum magnitude of a property to actuate an action or a warning for the action.

In addition to the detection of the property, if a device provides a visible or audible evidence of the detection then it is classified as an indicator. A breathalyser, for example, is

†On leave from University of Bristol, H. H. Wills Physics Laboratory, Bristol, U.K.

an indicator because it provides the visible evidence of the presence of alcohol in a person's breath.

Generally, detectors and indicators are not measuring devices (although in principle they may be converted into measuring devices). In order for a device to be classified as a measuring device it has to be calibrated so that it indicates the magnitude of the property it detects. Naturally, the design of a measuring device requires considerations of parameters, such as accuracy of the measured magnitude, discussed in section 4.1.2.

If an instrument, in addition to the detection or measurement, is designed to cause other desired effect(s) then it forms a part of the whole of a control system. A control system may be designed to cause prearranged effects (e.g. an alarm clock) or it may be designed to respond and change some parameters in proportion to other signal variables. The degree of control in the above systems may be inherently dependent upon the mode and accuracy of the actuating signal measurements.

It should also be remembered that in practice very few instruments provide an *output* directly in terms of the property under observation. The majority of measuring instruments employ a transducer which changes another property (output) in proportion to the property under observation (input).

4.1.2. DEFINITIONS IN MEASUREMENT

For a given instrument and its range of measurements the following parameters are used to indicate its property with respect to its usefulness for the purpose of measurement:

1. *Specific variance or variancy.* This term is used to measure the extent to which the response of an instrument, to an input signal of a given magnitude, remains constant. Numerically the extent of the consistency of an instrument may be expressed in the term of variance which is defined, at a given value of the input, as the range of variations of the steady response of an instrument for the repeated application of the input under constant external environment. Now the specific variance can be defined as follows:

Specific variance = (variance at a given input)/(the magnitude of the input). (4.1)

Specific variance accounts for the variations in response due to such phenomena as hysteresis, backlash, etc.

2. *Passivity.* All instruments require a minimum change in the strength of an input signal to produce an observable change in the response. This characteristic (sluggishness) of an instrument is described by the term passivity which is defined as

Passivity = (least change in the magnitude of the input to produce the observable change in response)/(the magnitude of the input). (4.2)

3. *Sensitivity.* This term describes the proportion of the response with respect to the input. It is simply defined as

Sensitivity = (the magnitude of the instrument response)/(the magnitude of the input). (4.3)

4. *Accuracy.* The fidelity of the response of an instrument is stated by its maximum inaccuracy (generally for a range of measurement) by the following ratio expressed in percentage:

$$\text{Accuracy} = \{(\text{the error in the magnitude of response})/(\text{the range of measurement,}$$
$$\text{e.g. full scale deflection})\} \times 100\,\%. \tag{4.4}$$

The errors giving use to the inaccuracy in the measurement are of two kinds, namely, (1) "scale" errors, which are constant for a given range of inputs and may be known and eliminated by a careful calibration of the response of an instrument, and (2) random and variable, which are not constant and vary with each input. These errors are very difficult to remove. They may be estimated by standard statistical procedures described by Topping (1957) and Ku (1968).

The prediction of errors in a function of measurement and their significance from the statistical point of view are discussed by Illig (1968) and Nelson (1969).

It is apparent from the above discussion of parameters, that whilst a sensitive instrument is not necessarily accurate, an accurate instrument has to be sensitive. The sensitivity and the accuracy of instruments are also of fundamental importance in the design of control systems.

4.2. Temperature

In the majority of crystal growth processes it is necessary to effectuate and control temperatures above room temperature. In fact temperature is unquestionably a universal parameter which is manipulated by crystal growth process operators to achieve their experimental objectives. It is, therefore, useful to consider methods of production of elevated temperatures, their measurement and their control. The production of lower temperatures necessary for crystal growth of some materials will not be discussed here.

4.2.1. METHODS OF HEATING

True operational temperatures such as the temperature of a melt or that of a solution, etc., are primarily dependent upon the heating methods, the elements of thermal transfer and the heat transfer geometry utilized in a crystal growth system.

The selection of a heating method for the purpose of crystal growth is influenced by the following factors:

1. The magnitude of the operating temperature.
2. The principal method of heat transfer from the heat sources to heat dissipants, i.e. radiation, conduction, or convention.
3. The rate of thermal transfer or the rate of heating.
4. The degree of uniformity of temperatures, the volume (or the area) of uniformity of temperatures, and the shape of temperature gradients.
5. The requirement of programming temperatures and temperature gradients.
6. The "operation" environment (e.g. furnace atmosphere).
7. The cost and convenience of the utilization of heating equipment.

The variety of heating methods available for crystal growth may be categorized as follows:

1. Flame heating.
2. Indirect and direct electrical resistance heating.
3. High frequency heating.
4. Electron beam heating.
5. Plasma heating, i.e. glow discharge and arc heating.
6. Radiation imaging or focusing (i.e. photon heating).

Although a detailed description of each of the heating methods is too lengthy to be included here, it is appropriate to discuss in brief the principal area of application and the limitations of the above listed methods.

4.2.1.1. *Flame Heating*

Air–fuel flames (i.e. air + lower hydrocarbon gases and coal gases) are utilized for creating temperatures below 2000 K. Oxy-coal gas, oxy-hydrogen and oxy-acetylene flames, on the other hand, are used for obtaining temperatures in the range 2000–5000 K. By far the most recurrent use of a gas flame for crystal growth is in the Verneuil apparatus for melting oxides and other high melting point materials. It has been shown by Adamski (1965), Adamski *et al.* (1968), and Lipson *et al.* (1968) that by a careful design of the burner it is possible to obtain a radially uniform temperature zone for the melt and well-defined longitudinal thermal gradient and grow good quality crystals. Recently Khambatta *et al.* (1971) have calculated optimum growth rates using a relatively simple mathematical model of a high temperature flame fusion crystal growth system, and indicating the importance of the gas temperature and the temperature distribution in the annulus surrounding the growing crystal. Outside the melt region in the above system the radiative transfer is shown to be an order of magnitude greater than the convective transfer. It should, however, be remarked that generally speaking it is rather difficult to control the temperature and temperature gradients to a close proximity.

4.2.1.2. *Electrical Resistance Heating*

The indirect electrical resistance heating furnaces are widely used for melting and heating in the range 300–3000 K. The construction of an indirectly heated resistance furnace is straightforward.† In most designs (for temperatures up to 2000 K) the windings of a suitable metal wire or a ribbon are incorporated either on the exterior surface of a refractory shape, such as a muffle or a tube, or in contact with the interior surface of the refractory. To avoid short circuiting of the turns of the windings at an operating temperature each turn is placed in a groove or a slot on the refractory. Alternatively, each turn is separated by protrusions or "pips" on the refractory surface and/or embedding the windings in a refractory cement. It is important to ensure that the windings are not subjected to undue mechanical strains as often the metallic heating elements become

---

†Most manufacturers and major suppliers of heating elements normally provide useful information on the design of furnaces using their heating elements.

brittle during use, due to oxidation and/or recrystallization. The main design parameters of the windings are resistivity, temperature coefficient of resistivity, current carrying capacity, maximum and operating surface temperatures and surface emissivity, and the operating atmospheric condition for long life. Additionally, coefficient of linear expansion and specific gravity are of importance, especially for the design of self-supporting (i.e. without any refractory support) heating elements. In Table 4.1 most of the above properties of more common heating element materials are listed. It can be seen from the table that non-metallic heating elements made from silicon carbide, molybdenum disilicide, and graphite may be used with a greater ease in the range 1500–2000 K as they can be operated in air and oxidizing atmospheres. Graphite heating elements require a protective inert atmosphere, but they may be used to attain much higher temperatures. For example, an application of graphite to achieve the temperature of 3670 K is reported by Moore *et al.* (1962). Recently Winsemius and Lengkeek (1973) have described a "paint-on" heater in vacuum systems. A thin layer of colloidal graphite can be painted either inside or outside a glass or a silica container and baked at a temperature of 573 K to produce a thin, hard coating. An electric current is passed through the coating to produce heat. In principle such a heater assembly could be operated to obtain temperatures in the vicinity of 1700 K. Its principal advantage lies in space saving and ease of heating in inaccessible places.

Gradient furnaces and multiple temperature zone furnaces incorporating resistance heating are increasingly used in crystal growth processes. There are basically three design concepts which can be utilized in construction of the special furnaces of the kinds mentioned above.

1. Provision of a heat source and a heat sink or forced cooling at the extremities of the furnace or temperature zones. Heat sink is often provided by an insulating space between two separate heaters or parts of the same heaters. This is often achieved simply by leaving a gap on the heater; former temperature gradients of the order of 20 % can be commonly generated this way (see Lopez *et al.*, 1978). Iseler (1977) used air space to produce higher temperature gradient. Recently Shah and Galindo (1979) constructed a Bridgman apparatus. They achieved a temperature gradient of the order of $60 \, \text{K cm}^{-1}$ by using circulating water as a heat sink. In this arrangement temperature gradient could be varied by altering the flow velocity of water.

2. Independent control of the power input in the heating elements situated in different regions of the furnace.

3. A modular tapping arrangement of heating elements in an electric circuit incorporating one power source.

Electrically heated furnaces in general are easily adaptable for sophisticated temperature control procedures, although incorporation of a massive heat capacity, due to the use of a refractory for heat insulation and/or support, may pose problems. Recently, furnaces using a thin transparent gold film insulation (gold film reflects 95 % of the infrared energy) have become commercially available. These have very low capacity and fast heating-up time. They are relatively inexpensive with regard to construction and maintenance.

Direct Joule heating may be profitably utilized in a "molten zone" crystal growth process. In creation of a solid–liquid zone surface by Joule heating, however, other thermoelectric phenomena such as the Peltier and the Thompson effect concurrently

TABLE 4.1. *Some Useful Properties of Common Electrical Resistance Heating Elements*

Material and Trade Name	Approximate Maximum Operating Temperature (K)	Permissible Environment	Resistivity ($\mu\Omega\,m^{-1}$)	Temperature Coefficient of Resistivity	Suitable Refractory Support	Coefficient Linear Expansion ($K^{-1} \times 10^6$)	Thermal Conductivity (W cm^{-1} K^{-1})	Form of the Elements
Globar "hot rod" (Sintered SiC)	1900 1600 1650	Oxidizing Reducing Neutral	$1.0 + 10^2$	—	Alumina based	0–1000	0.15 at 1973 K	Rods
Gold	1273	Neutral and reducing	2.06 at 273 K	0.004 at 323 K	Silica and alumina based	14.38 at 323 K	3.1 at 323 K	Wire Ribbon Thin film
Graphite (C)	2850 2700	Reducing and neutral Vacuum	8.7–10	—	Water-cooled copper and other refractory materials at lower temperatures	2–2000	0.6	Rods Grooved Tubular
Graphite–pyrolytic	3650	CO	8.7–10	—	Water-cooled copper	2–2000	—	Grooved Tubular
Iridium	2450	Neutral and reducing	6.1×10^{-2}	0.004	Zirconia based	6	1.48 at 323 K	Wire Ribbon
Kanthal A-1 (22% Cr, 5.5% Al, Co + Fe)	1600	Oxidizing and reducing	1.45	3.2×10^{-5}	Magnesia and alumina based	11–15	—	Wire Ribbon

	Temperature	Atmosphere	$(2–40) \times 10^4$	2.4×10^2	Compatibility			Form
Kanthal super (MoSi$_2$)	1950 1650–1850 1650 1350	Oxidizing Reducing Neutral Vacuum (1.33 × 10^{-2} Pa)	—	—	Alumina 60–70% Silica 30–40% Fe$_2$O$_3$ (maximum) <1%	—	—	—
Molybdenum	2650	—	0.21	0.004	—	5	1.34	Wire Ribbon
Nichrome Nichrothal 80% Ni 20% Cr	1400	Oxidizing and neutral Reducing	1.35	0.0002	Alumina Silica based	13.2	0.13	Wire Ribbon
Platinum	1850	Oxidizing and neutral	0.47	0.004	Zirconia Alumina based	9.56	0.73	Wire Ribbon
Pt 90% Ir 10%	1973	Oxidizing and neutral	24 at 273 K	0.0012	Zirconia Alumina based	15	0.31 at 290 K	Wire Ribbon
Rhodium	2073	Oxidizing and neutral	4.33 at 273 K	0.0046	Zirconia Alumina based	8.3	15 at 323 K	Wire Ribbon
Tantalum	1973	Reducing and neutral	44 at 973 K	0.003	Silica at low temperature	7	0.55 at 290 K	Wire Ribbon
Tungsten	3250	Neutral and reducing	5 × 10^{-2}	0.006	Silica at low temperature	5	0.2	—
Vitreous carbon	3250	Neutral and reducing	10	—	—	2	0.06	Tubular Grooved

occur. The release of Peltier heat, especially, plays an important role in the determination of the effective growth velocity. Joule heating in the above process actually provides heat input to balance the heat losses in the system. For a concise review of the zone melting processes in an electric field and associated thermoelectric effects the reader is referred to Shah (1981).

4.2.1.3. *High Frequency Heating*

High frequency (h.f.) heating is of major importance in the crystal growth procedures because it can achieve, easily, a large range of temperatures with a reasonably high efficiency of energy transfer and is capable of operation in a variety of process ambients and can, in principle, make the crystal growth process free of contamination. High frequency heating is also versatile for direct and indirect heating.

Heating with high frequencies could be due to one of the two entirely different processes, namely, (1) induction heating, or (2) capacitive or dielectric heating. Capacitive heating has not been used to a great extent in crystal growth. It does, however, offer an alternative for some materials.

Induction heating normally occurs in a conducting material (and in ferromagnetic materials on loss of magnetization). It is in fact a kind of "resistance heating" due to the induced eddy currents (also known as Foucault currents) in the conductor by the electromagnetic field of a h.f. current carrying coil (the h.f. work coil) which surrounds the charge. The useful h.f. band is typically 100 kHz to 10 MHz, although for relatively large metallic ingots frequencies in the medium frequency (m.f.) band (0.5–10 kHz) is useful and advantageous.

The eddy current density distribution across the cross-section of a conductor is a function of the frequency. At high frequencies the impedance at the centre of the conductor is high due to a higher flux density. Thus the current density in the central core is negligible and most of the current flows in a thin layer (skin) near the surface. It can be shown that the current at a depth r below the surface of a conductor is given by the expression

$$I(r) = I_0 \exp(-r/\delta), \qquad (4.5)$$

where $I(r)$ is the current at the depth r below the surface of the conductor, I_0 is the current at the surface, and δ is a parameter called the skin depth or the depth of penetration.

From eqn. (4.5) it is clear that the skin depth is defined as the thickness of a layer, at the surface, over which the current diminishes by the value $1/e$ of the surface current. The above definition is diagrammatically shown in Fig. 4.1. The skin depth (in mm) may be calculated from the expression

$$\delta = \frac{1}{2\pi} \left(\frac{\rho \times 10^7}{v\mu} \right)^{1/2}, \qquad (4.6)$$

where ρ is the resistivity of the conductor (in $\mu\Omega$), v is the frequency (in Hz), and μ is the effective relative permeability of the material.

It can be seen from eqn. (4.6) that the skin depth rapidly decreases with increasing frequency. The relationship of the skin depth with the frequency is more clearly illustrated

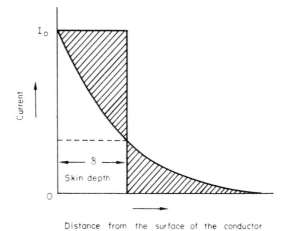

FIG. 4.1. Diagrammatic representation of the skin depth of a conductor in induction heating.

in Fig. 4.2, where the current distribution in a 5 mm diameter copper conductor at various frequencies is shown. It should, of course, be appreciated that most of the heat is generated in the skin layer over the coupled flux volume of the conductor. The heat then is conducted inwards. The temperature achieved therefore for a given power input would depend upon the frequency, the rate of the heat conduction to the centre of the bar, other heat losses in the system, and, of course, the degree of coupling between the h.f. work coil and the conductor.

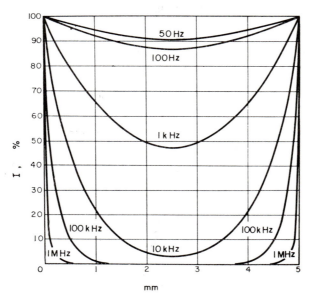

FIG. 4.2. The current distribution in a 5 mm diameter copper conductor at various frequencies. (From *Industrial H.F. Generators* by Sobotka, by the courtesy of Philips Technical Library, Eindhoven.)

The choice of frequency will depend upon the degree of penetration required in accordance with eqn. (4.6) and the efficiency of power transfer. (In many cases the power delivered to the conductor depends upon the frequency used.) For example, the use of a frequency in the m.f. band offers a greater depth of penetration and thus a larger volume in which heat is generated. It should be remembered that the heating efficiency can reach a maximum at a certain frequency (see Shah, 1981). Also at high frequencies the power losses via lead connections, etc., become appreciably more serious. The reader is reminded here that the working frequencies (especially above 10 MHz) are tied to the international agreement to cope with the danger of interference with other radio services. For further details see the latest CISPIR† recommendations.

Modern high and medium frequency generators use resonant oscillator circuits tuned to the working frequency. A block diagram of a typical generator is shown in Fig. 4.3. For an efficient power transfer to the "load" to be heated, its impedance has to be matched with the resonant resistance of the oscillator circuit. This is given by

$$R_{(R)} = 530Q/Cv_0, \tag{4.7}$$

where Q is the circuit quality factor (Q factor = reactance/resistance), C is the circuit capacitance (in pF), and v_0 is the resonant frequency given by

$$v_0 = 1/\{2\pi(LC)^{1/2}\}, \tag{4.8}$$

where L is the inductance in the tuned resonant circuit.

The procedure to match the load is described in detail by Sobotka (1963).

For cylindrical material, of diameter d and axial length l, being heated with a small skin depth ($\delta \ll d$), the skin layer can be treated as a single turn secondary coil of dimensions: πd = width, δ = thickness, and l = length, forming a transformer with the work coil as the primary winding. Consequently, an approximate expression for the power induced for dissipation is given by

$$P = 4 \times 10^2\pi^2 N^2 I^2 ld(\vartheta\mu v \times 10^{11})^{1/2} \quad \text{(in watts)}, \tag{4.9}$$

where I is the work coil current (in amperes r.m.s.), N is the axial length of the work coil (in turns cm^{-1}), l is the axial length of the material (in mm), d is the diameter of the material (in mm), ϑ is the resistance (in Ω), μ is the relative permeability of the material, and v is the frequency (in Hz). For a more rigorous expression of the power dissipated in the material the reader is referred to Simpson (1960).

It was stated above that induction heating is applicable to conducting materials. However, Warren (1962) showed that it was in fact possible to sustain a molten zone in relatively poor conducting materials. At an operating frequency ($\sim$ few MHz) the skin depth for a poor conductor such as liquid NaCl is much greater than the small dimensions of the zone containing liquid, and under this condition it is possible to achieve a

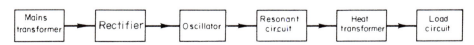

FIG. 4.3. A block diagram of a modern h.f. and m.f. generator for heating.

†Comité International Spécial des Perturbations Radio-électriques.

reasonably large power transfer in the material at the operating frequency. Thus with a suitable choice of coil geometry it should become possible to deliver enough power to sustain the molten zone by an h.f. generator output of only a few kW.

Capacitive heating due to the application of an h.f. field occurs in non-conductors or insulators. It is a result of the movement of atoms or molecules in a material caused by a h.f. electric field. Therefore, a high efficiency of heating may be brought about by nearly matching the frequency of the oscillatory electric field to that of the natural frequency of vibration of the molecules/atoms in the substance. A useful frequency band for capacitive heating is typically 30–50 MHz.

If a material to be heated is placed filling the space between two parallel plate electrodes as shown in Fig. 4.4, then the capacitance C (in pF) of the capacitor thus formed is given by

$$C = 11.1\varepsilon A/(4\pi d),\tag{4.10}$$

where ε is the dielectric constant of the material, A is the surface area of each of the plates (in mm^2), and d is the distance between the plates (in mm).

The leakage Σ of the above capacitor is given by

$$\Sigma = 10^{-1}\sigma A/d \quad \text{(in Siemens, i.e. } S = \Omega^{-1}),\tag{4.11}$$

where σ is the conductivity of the material (in $\mathrm{S\,m}^{-1} = \Omega^{-1}\mathrm{m}^{-1}$). If now an alternating voltage, $V = V_0 \cos \omega t$ (where $\omega = 2\pi v$ and where v is the frequency in Hz), is applied to the capacitor, then the ohmic current i_r and capacitive current i_c flowing through the capacitor are given by

$$i_r = \Sigma V_0 \cos \omega t\tag{4.12}$$

and

$$i_c = \frac{\mathrm{d}q}{\mathrm{d}t} = \frac{\mathrm{d}}{\mathrm{d}t}(CV) = \omega C V_0 \cos(\omega t + \tfrac{1}{2}\pi),\tag{4.13}$$

where q is the charge in the capacitor.

Taking into account the phase angle between the total current and capacitive current, the total current i_T is given by

$$i_T = i_c(1 - j\theta),\tag{4.14}$$

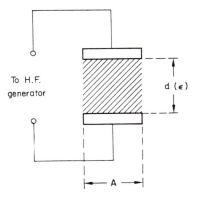

FIG. 4.4. A schematic h.f. capacitive heating arrangement.

where $j = \sqrt{(-1)}$ and θ is called the loss factor of the capacitor. The expression for the loss factor is given by

$$\theta = \Sigma/(\omega C). \tag{4.15}$$

Alternatively, θ may be given by the expression

$$\theta = 2\sigma/(\varepsilon v). \tag{4.16}$$

The dielectric constant is not a very sensitive function of the frequency, therefore it is inferred from eqn. (4.16) that the loss factor increases with lower frequencies.

The power loss available for dissipation can be shown to be given by

$$W = V^2\Sigma = V^2\omega C\theta. \tag{4.17}$$

It is fairly clear from eqn. (4.17) that the thermal efficiency in capacitive heating rises with increasing frequency. The limit of the electric field strength applicable in the method is governed by the electric breakdown strength of the material.

High frequency heating is initially expensive due to the high cost of installation of a generator. There are, however, a number of advantages, some of these having been listed earlier, which offset this and make it an attractive method of heating. Attention is drawn here to further benefits that can be gained by h.f. heating and induction heating in particular:

1. When induction heating is used with a water-cooled metal container, for instance, as in the silver boat technique by Sterling and Warren (1963), forces of levitation due to inductive forces lift a molten charge (the temperature of which may be very high) and completely eliminate the risk of contaminating.
2. Electromagnetic induction forces can provide automatic stirring to a molten charge. (Stirring in many crystal growth processes is beneficial and therefore desirable.)
3. Another major feature of h.f. heating is the accuracy with which the generation of heat is localized.
4. Sophisticated control of other crystal growth parameters becomes possible with the use of h.f. heating.

For further information on h.f. heating the reader is referred to Brown *et al.* (1947), Simpson (1960), Sobotka (1963), and Brown (1965).

4.2.1.4. *Electron Beam Heating*

The earliest observation of electron beam heating appears to be made by Grove (1842) who in his paper on "The electrochemical polarity of gases" described an experiment in which two electrodes, a metal plate, and a thin platinum wire were mounted in a chamber enclosing various gases at low pressure. He observed that on the application of the output of an induction coil, the plate electrode being negative, the plate rapidly developed a circular "oxidized" spot beneath the platinum wire tip. The effect he claimed was due to the attraction of negative ions from the discharge towards the plate. Many years later Sir William Crookes (1879) became the first man to demonstrate electron beam melting by fusing a platinum anode in a cathode-ray tube by the bombardment of cathode rays (which were not at that time established as electron streams). The first patent for an electron beam melter was granted to Pirani (1907), who actually produced homogeneous beams of metals like tantalum. It is interesting to note that Pirani had pointed out the potential of vacuum-melting purification by electron beam heating. In the period 1907–1940 a handful of experimental electron beam furnaces were constructed. An interesting account of the pre-war efforts in this field is provided by Trombé (1950). It is only in the last two decades that electron beam heating has developed into a very sophisticated technique and is now widely used in a variety of processes in laboratory and industry.

Electron beam heating is accomplished by subjecting a material to bombardment by a controlled stream of high energy electrons. Upon impact the kinetic energy of the electrons is dissipated in the material producing heat. The production of high energy electrons may be achieved by emission of electrons from an electron emitter (e.g. a heated filament) and then accelerating the electrons towards the material to be heated (target) by the application of a potential between the target and the filament. Alternatively, the target can be posed in the path of an accelerated beam of electrons. This process is only feasible at pressures below 1.333×10^{-2} Pa (Pa = pascal = N m^{-2}, 1.333×10^2 Pa = 1 torr). A very simple arrangement for electron beam production (i.e. an electron gun) may only consist of a filament (cathode) sitting at a high $-$ve potential and the target, as an anode, at the earth potential. A more advanced electron gun and associated equipment may consist of a cathode, a modulator grid (i.e. a metal screen with an aperture held at the same potential as or slightly $-$ve to the filament), an anode, a focusing lens and a deflection coil as shown in Fig. 4.5. The area of heat production on the target can be varied by a factor of one hundred simply by variation of a current through the focusing lens. In contrast to the heating method so far described, electron bombardment is capable of concentrating large power densities ($\sim 5 \times 10^5$ kW cm^{-2}) in a very small area ($\sim \mu^2$) because of the ability to focus the electron beam. The direction of the beam can be varied as well by manipulation of current through the deflection coil.

It is apparent from the foregoing description that the constituents of an electron gun may be conveniently classified into two categories, namely (1) cathode element(s), and (2) shaping elements, which include modulating electrodes, anode, and screen electrodes. The detailed design of a gun will depend upon factors such as:

1. Required furnace power—laboratory furnaces typically can deliver 2–30 kW.
2. The availability and suitability of the accelerating voltage.

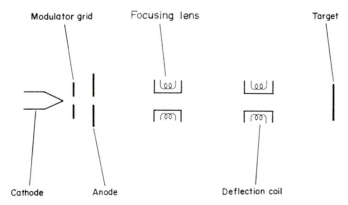

FIG. 4.5. A versatile electron gun arrangment for a heating device.

3. The filament material—its emission characteristics and the value of maximum working temperature ($0.9 \times$ melting point).
4. The area of heat generation and beam shaping requirements.

The saturated current density J_{sat} for a thermionic emission is given by the Richardson–Dushman equation

$$J_{(sat)} = A T_f^2 \exp\{-e\psi/(kT_f)\}, \qquad (4.18)$$

where A is a constant dependent upon the filament material, T_f is the filament operating temperature, ψ is the work function of the filament material, and k is Boltzmann's constant.

In practice, however, the gun operation is "space charge limited", i.e. it is independent of the filament temperature. The operational current density J_{op} therefore is given by the Childs law equation (for two parallel electrodes)

$$J_{op} = \tfrac{4}{9}(2\varepsilon_0)^{1/2}(e/m)^{1/2}(V^{3/2}/D^2), \qquad (4.19)$$

where ε_0 is the permittivity of free space, e/m is the electronic charge to mass ratio, V is the voltage between the electrodes, and D is the distance between the electrodes.

It is emphasized here that Langmuir and Compton (1931) have shown that the anode voltage power law (i.e. 3/2-power dependence) is valid for all electrode configurations, though different constants are involved for different configurations.

The choice of accelerating voltage is dependent upon the required minimum beam diameter and the beam current. The upper limit, however, is set by the X-ray generation threshold (which is dependent upon the voltage, current, and the target). The gun can, of course, be operated at almost any voltage with proper shielding of the equipment. Potential designers and operators are strongly urged to consult the latest MPD (minimum permissible dose) limits in operation.

The beam current is determined by the total beam power required. In actual fact the relationship between the beam current and the accelerating field defines one of the basic parameters called perveance G as follows:

$$G = I/V^{3/2}. \qquad (4.20)$$

The design analysis of the shaping electrodes is beyond the scope of this article. However, it is pointed out that the beam-shaping requirements determine the choice of the type of gun required. For further details the reader is recommended to refer to Leonard (1962).

The designer of complete electron beam systems for crystal growth purposes, such as melting and float-zone melting, in addition to the choice of an electron gun needs to consider a suitable vacuum system, work chamber, traverse mechanisms and temperature control. Some of these aspects have been discussed by Barber and Bakish (1962).

Electron beam heating is widely used in float zone melting. A brief discussion on the electron beam float zone melting is given by Shah (1981). A major source of instability in electron beam heating is due to outgassing of the material when it is being heated. Two basic methods of emission control are used to overcome this: (1) adjustment of the filament control and (2) fire control. Kanaya et al. (1968) used separate circuits to stabilize accelerating voltage and beam current. Kamm (1977) has also described a system for emission control which uses a regulated d.c. power supply and operational amplifier control circuits to regulate the emission current and the filament power supply.

However, electron beam heating is not popularly utilized in Czochralski crystal pulling systems, although an elegant crystal pulling furnace, utilizing four symmetrically mounted 5 kV–0.25 A electron guns, was described by Eaton *et al.* (1961). Another similar technique, namely the pedestal technique recently employed to grow large diameter (40 mm) silicon crystals, may be regarded as a modified pulling process using electron beam heating.

The main advantages and/or disadvantages of electron beam heating are summed up as follows:

1. Very rapid heating (and cooling) is inherent.
2. Contamination can be kept to a minimum by provision of an ultra high vacuum system. The necessity of working in a vacuum means that the dissociating materials cannot be processed.
3. Very high temperatures can be achieved due to high power densities.
4. Very high efficiency of energy transfer. For instance, in some gun designs less than 0.1 % of the cathode current is lost to the electrodes.
5. Well-defined and sharp thermal gradients inherent.
6. Temperature control in certain circumstances could be difficult.
7. Processed materials could be contaminated by the evaporation of the filament although it can be avoided by deflection of electron beam and concealing the target from evaporating atoms from the filament.
8. Higher defect densities in the processed crystals due to high thermal stresses.
9. Possibility of radiation damage.
10. Flexibility due to the utilization of electron beam shaping and deflection.
11. Very high installation costs.

4.2.1.5. *Plasma Heating*

The word plasma seems to have been first used by Langmuir (1929) in connection with glow and arc discharges to describe the region where positive and negative space charges are nearly balanced. It is difficult to give a precise definition of plasma, although the word (or/and the word plasmoid) is now commonly used to describe a "gaseous cloud" consisting of molecules, atoms, ions, electrons, and photons, which is capable of electrical conduction but in aggregate may be electrically neutral. Among the earliest investigators contributing to plasma physics, Geissler, Crookes, Thomson, Franck, and Hertz have a prominent place.

The different kinds of plasmas can be characterized by the magnitude of the following properties:

1. Degree of ionization: a few percent ionization is counted as a low degree of ionization while the ionizations of the order 30 % and above are regarded as high.
2. Pressure: plasmas in the pressure range 1–130 Pa are low pressure plasmas, while those at the pressures 1×10^4 and above (note: 1 atmosphere pressure = 101.8×10^2 Pa) are regarded as high pressure plasmas.
3. Temperature: a low temperature plasma does not exceed temperatures over 10^5 K. A high temperature plasma, on the other hand, has a temperature in the region of 10^6 K and above.

The internal pressure of a plasma is determined by the mutual impact and energy transfer of the particles. In the steady state the external pressure equals the internal pressure. Therefore, assuming elastic collisions predominate, the plasma pressure P is given (to a first approximation) by the famous equation in the kinetic theory of gases

$$P = \tfrac{1}{3}Nmv^2, \tag{4.21}$$

where N is the number of particles per unit volume, m is the mass of the particle, and v is the quadratic mean value of the particle velocities as determined from the Maxwell-velocity distribution curve.

The degree of ionization in a plasma depends upon the energy supply to the plasma, i.e. the energy available for ionization. The ionization energy may be supplied in various ways. In gas discharge the energy is supplied by the electrical energy.

The electrical conductivity of a plasma may be expressed by the equation

$$\sigma = n\mu_n q + p\mu_p Q, \tag{4.22}$$

where σ is the conductivity of the plasma (in $S = \Omega^{-1}\,m^{-1}$), n is the number of electrons per unit volume (m^3), μ_n is the electron mobility (in $m^2\,V^{-1}\,s^{-1}$), q is the elementary electron charge (in C), p is the number of ions per unit volume (per m^3), μ_p is the ion mobility (in $m^{-2}\,V^{-1}\,s^{-1}$), and Q is the elementary ion charge (in C).

It is pointed out that Ohm's law does not apply to plasma as σ is not a constant but a function of the number of particles.

The energy transfer and therefore the temperature of plasmas is predominantly dependent upon the electrons, as electron mobility $\gg$ ion mobility. Electronic collisions with molecules delivering the energy for absorption could be classified into four different categories according to their consequences:

1. Elastic collisions: these lead to low energy losses of electrons and consequently the temperature of the plasma would increase.
2. Dissociating collisions: these lead to the dissociation of polyatomic gases into atoms.
3. Ionizing collisions: these ionize the atoms in the plasma.
4. Excitation collisions: these lead to raising of energy state of the atoms.

Conductivity plasma may exist in a state of internal equilibrium. Under the influence of an external electric field the plasma electrons acquire a resultant velocity (i.e. drift velocity) in the direction of the field.

When a plasma is employed as a heating source, the kinetic energy of the particles contained in the plasma is transferred to the material to be heated. The evaluation of the plasma as a heating source could be facilitated by the consideration of the following four parameters.

1. Total energy content of the plasma: this can be calculated. If one neglects the rotational and vibrational energy of the plasma and only considers the translational kinetic energy, then by the application of Boyle's law the energy of the plasma may be calculated from

$$\tfrac{1}{2}mv^2 = \tfrac{3}{2}kT, \tag{4.23}$$

where k is Boltzmann's constant. The total energy of the plasma available for heating cannot exceed the power of the energy source feeding the plasma.

2. Flow of energy E, i.e. the energy transfer per unit area per unit time.
3. Heat transfer density Q, i.e. the heat quantity developed per unit area per unit time.
4. Efficiency η: this is defined as the ratio of heat transfer density: flow of energy

$$\eta = Q/E. \tag{4.24}$$

Two distinct types of plasma have been usefully employed in crystal growth processes. They are:

1. *Gas discharge plasmas.* These are generally low pressure and poorly ionized plasmas. For high temperature crystal growth and material processing hollow cathode and hollow anode discharges have been successfully used by Class *et al.* (1966), Dugdale (1966), Class (1968), and Storey and Laudise (1970).

A glow discharge occurs on application of a suitable voltage across two electrodes enclosed in a gas at low pressures (13.3–133 Pa). The exact nature of the discharge depends upon the gas, its pressure and the geometry of electrodes. Most glow discharges, however, are considered to consist of two spatial regions, namely:

1. The anode plasma region: this contains gas ions and electrons with near thermal energies. It is therefore conducting at anode potentials and consequently only a small voltage drop occurs across this region.
2. The cathode fall region: this region has a high density of ions, moving towards the cathode, and it is due to the multiple electron gas atom collision occurring in the regions. The ions contained herein give rise to electrons by two processes, namely the cathode emission by ion bombardment, and all the collisions in the cathode fall region. It has been shown by van Paasen *et al.* (1962) and Boring and Stauffer (1963) that 10–200 eV can be generated in the glow discharge.

Generally, electrons which do not take part in ion-electrons generation retain their high energy. They are often distinguished as the "run-away" electrons. The heating of a body placed in the plasma actually results from the bombardment of the high energy electrons. In a hollow cathode apparatus, schematically shown in Fig. 4.6, the electrons emanating from the cathode walls are accelerated and scattered back from the opposite wall in the space at the middle of the hollow cathode. It is for this reason that the focusing of heating may be attained by shaping the cathode. In a hollow cathode arrangement the position of the anode is not important as a very small potential drop occurs across the anode path. The formation of the run-away electrons is encouraged by higher voltages. It is also evident that the energy of the bombarding electrons is dependent upon the voltage drop across the cathode fall region which is nearly equal to the externally applied voltage. In fact typical operation parameters for a hollow cathode discharge in melting application lie in the following range: pressure 13.3–133.3 Pa, voltage 1.5–5 kV, and current 0.1–1.5 A. The approximate estimate of the energy density available for heating is given by

$$E = E_e j, \tag{4.25}$$

where j is the flux of electrons impinging on the surface of the heating material and E_e is the average electron energy given by

$$E_e = Ve, \tag{4.26}$$

where V is the externally applied voltage and e is the electronic charge. For melting of small volumes the energy for heating is preferred to be delivered by a relatively small flux

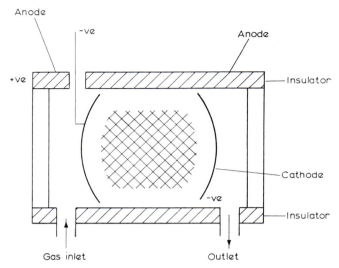

FIG. 4.6. A schematic diagram of a hollow cathode apparatus.

of high energy electrons because under this condition the space charge dispersion of electrons in the plasma is smaller. The above situation is created by operating the plasma at a high voltage and a low current. It should, however, be remembered that an important advantage of the hollow cathode heating stems from the fact it can be operated at a relatively high voltage to minimize sputtering of the cathode and thereby reduce contamination. (Behrisch, 1964, has pointed out that the sputtering rate may be kept to a minimum by an appropriate choice of cathode material.) The other main advantage of the hollow cathode process is that it can operate relatively higher pressures than those for electron beam heating.

Dugdale (1966) showed that it was possible to produce well-defined beams of electrons and ions at relatively high voltages ($\sim 20\,kV$ and over) by hollow anode geometries. A simple hollow anode geometry is shown in Fig. 4.7. It consists of a plane cathode and a hollow anode of either cylindrical or rectangular shape. The boundary of the cathode fall region normally follows the shape of the equipotential lines for the geometry as in Fig. 4.7; here it is shown that the ions moving towards the cathode tend to funnel at the centre of the cathode generating electrons. Consequently, the fast electrons crossing the cathode fall region tend to be directed along the axis of the anode. At voltage around $15\,kV$ 40% of the input power can be transmitted to the end of the anode furthest from the cathode. Furthermore, electrons can be brought out of the anode plasma through an aperture in the anode. The electron beam thus obtained can be focused and shaped in the usual manner. An ion beam can also be obtained outside the electrode assembly by making a hole in the central region of the cathode. The use of a second electrode, as shown in Fig. 4.7b, which is biased positively with respect to the cathode, is necessary to achieve a reasonable energy density in the ion beam. With the above arrangement the power density normally obtained is of the order of 15% of the input power.

Finally, it is pointed out that the power density in the above-mentioned glow-discharge devices will be limited due to the factors, such as maximum temperature limits of the

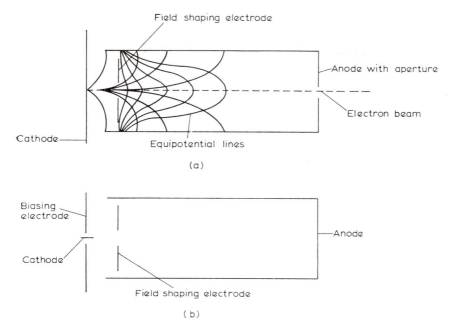

FIG. 4.7. Examples of hollow anode geometries: (a) for obtaining an electron beam, and (b) for obtaining an ion beam.

design, erosion due to sputtering, glow to arc transition conditions, and thermal characteristics (including thermal fatigue) of the cathode material.

2. *Induction plasmas*. These are relatively high pressure plasmas which can be used for crystal growth as demonstrated by Reed (1961) and Alford and Bauer (1966). An induction plasma is energized by a h.f. current (over 4 MHz). It is widely used commercially in the form of a torch (i.e. an open plasma torch) for welding and other purposes. A plasma torch in many respects is very similar to a flame torch and therefore can be used as a heating source in the Verneuil technique. The advantages of a h.f. plasma torch over a flame torch are that (1) they can be operated with inert or active gas ambient, and (2) the crystals can be kept free of contamination from the products of combustion.

A simple design of a Verneuil apparatus using an open plasma torch is shown in Fig. 4.8. The plasma torch utilized in the apparatus essentially consists of a tube through which a continuous stream of gas is passed in to the open. A h.f. field is posed into the path of the stream. The torch is "ignited" by thermal ionization of the gas by inserting a conducting rod in the h.f. field. The heating of the rod causes the reduction in ionization potential of the gas and the plasma is excited. For practical purposes the temperature attained depends upon the rate of the supply of gas. The stabilization of plasma for crystal growth purposes is essential and it is achieved in practice, thermomechanically, by a suitable design of the gas path (i.e. by utilizing the tangential gas supply and causing the gas to travel in a spiral course along the walls of the tube).

A stationary h.f. plasma (low pressure 1.33 Pa) may be utilized for zone melting purposes as shown in Fig. 4.9. Focusing of heat to obtain a small melt zone is possible by pinching (and incidentally stabilizing) the plasma by an electrical or magnetic field. The

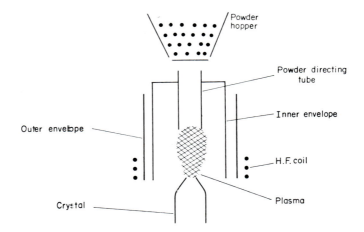

FIG. 4.8. A Verneuil apparatus using an open h.f. plasma torch.

energy transfer efficiency of the stationary plasma is, understandably, much higher than that of the open plasma. In practice it is found that impurity contamination can be avoided by supplying a very weak gas stream around the molten zone without sacrificing the advantage of the stationary plasma. Recently Savitsky and Burkhanov (1978) showed that temperatures higher than 4273 K can be obtained. They have zone-refined carbides of tantalum and hafnium and have claimed to achieve higher efficiency due to the achievement of high temperature.

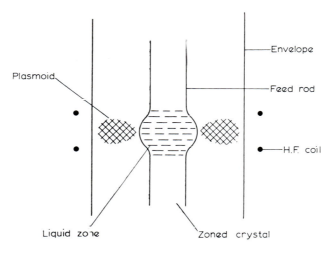

FIG. 4.9. A zone melting arrangement using a stationary h.f. plasma.

4.2.1.6. *High Intensity Arcs*

The first description of an electric arc appears to have been given by Davy (1809). It is now recognized that an arc is a high pressure, low temperature plasma. The distinction between a glow discharge and an arc is that the arc carries a higher current density ($0.1 \, \text{A cm}^{-2}$ and above) at a much lower voltage than the glow discharge. In fact the current density j in the glow discharge varies as follows:

$$j \propto p^2, \tag{4.27}$$

whereas the current pressure variation in the discharge (arc) above pressures of a few thousand Pa (i.d. few cm of mercury) was theoretically shown by Meek and Craggs (1953) to be

$$j \propto p^{4/3}. \tag{4.28}$$

Although in practice the relationship eqn. (4.28) is not exactly obeyed, it can be experimentally observed that at flow to arc transition the voltage between the electrode falls and the current increases. The glow discharge to arc transition is further characterized by the fact that the cathode fall region becomes invisibly small and the space between the electrode is filled by the much brighter positive column (i.e. the anode plasma) which carries a higher current.

The electric arc in its early days was used by Hare (1839)—cited by Doremus (1908)—for melting purposes. Just before the turn of the century Moisson (1897) constructed an arc furnace with carbon electrodes in air to achieve temperatures over 4000 K. It should, however, be pointed out that the arc sources were more popular for high intensity illumination rather than for use as heat sources. With the advent of cold hearth arc melting by Kroll (1940), the arc began to be frequently used as a direct heat source for crystal growth. The melting techniques and furnaces using the electric arc have recently been reviewed by Weatherly and Anderson (1965) and Reed (1967a, b). It would suffice to say here that the electric arc has been successfully used for Verneuil technique by Bartlett *et al.* (1967), arc fusion by Rabenau (1964), electron beam zone refining by Hulm (1954), Cabane (1958), and Geach and Jones (1959), and Czochralski growth by Reed and Pollard (1968). Recently Jordon and Jones (1971) have described a relatively inexpensive and versatile arc furnace which can be used for crystal growth under totally controlled atmospheric conditions.

Single crystals of some iron group metals and their oxides have been prepared by a novel d.c. arc transfer process initiated by Drabble and Palmer (1966). The process essentially consists of striking a d.c. arc between conducting electrodes of the material to be processed. (For growing a metal oxide crystal the arc is struck between the metal electrodes in air or oxygen.) The sustainment of the arc is accompanied by the transport of the material from one electrode to the other. The direction of the transport depends upon the polarity of the material. The actual growth of the crystal, however, takes place due to a unidirectional freezing of a molten pull on the tip of the growing electrode. A review of the materials grown by the above technique is given by Drabble (1968).

The main advantage of the arc heating is that it is a relatively inexpensive mode of attaining very high temperatures. However, it is very difficult to incorporate temperature control in the arc processes. The stability of the arc can often pose a problem in its application to crystal growth. The contamination possibilities from the electrodes are high in the above process. It is claimed by Reed (1967) that at least in cold-hearth arc melting, the processed materials were found free of contamination.

4.2.1.7. Radiation Imaging

One of the earliest applications of imaging radiant energy could be attributed to Archimedes, who employed solar energy to heat the objects with the help of mirrors. Towards the end of the seventeenth century and during the eighteenth century there were a number of solar furnaces in existence. The famous phlogiston theory was demonstrated to be "proved" with the help of solar energy in Italy during the seventeenth century. Platinum was first partially fused in a solar furnace by Macquer (1758). Lavoisier had performed many of his heating experiments in vacuum and controlled atmospheres with a solar furnace using a combination of 1170 mm and 200 mm diameter alcohol filled glass envelope lenses.

Any heating system using radiant energy imaging normally consists of at least three distinct components, namely

1. A radiation source of adequate intensity.
2. An optical image forming system comprising lenses and mirrors.
3. A material holder or a material mounting arrangement.

In addition the system incorporates heat flux attenuators in the form of slits and shutters to achieve a form of control on heating.

The oldest and perhaps the most fascinating source of radiant energy is the sun. Solar radiation is of course free and abundant. The intensity of normally incident solar radiation on the earth surface can be as high as 1.45×10^{-3} W mm^{-2} at some sites. Furthermore, the intensity of solar radiation is remarkably stable. Foex (1964), for instance, has reported that on a clear day the intensity was found to vary well within 5 % during a period of 6 hours. The average flux density obtained within the image area is high enough to melt most of the high melting point materials. With the normal incidence radiation of intensity 1×10^{-3} W mm^{-2}, Le Phat and Vinh (1961) were able to obtain the power density of 43 W mm^{-2} using a perfect paraboloid mirror (reflection coefficient 0.93). The relationship between the radiant intensity and heating time in a solar furnace has been considered by Cobble (1964) and Noguchi and Kozuka (1966). They were, however, not able to account for heat losses by convection, radiation, and conduction to obtain meaningful results. Experimentally, Noguchi and Kozuka (1966) found that Y_2O_3 (m.p. 2686 K) could be melted in 2.93 s, using flux density of 7.41 W mm^{-2} or in 2.3 s with the flux density of 10.89 W mm^{-2}. The size of the solar image (i.e. heating area) may be varied by the size of the aperture of the concentrator up to the image diameter ~ 125 mm without an appreciable decrease in the maximum flux across the image. It is pointed out that uniform large diameter image areas are expensive to produce. Amongst the major disadvantages of the sun as a radiation source are (1) it is not always available when needed, and (2) sophisticated temperature control is difficult to achieve. A solar furnace is particularly well suited for small scale fusion crystal growth as recently demonstrated by Sakurai and Ishigame (1968) and Sakurai et al. (1968) in successfully growing crystals of NiO and UO_2. Noguchi (1978) has also described a design of a high temperature solar furnace suitable for small scale crystal growth. Digital pyrometric techniques for measuring temperatures in solar furnaces have been described by Yamada and Noguchi (1978). They essentially use a brightness pyrometer as a signal source which is then amplified, converted into a digital form by an analogue to digital converter and fed into the central processor of a computer. Conversion errors through the system are of the order ± 1 K at 3273 K and ± 7 K at 2273 K.

One of the earliest and widely used manmade radiation sources used for heating is the familiar arc. The density of the current flowing through the arc, in the main, depends upon the pressure and the rate of consumption of the electrode carbon; electrode arcs are operated in vacuum or a low pressure ($\sim 1.013 \times 10^5$ Pa) inert atmosphere. Due to the high electrode consumption rates carbon arc cannot be operated continuously for more than a few hours. Laboratory carbon arc image furnaces are capable of giving power density of 1 W mm^{-2} over an area of 2.5 cm. The flux density, however, is spread over the spectrum of 300–900 µm. Furthermore, the flux stability is very poor due to rapid and slow fluctuations in the flux density. Faster variations are due to changes in the solid particle content in the arc gap while slower changes are due to the variations in the electrode gap and the electrode resistance. Tungsten electrodes are normally used for high pressure arcs such as xenon or xenon–mercury arcs. The arc is normally operated in the pressure range 6.078–30.49 $\times 10^5$ Pa (6–30 atm). Due to high pressure the arc discharge is confined in a small volume between the tips of electrodes and thus is a highly concentrated source. The arc operation is easy to maintain and the life of the electrodes extends over a few hundred hours. The serious disadvantage, however, is that flux distribution in the source itself has high radiant non-uniformity. In general therefore arc images are not the ideal sources for crystal growth.

High refractory electrical resistance can serve the purpose of a radiation source for heating. With heated tungsten filament lamps (e.g. a projector lamp) with suitable reflector or concentrator it is easy to obtain flux density of 0.25 W mm^{-2}. Recently, high wattage tungsten–halogen lamps have become available. They operate at a higher temperature than the ordinary tungsten bulbs, in the range 3000–3400 K. Eighty per cent of their output is in the infrared (slightly less than the tungsten lamp) and they have a reasonable lifetime of a few hundred hours. They have been used recently with elliptical focusing arrangements in float-zone growth by Okada et al. (1971), Takahashi et al. (1971), and Takai (1978).

The availability of high power continuous wave (c.w.) output carbon dioxide (CO_2) type lasers has opened up the potential of radiation imaging technique of heating. The carbon dioxide type laser is the most efficient molecular gas laser, first discovered by Patel (1964). Its power generation is in fact a stimulated emission of the energy due to transitions between the rotational–vibrational energy states of CO_2. Practically, a mixture of carbon dioxide, nitrogen, and helium is electrically excited in a laser cavity. The excited molecules return to the lower ground state emitting a powerful monochromatic radiation of wavelength 10.6 µ or 9.6 µ. The device due to inherent high quantum efficiency operates at a working efficiency in excess of 20%. Lasers, giving outputs in the range 500 W to 1 kW, are commercially available. Eickhoff and Gurs (1970) and Cockayne and Gasson (1970) have successfully used laser devices for crystal growth. One of the advantages of the laser radiation is that one can obtain sufficiently high power with relatively simple and inexpensive optics.

The design and the quality of the concentrators is of great importance in an image furnace as they determine the performance of the furnace. The design is dictated partially by the source and partially by the desired energy density. Single, double, and compound reflective systems using paraboloidal and elliptical mirrors may be used. Refractive systems with lenses can also be used. Details of different designs are given by Laszlo (1965).

Basically two types of control are incorporated in the imaging technique: (1) control of the intensity of the image area (i.e. energy density) by attenuation, and/or (2) the time of "exposure" by the shutter mechanism. Obviously the above mechanisms offer only limited flexibility.

On the whole imaging techniques are better suited for small melt/solution volume techniques, such as zone-melting and fusion, and therefore have been successfully used in high temperature float zone-melting and Verneuil type fusion processes.

4.2.1.8. *Thermal Transfer*

In many cases the principal mode of heat transfer is dictated by the selection of the source. An efficient design of the rest of the apparatus should ensure the minimum of losses due to other modes of transfer because heat transfer together with the thermal geometry determines all the inherent losses in the system. Consideration of the above features also enables one to estimate the time lag between the energy input and the result(s) of application. It is impossible in the space of this article to discuss the laws of heat transfer. The interested reader is referred to Kutz (1968) for an exposition of the mechanics of thermal transfer with reference to temperature.

4.2.2. TEMPERATURE MEASUREMENT

The absolute scale of temperature (also referred to as the thermodynamic scale) was defined by Kelvin (1848) on the basis of the efficiency of a reversible Carnot cycle. As the name implies, the above scale is independent of the properties of thermometric substances. All the practical methods of temperature measurement are related to the absolute scale through the International Practical Temperature Scale (IPTS) which formally came into being, in 1927, under the auspices of the CGPM.† The IPTS is defined in terms of the temperatures of a number of fixed points. The most important and fundamental single fixed point is the triple point of water. The practical definition of the unit Kelvin of temperature was reasserted in the resolution of the Thirteenth General Conference (1968) as the fraction 1 273.16 of the thermodynamic temperature of the triple point of ice. Kelvin is incidentally one of the basic units of the Système International d'Unités (SI); it is represented by the symbol K (and *not* °K), and its "size" is known to about ± 1 in 27,000. The IPTS (1968) covers the range from 13.81 K (triple point of hydrogen) upwards in terms of the specified primary and secondary fixed points given in Tables 4.2 and 4.3. The latest revision of the IPTS was effected in 1975. The differences between IPTS (1968) and IPTS (1975) are however minor and should not affect the practice of temperature measurement. Differences between IPTS 1968 and the thermodynamic scale have been discussed by Quinn *et al.* (1978). The interpolation of the scale between the fixed points in the IPTS definition is specified by the use of the temperature transducer listed in Table 4.4. It should, however, be emphasized that the transducers listed therein can be (and indeed

† Conférence Générale des Poids et Mesures.

TABLE 4.2. *IPTS* (1968) *Definition of the Primary Fixed Points*

Fixed Point (Defined by Equilibrium State)	Assigned Value of International Practical Temperature
Triple point of equilibrium hydrogen (Equilibrium between solid, liquid and vapour phases of equilibrium hydrogen)	13.81 K ($-259.34°$C)
Equilibrium between the liquid and vapour phases of equilibrium hydrogen at a pressure of 33330.6 N m^{-2}	17.042 K ($-256.108°$C)
Boiling point of equilibrium hydrogen (Equilibrium between the liquid and vapour phases of equilibrium hydrogen)	20.28 K ($-252.87°$C)
Boiling point of neon (Equilibrium between the liquid and vapour phases of neon)	27.102 K ($-246.048°$C)
Triple point of oxygen (Equilibrium between the solid, liquid and vapour phases of oxygen)	54.361 K ($-218.789°$C)
Boiling point of oxygen (Equilibrium between the liquid and vapour phases of oxygen)	90.188 K ($-182.962°$C)
Triple point of water (Equilibrium between the solid, liquid and vapour phases of water)	273.16 K (0.01°C)
Boiling point of water[a] (Equilibrium between the liquid and vapour phases of water)	373.15 K (100°C)
Freezing point of zinc (Equilibrium between the solid and liquid phases of zinc)	692.73 K (419.58°C)
Freezing point of silver (Equilibrium between the solid and liquid phases of silver)	1235.08 K (961.93°C)
Freezing point of gold (Equilibrium between the solid and liquid phases of gold)	1337.58 K (1064.43°C)

[a] Freezing point of tin (505.1181 K *or* 231.9681°C) may be used as an alternative to the boiling point of water.

are) used outside the interpolation ranges. It is often desirable for correct operation and greater accuracy to calibrate a temperature measuring instrument. The calibration, however, need not necessarily be carried out directly against the standard used to produce the IPTS. In fact it can be conveniently carried out using a substandard which is traceable to a calibration on the IPTS. Alternatively, the primary and secondary fixed points of the IPTS may be used directly for calibration. For calibration procedures for practical thermometers the reader is referred to Barber (1971). It is, however, desirable to describe here, briefly, the principal transducers used in temperature measurement in crystal growth processes.

4.2.2.1. *Expansion Thermometry*

Two main types of thermometers in this category are in use:

1. *Gas thermometers.* These rely on the fact that the change in the volume of a perfect gas at a constant temperature is proportional to the change in the temperature. Gas

TABLE 4.3. *IPTS* (1968) *Definition of the Secondary Fixed Points*

Fixed Point	International Practical Temperature 1968 (K)
Boiling point of normal hydrogen	20.397
Triple point of nitrogen	63.148
Boiling point of nitrogen	77.348
Sublimation point of carbon dioxide	194.674
Freezing point of mercury	234.288
Ice point	273.15
Triple point of phenoxybenzene	300.02
Triple point of benzoic acid	395.52
Freezing point of indium	429.784
Freezing point of bismuth	544.592
Freezing point of cadmium	594.258
Freezing point of lead	600.652
Boiling point of sulphur	717.824
Freezing point of antimony	903.89
Freezing point of aluminium	933.52
Freezing point of copper	1357.6
Freezing point of nickel	1728
Freezing point of cobalt	1767
Freezing point of palladium	1827
Freezing point of platinum	2045
Freezing point of rhodium	2236
Freezing point of iridium	2720
Melting point of tungsten	3660

thermometers are in fact experimentally used as the mainstay of the thermodynamic temperature scale. They are often used in the range 2–90 K and are capable of achieving a very high degree of precision. At higher temperatures, however, they are rarely used as routine temperature measuring apparatus, mainly because they are cumbersome. They will not be described further.

2. *Liquid-in-glass thermometers.* The liquid-in-glass thermometer is probably the most widely used instrument for measuring temperature. Its relative simplicity in use makes it highly attractive where reliability is required without an excessive demand on the accuracy

TABLE 4.4. *Instruments of Interpolation between the Fixed Points Specified in the IPTS* (1968)

Fixed Points and the Temperature Range	Interpolation Instrument
Triple point of equilibrium hydrogen and freezing point of antimony (13.81–903.89 K)	Platinum resistance thermometer
Freezing point of antimony and freezing point of gold (903.89–1337.58 K)	Ten per cent rhodium–platinum/ platinum thermocouple
Above freezing point of gold (i.e. above 1337.58 K)	By Planck's law of radiation, i.e. by radiation pyrometer

in the temperature range 250–800 K. Generally, a typical liquid-in-glass thermometer consists of a reservoir of thermometric liquid in a bulb to which is attached a long capillary stem so that the liquid can expand in the stem. The temperature of a medium is measured by totally "immersing" the liquid (the whole of the liquid column) in the medium and reading the length of the liquid in the calibrated stem. The actual construction and method of measurement of a particular thermometer slightly differs due to the particular application(s) for which it is designed. ASTM methods for instance require the use of well over forty different types of thermometer giving complete specifications for each type.

The sensitivity of a liquid-in-glass type thermometer may be expressed by the equation

$$\frac{\mathrm{d}l}{\mathrm{d}T} = \frac{4\beta_{LB}V_{LB}}{\pi D_{\mathrm{cap}}^2}, \tag{4.29}$$

where l is the length of the liquid in the stem, T is the temperature, β_{LB} is the net expansion coefficient of the liquid in the glass bulb, V_{LB} is the volume of the liquid in the bulb, and D_{cap} is the inside diameter of the capillary stem. It can be seen that for minimum values of V_{LB} = 1000 mm^3 and D_{cap} = 0.06 mm, as recommended by Hall and Leaver (1962), and mercury as the thermometric liquid ($\beta_{LB} = 1.81 \times 10^{-4}\,\mathrm{K}^{-1}$), a sensitivity of 64 mm K^{-1} can be obtained. Indeed, "limited range" Beckmann-type thermometers do have a sensitivity around ± 0.002 K. The accuracy of such thermometers is limited by the changes in the glass bulb dimensions as they are tremendously magnified in terms of the calibrated stem length. For instance, in a typical high sensitivity mercury thermometer the accuracy of 0.01 K requires the volume of the glass bulb to remain constant to 0.0016 %. Other causes of inaccuracies are due to inability of the glass to recover on sudden temperature changes and irreversible changes in glass investigated by Thompson (1962). Within the limits of instabilities listed here the accuracy would depend upon the calibration. The typical accuracies available are from 0.01 to 0.03 K in the temperature range 273.15–373.15 K and to 0.2–0.5 K in the range 573.15–773.15 K. Thermometers specially designed as the working standards are described in the British Standards specifications BS 593 and BS 1900.

4.2.2.2. Resistance Thermometry

Electrical resistance of most solids is a sensitive function of temperature. In metals the resistivity increases with the temperature while in semiconductors it decreases with temperature. Therefore, it is possible to use the above materials as transducers for temperature measurement.

1. *Metal resistance thermometry.* Metal resistance thermometry, using platinum elements, was first put on a sound basis by Callander (1887)† who produced an accurate equation for the temperature dependence of the resistance, namely

$$R_T = R_0(1 + AT + BT^2), \tag{4.30}$$

---

† The first practical platinum resistance thermometer was constructed by Siemens (1871), but the test report of the committee of British Association produced in 1874 on the "Siemens Pyrometer" was unfavourable. Subsequently a satisfactory construction was produced by Siemens. His method of calibration and his equation for the relationship between the temperature and resistance, however, have proved wrong in the light of present-day knowledge.

where R_T is the resistance at temperature T, R_0 is the resistance at $0°C$ (273.15 K), and A and B are coefficients (constants).

The resistivity of any non-ferromagnetic metal is described by the well-known Matthiessen's rule, which states that the observed resistivity $\rho(T)$ at a temperature T consists of a temperature independent part ρ_0, known as the residual resistivity, which is determined by impurity and defect scattering, and a temperature dependent part $\rho_i(T)$, known as ideal resistivity and determined by the lattice vibration

$$\rho(T) = \rho_0 + \rho_i(T). \tag{4.31}$$

Qualitatively the approximate linear dependence (above the Debye temperature)[†] is shown in the Bloch expression for $\rho_i(T)$ derived from one electron single band theory, using the Debye approximation

$$\rho_i(T) = (T/\theta_D)^5 \tfrac{1}{4}(\theta_D/T)^4 \rho(\theta_D) \quad \text{for} \quad T \gg \theta_D, \tag{4.32}$$

where θ_D is the Debye temperature and $\rho(\theta_D)$ is a constant for a given metal. It is, however, not possible to predict accurately and quantitatively from the theory the exact temperature dependence of the resistance with temperature as Matthiessen's rule is not strictly obeyed in many metals and ρ_0 does vary with the temperature. In practice, therefore, resistance thermometry can be effected by actually calibrating the resistance of a metal wire and specifying the impurity content. This is conveniently done by specifying the so-called alpha value ratio

$$\text{alpha value ratio} = R_{273.15\,K}/R_{373.15\,K} \tag{4.33}$$

where $R_{273.15\,K}$ and $R_{373.15\,K}$ are the values of the resistance at the freezing point and boiling point of water.

Callander (1899) had, of course, realized the importance of the impurity content and had insisted on using platinum wires with alpha values > 1.385. Since then the purity of platinum has improved considerably and the IPTS (1968) recommendations are that the alpha value for the "standard" platinum wire for thermocouple use should not be less than 1.39250.

In principle any metal can be used for resistance thermometry but in practice in the temperature range 300–1300 K indium, platinum, nickel, and copper are found to be the most satisfactory. By far the widest used metal is platinum as its resistance versus temperature curve is nearly linear over a broad temperature range and to some extent it is chemically inert. The IPTS (1968) had adopted platinum resistance thermometers (PRT) for interpolation up to the antimony point, but with a high purity platinum wire and highest grade alumina sheaths and supports (taking some additional precautions) its use can be extended up to 1300 K. The causes of errors in the higher temperature ranges arise from evaporation, contamination, and imperfect electrical insulation. Barber and Blanke (1961), Evans and Burns (1962), and Nakaya and Uchiyama (1962) have shown that very low drift rates ~ 0.001–$0.002\,K\,hr^{-1}$ and good stability around the gold point can be achieved. Callander's equation (4.30) was originally shown to be valid in the range 273.15–373.15 K, but Diamond (1969) has shown that for practical use eqn. (4.30) can be generalized to permit interpolation between any arbitrary fixed points T_1 and T_2 (including the high temperature range). The coefficients A, B should, of course, be

---

[†]For a detailed treatment on the resistance of metals the reader is referred to Ziman (1963).

determined to give a correct fit for the above temperature range. Evans (1977) has shown that in relation to thermocouple and radiation temperature scales simple quadratic interpolation for PRT in the range 903–1341 K is good. For a direct read out of temperatures a linearization circuit for PRT has been described by Monk (1977). It uses an operational amplifier with feedback for linearizing resistance–temperature characteristics. Additionally, the circuit is arranged in such a way that the measuring current required in the platinum element for a given sensitivity is reduced. This decreases errors in the temperature measurement arising due to the power dissipated in the element.

For basic guide lines for equipment selection for PRT the reader is referred to Monk (1977a). Information on the time constants of PRT is given by Jarusek (1976). Recently thick films of platinum are being industrially used for PRT. Chattle (1977) has evaluated procedures for evaluating their resistance–temperature relationship. In order to achieve the maximum degree of precision it is a good practice to anneal PRT at a temperature above the range in which it is used. Rapid heating and rapid cooling should be avoided as the defects and strain introduced due to thermal shock alter the resistance–temperature characteristics. The ultimate sensitivity and accuracy achieved with a PRT obviously depends upon the method of resistance measurement (see below), but it is possible to determine the temperature with an accuracy of ± 0.01 K at 1000 K.

2. *Semiconductor resistance thermometry* (*thermistors*). Electrical conduction in semiconductors can be due to two types of charge carrier, namely, electrons (in the conduction band) and "holes" (in the valence band). Therefore, the electrical conductivity σ of a semiconductor may be described by the equation

$$\sigma = 1/\rho = e(n_e\mu_e - n_p\mu_p), \tag{4.34}$$

where e is the electronic charge, n_e and n_p are the number of electrons and holes per unit volume respectively (i.e. carrier concentrations), and μ_e and μ_p are electron and hole mobilities.

The electrical conduction may further be distinguished in two different categories, namely:

(i) Extrinsic conduction: this is predominantly due to carriers excited from the impurities in the semiconductor. It occurs in a relatively low temperature range where thermal energy is predominantly expended to ionize impurities. Depending upon the nature of a major impurity present one may have $n = n_e \gg n_p$ or $n = n_p \gg n_e$. The number of carriers taking part in conduction processes can be expressed by an exponential function of the form

$$n = \exp(-A_1/T), \tag{4.35}$$

where A_1 primarily depends upon the ionization energy of the impurity level and also on the fraction of impurity levels from which the carriers are excited. From eqn. (4.35) it can be clearly seen that the sensitivity of an extrinsically conducting thermistor as defined by $d\rho/dT$ increases with decreasing temperature. Extrinsic thermistors are therefore used in the low temperature ranges. It can also be seen from eqn. (4.35) that the exact temperature dependence of resistance in the extrinsic region would be determined by the impurity content of the semiconductor.

(ii) Intrinsic conduction: This is due to the charge carriers thermally excited from the valence band to the conduction band and one may write $n_e = n_p = n$. The number of carriers in the intrinsic conduction is given by

$$n = A_2 T^{3/2} \exp\{-E_g/(2kT)\},$$ (4.36)

where A_2 is a constant, E_g is the energy gap between the valence band and the conduction band, and k is Boltzmann's constant. Again it can be seen that the sensitivity $d\rho/dT$ decreases with increasing temperature. The temperature dependence of the resistance, however, is not dependent upon the impurity concentration.

Doped germanium resistors are extensively used in the range 2–35 K, as in this range resistance decreases by about 1 % for a 1 % increase in temperature. Carbon resistors can also be used in the temperature range 1–20 K. For higher temperatures intrinsic silicon and germanium (up to 1650 K for silicon and 1200 K for germanium) can be used.

Commercial thermistors are generally made from polycrystalline oxides of iron, nickel, cobalt, and manganese as well as silicates and sulphites of aluminium, copper, and iron. In the specified limited temperature range the resistance R of a thermistor can be given by an equation of the form

$$\ln R = A + B/(T + C),$$ (4.37)

where A, B, and C are calibration constants. Commercial thermistors are available for measuring temperatures 4.2–1500 K. The stability of these devices for use at high temperatures is not, however, very good, and therefore the practical use is confined to below 500 K. Hannenmann (1977) has described fabrication of a thin film device of germanium which can be used for point to point measurement. It has a rapid transfer response. Using microelectronic technology he has also fabricated thin film of titanium in meandering pattern for the measurement of average surface temperatures. The possibility of the use of refractory oxides such as Al_2O_3, MgO, etc., for temperatures above 1700 K and commercial ceramics has been suggested by McElroy and Fulkerson (1968) and Anderson and Stickney (1962), respectively. Wolff (1969) has recently reported a thermistor utilizing 85 % mole zirconia and 15 % mole yttria as the sensor element using iridium–rhodium lead wires and beryllium oxide as the insulating material. The above thermistor is capable of measurement in the range 1000–2500 K with an accuracy of ± 2.5 K at 2000 K. Unlike other high temperature resistive elements both thermal stability and shock resistance of the above device are claimed to be excellent.

3. *Methods of resistance measurement.* It is evident from sections 1 and 2 above that the order of magnitude of resistance would depend upon the type of thermometer used. The requirement of sensitivity for one thermometer may vary according to the temperature being measured. For instance, for a PRT the temperature coefficient $(1/R)dR/dT$ changes from 5.8×10^{-4} K^{-1} at 1275.15 K to 2.1×10^{-1} K^{-1} at 20 K. Therefore, to obtain 0.001 K sensitivity at 1275 K one must be able to detect resistance changes up to 0.6 part per million while at 20 K the same sensitivity may be obtained by being able to detect resistance changes of the order two parts in ten thousand. The degree of accuracy obtainable, however, depends upon experimental variations in lead resistance, stray thermal e.m.f. due to inhomogeneities in the wire contact resistance of switches and connections, and the stability of other standard resistances in the circuit. As the actual resistance of a metal thermometer is low, the overall error due to the above causes can be large. A refined measurement circuit therefore should ensure the elimination of causes of stray voltages and instabilities to achieve a high degree of accuracy. Thermistors, on the

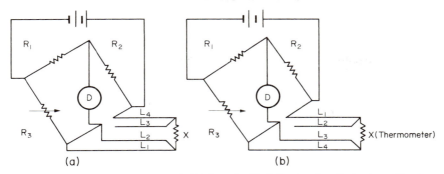

FIG. 4.10. Mueller bridge: (a) and (b) show double balancing for elimination of lead resistance.

other hand, exhibit large resistance changes with temperature; therefore the error due to the above-mentioned causes can be relatively smaller.

There are basically two approaches adopted for resistance thermometry:

1. *Bridge circuit methods*: the bridge circuits employed in high precision resistance thermometry are modifications of either the Wheatstone or the Kelvin double bridge and follow the four variants: two based on the Wheatstone bridge and two based on the Kelvin double bridge published by Smith (1912). The above modifications and subsequent refinements are primarily designed to eliminate inaccuracies due to the lead resistance of a thermometer. In the famous Mueller bridge designed by Mueller (1916–17) the lead resistance is eliminated by twice balancing the bridge with reversal of the leads as shown in Fig. 4.10a and b. Recently Evans (1962) has shown that by using mercury wetted lead switching contact in the Mueller bridge it is possible to achieve discrimination of 1 μΩ. Also van der Wall and Struik (1969) have described a simple and inexpensive modification of the Wheatstone bridge circuit to enable direct reading in degrees Celsius with a platinum resistance thermometer. Reading accuracy of $\pm 0.01°C$ is claimed for the temperature range 0 to $+500°C$.

A Smith bridge based on the Kelvin double bridge, and shown in Fig. 4.11, was officially adopted by the NPL, UK, in 1923. With a bridge of this type only a single balance is required. The error due to the lead resistance in the above method is essentially rendered insignificant by arranging the thermometer leads in series with a large resistance. A major source of inaccuracy is due to the instability of large standard resistance coils, but a coil designed by Barber *et al.* (1952) overcomes this problem.

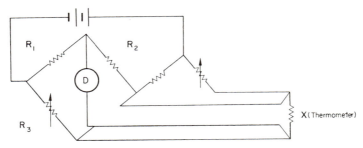

FIG. 4.11. A Smith bridge design adopted by the NPL, UK.

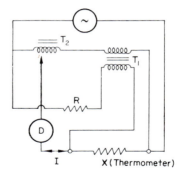

Fɪɢ. 4.12. An a.c. bridge using a transformer ratio arm.

An a.c. bridge may also be used for accurate measurement provided a precise voltage source can be incorporated. Leslie *et al.* (1960) have pointed out that a highly accurate voltage ratio source can be incorporated in a bridge by the use of a "transformer ratio arm" as shown in Fig. 4.12. The voltage ratio thus obtained depends only on the ratio of the turns and is not sensitive to vibrations and ageing. Hill and Miller (1963) have produced a highly accurate bridge using the above principle.

Recently, Ekin and Wagner (1970) have described an a.c. bridge suitable for use with four terminal resistance thermometry (see below) and claimed to have achieved 0.02% accuracy with a commercial germanium resistance thermometer.

2. A review of potentiometric methods for resistance thermometry is given by Dauphinée (1962). The main advantage of the potentiometric methods over the bridge methods is that the lead resistance effects can be eliminated by using four leads (i.e. two current leads and two potential leads). Voltages of the order of $n\,\text{V}$ can be potentiometrically measured by commercially available instruments. The chief disadvantage, however, is that the accuracy of the measurement could be considerably reduced due to stray thermal e.m.f.s and the current drift. The effect of thermal e.m.f. may be eliminated by reversal of the current through the circuit. The procedure, however, is slow and the thermal e.m.f.s are not always constant. Moreover, reversal of the current may be accompanied by instability in the current. The current drift can be stabilized by low drain current and utilizing stable mercury batteries or a low drift chopper stabilized power supply. Dauphinée (1962) has devised an elegant technique of "isolating potential comparison" which largely overcomes the above-mentioned drawbacks. The basic circuit of this method is shown in Fig. 4.13. Here two resistances are compared, essentially by

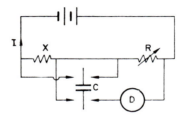

Fɪɢ. 4.13. The isolating potential comparison method.

switching a capacitor, using a low frequency chopper (20–80 Hz), to transfer a voltage V_R, across a known resistance R, into series opposition with the voltage V_X across the thermometer. A null reading on the detector, placed in series with a potential lead, is obtained when the two resistances are exactly equal. Alternatively the small difference ΔR in the two resistances can be measured by recording the equation.

$$V = I\Delta R, \tag{4.38}$$

where I is the constant current flowing in the circuit. By using a highly stable (low leakage) capacitor, reversal of current and a suitable modification of the circuit, Dauphinée (1962) was able to obtain an accuracy of 0.001 K. Moreover, by using a quadratic resistance network the above instrument could be converted to read the temperature directly.

4.2.2.3. Thermocouple Thermometry

Following Seeback's (1823, 1826) production of an e.m.f. by heating a junction between two metals, the first applications of the thermoelectric phenomenon for temperature measurement was considered by Becquerel (1830) and Pouillet (1836). Since then metal thermocouple thermometry has been perfected and widely applied in the range 100–3000 K.

Thermoelectric effects arise from the non-equilibrium distribution of current carriers (electrons and/or holes) in a material when placed in a temperature gradient. Rigorous phenomenological treatment of the various thermoelectric phenomena in metals and semiconductors is given by Ioffe (1957), MacDonald (1962), Ziman (1963), and Goldsmid (1964). A satisfactory expression for the temperature dependence of the thermoelectric power of metals and metal junctions, however, cannot be derived theoretically. It is therefore a common practice to fit a polynomial expression of the form†

$$E = aT + bT^2 + cT^3 + dT^4 + eT^5 + \ldots, \tag{4.39}$$

where E is the e.m.f. produced by a thermocouple junction at a temperature T and a, b, c, d, and e are the coefficients calculated from the experimental data.

Alternatively, a smooth calibration curve may be drawn through the temperature–e.m.f. data of a number of fixed points. For digital instrumentation and the direct read-out of temperatures eqn. (4.39) has to be solved *or* linearized. Monday (1977) has approximated the non-linear voltage–temperature relationship curve in a relatively large number of linear sections by a dual shape analogue to digital converter. As demonstrated by Shah (1967) it may be possible to fit an inverse expression of the form

$$T = AE + BE^2 + CE^3 + \ldots, \tag{4.40}$$

where $A, B, C, \ldots$, are coefficients, without any loss in the accuracy. The advantage of eqn. (4.40) is that once the coefficients $A, B, C, \ldots$, are known it is easier to compute T using eqn. (4.40) than from eqn. (4.39). The absolute accuracy and the precision in measurement is, in any case, affected by the purity or the composition of the thermocouple materials. Close control of the impurity and composition at the manufacturing stage has contributed to a

---

† The use of only two or three terms in eqn. (4.39) may suffice for an accuracy of ± 1 K.

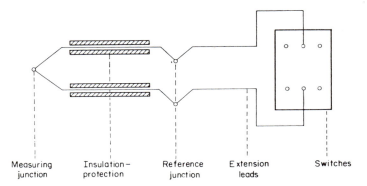

FIG. 4.14. Some error contributing factors in a thermocouple circuit.

greater reproducibility in temperature measurement. In addition other critical factors affecting the accuracy in a typical thermocouple circuit shown in Fig. 4.14 are as follows:

1. The measuring junction and the reference junction.†
2. Extension wires and the presence of thermoelements in temperature gradient.
3. Purity of the insulating and/or protecting refractory material.
4. Switching.

Junctions between two thermocouple metals are produced by either fusing the two wires in a gas flame, using a suitable flux which can be easily removed after the fusion, or by spot welding. Mechanical or brazed connections can also be used. Descriptions of the codes of practice for the junction manufacture have been provided by the Instrument Society of America (1964) and the British Standards Institution (1966).

It was shown by Mortlock (1958) that the major source of error in a thermocouple junction arises if compositional changes by interdiffusion are accompanied by the presence of thermal gradients. For a greater accuracy, unwanted heat transfer at the measuring junction should be avoided. This can be done by ensuring that the wires adjacent to the measuring junction are not in a temperature gradient by simply immersing the wires to a greater depth and/or by insulating the wires at the thermal exit.

It should also be ensured that the reference junction is at a correct temperature. The reference junction practices have been reviewed by Caldwell (1965). In brief, the reference junction can be maintained at 273.15 K by immersing the junction in the centre of a Dewar containing a mixture of shaved ice and water. As the impurities of the ice water mixture can alter the equilibrium temperature, due regard should be paid to maintaining the purity of the mixture. For the reference junction temperatures above the ice point, commercially available "zone boxes" can be used. These are fully described by Roeser (1941). Alternatively an accurately controlled miniature oven may be used for the above purpose.

Recently electrical solid state reference junctions have become available. Basically, these provide a zero suppression in accordance with the e.m.f. generated by the cold junction. The automatic compensation in the e.m.f. is simply provided by a basic bridge design shown in Fig. 4.15. Here one of the resistances R_2 is a temperature sensitive

† In practice temperature is measured by observing the e.m.f. across a two-junction differential thermocouple. One junction (the reference junction) is kept at a known reference temperature (usually the ice point) while the other proves to be the measured temperature region.

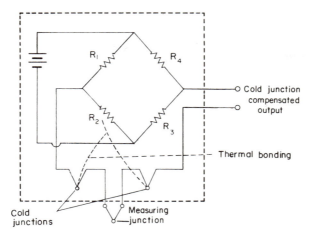

Fig. 4.15. A basic bridge circuit for an electrically compensating reference junction.

component, thermally bonded to the cold junction and its e.m.f.–temperature characteristic is matched to that of the thermocouple. The circuit is so arranged that the potential across R_2 is equal and opposite to that of the cold junction for a wide ambient temperature range. More sophisticated electrical reference junctions are available and are increasingly incorporated into temperature recording and measuring devices commercially available. A review of electrical reference junctions together with others is given by Muth Jr. (1967).

Thermocouple wire and extension leads which pass through a temperature gradient should be highly homogeneous, as the inhomogeneity either in the impurity content or the composition usually results in the generation of unwanted thermal e.m.f. Compositional inhomogeneities may be present due to faulty manufacturing, but often the causes occur because of faulty installation and service. Coldwork, recovery and recrystallization, etc., often contribute to the inaccuracies in the measurement. It was shown by Starr and Wang (1963) that the chemical inhomogeneities in thermocouples may arise during service due to oxidation, surface reactions, and interdiffusion effects, and change the temperature–e.m.f. characteristic of a thermocouple. Another source of error is impurity contamination. It was shown by Nielson (1970) that Pt–Pt 13% Rh thermocouples made from wires of both standard and reference grade showed a degradation in calibration of 5–10 K at 1500 K, after only 15 days heating, due to impurity contamination. It is, therefore, a sound practice to anneal the thermocouple wires at a temperature slightly above the range in which they are used, as this removes the mechanical inhomogeneities and stabilizes the calibration before use. Another source of impurity contamination is insulation protection. Anderson (1978) showed that after 72 h 293 K inconel sheathed and MgO insulated $Pt_{90}Rh_{10}/Pt$ compacted thermocouples exhibited 47% decalibration. Metalographic and ion microprobe analyses showed that the reaction had occurred at Pt–MgO and $Pt_{90}Rh_{10}$–MgO interface. Selman and Rushforth (1971) have also investigated the protective sheaths for $Pt/Pt_7Rh_{(100-x)}$ thermocouples. It appears that the best results are obtained when a recrystallized alumina sheath is used with $Pt/Pt_{87}Rh_{13}$ thermocouple. The selection and use of sheaths have been elaborated by Brown (1967).

Thermocouples are often connected to extension leads to transfer the reference junction to a position where the ambient temperature does not vary. Such leads should ideally have the same e.m.f. temperature characteristics as the thermocouple between the interval of 273–323 K (i.e. they are compensating). It is important to ensure that all compensating lead–thermocouple connections are at the same temperature. There is, however, a danger of additional errors arising from the above practice which is analysed by Moffat (1962). For more accurate measurement therefore it is advisable to dispense with the use of compensating leads.

An electrical insulation has to be provided to prevent short circuiting of the thermocouple element. Additionally, at high temperatures thermocouple wires may require protection against chemical attack in the furnace environment. Both insulation and protection may be provided by refractory ceramic in the form of beads or sheaths. The sheath material obviously has to be chosen on the basis of high electrical resistivity. It is often not appreciated that impurity contamination of the thermocouple may occur because of impurity diffusion from the sheath material. Therefore, it is of the utmost importance to ensure that the sheath material is of high purity and is itself kept clean.

Finally, switches used in the e.m.f. measuring system should be free from all stray thermal e.m.f.s. Therefore, it is necessary to shield the measuring circuit from the thermal field. The connections to the switches and terminals should be made by thermal free and flux free soldering.

Table 4.5 gives a summary of the characteristics of various thermocouples in use for the reader's reference. It should be noted that for most materials e.m.f.–temperature conversion tables are available. Various British and American standards may be referred to for the selection of the thermocouple materials for reliable accuracy and relation to the IPTS (1968).

It is worth mentioning that the crystal/melt interface can be utilized as a thermocouple hot junction for measuring differential temperature across the interface during growth. Owen and White (1977) modified a conventional crystal pulling system in the following way.

The bottom of the crucible was connected to $Pt/Pt_{87}Rh_{13}$ thermocouple. At the other end the seed was fused in a large tin block lagged with zirconia felt. This seed holder arrangement was in electric contact with the pulling shaft and a rotating mercury bath. The potential difference V between the crucible and the mercury bath was measured. The authors argue that under the experimental conditions used $dV/dt \simeq dT/dt$, where T is the temperature of the solid–liquid interface. Their measurements were done with InSb and showed that the periodic variations in the potential V were in correlation with different rotation rates. The peak to peak differences are in the range 0–0.6°C.

4.2.2.4. Radiation Thermometry

Extrapolation of the IPTS (1968) above the gold point is defined by the use of a radiation pyrometer and Planck's law of black-body† radiation, namely,

$$J_{b\lambda}\left(\frac{c_1\lambda^{-5}}{\pi}\right)\left\{\exp\left(\frac{c_2}{\lambda T}\right) - 1\right\}^{-1}, \tag{4.41}$$

† A black body is a hypothetical medium with the property that it absorbs all radiation incident upon it, i.e. its absorptance is unity.

TABLE 4.5. *Summary of Various Thermocouples*

Type and Trade Name	Thermocouple Elements		Sensitivity ($\mu V\,K^{-1}$)	Temperature Range (K)	Environment for Stability	
	+ve	−ve			Operating Atmosphere	Suitable Refractory
Base metal (K[a] type) Chromel P/alumel[b] Kanthal P/N Tophel/Ni Al Ni Cr[a]/Ni Al– Ni Cr/Ni	$Ni_{80}Cr_{20}$	$Ni_{95}(Al,Si,Mn_5)$	40	273–1470	Oxidizing	Alumina, silica
Base metal	$Ni_{80}Cr_{20}$	$Ni_{97}Si_3$	23–45	273–1470	Oxidizing and reducing	Alumina, silica
Base metal (Y.J. type) Iron/copnic Iron/constantan	Fe	$Cu_{60}Ni_{40}$	50–64	273–930	Oxidizing and reducing	Alumina, silica
Base metal (T type) Copper/constantan	Cu	$Cu_{60}Ni_{40}$	38–61	273–930	Oxidizing	Alumina, silica
Base metal (E type) Chromal/constantan	$Ni_{90}Cr_{10}$	$Cu_{60}Ni_{40}$	58–75	273–1273	Oxidizing	Alumina, silica
Noble metals (S type)	$Pt_{90}Rh_{10}$	Pt	6–11	273–1273	Oxidizing	Sapphire–recrystallized alumina
Noble metals (R type)	$Pt_{87}Rh_{13}$	$Pt_{90}Rh_{10}$	13.2	273–1273	Oxidizing	Sapphire–recrystallized alumina
20/5 Rh	$Pt_{80}Rh_{20}$	$Pt_{95}Rh_5$	10–11	273–1273	Oxidizing	Sapphire–recrystallized alumina
	$Ir_{60}Rh_{40}Ir$		3.5–6.2	273–2173	Reducing	Thoria (He atmosphere) Boron nitride
Platinel	$Au_3Pd_{83}Pt_{14}$	$Au_{65}Pd_{35}$	40	273–1573	Oxidizing and reducing	Alumina, silica
Refractory metal	W	Ta	16	273–1773	Inert and vacuum	Thoria
Refractory metal	W	Mu	8.4	273–2373	Inert	Thoria
Refractory metal	W	Re	5.5–7.2	300–2373	Reducing	Thoria
Refractory metal	W	$W_{74}Re_{26}$	2.3–15.7	300–3000	Inert and vacuum	Thoria

[a] For a recent review of the behaviour of this type of thermocouple see Campari and Garriba (1971).
[b] Investigations of the errors in chromal/alumal thermocouples have been carried out by Prewbrazhenskii and Letskas (1978).

where $J_{b\lambda}$ is the spectral radiance at wavelength λ (i.e. the energy radiated by a black body, at wavelength λ in a particular direction, per unit time, per unit wavelength, per unit projected area of the body, and per steradian solid angle), T is the temperature in K, and c_1 is a constant called the first radiation constant which is given by

$$c_1 = 2\pi h c^2, \tag{4.42}$$

where h is Planck's constant and c is the velocity of electromagnetic radiation (light) in vacuum so that

$$c_1 = 3.7413 \times 10^{-16}\,\mathrm{W\,m^2}.$$

c_2 is a second constant called the second radiation constant. It is defined by

$$c_2 = hc/K, \tag{4.43}$$

where K is Boltzmann's constant, so that

$$c_2 = 1.4388 \times 10^{-2}\,\mathrm{mK.}†$$

The definition of the IPTS above the gold point T_{Au} can now be expressed by the ratio

$$J_{b\lambda}(T)/J_{b\lambda}(T_{\mathrm{Au}}) = \exp\{c_2\lambda^{-1}(T_{\mathrm{Au}} + T_0)^{-1}\} - 1/\exp\{c_2\lambda^{-1}(T + T_0)^{-1}\} - 1, \tag{4.44}$$

where T is the temperature to be defined and T_0 is 273.15 K.

In practical measurements the radiation from a real body at any temperature is different from that from the black body. The difference in radiance between a real body and the black body can be expressed in terms of the emissivity of the real body (ε):

$$\varepsilon = J_{\mathrm{real\,body}}(T)/J_{\mathrm{black\,body}}(T), \tag{4.45}$$

where $J_{\mathrm{black\,body}}(T)$ and $J_{\mathrm{real\,body}}(T)$ are the radiance of the black body and real body at temperature T. The emissivity is a function of wavelength, temperature, and direction. Therefore, measurement of the temperature of a real body by a radiation pyrometer working at an effective wavelength λ will have an error ΔT given by

$$\Delta T = T^2\lambda \ln \varepsilon_{\lambda T}/c_2 \tag{4.46}$$

where c_2 is the second radiation constant and $\varepsilon_{\lambda T}$ is the emissivity of the real body at the temperature T and wavelength λ.

If the value of $\varepsilon_{\lambda T}$ is known then one can, in principle, accurately determine a temperature, using eqns. (4.44) and (4.46) with the aid of a "monochromatic" radiation pyrometer measuring the radiance at the effective wavelength λ.

Alternatively, it is also possible to measure temperature by gauging total radiation emitted from a body by means of a total radiation pyrometer. One can approximate that the total radiation being emitted is at an effective wavelength λ_{eff} which is given by the formula

$$\lambda_{\mathrm{eff}} = 0.37hT^{-1},$$

where h is Planck's constant. However, Rozhdestvenskii (1977) has shown that this can lead to considerable error. Total radiation emitted from a black body also obeys the fourth power law of radiation (also referred to as the Stefan–Boltzmann law),

$$J_b = \sigma(T_1^4 - T_2^4), \tag{4.47}$$

†The IPTS (1968) revised value.

where J_b is the total energy radiated by the body per unit surface area, per unit time, σ is the total radiation constant (or Stefan's constant) $= 5.670 \times 10^{-8}\,\mathrm{W\,m^{-2}\,K^{-4}}$, and T_1 is the temperature of the black body and T_2 the ambient temperature.

The total radiation from a real body will obey a similar law, namely,

$$J = \sigma \varepsilon_T(T_1)(T_1^4 - T_2^4), \tag{4.48}$$

where $\varepsilon_T(T_1)$ is the total radiation emissivity at temperature T_1. The correction for the emissivity in the temperature T as determined by a total radiation pyrometer is given by

$$\Delta T = T\{1 - \varepsilon_T^{1/4}\}. \tag{4.49}$$

Yet another modification of a radiation pyrometer is based on the belief that the ratio of the radiances of a grey body† is the function of only the temperatures and the wavelengths (i.e. the ratio is independent of the emittance). Consequently a multicolour pyrometer (also called ratio pyrometer) measuring radiance at two or more wavelengths can determine temperatures more accurately as errors due to emissivity do not arise or, at any rate, are considerably reduced. Emslie and Blau Jr. (1959) have, however, demonstrated the inherent fallacy of the above statement and have shown that the error in temperature determined by a multicolour pyrometer is of the same magnitude as that in the temperature determined by a single colour pyrometer. Therefore, it seems that there is an additional disadvantage, in using a multicolour pyrometer, namely, the need for much more precise information (which is not always available) without any apparent gain in the accuracy. For the above reasons a multicolour pyrometer is not recommended for precision measurement. Although less expensive it will not be described any further.

A radiation pyrometer consists of the following components, as shown in Fig. 4.16.

1. A standard source calibrated in terms of temperature and the brightness or radiation output or an electric input.
2. An optical system for collecting radiation from the "to be measured" temperature region.
3. A radiation detector.
4. Filters or/and attenuators.

It is normal practice to use a tungsten strip lamp as a standard source for the temperature calibration of visual or photoelectric pyrometers. The normal width of the strip filament (in the UK) is 1.3 mm, although sometimes it may be necessary to use a wider strip of width 4 mm. Normally, vacuum lamps are used in the range 973–1823 K while gas-filled lamps may be employed in the range 1573–2573 K. The lamp is normally calibrated to give lamp current in terms of the reference temperature at 0.66 μ. Hall (1966) has recently stated that an accuracy of 0.5 K at the gold point is achievable with a standard pyrometer using a tungsten strip lamp. It should be a normal practice to run the lamp for more than 30 min at temperatures above 1000 K to allow it to reach current temperature equilibrium. Longer time should be allowed for measuring lower temperatures. A constant d.c. current source with a constancy of 0.1 % used in conjunction with the lamp enables accuracy around 1 K. As quoted above, the tungsten lamps are recommended for use below 2573 K. Above this temperature a black-body lamp such as the one designed by Quinn and Barber (1967) should be used. The black-body lamp incidentally enables one

---

† A grey body is a real body whose emissivity is constant through the whole radiation spectrum but may vary with the temperature.

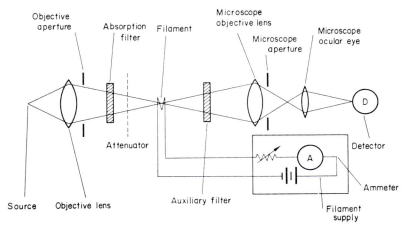

FIG. 4.16. A schematic representation of a pyrometer showing its components.

to obtain accuracy in the region of ± 0.2 K. Above 2573 K and up to 3793 K a carbon arc using a 6.35 mm ($\frac{1}{4}$ in.) diameter spectrographic grade graphite rod as the positive electrode and a similar grade rod of diameter 3.175–4.742 mm (i.e. $\frac{1}{8}$–$\frac{3}{16}$ in.) as the negative electrode should be employed.

It is of the utmost importance that every optical component used in a pyrometer should have unique spectral absorption, transmission, reflection, and diffraction characteristics, so that the proportionality of the pyrometer response output is uniquely dependent upon the unknown temperature of the source only. It is obvious that to obtain higher sensitivities, transmittance of the refracting components and reflectance of the reflecting components should be as high as possible.

In order to obtain reasonably large sensitivity at low temperatures it may be necessary to design a pyrometer with large entrance aperture and it may be required to be used with a large target area of the source. For high temperatures, however, both visual and photoelectric pyrometers are designed with a very small entrance angle. For fixed focus type pyrometers it is necessary to specify a minimum target area at different distances from the pyrometer. Ideally, the ratio of the brightness of the filament to that of the source to be measured should be independent of the source distance from the pyrometer. This may be achieved by suitable selection of objective apertures. In a disappearing filament type† pyrometer additionally it is necessary to match the colour of the source to that of the filament so that the image of the filament completely disappears. The complete disappearance, however, is not possible as in the near match conditions a thin dark line appears outside the edge of the filament and/or a bright line inside the edge (presumably due to diffraction effects). The above difficulty is avoided by a criterion due to Fairchild and Hoover (1923), who showed that if the ratio α/β as defined in Fig. 4.17 (which essentially is the ratio of the exit angle to the entrance angle at the pyrometer filament) is

---

† The disappearing filament type pyrometer is the most widely used monochromatic pyrometer. Briefly, to measure temperature the image of the source of unknown temperature is superimposed on the image of the filament of the standard tungsten lamp and the current through the lamp is adjusted so that the filament disappears.

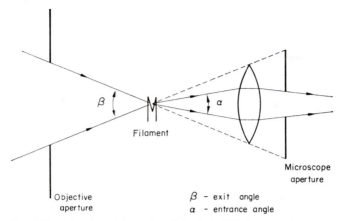

Filament

Objective
aperture

Microscope
aperture

β – exit angle
α – entrance angle

FIG. 4.17. Definition of the ratio α/β for Fairchild–Hoover criterion.

sufficiently small the above effect disappears. The size of the ratio required to eliminate the effect is, incidentally, a function of the width and the shape of the filament.

The ultimate sensitivity and the accuracy of a radiation pyrometer is determined by the detector it uses. Visual pyrometers, of course, rely on the human eye as the detector. It is in fact a highly sensitive detector. Jones (1953), for instance, quotes that the human eye can detect radiation energy of 1.12×10^{-17} W (noise equivalent power, i.e. r.m.s. noise level) with a typical response time of the order 0.1 s. The most important disadvantage of the eye as a detector is that it is highly subjective because its sensitivity, spectral sensitivity maximum, and accuracy vary from subject to subject. To make a pyrometer subject independent, one may use a variety of thermal, photoconducting, photovoltaic, photo-emission, and pyroelectric detectors. For details of the performance characteristics of the various detectors the reader is referred to Jones (1953), Lovejoy (1962), McElroy and Fulkerson (1968), and Smith *et al.* (1968). Briefly, the relevant properties of various types of detectors are listed in Table 4.6. A novel and potentially very useful application of a thermopile as a thermal detector in a high temperature crystal growth apparatus is described by Blum and Chicotka (1968). In their application a sapphire rod was used as a radiant energy "pipe" to deliver the radiance energy on to the detector outside the crystal growth chamber. Photoelectric pyrometers for precision work† have employed a photo-multiplier (a photo-emission detector) having a tri-alkali photocathode (i.e. S–20 type). For instance Quinn and Ford (1969) have described the design and performance of the NPL photoelectric pyrometer, using an S–20 type detector, which is used to establish the temperature scale above the gold point. The extrapolated scale of the above instrument is estimated to reproduce the IPTS to ± 0.1 K at the gold point ± 0.25 K at 1823 K, at 1 K at 2973 K. Additionally the S–20 type detector can be used in a high speed optical pyrometer (1 ms response time or roughly 1200 measurements per second) as demonstrated by Foley (1970). Nutter *et al.* (1967) have further shown that a pyrometer using the S–20 photomultiplier on automation retains its high precision resolution of the order ± 0.1 K at 1273.15 K. For measurement of lower temperatures, detection in the infrared is better suited. Infrared semiconductor photo-detectors, therefore, show a greater promise in the

†Also see Ricoli and Lanza (1977).

TABLE 4.6. *Summary of Performance Characteristics of Different Types of Radiation Detectors*

Detector Type	Noise Equivalent Power (NEP) (W)	Response Time (s)	Usable Wavelength Range (μ)	Peak Response Wavelength (μ)	Detector Operational Temperature Range (K)
Human eye	1.7×10^{-17}	~0.1	—	0.55	Room temperature and above
Thermal					
Thermal pile, bolometer Thermistors	1×10^{-8} to 1×10^{-11}	0.01–10	Broad wavelength band (limited by window and coating material)	—	Room temperature and above
Pyroelectric					
$BaTiO_3$	~10^{-8}	0.1–10^{-3}	Broad wavelength (limited by window material)	—	
Triglycine sulphate	2×10^{-9}	0.1–10^{-3}	Broad wavelength band (limited by window material)		300–313 K
Photo Emission					
Photo multipliers					
S_{20}	10^{-15}	~10^{-9}	0.3–0.7	0.14	Room temperature
S_4	10^{-15}	10^{-9}	0.3–0.7	~1	
S_1	10^{-12}	10^{-9}	0.36–1.1		
Photo cells					
Vacuum S_1	10^{-11}	10^{-9}	1.1	0.8	
Gas S_4	10^{-11}	10^{-9}	1.1	0.8	
Photo Voltaic					
Si photo transistor	10^{-13}	10^{-7}	1.1	0.85	210–400
Si photo diode	10^{-13}	10^{-7}	<1.1	0.85	210–400
Ge photo diode	10^{-13}		<1.8	0.85	
Ge photo transistor	10^{-10}		<1.8		
Photo Conductive					
CdS	10^{-16}	1–500×10^{-6} (dependent upon the level of illumination)	0.5–0.8	5.2	210–340
CdSe	10^{-6}	10^{-4}–5×10^{-3}	0.5–7.5	7.2	
PbS	10^{-10}	1–500×10^{-6}	<4	2.2	240–330
PbSe	10^{-9}	1–500×10^{-6}	1–7	3	
PbTe	10^{-10}	1–500×10^{-6}	>6		
InSb	10^{-9}		<8	6	

low temperature measurement. Hall (1966) has actually determined a radiation temperature scale, using a PbTe cell and measuring radiation in the 2–4 μ region with an accuracy of 0.05 K for the range 445.15–1033.15 K (the error at the gold point exceeded 0.8 K). Igras *et al.* (1977) have described a detector for the temperature range 273–973 K using a non-cooled semiconductor alloy $Hg_{(1-x)}Cd_x$ in the wavelength range 3.5–5 μm. Resolution claimed is of the order of 2×10^{-2} K. Semiconducting heterojunctions, with a larger lattice mismatch between the junctions, may turn out to be the detectors of the future as their internal quantum efficiency is reported to be high by Tansley and Newman (1967). Hampshire *et al.* (1970) have recently shown the usefulness of a n–n CdSe–Ge junction which can be operated as a null detector, because the photo-voltaic output produced by the radiation can be adjusted to zero by biasing the junction. They have described a two-colour type pyrometer in which the bias voltage to the junctions was calibrated directly in terms of temperature. The sensitivity of detection claimed was 0.6 mV K^{-1} at 1000 K, 0.1 mV K^{-1} at 2000 K, and 0.05 mV K^{-1} at 3000 K.

Filters and attenuators (the sectored discs) are mainly employed to extend the range of the pyrometer essentially by measuring the attenuated intensity from the source to be measured. Mechanical and electrical chopping devices are often incorporated in pyrometers to modulate the light beam(s) and eventually to obtain an a.c. electrical output which can be easily and inexpensively amplified without introducing substantial noise in the amplified signal.

A review in the laboratory high temperature measurement is given by Ruffino (1977).

4.2.2.5. *Other Less Widely Known Thermometries*

As pointed out earlier, any transducer which relationally changes its properties can potentially be used for thermometry. In the literature numerous forms of thermometers are reported and it is impossible to describe them all in the space available. In passing, however, one may list the following as "employables" in crystal growth processes:

1. Piezoelectric thermometers based on quartz crystals have a sensitivity of the order of 3×10^{-4} K in the range 230–500 K.
2. Capacitance thermometry based on polycrystalline oxides as the dielectric medium is capable of being used up to 2300 K.
3. Noise thermometry based on the thermal noise of an ordinary carbon resistor is potentially a simple technique capable of measuring temperatures up to 1700 K with an absolute accuracy of ± 1 K.

Gulskii (1977) has described a method where thermal noise of two temperature sensitive elements of differing resistances are measured without measuring their actual resistances. Thus the noise thermometer is insensitive to the temperature coefficient of the resistance.

4.2.3. TEMPERATURE CONTROL

It was stated at the outset of this article that the regulation and, indeed, manipulation of crystal growth processes is carried through the control of parameters of crystal growth environment. It is, therefore, useful to examine control processes and mechanics from the so-called systems viewpoint, and then describe the application of the systems principle to the control of crystal growth parameters such as temperature and pressure.

4.2.3.1. *Systems Description of Control Mechanisms*

A system is defined as an integrated group of functionally distinct constituent units coupled together to perform specific functions. The main aim of any control system is to regulate the magnitude of a desired variable (i.e. a controlled variable) as the output of the system, either (1) to maintain it constant within specified limits, or (2) to maintain a known relationship between the output and other parameters which may be changing or are being manipulated. Basically all control systems can be classified into two principal categories, namely:

1. Open loop systems.
2. Closed loop systems.

An open loop is schematically shown in Fig. 4.18. Here the corrective effort applied by the system to alter the controlled variable output does not depend upon the value of the output, i.e. the input command to the control system does not change according to the value of the output but is in fact determined by a kind of reference functional relationship (frequently linear) normally embraced in the "calibration" of the control equipment. Traffic control by sets of automatic lights is a typical example of this kind of control system. In general open loop control techniques are applicable to those processes where the controlled variable produces a consistent and corresponding state of the process conditions.

In a closed loop system, on the other hand, the input command is closely connected with the actual value of the output as shown in Fig. 4.19. The disturbances, which are beyond control of the system, are primarily responsible for the unwanted change in value of the controlled output. The error thus originated is measured and its magnitude then triggers a "feedback" into the input command to take the corrective action.

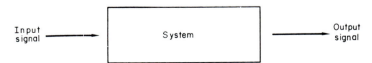

FIG. 4.18. An open loop control system.

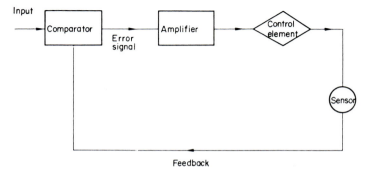

FIG. 4.19. A closed loop control system.

Analytical representation of a control system is governed by the mathematical formulation of the fixed laws of physics or chemical kinetics the system obeys. The presentation, however, has to include the relationships of the input and output functions with time. Most feedback control systems used in crystal growth processes are linear, i.e. they can be represented by a linear differential equation of the type

$$A_n D^n y + A_{(n-1)} D^n y + \ldots + A_0 y = B_m D^m x + B_{(m-1)} D^{(m-1)} x + \ldots + B_0 x, \quad (4.50)$$

where D^n, etc., are the operators $d^n\,dt$, $y = y(t)$ is the output or response function, $x = x(t)$ is the input or forcing function and $A_0 \ldots A_n$, $B_m, \ldots, B_0$ are constants. The theoretical dynamic analysis of a feedback control system is therefore possible by the solution and manipulation of eqn. (4.50). An important and significant advantage of a linear system is that it possesses the superposition property, i.e. the responsive characteristics of the linear system are independent of the magnitude of the input signal and the effect of several input (or disturbance) functions on the output can be obtained by the algebraic sum of the independent effects of each signal as if it were acting separately. In practice this means that any complex multi-variable system can be expressed as a combination of simple, easily manipulated input–output elements as shown in Fig. 4.20. The solutions of the control systems equations are beyond the scope of the present chapter, but they can be found in several textbooks on the analysis of control systems.

In most modern practical control systems a transducer is used to obtain an electrical error signal which is subsequently transmitted to trigger a controller action to modify the input function.

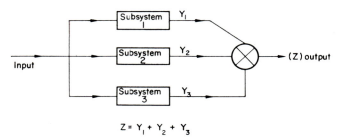

$$Z = Y_1 + Y_2 + Y_3$$

Fig. 4.20. A block diagram of a multivariable linear control system.

4.2.3.2. Elements of a Control System and Controller Actions

A closed loop automatic control system normally consists of components performing the following functions:

1. A transducer which senses or measures the instantaneous value of the controlled variable (referred to as the control point) or its deviation (called the error) between the control point and the set point (i.e. a desired value of the controlled variable). In an electric control system the transducer employed usually delivers a low-level electric signal proportional to the value being metered.
2. The low-level signal is then amplified either by electro-mechanical elements in the control loop (e.g. a self-balancing potentiometer or a Wheatstone bridge) or an electronic amplifier. The final sensitivity of the control naturally depends upon the properties of the amplifying units as well as those of the sensing elements.

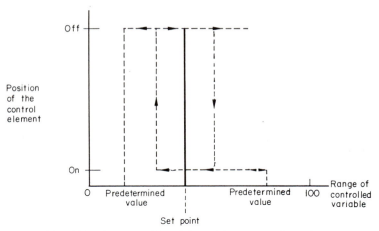

FIG. 4.21. The two-position differential gap controller action.

3. The amplified signal is transmitted into the feedback loop to drive elements (control elements) which take the corrective action to restore the control point to the set point. Various controller actions utilized in different systems can now be explained in terms of the manipulating actions of the control elements to adjust the output signals.

1. *Two-position actions.* The simplest two-position action is an "on–off" action of the control element at a fixed value of the control variable. A more popular two-position action is called the two-position differential gap action and it is illustrated in Fig. 4.21. Here there are two predetermined values of the controlled variable, one on each side of the set point. The control point, under normal controller action, remains in the "differential gap" between the two predetermined values. The control element is moved from one position to the other when the control variable reaches (say, the first) predetermined value from one direction. It remains in that position until the control point has passed in the opposite direction through a range of values to the second predetermined value, whence the control element is moved to the other position. Two-position controller actions suffer from an inherent disadvantage in that, due to the on–off action, the value of the control variable continuously oscillates in a range of values.

2. *Floating speed actions.* The floating speed actions are characterized by the fact that the adjustment of the control element is moved (floated) at one or more speeds during the movement of the control point between its extreme values, i.e. the corrective action of the control element is changed either in steps at fixed intervals throughout the range or continuously varied. The control element is said to have a single-speed floating action if it is moved at a single rate between the extreme values of the controlled variable as shown in Fig. 4.22. In a multispeed action, on the other hand, the control element is moved at multiple rates, each rate corresponding to each consecutive range of values as shown in Fig. 4.23.

3. *Continuously modulated actions.* In more sophisticated control systems the output signal from the control element is modulated in time and bears a specific relationship to

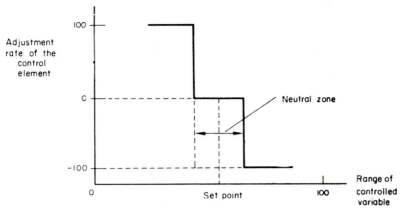

FIG. 4.22. The single-speed floating (controller) action.

the error signal. In the proportional action, it is proportionally related to the error signal defined by an equation of the type

$$\Delta y = -(100\Delta\theta)/A, \tag{4.51}$$

where Δy is the change in the controller output signal expressed as the percentage of the range, $\Delta\theta$ is the change in the input signal (transduced) expressed as the percentage of the range, and A is the range of the values of controlled variables in which eqn. (4.51) is a valid relationship. It is referred to as the proportional band. In the reset action the output of the control element is proportional to both the time and the magnitude of the error signal. It can be seen that the reset action is an integrating reaction and can be defined by an integral

$$\Delta y = 100 \frac{B}{A} \int \Delta\theta \, dt, \tag{4.52}$$

where Δy, $\Delta\theta$, and A are the same quantities as in eqn. (4.51) and B is a time constant called the reset rate.

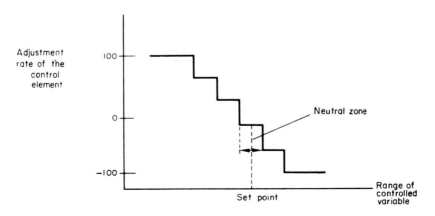

FIG. 4.23. The multispeed floating (controller) action.

In the rate action the change in the output of the control element is proportional to the time derivation of the input signal

$$\Delta y = 100 \frac{C}{A} \frac{\mathrm{d}(\Delta\theta)}{\mathrm{d}t}, \tag{4.53}$$

where C is a constant and is called the rate time.

Some controllers may have combinations of either proportional + reset action or proportional + rate reaction. The summary of the relationships between the time variation of the error signal and the time variation of the output control element adjustment is given in Fig. 4.24.

More modern control units employ solid state electronic circuits utilizing control elements described in section 4.2.3.4. The associated circuitry in the continuously modulated action controllers usually incorporates a facility to alter the width of the proportional band and the set point.

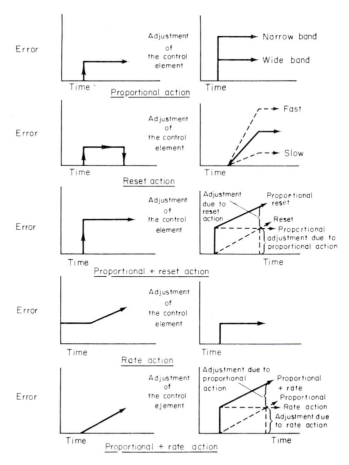

FIG. 4.24. Summary of the relationships of continuously modulated controller action and the error signal with time.

4.2.3.3. *Methods of Temperature Control*

The different methods which can be employed for a closed loop temperature control have the following physical elements:

1. Power supply for dissipation to create and maintain control point temperature (i.e. the heat source).
2. Heat sink (e.g. the environment of the controlled object, or mass).
3. Thermal resistance between the heat sink and the controlled object.
4. Heat capacity of the controlled object.

The different methods of temperature control can be shown to stem from the equation

$$qR = \Delta T, \tag{4.54}$$

where q is the rate of energy dissipation at the control point location, R is the thermal resistance between the heat sink and the control point location, and ΔT is the temperature difference between the control point and the temperature of the sink. To maintain the control point at the set point T has to be kept constant. The changes in ΔT can be compensated by adjusting either q or R or both. In Figs. 4.25a, b, and c three basic control loops are shown in which q, R, and both q and R are forcing functions respectively. In systems incorporating electrical power supply the adjustment in the power consumption can be achieved by varying either the current or the duration for which the current is on. The adjustments in thermal resistance can be incorporated on the variable leakage principle, e.g. by controlling an opening to alter natural convection or by adjusting the forced cooling of the sink.

For the detailed mathematical treatment of temperature control systems the reader is referred to Roots (1969). The practical considerations may be found in Kutz (1968).

4.2.3.4. *Common Control Elements in Temperature Control*

Contact relays and various thermostatic devices may be used in on–off type and other discontinuous control systems. They are described in detail by Coxon (1962) and Kutz (1968) and will not be dealt with here. For continuously modulated systems motorized variable transformers, grid controlled rectifiers, variable dummy load (in parallel with the real load), silicon controlled rectifiers (SCR), and magnetic amplifiers may be used. By far the most frequently used control elements are the SCR and the magnetic amplifier. They are discussed below.

1. *The silicon controlled rectifier or thyristor.* The SCR is a four-junction silicon device as shown in Fig. 4.26a. It is extensively used as a current control device as it is bistable either in a high impedance state or a low impedance state. It can be switched from one state to the other by a discontinuous impulse type control signal. The symbolic representation of the SCR in circuit diagrams is shown in Fig. 4.26b. The action of the SCR can be controlled by the signal applied to an electrode called the gate electrode (Fig. 4.26a). For zero gate current the SCR acts as an open circuit. It can be switched to a conducting state to pass large currents (a few tens of amperes) in the anode–cathode direction, by injecting a relatively small (few mA) gate current. The device can also be "fired" (i.e. made conducting) by increasing the forward voltage above a certain value, known as the

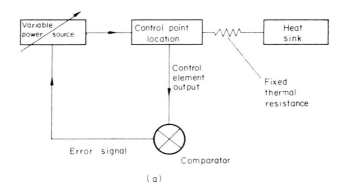

(a)

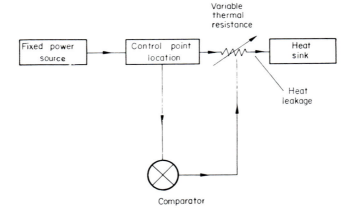

(b)

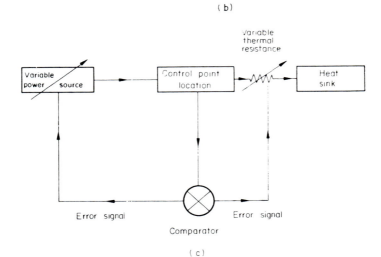

(c)

FIG. 4.25. The basic methods of temperature control: (a) by power adjustment, (b) by thermal resistance adjustment, and (c) by combined power and thermal resistance adjustment.

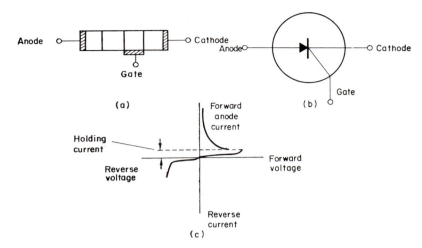

FIG. 4.26. The silicon controlled rectifier: (a) schematic construction, (b) circuit-diagram representation, and (c) current–voltage characteristics.

forward backover voltage. Normally, once the SCR is turned on the gate loses its control. Turning off the device requires a reduction in the forward anode–cathode current below a certain value called the holding current. The current–voltage characteristics of a typical SCR are shown in Fig. 4.26c. If the SCR is used in an a.c. circuit, turn-off is assisted by the reverse half-cycle of the applied voltage. In fact a continuous variability of the average current through the SCR can be obtained by the application of gate pulses at different phase angles of the waveform. When the SCR is used in d.c. circuits the turn-off may be achieved by reversing the anode–cathode voltage. The triggering pulses in the d.c. mode may be conveniently applied by a relaxation oscillator circuit employing unijunction transistors. A detailed discussion of thyristors, together with their associated devices and application circuits, may be found in Gentry *et al.* (1964) and Ankrum (1971).† In brief, the SCR is normally used to control power in one of the four modes shown in Fig. 4.27a, b, c, and d, namely:

(a) As a constant frequency on–off switch.
(b) As a pulsed switch.
(c) As a pulse width modulated switch with the aid of an additional circuit for turning off the SCR on command.
(d) As a time switch; i.e. the SCR is switched off after a fixed time interval with the aid of an external circuit.

It may be noted that the modes (c) and (d) can be used for controlling d.c. input power.

2. *The magnetic amplifier.* The heart of any magnetic circuit amplifier is a saturable reactor. It is in essence an adjustable inductor, providing high and low impedance states. Its current–voltage relationship can be adjusted by controlling the magnetomotive force,

† Designers' handbooks and manuals supplied by the manufacturers normally discuss common applications.

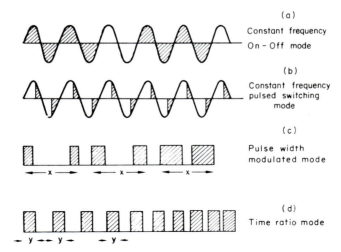

(a)
Constant frequency
On – Off mode

(b)
Constant frequency
pulsed switching
mode

(c)
Pulse width
modulated mode

(d)
Time ratio mode

FIG. 4.27. Basic modes of control using the SCR: (a) constant frequency on–off mode, (b) constant frequency pulsed switching mode, (c) pulse width modulated mode, and (d) time ratio mode.

i.e. the degree of magnetization of its core.† The basic construction of a saturable reactor and its circuit diagram representation are shown in Fig. 4.28a and b. Figure 4.28c and d shows simple forms of application of a magnetic amplifier in an a.c. and a d.c. circuit respectively. It may be noticed that in Fig. 4.28c and d the rectifying elements are connected in a series with the output to limit the flow in the windings in one direction only. In temperature control applications the error signal, derived from the temperature sensing transducer, is applied to the control winding to form a closed loop.

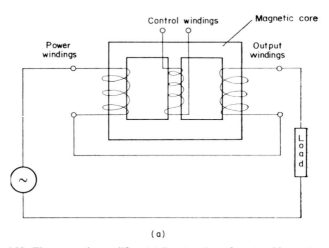

(a)

FIG. 4.28. The magnetic amplifier. (a) Construction of a saturable reactor.

† The variable impedance exhibited by the saturable reactor is due to the fact that during the increasing magnetization stage of magnetization cycle of the core, the device is in high impedance state, but as the core becomes saturated the device is switched to the low impedance state. For detailed understanding of the principles and workings of saturable reactors and magnetic amplifiers refer to Say (1954).

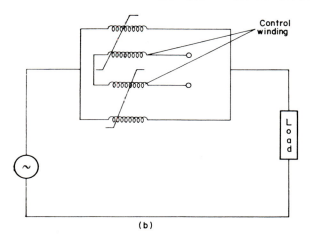

(b)

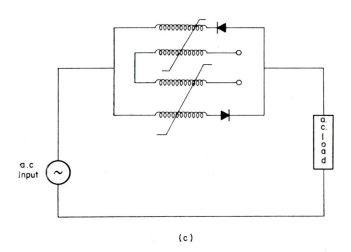

(c)

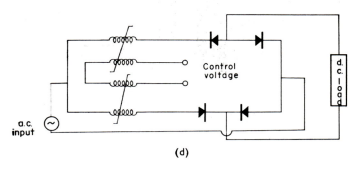

(d)

FIG. 4.28 (*cont.*). The magnetic amplifier: (b) circuit diagram representation of a saturable reactor, (c) an application of a magnetic amplifier in an a.c. circuit, and (d) an application of a magnetic amplifier in a d.c. circuit.

3. *Digital techniques.* Any analogue (continuous) signal, after suitable amplification, can be converted into a digitized signal by the use of a commercially available device as analogue to digital (A–D) converter. A digital signal in reverse can be converted into an analogue signal by another device known as digital to analogue converter. The mechanism on the mode of conversion can be found in any standard textbook on digital electronics. This capability can then be utilized in conjunction with microprocessors and/or minicomputers. A large number of functions can be manipulated in a complicated programme by the use of adaptive algorithm. Microprocessors and their accessories are increasingly becoming cheap and readily available so that they can be incorporated into various crystal growth processes.

4.2.3.5. *Temperature Control in Crystal Growth*

For most laboratory applications, cheap and reliable temperature controlling packages are commercially available. They can be incorporated in induction or resistance heating circuits in conjunction with PTR and thermocouple temperature sensors.† Control systems with optical pyrometers can be devised, but a little ingenuity is required to avoid problems due to variations in transmission of radiation from the window in crystal growth apparatus due to the deposition of the evaporated material onto the window. See, for instance, King (1970) and Webber and Hiscocks (1973).

It should, however, be stressed that the successful temperature control depends upon the design of the whole crystal-growing apparatus. It has been stated that a feedback system can be represented by an equation similar to eqn. (4.50). The solutions of the above equation tend to be oscillatory with increasing amplitude, decreasing amplitude, or constant amplitude of the periodic changes in the controlled variable. Oscillatory solutions with increasing amplitude are unstable. The onset of the above uncontrollable instability can be prevented by making it impossible for oscillations (with characteristic

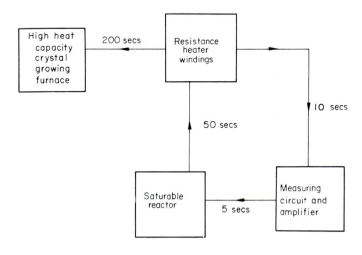

(a)

FIG. 4.29. Typical examples of temperature control sysgems in crystal growth. (*Continued opposite*).

† For information on time constants see Jarusek (1976).

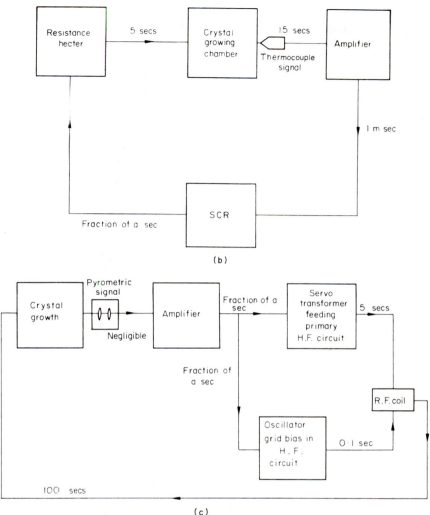

(b)

(c)

FIG. 4.29 (*cont.*).

frequency) from one stage of the feedback loop to the other. As the heat source, the load (i.e. crystal growing melt or solution) forms the parts of the control loop. Prevention of instability in the control system can be achieved readily by making the characteristic response times of all the stages different and by limiting the "gain"† of the heater stage. Figure 4.29a, b, c shows some practical layouts with typical characteristic response times of the different parts of the crystal growth systems.

It is advisable to check that no spurious signals leak in or out of the stages in the control loop; for example, one must guard against degradation of insulation which can induce signals often large enough to make the control system ineffective.

---

†A control loop can be essentially looked upon as a feedback amplifier and therefore its stability of operation can be achieved by negative feedback which effectively reduces the gain or by a differential feedback which limits the gain.

Time constants of the furnaces are often very long due to their large heat capacities. It is therefore often advisable to derive and control signals from a sensor at a remote point and/or separate the control element from the large thermal capacity source; for example, by the use of an auxiliary small capacity heater.

4.2.3.6. *Automatic Diameter Control of Crystals in Czochralski Growth*

Following the interest in minimizing waste in crystal cutting for devices a great deal of effort has been put into the development of the techniques for automatic diameter control of cylindrical crystals. Control signals for automatic controls have been derived in a variety of ways. (For a more detailed review see Hurle, 1977.)

Patzner *et al.* (1967) used a signal from an optical pyrometer focused on to the melt surface in contact with the growing crystal to modulate pulling rate. The signal was related to a bright ring around the crystal and it was necessary to use crucible lift to maintain melt at a constant level. Use of the bright ring and crucible lift has also been used by Domey (1971) in conjunction with computer control. Diggs *et al.* (1973) identified the appearance of the bright ring due to the reflection of light emitted from hot areas from the meniscus. Turovski *et al.* (1977) also used meniscus reflection for automatic control. The methods based on meniscus reflection are widely and successfully used for silicon because it has metal-like reflectivity and a high value of capillary (Laplace) constant, $L = 2\sigma_l/\rho_l g$ where σ_l and ρ_l are surface tension and the density of liquid silicon and g is gravitational acceleration. Gross and Kersten (1972) and van Dijk *et al.* (1974) used near normal incidence of a laser beam to obtain meniscus reflection to grow crystals of KCl and $Bi_{12}GeO_{20}$. Control signal was derived from the looped, reflected He–Ne laser radiation in conjunction with a pair of photodiodes. The signal via a PID controller modulated the power supplied to the melt. Diameter constant to about 1 % was grown.

Bachmann *et al.* (1970), Desaur *et al.* (1970), Gärtner *et al.* (1971), Gross and Kersten (1972), and O'Kane *et al.* (1971) used a visible–infrared TV camera to obtain crystal imaging in a closed loop control system. However, the techniques required the use of a large diameter crucible to view the solid–liquid interface and the definition of the diameter on the scan lines was poor. The constancy of the diameter achieved in these methods is somewhat less impressive than that achieved in the other methods.

van Dijk *et al.* (1974) used X-ray imaging system at normal incidence and in conjunction with liquid encapsulated growth of GaP. A small X-ray source and an image intensifier were used with aluminium windows in the pressure chamber. Image quality was enhanced by avoiding saturation of the image intensifier, with a pair of metal masks. Measurement of the diameter on TV line scan was used to control the temperature of the melt. The scan line corresponding to the interface was automatically selected by counting a predetermined number of lines from a reference line corresponding to the melt level well away from the crystal. The number of predetermined lines was continuously monitored in practice and was in actual fact a little less than the true meniscus height. The constancy in diameter achieved was 0.5 mm in 20 mm.

Bardsley *et al.* (1972) used a crystal weighing technique for the control of the diameter. The control signal was essentially derived as an error signal, i.e. the difference between the observed weight and the desired weight. "Desired weight" signal was generated from a linear potentiometer coupled to a pull rod. This system was, however, unstable at low growth speeds (of a few centimetres per hour). Reinhart and Yatsko (1974) used a similar technique with a lever mechanism to obtain greater sensitivity. Crucible weighing has also been used by van Dijk *et al.* (1974) and Kyre and Zydzik (1973). Zinnes *et al.* (1973) used crucible weighing in conjunction with a minicomputer. Valentino and Brandle (1974) have improved noise-to-signal ratio by providing dashpot damping to the weighing system. They used differentiated weight signal to compare it with desired rate of weight increase (i.e. differentiated weight error signal). Bardsley *et al.* (1974, 1975) applied a crucible weighing system for liquid encapsulation growth of InP, GaP, and GaAs. Protection of the load cell from the corrosive evaporants and convective heat transfer was necessary. In crucible weighing techniques when used with r.f. heating it is necessary to provide compensation for the vertical force induction in the susceptor. Bardsley *et al.* (1972) have developed an analogue controller which has been used to grow a variety of materials. This can be used with either weight error signal or differentiated weight error signal.

Bardsley *et al.* (1974, 1977) have shown that the apparent weight of the crystal contains contributions arising from the surface tension force which changes with the diameter of the crystal. For materials which expand on solidification and/or are only partially "wettable" in their own melt, the sign of the change in rate of weight increase with changing diameter can be anomalous. Bardsley *et al.* (1974, 1975) have included elements to generate quantified estimates of the anomalous component in their control systems. With these modifications they have been able to grow Si, Ge, GaAs, GaP, and InP.

In the above methods servo control loops have to cope with long time constants involved with the crystal growth process. For improving the stability against drift of phase advanced networks have been used. One has also to remember that in the system there are two dominant sources of noise, namely mechanical noise from pulling and rotation mechanisms.

4.3. Atmosphere

The atmosphere in a crystal growth apparatus (i.e. the ambient) may be defined by specifying:

1. The composition of the ambient.
2. The pressure of the ambient.
3. The purity or the impurity content of the ambient.†

The composition of the ambient is dictated by the reaction(s) involved in the growth processes. The magnitude of the pressure is determined by the need:

1. To prevent unwanted and undesirable reactions.
2. To avoid impurity contamination.
3. To arrest dissociation reactions associated with the growth process.

Subatmospheric pressures in the high vacuum (h.v.) and ultra high vacuum (u.h.v.) ranges help to avoid unwanted reaction and impurity contamination. Pressures higher than atmospheric pressure are necessary to stop dissociation reactions involved in the crystal growth process. The purity of the ambient is achieved by prepurification of the gases

---

† Purity of the gases available in cylinders is normally available from the suppliers.

before insertion in the apparatus and/or by adjusting the flow conditions of the gases in the apparatus.

It is clear from the foregoing that the ambient conditions will vary from material to material and from process to process. It is therefore desirable to discuss the techniques for creating and controlling vacuum and high pressure.

4.3.1. VACUUM TECHNIQUES

There is a wide variety of pumping equipment available for creating a vacuum environment and their choice in the design of a vacuum system for a particular process depends upon the following factors:

1. The desired operational pressure.
2. The pumping speed to maintain the desirable pressure.
3. The permissible amount of the residual components in the ambient (i.e. in the vacuum vessel).

One can, however, divide the vacuum technology into three distinct pressure regions, namely,

1. Medium vacuum.
2. High vacuum (h.v.).
3. Ultra high vacuum (u.h.v.).

Some important properties associated with the above regions are compared with those at atmospheric pressure in Table 4.7.

Potential rate of contamination or surface reaction can be estimated from the rate of impingement of gas molecules per unit surface area or the impingement frequency v. It is given by

$$v = 1/4 n v_{av}, \tag{4.55}$$

where n is the number of molecules per unit volume and v_{av} is the average velocity of molecules. It may be shown that v is related to the pressure by the formula

$$v = 4.695 \times 10^{24} P/(MT)^{1/2}, \tag{4.56}$$

where v is the impingement frequency expressed in $m^{-2} s^{-1}$, P is the pressure in Pa, M is the relative molecular mass, and T is the temperature in K.

The monolayer formation times listed in Table 4.7 are calculated from eqn. (4.56) by assuming that the average number of molecules per m^2 area is 10^{19} and that the sticking coefficient of the molecules to the surface is 1 (i.e. all the molecules which hit the surface stick to it). Of course, in practice the value of sticking coefficients is less than unity and they depend upon the nature of the gas, the surface, and the temperature. The values listed in Table 4.7, however, should provide a rough idea of the rate of contamination in different vacuum conditions. Figure 4.30 shows the residual impurities in a gas expressed in parts per million (ppm) and in the units of pressure. It can be immediately appreciated from this that a relatively poor vacuum of 1.33×10^{-1} Pa corresponds to an ultra high purity gas having total impurity concentration of 3 ppm. It is for this reason one endeavours to grow crystals in u.h.v. or h.v. whenever possible.

TABLE 4.7. *Comparison of Physical Properties of Gases in Different Vacuum Ranges*

	Pressure < Pa (Nm^{-2})	Density of Molecules per m^3	Mean Free Path (m)	Rate of Impingement in no. of Molecules $(m^{-2} s^{-1})$	Monolayer Formations Time (s)	Type of Flow	Range of Pumps	Common Pressure Gauges
Atmospheric	1.013×10^5	2.5×10^{25}	6.6×10^{-8}	2.9×10^{27}	3×10^{-7}	Viscous		McLeod
Medium vacuum	1.33×10^2 – 1.33×10^{-1}	3.3×10^{22} , 3.3×10^{19}	5×10^{-5} , 5×10^{-2}	3.8×10^{24} , 3.8×10^{21}	2.5×10^{-6} , 2.5×10^{-3}	Viscous	Mechanical pumps Cryosorption pump Molecular drag pumps	McLeod Diaphragm, thermal, conductivity gauge, thermocouple
High vacuum	1.33×10^{-1} – 1.33×10^{-6}	3.3×10^{19} , 3.3×10^{12}	5×10^{-2} , 5×10^3	3.8×10^{21} , 3.8×10^{16}	2.5×10^{-3} , 2.5×10^2	Transition Viscous–molecular	Diffusion (or mercury) sputter-ion pump Getter-ion pump (in combination with other medium vacuum pumps)	McLeod Hot cathode ionization gauge Cold cathode
Ultra high vacuum	1.33×10^{-6} – 1.33×10^{-11}	3.3×10^{12} , 3.3×10^9	5×10^3 , 5×10^8	3.8×10^{16} , 3.8×10^{11}	2.5×10^2 (4 min) 2.5×10^7 (290 days)	Molecular	Oil–mercury diffusion pumps; sublimation, cryogenic, condensation, and sputter-ion pumps (in combination with other medium vacuum pumps)	Bayard–Alpert ion gauge Penning Cold cathode magnetron gauge Hot filament magnetron gauge

Notes: $Pa = 1.33 \times 10^2$ torr (or mmHg).
Monolayer formation times are calculated assuming that the sticking coefficient $= 1$.

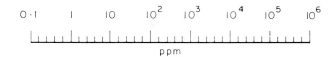

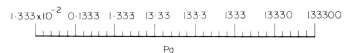

FIG. 4.30. The residual impurities in a gas ambient expressed in units of pressure.

4.3.1.1. *Design of Vacuum Systems*

It is impossible to give even an abridged account of the principles and description of the various vacuum pumps and associated equipment within the space of this chapter. Table 4.8, however, provides a summary of the commercially available pumping equipment. For a detailed treatment and discussion on vacuum equipment the reader is recommended to refer to Pirani and Yarwood (1951), Dushman and Lafferty (1962), Roberts and Vanderslice (1963), Beck (1964), Lewin (1965), Diels and Jaeckel (1966), Redhead *et al.* (1968), and Bunshah and Batzer (1968). For the latest advances on diffusion pumps, getter–ion pumps, and cryogenic pumps, readers are recommended to refer to Hablanian and Maliakal (1973), Bills (1973), and Hobson (1973). Generally, in designing a vacuum system the knowledge of the following operational parameters is highly desirable:

1. The working pressure or operational pressure.
2. The volume of the work chamber.
3. The rate of outgassing of various surfaces in the vacuum vessel at the operational temperatures.
4. The permissible levels of residual components of the ambient.

It is pointed out here that it is extremely difficult to measure the working pressure at the crystal growing interfaces. Therefore, at the point of measurement the measured pressure may be considerably lower than that at the interface (particularly if there is excessive outgassing or evaporation of the volatile components). Account of the above fact should be taken at the design stage. Often the rate of outgassing or volatilization is very high at elevated temperatures and may dictate the choice of pumping speeds in the system.

The extent of the pumping action of a vacuum pump can be expressed in terms of its throughput Q defined by

$$Q = SP, \tag{4.57}$$

where S is the pumping speed in volume (extracted) per unit time and P is the pressure. The pumping speed is the volumetric flow generated by the pumping action and is therefore given by

$$S = \frac{dV}{dt}, \tag{4.58}$$

TABLE 4.8. *Summary of the Information on Various Vacuum Pumps*

Type of Vacuum Pump	Principle of Operation	Type of Fore Pump needed for Backing and Roughing	Pressure Range of Operation (Pa)	Range of Pumping Speed (litres s^{-1})	Other Remarks
Mechanical, oil sealed	Extraction of gas by rotary vanes or pistons		1.33×10^{-1} – 1.33×10^{5}	1–400	Used as fore-backing pumps. Possibility of oil contamination which can be avoided by fore line/backing line traps
Cryosorption	Sorption of gases on cooled zeolite or charcoal		1.33 – 1.33×10^{5}	50	For initial pumping of clean systems; may be used in series with mechanical pump down to 1.33×10^{-2} Pa
Oil or mercury diffusion pumps	Vapour jet trapping air molecules	Mechanical pumps	1.33×10^{-8} – 1.33	1–10^{5}	Used in the h.v. and u.h.v. Needs traps and baffles
Molecular drag pump	Extraction by high speed turbine rotor	Mechanical pumps	1.33×10^{-7} – 1.33	140	Does not need traps, low compression ratios for light gases (i.e. higher proportion of light gases in the residual pressure). Expansive
Sputter–ion pumps	Catchment of ionized gas atoms in metal cathode and by chemical gettering of sputtered metal on an electrode	Cryosorption	1.33×10^{-10} – 1.33	1–1000	Possibility of sputtering metal contamination exhibits a kind of "memory" effect from the prior usage. Pumping speed decreases rapidly at lower pressures due to lower ionization
Getter–ion pumps	Catchment of gas ion in titanium or other getter films and chemisorption of active gases	Cryosorption	1.33×10^{-10} – 1.33	1–1000	Small throughput Pumping speed for inert gases is usually low
Sublimation pumps	Chemical gettering of active gas on renewed metal films	Needs cryosorption and sputter–ion pumps as backing pumps	1.33×10^{-11} – 1.33×10^{-4}	1–10^{4}	Low speeds for inert gases
Cryogenic condensation pumps	Condensation on large surface at cryogenic temperatures		1.33×10^{-11} – 1.33	Medium high	High throughput, pump speeds depend upon conductance of radiation shield surrounding cryogenic surface

where V is the volume of the gas and t is the time. The mass flow rate corresponding to eqn. (4.57) is given by

$$n\frac{\mathrm{d}V}{\mathrm{d}t} = Q/(kT),\qquad(4.59)$$

where n is the number of molecules per unit volume (i.e. molecular density), k is Boltzmann's constant, and T is the temperature in K.

Normally, the calculations of mass flow are carried out at one temperature (i.e. room temperature); $Q/(kT)$ in eqn. (4.59) is often approximated by Q. In the absence of any gas flow due to a work process (e.g. crystal growth) the net mass flow due to a pump in the system is given by

$$-V\mathrm{d}P = \mathrm{d}t(Q - Q_I),\qquad(4.60)$$

where Q_I is the "inherent" flow due to the internal gas evolution in the system. Q_I is therefore given by

$$Q_I = Q_L + Q_G + Q_B,\qquad(4.61)$$

where Q_L is the flow due to leakage, Q_G is the flow generated by outgassing of the surfaces in the vacuum system, and Q_B is the mass flow due to backstreaming or back diffusion of the gas from the pump into the system. The ultimate pressure P_u of the system is reached when $\mathrm{d}P/\mathrm{d}t = 0$, thus

$$P_u = Q_I/S.\qquad(4.62)$$

The working pressure P_W is given by

$$P_W = (Q_I + Q_W)/S,\qquad(4.63)$$

where Q_W is the gas flow generated by the work processes.

The pumping speed at the outlet of the work vacuum chamber is different from that at the inlet of the pump because the interconnecting components, pipes, valves, etc., between the two orifices offer resistance to the gas flow. This resistance is expressed in terms of the conductance C defined by

$$C = Q/\Delta P\qquad(4.64)$$

where Q is the throughput and ΔP is the difference across the section of the vacuum system, the conductance of which is defined by eqn. (4.64). The overall conductance of the system can be calculated from the conductance of individual components by the following laws for parallel and series connections:

$$C = C_1 + C_2 + \dots \qquad \text{(for parallel connections)},\qquad(4.65)$$

$$\frac{1}{C} = \frac{1}{C_1} + \frac{1}{C_2} + \dots \qquad \text{(for series connections)}.\qquad(4.66)$$

The pumping speed S_W at the outlet of the vacuum work chamber can be shown to be related to the conductance and the working pressure, namely

$$1/S_W = 1/S_{\text{pump}} - 1/C = (P_W - P_u)/Q_I,\qquad(4.67)$$

where S_{pump} is the rated speed of the pump and C is the conductance of the vacuum system.

In selecting the pumping speed of the pump, therefore, due account of the conductance should be taken. A general rule of thumb is that to conveniently handle normal crystal growth gas loads,

$$P_W > 100\,P_u. \tag{4.68}$$

It is stated in Table 4.7 that in different vacuum regions the gas flow is either viscous or molecular. The conductance of the vacuum system for each type of flow is different. For a straight tube of diameter D and length L it can be shown that:

$$C_{\mathrm{visc}} \propto D^4/L, \tag{4.69a}$$

$$C_{\mathrm{mol}} \propto D^3/L, \tag{4.69b}$$

where C_{visc} and C_{mol} are the conductance for viscous and molecular flow respectively.

In the transition region where viscous flow gradually changes to molecular flow with decreasing pressure the overall mass flow rate is due to the contribution (sum) of both types of flow.

Finally, it is reminded that the pump down in h.v. and u.h.v. regions is normally accomplished in two or more stages as the pumps operating in h.v. and u.h.v. are incapable of pumping at higher pressure ($> 1.33 \times 10^2$ Pa) and therefore they need to be backed up by medium vacuum pumps. They are used for evacuating the vacuum work chambers up to medium vacuum region in the first stage and are then operated in series with the h.v./u.h.v. pumps to keep the pressure at the outlets (of h.v./u.h.v. pumps) in the medium vacuum region.

In h.v. systems the ambient may be contaminated by diffusion pump fluids due to backstreaming phenomena. The "backstream poisoning" of the ambient can be reduced by interposing water-cooled and refrigerated trapping baffles or by "active metal" traps. In designing u.h.v. systems for crystal growth regard must also be paid to the possibility of contamination by the gettering or ionizing electrode material.

4.3.1.2. *Measurement and Control of Vacuum*

Pressure measuring transducers in the various vacuum region fall into two categories, namely:

1. *Total pressure gauges*. These measure total pressure in the system and are based on measurement of various properties proportional to the total pressure. The summary of various total pressure gauges is provided in Table 4.9. The detailed description of these gauges is included in the references cited in section 4.3.1.1. Recent comparative reviews of low pressure measurement have been provided by Steckelmacher (1965), Bunshah and Batzer (1968) and Lafferty (1972).
2. *Partial pressure gauges*. These are also known as residual gas analysers (RGA) and are in fact gas mass spectrometers.

The basic design of a RGA is shown in Fig. 4.31. It contains three functional subsystems, namely, the ion generator, ion analyser and ion detector.

TABLE 4.9. *A Summary of Information on Total Pressure Gauges*

Type of Gauge	Principle of Operation	Measurable Pressure (Pa)	Other Remarks
McLeod	Gas compression by a column of mercury	$1.33 \times 10^{-3} - 1.33 \times 10^{3}$	Fragile—may contaminate apparatus with mercury Unsuitable for vapour pressure measurements
Diaphragm	Displacement of a movable membrane diaphragm	$1.33 \times 10^{-2} - 6.5 \times 10^{3}$	Cannot be used with bakeable vacuum system, highly sensitive gauges difficult to obtain commercially
Thermal conductivity gauge (Pirani)	Resistance changes due to cooling of a heated wire due to the presence of gases	$1.33 \times 10^{-1} - 4 \times 10^{3}$	Cheap, reproducibility may be affected due to gas adsorption
High pressure hot cathode ion gauges	Ionization of gases by thermionically emitted electrons from a hot filament	$1.33 \times 10^{-3} - 1.33 \times 10^{2}$	Reproducibility may be affected by gas reaction with the hot filament
Bayard–Alpert ion gauge	Gas ionization by thermionic emission	$1.33 \times 10^{-8} - 1.33 \times 10^{-1}$	Improvement in reproducibility due to provision for ion bombardment of the gas adsorbed filament
Penning gauge	Cold cathode gas discharge in crossed electric and magnetic fields	$1.33 \times 10^{-10} - 1.33$	Rugged, simple to use. Pressure indication is non-linear
Trigger discharge gauge	Discharge triggered at low pressures	$1.33 \times 10^{-12} - 1.33 \times 10^{-1}$	
Hot cathode and cold magnetron gauges	Gas discharge or thermionic electrons in electromagnetic fields (in magnetron configuration)	$1.33 \times 10^{-12} - 1.33 \times 10^{-2}$	Pressure indication non-linear

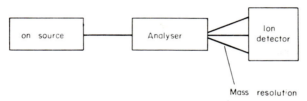

FIG. 4.31. A basic block diagram of a residual gas analyser (RGA).

In the ion generator the sampled gas from the vacuum environment is ionized by electron collision. Normally electron currents of the order of 10^{-4} A are required for sufficient ionization and therefore obtained by thermionic emission or a cold cathode discharge. The ions thus produced are accelerated and focused on to the entrance aperture/slit of the analyser. In the analyser the ions are dispersed according to their mass–charge ratio by electrostatic, magnetic, or electromagnetic fields. The working of an analyser is characterized by two quantities, namely, resolving power and transmission. The resolving power of an analyser expresses its ability to separate the mass number of the ions and a practical definition is given by

$$R_N = M/\Delta M, \tag{4.70}$$

where R_N is the resolving power (at a peak fraction N), M is the mass number of the mass peak which is separated, and ΔM is the "width" of the mass peak expressed as a function of M as a function M of the peak height.† Transmission T is defined as the ratio of the number of separated ions to the total number of ions entering into the analyser. It is, therefore, obvious that for an analyser with higher transmission, a higher ion current for each ion species (peak) would be obtained. It can be seen that the choice of a RGA should be dictated by the maximization of the product $R_N T$.

The ultimate sensitivity of a RGA depends upon the ion detection system. More modern ion detection systems incorporate ion collection accompanied by electronic amplification. The output from the ion detection system is normally in the form of a recorded spectrum showing the relative intensity of the ion current at different mass number.

All commercially available RGAs can be classified into three categories:

1. *Deflection type:* where dispersion is by deflection in a magnetic field or a combination of a d.c. electric field and a magnetic field.
2. *Radio frequency resonance type:* where ions are separated by resonance due to imposition of a r.f. field.
3. *Time of flight type:* here ions are first accelerated through a potential field and allowed to drift through a field-free space (of fixed length). The time of arrival of the ions depends upon m/e (e.g. ions with low values of m/e arrive faster than those with higher m/e).

The output from a RGA is in the form of a spectrum showing mass peak heights for each mass number. Most permanent gaseous substances partially decompose or dissociate and show a unique "cracking pattern" in terms of mass peak heights. It is

† The peak height is naturally proportional to the number of ions that are separated in the peak.

therefore possible to calibrate a RGA for quantitative analysis by admitting permanent gases at a known pressure into the RGA and measure its sensitivity. In principle a mass spectrometric RGA is capable of measuring the lowest gas density in the universe (roughly corresponding to pressure $\sim 1.33 \times 10^{-14}$ Pa). In practice, however, the sensitivity is statistically limited. It was shown by Huber (1964) that the statistical error Δ in a RGA is given by

$$\Delta = 100(nt)^{-1/2} \text{ per cent,} \qquad (4.71)$$

where n is the number of ions impinging on the analyser and t is the time in seconds. Thus for $t = 1$ s at a pressure 1.33×10^{-10} the statistical error of the instrument would be in the region of 4%.

Furthermore, the accuracy of a RGA when measuring a partial pressure is dependent upon the noise due to stray ion scattering which in turn is dependent upon the total pressure. The partial pressure sensitivity can, therefore, be defined by the ratio

Partial pressure sensitivity = (smallest detectable partial pressure)/(total pressure).
$$(4.72)$$

For most commercial RGAs the above ratio is the region 10^{-6}. Craig and Harden (1966) have found that with specific calibration it is possible to achieve accuracy of about 2% at 1.33×10^{-4} in the partial pressure measurement.

The automatic control of total pressure in a vacuum system is possible by suitable electronic amplification of the measurement signal obtained from either a total pressure gauge or a RGA. The control element in such a system is a servo-controlled vacuum valve as shown in Fig. 4.32. Automatic pressure controllers for pressures up to 1.33×10^{-10} Pa are commercially available. Annoni *et al.* (1968) have recently described an automatic pressure controlling system coupled with a safety protection circuit to shut off the processes in the case of an accidental leakage in the system.

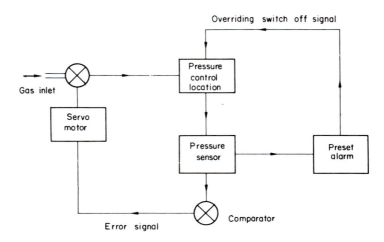

FIG. 4.32. A block diagram of an automatic pressure control in a vacuum system.

4.3.2. HIGH PRESSURE TECHNIQUES

As remarked earlier, the need for supra-atmospheric pressures in a crystal growth arises from the necessity of suppressing the dissociation of the compound materials used in the crystal growth process. The equilibrium vapour pressure of a highly volatile constituent component of a compound at the crystal growing temperature T could be as high as 10^7 Pa. One method of attaining a high equilibrium vapour pressure within an enclosed system is by enclosing the pure volatile component in one part of the crystal growing apparatus and then maintaining the whole of the crystal growing vessel at an elevated temperature T', so that the vapour pressure of the volatile component is equal to the equilibrium vapour pressure of the dissociating compound at the growing temperature T. Another method, originally devised by Mullin *et al.* (1965), has become available for high pressure crystal growth. It is referred to as the liquid encapsulation technique. In this method the dissociating liquid (melt) is encapsulated by an inert, optically transparent layer of a liquid.† The loss of the volatile component from the dissociating liquid is suppressed by providing an inert gas ambient over the encapsulating liquid at a pressure greater than the equilibrium dissociation vapour pressure of the volatile component. The chief advantage of the liquid encapsulation method is that it dispenses with the need to maintain the whole of the crystal growing chamber at an elevated temperature. Czochralski crystal pulling apparatus, using the above techniques, has been described by Mullin *et al.* (1968), Bass and Oliver (1968), and Wenckus and Doherty (1968). Nygren (1973) has grown 35 mm diameter GaP single crystals by the above technique. High pressure crystal pullers with a remote television viewing system are now commercially available. The main problem with liquid encapsulation is that contamination of the melt does take place due to the presence of the encapsulant. Moreover, encapsulants which do not react with the melt are difficult to find. B_2O_3, which is a very successful encapsulant, can be used only around and above 1000°C. Also other encapsulants have been used (see Shah, 1981); at lower temperatures they are not universally successful. To overcome the difficulties of contamination, Moulin *et al.* (1974) have described an indirect encapsulation technique to grow GaAs. It uses a rotating liquid B_2O_3 seal remote from the surface of the melt, actually in contact with the shaft of the rotating seed holder. The growth ampule made of silica was, however, complicated and consisted of two parts, namely the crucible and the sealing system, which can accept a difference of pressure of argon with the arsenic pressure. In the system the use of an afterheater was also made to prevent arsenic condensation and to ensure freedom from defects due to thermal stress. The system is claimed to be stable. The major source of contamination is silica.

Mullin *et al.* (1972) have described a pressure-balancing method in which the pressure over a liquid seal is dynamically balanced by sensing the pressure with a manometer and then controlling a valve which enables rapid equalization but does not allow diffusive losses of the vapour to escape. In another variant of the pressure-balancing method, namely partition balance, Mullin (1975) has described an incorporation of a two-way bubbler which allows the excess vapour pressure or gas pressure to flow and partition itself into the growth chamber and the pressure chamber according to the pressure differences between the chambers.

---

† Desirable properties of a liquid encapsulant are listed by Shah (1981).

For the technology of high pressure production and measurement the reader is referred to Bridgman (1949), Cornings (1956), Giardini and Lloyd (1963), Munro (1963), Deffet and Lialine (1965), and Bradley (1969).

4.3.3. DYNAMIC ATMOSPHERES

In many crystal growth processes it is necessary to provide a dynamic flow of gases to drive the necessary chemical reactions involved in the crystal growth. Where possible, high purity gases should be used to keep impurity contamination minimal. Most common high purity gases can be obtained commercially. Alternatively a prepurification unit (e.g. a palladium purifier for hydrogen) can be incorporated in the crystal growth apparatus flow system. In many cases unwanted constituents, like water vapour and oxygen from the gases, may be removed by standard traps. The sensing and controlling elements for flow control are cheaply available.

Jenson *et al.* (1972) have recently described a simple device for supplying a gas stream at a controlled temperature. This device essentially consists of splitting a gas supply into two streams. One of these passes through a fixed temperature bath. The relative flow of the streams is controlled by a single micrometer screw type valve. As the flow in one stream is increased the flow in the other decreases. The temperature constancy of the gas supply thus obtained is claimed to be of the order ± 0.01 K.

Dynamic atmospheres, and to a certain extent static atmospheres, may be actually used to remove certain trace impurities by promoting a chemical reaction between the trace impurities and constituents of the crystal growth ambient. The above technique has been recently employed to improve the resistance ratio of metals—see, for instance, Shah and Huntley (1970), Shah and Brookbanks (1972), and Reed (1972).

4.4. Container Materials

One of the most essential features of a practical crystal growth apparatus is the provision for the containment of a medium from which the crystal is grown. Thus, for a melt/solution growth it is necessary to provide a crucible to hold the liquid. Similarly, in a method in which the crystal is grown from the vapour phase it is essential to contain the vapours by a suitably designed enclosed vessel which does not hinder the process of crystal growth. Additional provision of a process enclosure is an absolute necessity in the design of a process apparatus to create and maintain a suitable environment for the crystal growth. It is, therefore, useful to consider the desirable features an ideal container material should have.

4.4.1. GENERAL CONSIDERATIONS

The first and foremost requirement is that the container material should be completely inert to the crystalline material and the crystal growing medium. The problem of finding an inert container for high temperature melting materials and highly reactive materials is often difficult and one is often forced into choosing a crucible free crystal growth process.

The container material should not contaminate the crystal growing medium, and eventually the crystal impurities on the surfaces within the interstices (particularly gaseous impurities), occluded impurities, and those in solid solution in the container materials can be potentially transferred to the crystal. Hence due regard should be given to the potential impurities in and on the container. The container design should facilitate cleaning. Chemical etchants and organic solvents used for cleaning purposes are often dangerous sources of contamination as they can leave impurities on the surface or in the interstices of the container material.

For both chemical and physical reasons the fluids used in the crystal growth process should not wet the container surface. Wetting is often the cause of the adhesion of the crystalline material which in turn may cause physical strain and, indeed, the cracking of containers and the crystals.

The surface of the container should not provide spurious nucleation sites, i.e. it should not have irregular features on its surface. In practice often the interior surfaces of a crucible are "sand blasted" so that the contamination is reduced by reduction in the contact area between the crystal growing medium and the crucible. It should, however, be pointed out that the problem of "keying" in these instances does not arise as the liquids contained in the crucible have a high surface tension.

The thermal shock characteristics of the material at crystal growing temperatures should be adequate to sustain thermal cycling. Furthermore, it is desirable for the crucible to have a lower coefficient of thermal expansion than the crystal growing medium, as this helps to eliminate thermal straining in the crystal. The reduction in thermal strain due to mismatch in thermal expansion may be achieved by the use of a displacement absorbing soft lining (e.g. a layer of loosely packed alumina powder) as demonstrated by Hurle (1959). In a crucible contained crystal growth, stray crystalline growth can be avoided if the thermal conductivity of the crucible is lower than that of the crystal because it was shown by Hurle (1962) that the lower thermal conductivity of the crucible helps in maintaining the solid/liquid surface convex (in the direction of the growth) when the crystals formed due to spurious nucleation on the surfaces of the crucible cannot grow into the melt.

Finally, the electrical properties of the container material should be compatible with the method of heating employed and they should not deteriorate with prolonged use at the operational temperatures.

Table 4.10 gives a brief survey of the properties of common container crucible materials. Generally metallic substances are grown from ionic or covalent containers.

4.4.2. MAINTENANCE OF CONTAINERS

It is recommended that every crystal grower should adopt the good practice of keeping crucibles and containers scrupulously clean. The following points should be borne in mind:

1. All the cleaning reagents (i.e. chemicals and organic solvents) should be of the highest purity and should be used once only.
2. Wherever possible, ultrasonic cleaning is recommended.
3. Chemical etching should be followed up by thorough rinsing in distilled–deionized water. In the course of cleaning with organic solvents it is easy to leave a dirty film on the cleaned surface due to rapid evaporation of the solvent. This can be avoided by repeated rinsing in fresh solvent or by following up rinsing with another solvent,

TABLE 4.10. *A Summary of the Properties of Common Container/Crucible Materials*

Material	Approximate Maximum Working Temperature (K)	Thermal Shock Resistance	Thermal Conductivity ($W\,m^{-1}\,K^{-1}$)	Coefficient of Linear Expansion ($K^{-1} \times 10^6$)	Other Remarks
Alumina	2170	Fair	16.748	8	Metallic element may react at temperatures above 1600°C
Aluminium nitride	2270	Fair		5.7	
Beryllia	2570	Good	1.6×10^3	8.4	Reaction possible above 1600°C
Boron nitride	1970	Very good	5.02	0.2-3	Oxidizes in air above 970 K
Calcium fluoride	1420	Fair		24	
Iridium	2600	Very good	148	6.8	—
Magnesia	2870	Fair	4.19-8.38	25	Possesses appreciable high vapour pressure
Platinum	1950	Very good	73	9.11	Becomes plastic at high temperatures
Pyrex	770	Good	1.13	3.2	Permeable to constituents of air at high temperature
Silica	1530	Very good	1.38-2.67	0.5-0.6	Permeable to air. Devitrification if continuously used above 1670 K and also if contaminated with organic films due to handling
Silicon nitride	1770	Fair		6.4	
Thoria	3070	Fair	4.19	6	Reacts with carbon and other refractory materials above 2290 K
Vitreous carbon	2070	Good	4.19-8.37	2-3.5	Oxidizes in air above 900 K
Zirconia	2570	Good	1.97	4.5	—

Note: The values listed here should be taken as an approximate guide for design purposes. The actual values are temperature dependent.

which is miscible in the first solvent and water, so that the final rinsing can be done in distilled–deionized water.

4. Cleaned crucibles should not be handled with bare hands; clean disposable plastic gloves are ideal for post-clean handling.
5. The cleaned crystal growth ware should be stored in dust-free pressurized cabinets or where possible in a clean oven at elevated temperatures.

4.5. Growth Velocity

4.5.1. MACROSCOPIC GROWTH VELOCITY

In most practical crystal growth methods a crystal is rarely grown under equilibrium conditions. Therefore, the growth velocity in a practical method is determined by the imposed gradient of the thermodynamic potential at the crystal growing medium boundary. The thermodynamic potential is a function of pressure, temperature, electrostatic potential, and chemical potential (which itself is a function of phase components in the growth system). The growth velocity, therefore, in principle can be influenced by imposing a gradient of one or more of the above-mentioned entities. Thus at least on a macroscopic scale the crystal growth velocity is dependent on the external conditions imposed by the crystal grower and his equipment. Due attention, therefore, should be paid at the equipment design stage to incorporate a required degree of variability and control on the imposed driving force, which is responsible for crystal growth. By far the most popular driving gradient is the temperature gradient.

There are several ways of moving a temperature gradient for crystal growth. The most convenient methods incorporate the movement of the temperature furnace away from the crystal growing medium boundary or the movement of the crystal medium boundary away from the furnace. The above mentioned movement can be created by the use of an electric motor and other mechanical appliances. The actual speed of the movement is often dictated by the necessity of avoiding micro-inhomogeneities in the crystal. It will suffice to mention here that whatever the actual magnitude of the movement, the movement itself should be smooth enough to avoid any abrupt change of conditions at the growth boundary. It has been recently demonstrated by Bachmann (1973) that by a careful design the background "noise level" of movement in the traverse system can be reduced to a level of 2 μm r.m.s. Alternatively, the driving of crystal growth by temperature may be accomplished by a programmed lowering of the temperature profile and thus altering the value of the temperature gradient.

Both chemical potential and pressure gradients under certain conditions may be influenced indirectly by controlling the temperature and temperature gradients. The chemical potential can be directly altered by altering the phase component composition, i.e. by altering the flow conditions, reaction rates, etc.

4.5.2. MICROGROWTH FLUCTUATIONS

Microgrowth fluctuations in a crystal growth process are caused by the instability of the crystal growing medium boundary. The reasons for the presence of a stability at the crystal medium interface may be manifold. But they can be divided into two main categories, namely:

1. Mechanical instabilities caused by the uneven mechanical movement to which the boundary may be subjected.
2. Chemical or thermodynamic instabilities caused by fluctuations of thermodynamic parameters at the interface. The fluctuations may be due to the presence of impurities and temperature gradients.

Mechanical instabilities can be easily avoided by a careful design of the equipment and by ensuring that the motors, gears, lead screws, etc., used in the equipment are of a high quality and are not subjected to uneven and excessive wear.

The chemical instabilities, on the other hand, are not so easy to eliminate. They depend upon the level of impurities and their diffusion coefficients. The theory of interface morphology is given by Delves (1975). It may, however, be pointed out here, briefly, that constitutional supercooling type instability can be successfully avoided by simultaneous adjustment of the growth velocity and the temperature gradient at the interface. The phenomenon of temperature fluctuation first observed by Hurle (1966) commonly occurs in crystal growth. The growth fluctuations in this case may be avoided by imposition of a magnetic field as demonstrated by Hurle (1966) or by suitable incorporation of baffles as demonstrated by Brice *et al.* (1971).

4.6. Conclusion

It is hoped that the foregoing pages have given the reader a sufficient insight into the creation, measurement, and control of crystal growth environment to spur him into becoming an enthusiastic and efficient apparatus designer, for, finally, he is reminded that the secret of successful crystal growth lies in the skill, craftsmanship, and care with which the apparatus is designed and executed.

References

ADAMSKI, J. A. (1965) *J. Appl. Phys.* **36**, 1784.
ADAMSKI, J. A., POWELL, R. C., and SAMPSON, R. L. (1968) *ICCG Birmingham*, p. 246.
ALFORD, W. J. and BAUER, W. H. (1966) *ICCG Boston*, p. 71.
ANDERSON, A. R. and STICKNEY, T. A. (1962) *Temperature—Its Measurement and Control in Science and Industry*, Vol. 3, Part 2 (ed. Herzfeld, C. M.), Reinhold, New York.
ANDERSON, R. L. (1978) *J. Less Common Metals* **59**, 517.
ANKRUM, P. D. (1971) *Semiconductor Electronics*, Prentice-Hall, New Jersey.
ANNONI, S., SALARDI, G., VERRAZZANI, L., and UCCELLI, F. (1968) *Nucl. Instr. Methods* **63**, 279.
BACHMANN, K. J. (1973) *J. Cryst. Growth* **18**, 13.
BACHMANN, K. J., KIRSCH, H. J., and VETTER, K. J. (1970) *J. Crystal Growth* **7**, 290.
BARBER, C. R. (1971) *The Calibration of Thermometers*, HMSO, London.
BARBER, C. R. and BLANKE, W. W. (1961) *J. Sci. Instr.* **38**, 17.
BARBER, C. R., GRIDLEY, A., and HALL, J. A. (1952) *J. Sci. Instr.* **29**, 65.
BARBER, G. F. and BAKISH, R. (1962) *Introduction to Electron Beam Technology* (ed. Bakish, R.), Wiley, New York, p. 96.
BARDSLEY, W., GREEN, G. W., HOLLIDAY, H., and HURLE, D. T. J. (1972) *J. Cryst. Growth* **16**, 277.
BARDSLEY, W., COCKAYNE, B., GREEN, G. W., HURLE, D. T. J., JOYCE, C. G., ROSLINGTON, J. M., TUFTON, P. J., WEBBER, H. C., and HEALY, M. (1974) *J. Cryst. Growth* **24/25**, 369.
BARDSLEY, W., GREEN, G. W., HOLLIDAY, C. H., HURLE, D. T. J., JOYCE, C. G., MACEWAN, W. R., and TUFTON, P. J. (1975) *Inst. Physics (London) Conference*, Series No. 24, 355.
BARDSLEY, W., HURLE, D. T. J., and JOYCE, C. G. (1977) *J. Cryst. Growth* **40**, 33.

BARTLETT, R. W., HALDEN, F. A., and FOWLER, J. W. (1967) *Rev. Sci. Instr.* **38**, 1313.
BASS, S. L. and OLIVER, P. E. (1968) *ICCG Birmingham*, p. 286.
BECK, A. H. (ed.) (1964) *Handbook of Vacuum Physics*, Vol. 1 and Vol. 3, Pergamon Press, Oxford.
BECQUEREL (1830) cited by WEBER, R. L. (1950) *Heat and Temperature Measurement*, Prentice-Hall, New Jersey.
BEHRISCH, R. (1964) *Ergebnisse der exakten Naturwissenschaften*, **35**, 295.
BILLS, D. G. (1973) *J. Vac. Sci. Technol.* **10**, 65.
BLUM, S. E. and CHICOTKA, R. J. (1968) *Rev. Sci. Instr.* **39**, 277.
BORING, K. L. and STAUFFER, L. H. (1963) *Proc. Nat. Elelectronics Conf. USA*, p. 535.
BRADLEY, C. C. (1969) *High Pressure Research in Solid State Research*, Butterworth, London.
BRICE, J. C., HILL, O. F., WHIFFEN, P. A. C., and WILKINSON, J. A. (1971) *J. Cryst. Growth* **10**, 133.
BRIDGMAN, P. W. (1949) *The Physics of High Pressure*, G. Bell, London.
BRITISH STANDARDS INSTITUTION (1966) *Code for Temperature Measurement*, Part 4—Thermocouples (*British Standard 1041*), London.
BROWN, D. W. (1965) *Induction Heating Practice*, Odham, U.K.
BROWN, D. W. (1967) *J. Inst. Met.* **95**. 12.
BROWN, G. H., HOYLER, C. N., and BIERWORTH, R. A. (1947) *Radio Frequency Heating*, van Nostrand, New York.
BUNSHAH, R. F. and BATZER, T. H. (1968) *Techniques of Metals Research* (ed. Bunshah, R. F.), Interscience, New York.
CABANE, G. (1958) *J. Nucl. Energy* **6**, 269.
CALDWELL, F. R. (1965) *J. Res. Nat. Bur. Stds.* **69**C, 95.
CALLANDER (1887) *Phil. Trans.* **178**, 160.
CALLANDER (1899) *Phil. Mag.* **47**, 191 and 519.
CAMPARI, M. and GARRIBA, S. (1971) *Rev. Sci. Instr.* **42**, 644.
CHATTLE, M. V. (1977) Report, National Physical Laboratory, UK, No. Qu 42.
CLASS, W. (1968) *ICCG Birmingham*, p. 241.
CLASS, W., NESOR, H. R., and MURRAY, G. T. (1966) *ICCG Boston*, p. 75.
COBBLE, M. H. (1964) *Solar Energy* **8**, 63.
COCKAYNE, B. and GASSON, D. B. (1970) *J. Materials Sci.* **5**, 837.
CORNINGS, E. W. (1956) *High Pressure Technology*, McGraw-Hill, New York.
COXON, W. F. (1962) *Temperature Measurement and Control*, Macmillan, London.
CRAIG, R. D. and HARDEN, E. H. (1966) *Vacuum* **16**, 67.
CROOKES, W. (1879) *Phil. Trans. Roy. Soc. (London)*, Part I, 135; Part II, 641.
DAUPHINÉE, T. M. (1962) *Temperature—Its Measurement and Control in Science and Industry*, Vol. 3, Part 1 (ed. Herzfeld, C. M.), Reinhold, New York, p. 269.
DAVY, H. (1809) *Phil. Trans. Roy. Soc. London*, A, **97**, 71.
DEFFET, L. and LIALINE, L. (1965) *The Physics of High Pressures and the Condensed Phase* (ed. van Itterbeek), North-Holland, Amsterdam.
DELVES, R. T. (1975) This book (1st edition), Chapter 3.
DESSAUR, R. G., PATZNER, E. J., and POPONIK, M. R. (1970) US Patent 3,493,770.
DIAMOND, J. M. (1969) *Rev. Sci. Instr.* **40**, 1477.
DIELS, K. and JAECKEL, R. (1966) *Leybold Vacuum Handbook*, Pergamon Press, Oxford.
DIGGS, T. G., HOPKINS, R. H., and SEIDENSTICKER, R. G. (1973) *J. Cryst. Growth* **29**, 326.
DOMEY, K. E. (1971) *Solid State Tech.* (Oct. 1971) **41**.
DOREMUS, C. A. (1908) *Trans. Am. Electrochem. Soc.* **13**, 347.
DRABBLE, J. (1968) *ICCG Birmingham*, p. 804.
DRABBLE, J. and PALMER, A. (1966) *J. Appl. Phys.* **37**, 1778.
DUGDALE, R. E. (1966) *J. Materials Sci.* **1**, 160.
DUSHMAN, S. and LAFFERTY, J. (1962) *Scientific Foundations of Vacuum Techniques*, 2nd edn., Wiley, New York.
EATON, N. F., GASSON, D. B., and JONES, F. O. (1961) *AEI Engineering J.*, p. 281.
EICKHOFF, K. and GURS, K. (1970) *J. Cryst. Growth* **6**, 21.
EKIN, J. W. and WAGNER, D. K. (1970) *Rev. Sci. Instr.* **41**, 1109.
EMSLIE, A. G. and BLAU, H. H. JR. (1959) *J. E. Chem. Soc.* **106**, 877.
EVANS, J. P. (1962) *Temperature—Its Measurement and Control in Science and Industry*, Vol. 3, Part 1 (ed. Herzfeld, C. M.), Reinhold, New York, p. 285.
EVANS, J. P. (1977) *Metrologia* 171.
EVANS, J. P. and BURNS, G. W. (1962) *Temperature—Its Measurement and Control in Science and Industry*, Vol. 3, Part 1 (ed. Herzfeld, C. M.), Reinhold, New York, p. 313.
FAIRCHILD, C. O. and HOOVER, W. H. (1923) *J. Opt. Soc. Am.* **7**, 543.
FOEX, M. (1964) *Proc. Internat. Conf. on Image Furance Technique, Cambridge, Mass.*, Plenum, New York.
FOLEY, G. M. (1970) *Rev. Sci. Instr.* **41**, 827.
GÄRTNER, K. J., RITTINGHAUS, K. F., SEEGER, A., and UELHOFF, W. (1971) *ICCG Marseille*, p. 619.
GEACH, G. A. and JONES, P. O. (1959) *J. Less Common Metals* **1**, 56.

GENTRY, F. E., GUTZWILLER, F. W., HOLONYAK, N., JR., and VON ZASTROW, E. E. (1964) *Silicon Controlled Rectifiers. Princples and Applications of p–n–p–n Devices*, Prentice-Hall, New Jersey.

GIARDINI, A. A. and LLOYD, E. C. (eds.) (1963) *High Pressure Measurement*, Butterworth, London.

GOLDSMID, H. J. (1964) *Applications of Thermoelectricity*, Methuen, London.

GROSS, U. and KERSTEN, R. (1972) *J. Cryst. Growth* **15**, 85.

GROVE, W. R. (1842) *Phil. Trans. Roy. Soc. London* **142**, 87.

GULSKII, B. I. (1977) *Izv. Vuz. Priborostz* **20**, 110.

HABLANIAN, M. H. and MALIAKAL, J. C. (1973) *J. Vac. Sci. Technol.* **10**, 58.

HALL, J. A. (1966) *J. Sci. Instr.* **43**, 541.

HALL, J. A. and LEAVER, V. M. (1962) *Temperature—Its Measurement and Control in Science and Industry*, Vol. 3, Part 1 (ed. Herzfeld, C. M.), Reinhold, New York.

HAMPSHIRE, M. J., PRITCHARD, T. I., TOMLINSON, R. D., and HACKNEY, C. (1970) *J. Phys. E.* **3**, 185.

HANNENMANN, R. J. (1977) *Trans. ASME*, Ser. A, **99**, 385.

HILL, J. J. and MILLER, A. P. (1963) *Proc. Inst. Elect. Engrs* **110**, 453.

HOBSON, J. P. (1973) *J. Vac. Sci. Technol.* **10**, 73.

HUBER, W. K. (1964) *Vacuum* **13**, 399.

HULM, J. K. (1954) *Phys. Rev.* **94**, 1390.

HURLE, D. T. J. (1959) Ph.D. thesis, Southampton University.

HURLE, D. T. J. (1962) *Prog. Metals Sci.* **10**, 79.

HURLE, D. T. J. (1966) *Phil. Mag.* **13**, 305.

HURLE, D. T. J. (1977) *J. Cryst. Growth* **42**, 473.

IGRAS, E., NOVAK, Z., PIOTROWSKI, J., PIOTROWSKI, T., and KILLAS, J. (1977) *Autom. Kontrola* **23**, 441.

ILLIG, W. (1968) *Microtechnic.* **22**, 531.

INSTRUMENT SOCIETY OF AMERICA, PITTSBURGH (1964) *American Standard for Temperature Measurement* C. 91–1.

IOFFE, A. F. (1957) *Semiconductor Thermoelements and Thermoelectric Cooling*, Infosearch, London.

ISELER, G. W. (1977) *J. Cryst. Growth* **41**, 146.

JARUSEK, J. (1976) *Mereni Regulace* **24**, 128.

JENSON, F. R., BUSCHWELLER, C. H., and BECK, B. H. (1972) *Rev. Sci. Instr.* **43**, 145.

JONES, R. C. (1953) *Advances in Electronics*, Vol. 5 (ed. Marton, L.), Academic Press, New York.

JORDON, R. G. and JONES, D. W. (1971) *J. Phys. E.* **4**, 245.

KAMM, G. K. (1977) *Rev. Sci. Instrum.* **48**, 463.

KANAYA, K., YAMAZAKI, H., KAWAKATSU, H., OKAZAKI, I., and SHIMZU, T. (1968) *J. Electron Microscopy* **17**, 229.

KELVIN (1848) *Math. Phys. Papers* **1**, 100.

KHAMBATTA, F. B., GIELISSE, P. J., WILSON, M. P., JR., ADAMSKI, J. A., and SAHAGIAN, C. (1971) *ICCG Marseille*, p. 710.

KING, G. D. (1970) *J. Phys. E.* **3**, 730.

KROLL, W. (1940) *Trans. E. Chem. Soc.* **78**, 35.

KU, H. H. (ed.) (1968) *Precision Measurement and Calibration—Statistical Concepts and Procedures*, National Bureau of Standards, USA.

KUTZ, M. (1968) *Temperature Control*, Wiley, New York.

KYRE, T. R. and ZYDZIK, G. (1973) *Mater. Res. Bull.* **8**, 443.

LAFFERTY, J. M. (1972) *J. Vac. Sci. Technol.* **9**, 101.

LANGMUIR, I. (1929) *Phys. Rev.* **33**, 954.

LANGMUIR, I. and COMPTON, K. I. (1931) *Rev. Modern Phys.* **13**, 191.

LASZLO, T. A. (1965) *Image Furnace Techniques*, Interscience, New York.

LEONARD, L. H. (1962) *Introduction to Electron Beam Technology* (ed. Bakish, R.), J. Wiley, New York.

LE PHAT and VINH, A. (1961) *J. Res. C.N.R.S.* **57**, 265.

LESLIE, W. H. P., ZUNTER, J. J., and ROBB, D. (1960) *Research* **13**, 250.

LEWIN, G. (1965) *Fundamentals of Vacuum Science and Technology*, McGraw-Hill, New York.

LIPSON, H. G., KAHAN, A., ADAMSKI, J. A., FARRELL, E., REDMAN, M. J., and KAWAMURA, J. (1968) *ICCG Birmingham*, p. 250.

LOPEZ, A., MARTINEZ, L., and SHAH, J. S. (1978) *Acta Scientifia Venezolana*, Proc. Annual Convention AsoVAC, Maracai, Venezuela.

LOVEJOY, D. R. (1962) *Temperature—Its Measurement and Control in Science and Industry*, Vol. 3, Part 1 (ed. Herzfeld, C. M.), Reinhold, New York, p. 487.

MACDONALD, D. K. C. (1962) *Thermoelectricity—An Introduction to the Principles*, J. Wiley, New York.

MACQUER (1758) *Mém. Acad. Sci. Paris* **119**, 133.

McELROY, D. L. and FULKERSON, W. (1968) *Techniques of Metals Research*, Vol. 1, Part 1 (ed. Bunshah, R. F.), Interscience, New York.

MEEK, J. M. and CRAGGS, J. D. (1953) *Electrical Breakdown in Gases*, Oxford University Press.

MOFFAT, R. J. (1962) *Temperature—Its Measurement and Control in Science and Industry*, Vol. 3, Part 2 (ed. Herzfeld, C. M.), Reinhold, New York.

Moisson, H. (1897) *Le Four Électrique*, Steinheil, Paris.
Monday, M. (1977) *Mess Preuf.* **12**, 809.
Monk, R. L. (1977) *Elektronik* **26**, 64.
Monk, R. L. (1977a) *Connol Instrum.* **9**, 28.
Moore, A. W., Ubbelohde, A. R., and Young, D. A. (1962) *Br. J. Appl. Phys.* **13**, 393.
Mortlock, A. J. (1958) *J. Sci. Instr.* **35**, 283.
Moulin, M., Faure, M., and Bichon, G. (1974) *J. Cryst. Growth* **24/25**, 376.
Mueller, E. F. (1916–17) *Bull. Bur. Std.* **13**, 547.
Mullin, J. B. (1975) In *Crystal Growth and Characterisation* (eds., Ueda, B. and Mullin, J. B.), North-Holland, Amsterdam.
Mullin, J. B., Heritage, R. J., Holliday, C. H., and Straughan, B. W. (1968) *ICCG Birmingham*, p. 281.
Mullin, J. B., MacEwan, W. R., Holliday, C. H., and Webb, A. E. V. (1971) *ICCG Marseille*, p. 629.
Mullin, J. B., MacEwan, W. R., Holliday, C. H., and Webb, A. E. V. (1972) *J. Cryst. Growth* **13/14**, 629.
Mullin, J. B., Straughan, B. W., and Brickell, W. S. (1965) *J. Phys. Chem. Solids* **26**, 782.
Munro, D. C. (1963) *High Pressure Physics and Chemistry*, Vol. 1 (ed. Brandley, R. S.), Academic Press, New York.
Muth, S. Jr. (1967) *Instruments and Control Systems* **40**, 133.
Nakaya, S. and Uchiyama, H. (1962) *Comptes rendus des séances du Comité Consultatif de Thermométrie*, Comité International des Poids et Mesures, 6ᵉ Session, Sèvres.
Nelson, H. (1969) *Electrotechnology* **183**, 43.
Nielson, I. O. (1970) *Solid State Technology* **13**, 33.
Noguchi, T. and Kozuka, T. (1966) *Solar Energy* **10**, 203.
Nutter, G. D., Wike, R. B., and Bollerman (1967) *Instruments and Control Systems* **40**, 96.
Nygren, S. F. (1973) *J. Crys. Growth* **19**, 21.
Okada, T., Matsumi, K., and Makino, H. (1971) *Solid State Physics Japan* **6**, 170.
O'Kane, D. F., Kwap, T. W., Gulitz, L., and Bednowitz, A. L. (1971) *ICCG Marseille*, E3–8.
Owen, J. R. and White, E. A. D. (1977) *J. Cryst. Growth* **42**, 499.
Patel, C. K. N. (1964) *Phys. Rev. Letters* **12**, 588 and **13**, 617.
Patzner, E. J., Dessaur, R. G., and Poponik, M. R. (1967) *SCP Solid State Tech.* (Oct. 1967), **25**.
Pirani, M. (1907) US Patent 848,600.
Pirani, M. and Yarwood, J. (1951) *Principles of Vacuum Engineering*, Reinhold, New York.
Pouillet (1836) *Comptes Rendus Acad. France* **3**, 786.
Prewbrazhenskii, V. P. and Letskas, V. G. (1978) *Thermal Engng (GB)* **24**, 73.
Quinn, T. J. and Barber, C. R. (1967) *Metrologia* **3**, 19.
Quinn, T. J. and Ford, M. C. (1969) *Proc. Roy. Soc. A.* **312**, 31.
Quinn, T. J., Guildner, L. A., and Thomas, W. (1978) *Metrologia* **13**, 177.
Rabenau, A. (1964) *Chem. Ingn. Tech.* **36**, 542.
Redhead, P. A., Hobson, J. P., and Kornelsen, E. V. (1968) *The Physical Basis of Ultra High Vacuum*, Chapman and Hall, London.
Reed, R. E. (1972) *J. Vac. Sci. Technol.* **9**, 141.
Reed, T. B. (1961) *J. Appl. Phys.* **32**, 821 and 2534.
Reed, T. B. (1967a) *Advances in High Temperature Chemistry*, Vol. 1 (ed. Eyring, A. P.).
Reed, T. B. (1967b) *Materials Res. Bull.* **2**, 349.
Reed, T. B. and Pollard, E. R. (1968) *J. Cryst. Growth* **2**, 243.
Reinhart, R. C. and Yatsko, M. O. (1974) *J. Cryst. Growth* **21**, 283.
Ricoli, T. and Lanza, F. (1977) *High Temp. High Pressure* **9**, 483.
Roberts, R. W. and Vanderslice, T. A. (1963) *Ultra High Vacuum and Its Applications*, Prentice-Hall, New Jersey.
Roeser, W. F. (1941) *Temperature—Its Measurement and Control in Science and Industry* (eds. Fairchild, C. O., et al.), Reinhold, New York.
Roots, W. K. (1969) *Fundamentals of Temperature Control*, Academic Press, New York.
Rozhdestvenskii, A. B. (1977) *High Temp.* **15**, 377.
Ruffino, G. (1977) *High Temp. High Pressure* **9**, 253.
Sakurai, T. and Ishigame, M. (1968) *J. Cryst. Growth* **2**, 284.
Sakurai, T., Kamada, O., and Ishigame, M. (1968) *J. Cryst. Growth* **2**, 326.
Savakova, A. (1978) *J. Soc. Instrum. Control Engng* **17**, 81.
Say, M. G. (ed.) (1954) *Magnetic Amplifiers and Saturable Reactors*, G. Newnes, London.
Seeback, T. J. (1823) *Gilb. Ann.* **73**, 115.
Seeback, T. J. (1826) *Pogg. Ann.* **6**, 133.
Selman, G. L. and Rushforth, R. (1971) *Platinum Met. Rev.* **15**, 82.
Shah, J. S. (1967) PhD thesis, University of Bath.
Shah, J. S. (1974), This book (1st edition), Chapter 4.

SHAH, J. S. (1981) Zone refining and its applications, this book, Chapter 8.

SHAH, J. S. and BROOKBANKS, D. M. (1972) *Platinum Metals Rev.* **16**, 94.

SHAH, J. S. and GALINDO, H. (1979) *Proc. 5th Latin American Conference on Solid State, Bogota, Colombia.*

SHAH, J. S. and HUNTLEY, D. A. (1970) *J. Cryst. Growth* **6**, 216.

SIEMENS, W. (1871) Bakerian Lecture, *Proc. Roy. Soc.* **19**, 351.

SIMPSON, P. G. (1960) *Induction Heating*, McGraw-Hill, New York.

SMITH, F. E. (1912) *Phil. Mag.* **24**, 541.

SMITH, R. A., JONES, F. E., and CHASMAR, R. P. (1968) *The Detection and Measurement of Infra-Red Radiation*, (2nd edn.), Oxford University Press.

SOBATKA, H. (1963) *Industrial HF Generators*, Philips Technical Library, Eindhoven.

STARR, C. D. and WANG, T. P. (1963) *Proc. Am. Soc. Testing Materials* **63**, 1184.

STECKELMACHER, W. (1965) *J. Sci. Instr.* **42**, 63.

STERLING, H. F. and WARREN, R. W. (1963) *Metallurgia* **67**, 301.

STOREY, R. N. and LAUDISE, R. A. (1970) *J. Cryst. Growth* **6**, 261.

TAKAHASHI, M., NANAMUTSU, S., and KIMURA, M. (1971) *ICCG Marseille*, E1–3.

TAKAI, H. (1978) *J. Cryst. Growth* **43**, 463.

TANSLEY, T. L. and NEWMAN, P. C. (1967) *Solid State Electron.* **11**, 497.

THOMPSON, R. D. (1962) *Temperature—Its Measurement and Control in Science and Industry*, Vol. 3, Part 1 (ed. Herzfeld, C. M.), Reinhold, New York.

TOPPING, J. (1957) *Errors of Observations and their Treatment*, The Institute of Physics, London.

TROMBÉ, F. (1950) *Les Hautes Températures et Leur Utilisation en Chimie* (ed. Lebeau, P.), Masson et Cie, Paris.

TUROVSKI, B. M., SHENDROVICH, L. L., and SHUBUKII, G. I. (1977) *Inorgan. Mater.* **13**, 948.

VAN DER WALL, C. W. and STRUIK, L. C. E. (1969) *J. Phys. E.* **2**, 143.

VAN DIJK, H. J. A., JOCHREN, C. M. G., SCHOLL, G. J., and VANIDER WERF, P. (1974) *J. Cryst. Growth* **21**, 310.

VAN PAASEN, L. L., MULEY, E. C., and ALLEN, R. J. (1962) *Proc. Nat. Electron. Conf. USA*, p. 590.

WARREN, R. W. (1962) *Rev. Sci. Instr.* **33**, 1378.

WEATHERLY, M. H. and ANDERSON, J. E. (1965) *Electrochem. Technol.* **3**, 80.

WEBBER, H. C. and HISCOCKS, S. E. E. (1973) *J. Mat. Sci.* **8**, 294.

WENKUS, J. F. and DOHERTY, P. R. (1968) *ICCG Birmingham*, p. 301.

WINSEMIUS, P. and LENGKEEK, H. P. (1973) *Rev. Sci. Instr.* **44**, 229.

WOLFF, E. G. (1969) *Rev. Sci. Instr.* **40**, 544.

YAMADA, T. and NOGUCHI, T. (1978) *Rept. Govt. Ind. Res. Inst., Nagoya, Japan* **26**, 302.

ZIMAN, J. M. (1963) *Electrons and Phonons*, Clarendon Press, Oxford.

ZINNES, A. E., NEVIS, B. E., and BRANDT, C. D. (1973) *J. Cryst. Growth* **19**, 187.

CHAPTER 5

Vapor Phase Growth

G. B. STRINGFELLOW
Hewlett-Packard Labs., Palo Alto, California, USA

5.1. Introduction

In any crystal growth process nutrient must be supplied to the growing interface through some external phase. At the melting point of the crystal this external phase is the molten element or compound being grown. At lower temperatures the external phase may be either liquid, vapor, or even solid. This chapter will concentrate on growth processes where transport occurs through the vapor phase.

Deposition from the vapor phase is in most cases the preferred technique for the fabrication of thin layers of metal, insulator, and semiconductor materials. Vapor phase deposition is convenient for large scale operations (from the coating of turbine blades to growing epitaxial layers of semiconductor materials), can be used to coat irregular-shaped substrates, including growth on inside surfaces, and offers maximum control of materials properties such as thickness and composition. Another practical advantage is that it does not involve the contacting of the growing surface with a liquid or solid phase, thus avoiding numerous potential problems during and after the growth process.

Both *physical vapor deposition* (PVD) and *chemical vapor deposition* (CVD) are widely used for the deposition of polycrystalline oxide and metal films on parts for improvement of mechanical and structural properties, corrosion, and abrasion resistance.

The vapor phase process is used for the growth of bulk crystals only in special cases because of its inherently slow growth rate, normally $<10^{-4}$ cm min^{-1} for single crystal growth as compared to $\sim 10^{-2}$ cm min^{-1} for growth from the melt. Thus, for example, in the semiconductor industry, where the use of vapor phase growth is widespread, epitaxial layers of Si, GaAs, GaAs$_{1-x}$P$_x$, Al$_x$Ga$_{1-x}$As, etc., are grown by the vapor phase epitaxial (VPE) technique onto substrates sliced from large boules of material grown from the melt. Melt growth is more efficient for the production of large quantities of material but the superior perfection and control of thickness, composition, stoichiometry, and doping available using VPE frequently makes this technique superior for the growth of "active" layers in which the electronic functions are actually performed. Vapor phase techniques are also used for deposition of insulator (SiO$_2$, Si$_3$N$_4$, etc.) and metal films necessary for device and integrated circuit fabrication.

A partial listing of commonly used vapor phase growth processes is found in Fig. 5.1. This list is by no means complete but gives an idea of the variety of VP systems which have been developed. The two major categories are PVD and CVD. The former category

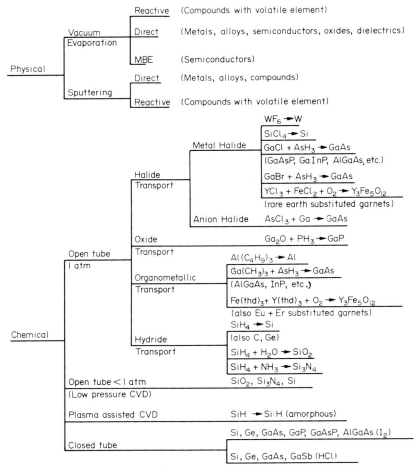

FIG. 5.1. Vapor phase growth systems.

involves both vacuum evaporation, where thermal energy is supplied by resistive or electron beam heating of the source, and sputtering, where source material is passed into the vapor phase by ion bombardment of a source electrode (target).

Traditional vacuum evaporation may involve either direct evaporation of the source material onto a substrate or reactive evaporation where the evaporated material reacts with a component intentionally added to the vapor phase to produce the desired deposit. This technique is frequently employed where direct evaporation of compounds yields deposits deficient in nonmetallic species. Thus, the evaporation is carried out in an ambient containing the deficient species in gaseous form. This topic has recently been reviewed by Bunshah (1974, 1976).

For the growth of semiconductor quality single crystalline epitaxial layers, the supply of evaporated species must be extremely well controlled. A relatively new and elegant evaporation technique, molecular beam epitaxy (MBE), has been developed for this purpose. In this technique molecular beams of the constituent elements are produced in

Knudsen cells and directed onto a carefully cleaned substrate under ultra high vacuum conditions. This technique can be used to grow high quality epitaxial layers of Si, GaAs, and other semiconductors at temperatures as low as 500–600°C (Cho, 1977). The slow growth rate, precise control of the supply of nutrients to the growing interface, and absence of appreciable solid state diffusion at such low growth temperatures makes this the most powerful technique yet developed for the growth of ultra thin ($\leq 10\,\text{Å}$) layers with controlled thickness, composition, and doping level (Gossard, 1976). The topic of vacuum evaporation, and in particular MBE, is the subject of a separate chapter in this volume (Chap. 6) and so will not be dealt with further here.

The control of composition, which is so difficult by vacuum evaporation, because of the more volatile species, is improved by sputtering of alloys or compounds. To achieve high deposition rates the ion source is generally a plasma discharge maintained adjacent to the target in an atmosphere of heavy ions such as Ar or Kr. Magnetic confinement of the plasma (using magnetrons) is commonly used to maintain sufficiently intense plasmas. In addition to superior control of composition, sputtering has the advantage over vacuum evaporation of being capable of coating larger areas. The adhesion is also frequently improved by the *in situ* cleaning of the substrate by the plasma. As in direct evaporation processes, sputtering may be carried out in an atmosphere containing one or more species in the gas phase (reactive sputtering). For more detailed information on sputtering the reader is referred to recent excellent review articles by Thornton (1977) and Francomb (1975).

The number of systems listed under CVD in Fig. 5.1 is much larger than for PVD. This does not reflect the relative importance of the two categories but is indicative of the variety of transport agents and systems available for CVD. A relatively few systems can be used to produce a great variety of metal, alloy, semiconductor, and insulator layers by PVD while the systems listed under CVD represent in some cases stages in the development of the important systems of today or systems developed for special needs where conventional techniques were unsuccessful. The list is by no means inclusive, but items from three important areas, deposition of metals, semiconductors, and insulators are included. Also examples are given for the most common transport agents, namely volatile halide, oxide, and organometallic compounds. A distinction is made between CVD, usually of polycrystalline or amorphous films, and VPE growth of single crystalline films.

At the present time the SiH_4 process for the growth of Si epitaxial layers is used almost exclusively in the semiconductor industry. Three types of systems—the horizontal reactor (Fig. 5.2a), the vertical pancake reactor (Fig. 5.2b), and the vertical barrel type reactor (Fig. 5.2c)—are used in production. The largest of these, the barrel type reactors, can accommodate 30 76 mm (3 in.) diameter Si substrates. This is an area of $> 1300\,\text{cm}^2$ of Si per run! High quality SiO_2 and Si_3N_4 are also produced in similar reactors using SiH_4 and CO_2 or NH_3 respectively.

In an effort to improve uniformity and volume, the low pressure CVD technique has been developed (Rosler, 1977) where growth takes place at a pressure of the order of 1 torr. Plasma-assisted CVD is also a recently developed process. The plasma is used to decompose the SiH_4 and energize adsorbed surface atoms to allow growth of hydrogenated amorphous Si films at very low temperatures (200–500°C) (Brodsky, 1978). Such films appear promising for solar cell applications (Carlson, 1977).

The growth of compound semiconductors and alloys, such as GaAs, $GaAs_{1-x}P_x$, and $Al_xGa_{1-x}As$ is another area where VPE growth is important. These materials are

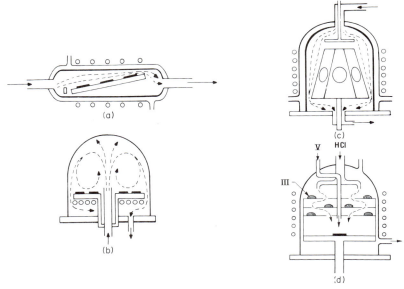

Fɪɢ. 5.2. Schematic diagrams of r.f. heated systems commonly used for VPE of semiconductors: (a) horizontal cold wall, (b) vertical (pancake) cold wall, (c) barrel cold wall, and (d) vertical hot wall.

important both for optoelectronic devices: solar cells, light-emitting diodes, injection lasers, radiation detectors, etc., and for high speed transistors. These III V compounds and alloys are usually grown by chloride transport techniques using either metal chlorides formed by HCl at high temperatures in the reactor, as shown in Fig. 5.2d, or using the group V chlorides. Especially for compounds and alloys containing Al, growth using simple organometallic sources such as trimethyl- or triethyl-metal compounds are being actively investigated (Dupuis and Dapkus, 1977; Stringfellow, 1978b).

The approach to vapor phase growth to be used in this chapter is to stress the fundamental aspects of the process. Specific systems discussed above, for which the fundamental growth mechanisms have been investigated, will be used as examples.

The fundamental aspects of vapor phase crystal growth might be divided into four major areas: (1) thermodynamics, which controls the driving force for the chemical reaction occurring at the vapor solid interface; (2) mass transport, by which the reactants reach the growing surface and products are removed; (3) surface kinetics including adsorption of reactants onto the surface, surface diffusion, step generation, surface chemical reactions between adsorbed reactants at steps, and desorption of products; (4) structural aspects of growth, especially epitaxial growth such as defect generation and surface morphology of homoepitaxial and heteroepitaxial layers.

A summary of the fundamental processes, which must be dealt with in any attempt to understand vapor growth, is shown in Fig. 5.3. The vapor growth process is extremely complex for practical systems. The fundamental concepts were introduced in Chapter 2. In this chapter we shall review the fundamental aspects of VPE and the simplified models which can be used as a basis for understanding and predicting the characteristics of specific vapor growth systems.

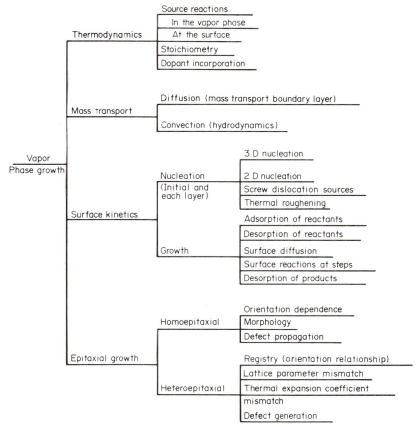

F:G. 5.3. Summary of fundamental processes in vapor phase growth.

5.2. Thermodynamics

Thermodynamics is the first important fundamental consideration for vapor phase growth because it defines the maximum rate of growth, determined as input rate of reactants multiplied by the equilibrium reaction efficiency. For example, for the process

$$GaCl(g) + AsH_3(g) = GaAs(s) + HCl(g) + H_2(g) \qquad (5.1)$$

the reaction efficiency is defined as

$$\eta = p_{HCl}/p_{GaCl}^{\circ}$$

where p° represents the input partial pressure. The growth rate may never exceed the thermodynamically limited growth rate and is often an order of magnitude smaller, limited by mass transport to the growing interface or surface kinetics. In addition, thermodynamics often determines composition (for the growth of alloys) and doping level. In many cases thermodynamics is also thought to control stoichiometry of compounds grown from the vapor phase.

5.2.1. SiCl$_4$ GROWTH OF Si

The VPE growth of Si using either SiCl$_4$ or SiH$_4$ is usually carried out in either a horizontal (Fig. 5.2a) or vertical (Fig. 5.2b, c) cold wall reactor. Only the substrate holder is heated since the endothermic deposition reaction would occur at heated walls. The reaction for deposition of epitaxial and polycrystalline layers of Si from SiCl$_4$ is

$$SiCl_4(g) + 2H_2(g) = Si(s) + 4HCl(g) \tag{5.2}$$

which occurs heterogeneously at the growing interface. One might expect the Si deposition rate to be proportional to $p_{SiCl_4}^{\circ}$, the silicon tetrachloride partial pressure entering the reactor, multiplied by the total flow velocity. In an early investigation of this system Theuerer (1961) discovered that this does occur at low partial pressures, but etching of the Si can actually occur for high partial pressure of SiCl$_4$. In Fig. 5.4 growth rate is plotted vs. p_{SiCl_4} including data from Theuerer (1961), Bylander (1962), and Alexander (1967). The data show the same general trend but parameters other than simply substrate temperature seem to affect the value of SiCl$_4$ partial pressure above which etching is observed. Such behavior is not consistent with reaction (5.2) being the only reaction occurring. Theurer (1961) pointed out that the 20% yield of SiHCl$_3$ for a 10% yield of Si indicated more complex reactions to be occurring.

Steinmaier (1963) first suggested that the reaction to form SiCl$_2$, i.e.,

$$SiCl_4(g) + Si(s) = 2SiCl_2(g), \tag{5.3}$$

was thermodynamically favorable and would in fact qualitatively explain the occurrence of etching at high SiCl$_4$ partial pressures. A complete thermodynamic calculation was carried out by Sirtl *et al.* (1974) using the best thermodynamical data compiled and critically analyzed by Hunt and Sirtl (1972). For a three-component system with two phases the phase rule allows three degrees of freedom. Thus specifying temperature,

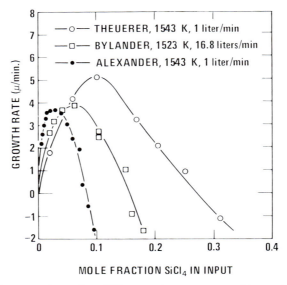

FIG. 5.4. Silicon growth rate from SiCl$_4$ versus input gas composition as determined by Theuerer (1961), Bylander (1962), and Alexander (1967). (From M. E. Jones, 1969.)

pressure (1 atmosphere for the data considered here) and the Cl/H ratio, which does not change as the reaction proceeds, completely specify the system at equilibrium. Sirtl et al. (1974) considered the gaseous species $SiCl_4$, $SiCl_3H$, $SiCl_2H_2$, $SiClH_3$, SiH_4, HCl, $SiCl_2$, SiCl, $SiCl_3$, Cl, Si, Cl_2 and Si_2Cl_6. The set of mutually independent mass action expressions for the formation of these compounds plus the condition for conservation of Cl and the constraint that total pressure equal 1 atmosphere gives a set of equations which when solved simultaneously yields a unique value of partial pressure for each of the species. Alternatively the free energy of the system could be minimized numerically (van der Putte, 1975) to give the same result. The resulting partial pressures calculated by Sirtl et al. (1974) for Cl/H = 0.10 are plotted versus temperature in Fig. 5.5. These partial pressures exist only in equilibrium with the substrate, thus a thermodynamic growth rate R can be calculated based on the assumption of equilibration with the substrate of the total volume of gas entering the reactor:

$$R = \eta \, p^\circ_{SiCl_4} F_T \tag{5.4}$$

where $p^\circ_{SiCl_4}$ is the input partial pressure of $SiCl_4$ before any reactions occur, F_T is the total flow rate, and η, the thermodynamic efficiency, is defined as

$$\eta = \frac{p^\circ_{SiCl_4} - \sum x p_{Si_xH_yCl_z} \text{ (eq.)}}{p^\circ_{SiCl_4}} \tag{5.5}$$

which is zero if the total amount of Si entering the reactor as $SiCl_4$ remains in the gas phase and unity if no Si remains in the gas phase at equilibrium. The growth rate versus Cl/H ratio curves calculated for three values of substrate temperature are plotted in Fig. 5.6. The magnitude of the growth rate calculated in this manner is generally higher than experimental growth rates by a factor ranging from 2 to 20. Sedgwick (1964) proposed a

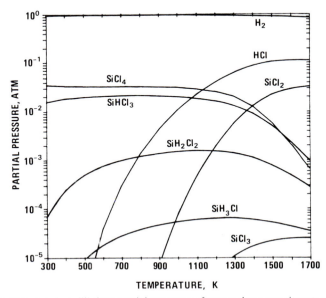

FIG. 5.5. Calculated equilibrium partial pressures of vapor phase constituents at 0.1 atm total pressure and Cl/H = 0.10 in the $SiCl_4/H_2$ system. (From Sirtl et al., 1974.)

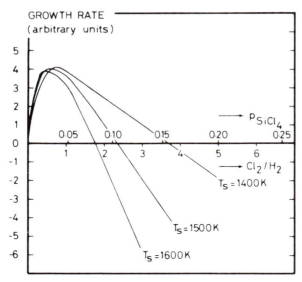

Fig. 5.6. Calculated silicon growth rate from $SiCl_4$ versus input gas composition at 1400 K, 1500 K and 1600 K. (From van der Putte *et al.*, 1975.)

quasi-equilibrium model where an additional term β (ranging from 1/2 to 1/20) is included in eqn. (5.4) to reflect the fact that not all of the flowing gas stream would be able to equilibrate with the substrate in most systems, due to relatively slow mass transport in the gas stream. These mass transport limitations will be dealt with more fully in the next section.

The calculated results shown in Fig. 5.6 are qualitatively in agreement with the experimental data of Fig. 5.4. A bothersome discrepancy is that the maximum growth rate peak and the value of Cl/H at which etching begins are considerably underestimated in the thermodynamic calculation.

Since the thermodynamic equilibrium is established at the Si surface the possibility exists that the surface reactions or mass transport are so sluggish as to impede the attainment of equilibrium in the gas phase. An important series of experiments was performed by Ban and Gilbert (1975a, b, 1978) in which a mass spectrometer was introduced into the system in order to sample and analyze the content of the gas stream. The initial experiments (Ban and Gilbert, 1975a) were carried out in a hot wall reactor at very slow gas velocities. The Si was deposited (or etched) at the tube walls. The vapor phase in this system would be expected to be nearer equilibrium than in conventional Si reactors because of the longer residence time in the reactor and the larger surface area of silicon. The results for inputs of $SiCl_4$, $SiCl_3H$, and $SiCl_2H_2$ were similar in that for temperatures above 1000 K the vapor phase composition was similar to that predicted from thermodynamics. The discrepancies were that the concentrations of HCl, $SiCl_2$, and $SiCl_2H_2$ tended to be higher and $SiCl_3H$ and $SiCl_4$ lower than predicted. Their data for $SiCl_2$ and $SiCl_4$ are plotted in Figs. 5.7 and 5.8. These discrepancies were interpreted as due to the formation of the various species at the Si interface being impeded by insufficient adsorption of HCl, thus inhibiting formation of species with more Cl and favoring species with fewer Cl atoms in the molecule.

In a system more typical of a commercial Si reactor, Ban and Gilbert (1975b, 1978) found the gas stream not to be homogeneous in temperature, composition, or flow

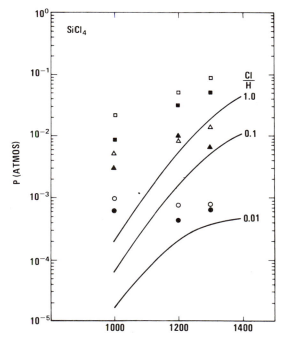

FIG. 5.7. Comparison between calculated and experimental (mass spectrometer) pressures of $SiCl_2$ versus temperature for the $SiCl_4/H_2$ system with $Cl/H = 1$ ($\square$), 10^{-1} ($\triangle$), and 10^{-2} ($\bigcirc$). Filled symbols indicate HCl initially present, open symbols HCl initially absent. (From Ban and Gilbert, 1975b.)

pattern. They observed a stagnant boundary layer 1–1.5 cm thick in which steep temperature and composition gradients were formed. In this case growth would be limited by mass transport through the boundary layer.

van der Putte (1975) extended his thermodynamic calculations to include diffusion across this boundary layer. The case where all diffusion coefficients are equal can be shown to be equivalent to the Sedgwick quasi-equilibrium model (Shaw, 1975). Using realistic diffusion coefficients for each species the calculated results are found to agree much better with experimental data. Taking into account thermodiffusion the calculations can be made to agree with the data very well. Discrepancies between the data of various authors are then attributed to small differences in the temperature gradients in the boundary layer which would be expected to change from reactor to reactor. The various aspects of mass transport will be dealt with in more detail in section 5.3.

In all of the above discussion it has been assumed that equilibrium is established only by heterogeneous reactions at the Si surface. Recent experiments by Smith and Sedgwick (1977) indicate that $SiCl_2$ may be formed by homogeneous reaction above the substrate presumably by the reaction

$$SiCl_4(g) + H_2(g) = SiCl_2(g) + 2HCl(g). \tag{5.6}$$

They found by inelastic light scattering measurements versus distance from the substrate that a maximum in p_{SiCl_2} occurred at a position in the boundary layer away from the substrate. This could occur only if homogeneous reactions were important. This finding

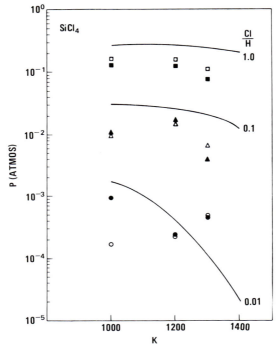

FIG. 5.8. Comparison between calculated and experimental (mass spectrometer) pressures of $SiCl_4$ versus temperature for the $SiCl_4/H_2$ system with $Cl/H = 1$ ($\square$), 10^{-1} ($\triangle$), and 10^{-2} ($\bigcirc$). Filled symbols indicate HCl initially present, open symbols HCl initially absent. (From Ban and Gilbert, 1975b.)

could have a significant impact on the correct treatment of mass transport which in this system is virtually always assumed to occur by diffusion of $SiCl_4$ from the main gas stream where pressure is $p_{SiCl_4}^{o}$ to the substrate where $p_{SiCl_4} \approx 0$, with no decomposition of $SiCl_4$ except at the substrate.

5.2.2. COMPOSITION OF III V ALLOYS

A much more complex VPE system thermodynamically is the growth of III V semiconductor alloys. For exploratory growth of these alloys, thermodynamics has proven to be a useful guide to the prediction of alloy composition versus growth variables. These calculations, particularly the treatment of the solid phase free energy of mixing, follow the successful prediction of composition for liquid phase epitaxial growth of III V alloys (Ilegems and Pearson, 1969; Stringfellow and Greene, 1969; Panish and Ilegems, 1972; Stringfellow, 1974).

A typical vertical III V growth system is shown in Fig. 5.2d. This particular system was developed for the growth of $GaAs_{1-x}P_x$ alloys. The basic growth reactions are:

$$GaCl(g) + AsH_3(g) = GaAs(s) + HCl(g) + H_2(g) \tag{5.7}$$

$$GaCl(g) + PH_3(g) = GaP(s) + HCl(g) + H_2(g). \tag{5.8}$$

The general method of performing thermodynamic calculations for the case of III V alloys is similar to that described above for the $SiCl_4$ analysis. For a system with n vapor species plus the solid composition as variables, $n + 1$ equations are necessary. The $n - 3$ equations representing mass action equilibria plus 4 conservation equations, one each for total pressure, Cl/H, solid stoichiometry, and solid composition, are used to calculate the equilibrium solid and vapor compositions. One difference between the III V systems and Si is that the input reactants to the deposition zone of the reactor are actually formed in the source zone where HCl reacts with the group III elements to form the volatile halides. Thus a thermodynamic calculation is first carried out for the source zone. The concentrations of group III chlorides, HCl, and group V elements formed by the source reactions are then used as the input variables for the calculation carried out at the vapor–solid interface.

A second difference is related to the solid solution. In the mass action expressions the activity coefficients in the solid phase are not unity and a simple model is used to calculate them. Two models have been used: the quasi-chemical equilibrium (QCE) (Stringfellow and Greene, 1969) and the regular solution (Ilegems and Pearson, 1969) models. These models treat the configurational free energy of the solid considering only nearest neighbor interactions (on the same sublattice). The simpler regular solution model assumes a random distribution of atoms on the sublattice where mixing occurs. This results in the following expressions for the activity coefficients:

$$\gamma_A = \exp\left(\frac{(1 - x)^2 \, \Omega_{AC-BC}}{RT}\right) \tag{5.9}$$

$$\gamma_B = \exp\left(\frac{x^2 \, \Omega_{AC-BC}}{RT}\right). \tag{5.10}$$

The values for the interaction parameters of various III V systems have been obtained by fitting the calculated liquid–solid phase diagrams to experimental data. For systems where data are not available a simple, fairly accurate method of calculating interaction parameters, the DLP model (Stringfellow, 1974), has been developed. This model is based on the Phillips–van Vechten (1970) dielectric theory of electronegativity (Phillips, 1973), which may be used to pedict bonding energy in semiconductor systems. The bond strength is strongly related to the lattice parameter a_0; thus the value of Ω is found to be a function of the difference in lattice parameters of the components Δa_0,

$$\Omega = 5.07 \times 10^7 \, \Delta a_0^2 / [(a_{AC} + a_{BC})/2]^{4.5}. \tag{5.11}$$

This expression is qualitatively in accord with what we expect for solid solutions, that for $\Delta a_0 \approx 0$ the solution is nearly ideal and for large values of Δa_0 immiscibility results. Listed in Table 5.1 are calculated and experimental values for several systems where experimental data exist.

Thermodynamic analyses of solid composition versus vapor composition have been published for the systems $GaAs_{1-x}P_x$ (Manabe and Gejyo, 1971; Ban, 1971; Mullin and Hurle, 1973), $InAs_{1-x}P_x$ (Mullin and Hurle, 1973), and $In_{1-x}Ga_xAs$ (Nagai, 1971; Mullin and Hurle, 1973; Kajiyama, 1976). Generally good agreement between calculated and experimental results is obtained. Here we shall use two systems $InAs_{1-x}P_x$ and $In_{1-x}Ga_xAs$ as examples, using the calculated results of Mullin and Hurle (1973) and the experimental data of Tietjen et al. (1969) for $InAs_{1-x}P_x$ and Conrad et al. (1967) for

TABLE 5.1. *Comparison of Experimental Values of Ω^s with Those Calculated Using the DLP Model*

System	$\Delta a_0/\bar{a} \times 10^2$	Ω (exp)[a]	Ω (Calculated DLP Model)
AlAs–GaAs	0.159	0	0
AlAs–InAs	6.76	2500	2814
AlSb–GaSb	0.654	0	23
AlSb–InSb	5.45	600	1456
GaP–GaAs	3.42	400, 1000[b]	985
GaP–InP	7.39	3500, 3250[c]	3630
GaAs–GaSb	7.54	4500, 4000[c]	3355
GaAs–InAs	6.92	1650,[c] 3000, 2000[d]	2815
GaSb–InSb	6.09	1475,[c] 1900	1846
InP–InAs	3.39	400	583
InAs–InSb	6.72	2900,[d] 2250	2289
GaP–GaN	18.9	23000[e, h]	28900[f]
GaP–BP	18.2	<30910[e, g, h]	26800[f]
GaP–GaBi	14.5	<28000[e, i]	14600[f]
Ge–Si	4.09	1200[j]	1190
Si–Sn	17.6	19530[k]	18500
Ge–Sn	13.5	7552[k]	10400

[a] Except where noted Ω obtained from M. B. Panish and M. Ilegems (1972) in *Progress in Solid State Chemistry*, Vol. 7 (H. Reiss and J. O. McCaldin, eds.), Pergamon Press, New York.
[b] G. A. Antypas (1970) *J. Electrochem. Soc.* **117**, 700.
[c] L. M. Foster (1972) *J. Electrochem. Soc.*, Extended Abstracts, Houston Mtg., p. 147.
[d] G. A. Antypas (1970) *J. Electrochem. Soc.* **117**, 1393.
[e] H* obtained from solubility limit.
[f] H* calculated: G. B. Stringfellow (1974) *J. Crystal Growth* **27**, 21.
[g] G. B. Stringfellow (1972) *J. Electrochem. Soc.* **119**, 1780.
[h] E. C. Lightowlers (1972) *J. Electron. Mat.* **1**, 39.
[i] F. A. Trumbore, M. Gershenzon, and D. G. Thomas (1966) *App. Phys. Lett.* **9**, 4.
[j] F. A. Trumbore, C. R. Isenberg, and E. M. Porbansky (1958) *J. Phys. Chem. Solids* **9**, 60.
[k] C. D. Thurmond and M. Kowalchik (1960) *Bell System Tech. J.* **39**, 169.

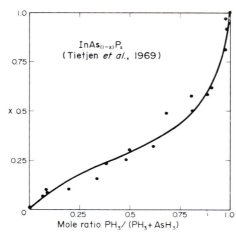

FIG. 5.9. Comparison of experimental and calculated compositions for the system $InAs_{1-x}P_x$. (From Mullin and Hurle, 1973.)

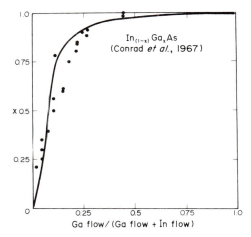

FIG. 5.10. Comparison of experimental and calculated compositions for the system $In_{1-x}Ga_xAs$. (From Mullin and Hurle, 1973.)

$In_{1-x}Ga_xAs$. The results are plotted in Figs. 5.9 and 5.10. The two systems have relatively small deviations from ideality with values of $\Omega_{InAs-InP} = 400\,cal\,mole^{-1}$ and $\Omega_{InAs-GaAs} = 2500\,cal\,mole^{-1}$ used in the calculations.

The GaP–GaN system is important from a commercial point of view because N doped GaP is used for green LEDs. Because of the large difference between the covalent radii of P and N the interaction parameter is $23{,}000\,cal\,mole^{-1}$. Thus N has only a limited solubility in GaP. The distribution coefficient is defined as x_{GaN}/p_{NH_3} since NH_3 is used in the vapor phase to incorporate N into the solid. In Fig. 5.11 the ratios of experimental to calculated values for the N distribution coefficient are plotted versus growth temperature (Stringfellow, 1975). In the high temperature range above 840°C the growth process is controlled by thermodynamics and mass transport, and as can be seen the N distribution coefficient appears to be thermodynamically controlled. For lower substrate temperatures the growth process is controlled by surface kinetics and the experimental distribution coefficient is significantly larger than that calculated thermodynamically.

To investigate the degree to which equilibrium thermodynamics truly predicts the vapor composition, Ban and coworkers (1971, 1973) used *in situ* sampling for mass spectrometric analysis in the $GaAs_{1-x}P_x$ and $In_{1-x}Ga_xP$ systems. The experiments were performed reacting HCl with Ga and/or In at $T > 900°C$ in the source zone and reacting the group III chlorides with the As and P species produced by the decomposition of AsH_3 and/or PH_3. They found several departures from equilibrium. In the source zone thermodynamic calculations predict nearly 100% conversion of HCl to the group III chlorides. They found considerably lower conversion efficiencies although this was probably related to mass transport limitations in their reactor since this quantity is routinely monitored by measuring weight loss at the source and found to be nearly 100%. They also found that PH_3, which should decompose nearly completely to P_2 and P_4, was present in the deposition zone at concentrations higher than thermodynamically predicted. In general these deviations have little effect on the equilibrium established in the deposition zone. In fact the calculations of solid composition agreed well with experimental data.

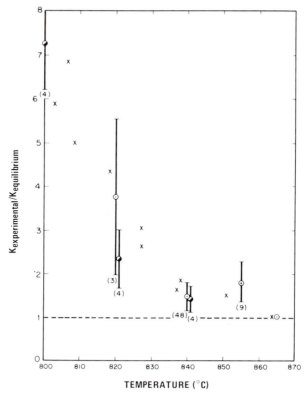

Fig. 5.11. Experimental nitrogen distribution coefficient versus temperature for VPE growth of $GaP_{1-x}N_x$ alloys. The data are from Weiner (●), Lindquist (⊕), and Burd (×). (From Stringfellow *et al.*, 1975.)

The accuracy of thermodynamic calculations of solid composition is somewhat surprising in view of the importance of mass transport and surface kinetics in determining growth rate. Mullin and Hurle (1973) speculate that even though kinetic terms may limit the rate at which atoms can be incorporated into the solid, the large energy terms involving composition determine the arrangement of these atoms on the surface and hence in the solid and thus control composition.

5.3. Mass Transport

As discussed in section 5.2, thermodynamics plays a key role in determining the composition of alloys grown by CVD and places an upper limit on growth rate. However, for most CVD growth processes, where the flow velocity is not slow enough to allow equilibrium to be established between the solid and the entire vapor phase, mass transport of reactant to the interface and products away from the interface plays an important role. This is well illustrated by comparing growth rates in two systems with identical concentrations of Si, Cl, and H in the input gas stream, one introduced as $SiCl_4$ and the other as $SiH_4 + HCl$. The resulting growth rate is plotted versus reciprocal temperature in

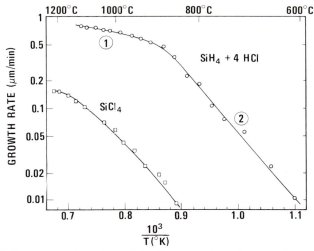

FIG. 5.12. Temperature dependence of the silicon growth rate for 0.1 mol % SiH_4 in H_2 and 0.1 mol % SiH_4 + 0.4 mol % HCl in hydrogen. (From Bloem, 1975.)

Fig. 5.12. The markedly faster growth rate obtained using SiH_4 is interpreted as being due to the more rapid diffusion of SiH_4 (Bloem, 1975).

The rigorous solution of mass transport problems in practical vapor phase growth systems is difficult. To treat thermal and forced convection in a flowing fluid with large temperature gradients it is necessary to solve the Navier–Stokes and continuity equations subject to the boundary conditions of the reactor. Such calculations have allowed reasonably accurate predictions of convection patterns and predict growth rates in agreement with experimental results; Sugawara (1972), Takahashi *et al.* (1972), Wahl (1977), and Manke and Donaghey (1977). However, such calculations are so complex as to fall more in the domain of the specialist in hydrodynamics than the crystal grower. The rigorous analysis of the hydrodynamics of vapor deposition systems is treated in more detail in Chapter 3 of this volume.

Based on hydrodynamic calculations, and more importantly specific experiments designed to reveal flow patterns using TiO_2 smoke, several simple models have been developed which accurately predict the macroscopic parameters important for real growth systems, such as efficiency and uniformity of growth rate. Because mass transport is so dependent on geometry and thermal gradients, this discussion will be divided to discuss closed tube, horizontal, and vertical systems separately.

5.3.1. CLOSED TUBE SYSTEMS

In the late fifties and early sixties the initial growth of group IV, III IV, and II VI semiconductor single crystals was accomplished in closed SiO_2 ampoules using I or Cl as the transporting agent. In these closed systems, diffusion plays a major role since the temperature is often relatively uniform and forced convection is of course not a factor. These are perhaps the easiest systems to analyze, but since these closed tube processes are no longer used extensively the results are of more historical than practical interest. Thus, they will not be discussed in detail here. The interested reader is referred to articles by Curtis and Dismukes (1972) and Reed and Lafleur (1972), which treat the mass transport in closed systems in some detail.

5.3.2. HORIZONTAL REACTOR

The simplest model for the treatment of mass transport is the quasi-equilibrium model of Sedgwick (1964) discussed in section 5.2.1. In this model the deposition rate is equal to the number of moles of the reactant entering the system multiplied by the thermodynamic reaction efficiency η and the reactor efficiency β, which represents the fraction of the gas stream which can equilibrate with the substrate. This is basically the equivalent of assuming a uniform stagnant layer of thicknesses δ through which mass transport occurs strictly by diffusion with the diffusion coefficients of all species being equal (Shaw, 1975b). In this case the flux of component i may be expressed:

$$J_i = D_0(p_i^\circ - p_i^{eq})/RT\delta. \tag{5.12}$$

This concept was advanced further by Bradshaw (1966). He applied the Burton *et al.* (1953) model developed for single crystal growth from the melt. In this model it is assumed that there is a boundary layer of relatively static gas adjacent to the growing solid surface within which the flow is essentially laminar. Equilibrium is established at the growing interface and convective mixing above the boundary layer maintains the partial pressures of input gases constant. It is tacitly assumed that no reactions take place homogeneously in the vapor phase. Bradshaw (1966) solves the one-dimensional continuity equation for each reactant. He then assumes that the thermal, mass, and momentum boundary layers are identical in thickness. Using these and other plausible assumptions he is able to treat the $SiCl_4$ system with reasonable accuracy, although, as discussed in section 5.2.1, the differences in diffusivities of individual molecular species and thermodiffusion effects must be taken into account for an accurate calculation.

Using TiO_2 smoke experiments, Eversteyn *et al.* (1970) observed a uniform stagnant layer in a SiH_4 horizontal reactor. Assuming that the temperature decreases linearly across the boundary layer, and taking into account the temperature dependence of the diffusion coefficient $[D = D_0(T/T_0)^2 \ (cm^2 s^{-1})]$, the growth rate in $\mu m/min$ was calculated to be

$$G = 7.23 \times 10^6 \frac{D_0 T_s p_0}{RT_0^2 \delta} \exp\left(\frac{-D_0 T_s x}{T_0 v_0 b\delta}\right) \tag{5.13}$$

where x is the distance along the susceptor, v_0 the mean gas stream velocity $(cm\,s^{-1})$, and b the free height above the susceptor (cm). Determining the boundary layer width from growth rate and direct observation using TiO_2 smoke, they found a $v_T^{-1/2}$ relationship with total gas stream velocity

$$\delta = A/\sqrt{v_T} - B \tag{5.14}$$

where $A = 7\,cm^{3/2}\,s^{-1/2}$ and $B = 0.2\,cm$. An important element of the analysis of Eversteyn *et al.* (1970) is that the depletion of $SiCl_4$ with distance in the direction of gas flow leads to a significant reduction in growth rate with distance along the reactor. This may be compensated for by tilting the susceptor, which causes an increase in gas velocity along the tube. This produces a decreasing boundary layer width which compensates for the decrease in p_{SiCl_4} and produces a uniform growth rate. Using this technique commercial horizontal reactors have specifications on growth rate of uniformity of $\pm \sim 5\%$ for growth runs containing $> 1300\,cm^2$ of Si.

Recently Berkman *et al.* (1978) have done a more detailed analysis of the boundary layer in the horizontal system with a tilted susceptor. The calculations are done without adjustable parameters and the results are relatively simple since they argue that the mass transport boundary layer thickness is approximately half of the tube height.

The actual flow pattern in a horizontal reactor with thermal and forced convection is a spiral (Ban, 1978). Further complications are due to entry conditions which must be overcome to achieve a fully developed boundary layer. Recent work of Takahashi (1972) and Ban (1978) indicate that in typical reactors the boundary layer is not fully developed. To date no realistic model has been developed to deal with the complications.

Another interesting and somewhat surprising recent finding is that homogeneous gas phase reactions apparently cannot be ignored in some cases. As discussed in section 5.2.1, Smith and Sedgwick (1977) found that the concentration of $SiCl_2$ detected by inelastic light-scattering measurements has a maximum value at a point in the vapor phase *above* the substrate. This can occur only if $SiCl_4$ reacts in the gas phase above the substrate to form $SiCl_2$. This may be a serious problem for all of the mass transport analysis described above which assume reactions only at the solid surface.

5.3.3. VERTICAL REACTOR

Another popular configuration for CVD reactors for the growth of semiconductor and insulator layers is the vertical reactor, of either the "pancake" variety shown in Fig. 5.2b or the barrel reactor shown in Fig. 5.2c. Considering first the pancake reactor, where gas impinges vertically on a horizontal heated pedestal, we would expect rather complex convection patterns. Wahl (1977) made a definitive study of this reactor geometry including hydrodynamic calculations and flow visualization experiments using TiO_2 smoke. His calculated results, which agree well with the experimental results, are shown in Fig. 5.13. Complex convection currents are set up as the cool gases flow downward and are heated by the pedestal, as shown in Fig. 5.13a. These can be avoided by simply putting the substrate at the top as in Fig. 5.13b. Buoyancy forces then make the currents more laminar. Wahl (1977) calculated growth rates and deposition profiles for the growth of Si_3N_4 and SiO_2, which were generally in good agreement with experimental results. In spite of these complications, uniformities of $\pm 5\%$ are obtained in commercial 305 mm (12 in.) diameter vertical pancake reactors with downward gas flow.

Manke and Donaghey (1977) have analyzed the hydrodynamics of barrel type cylindrical reactors. They propose that, similar to the horizontal reactors, the entry conditions of the gases into the reactor play an important role over most of the length of the deposition zone. They were able to obtain results for yield versus position in the reactor in good agreement with experimental results. This material is discussed in more detail in Chapter 3.

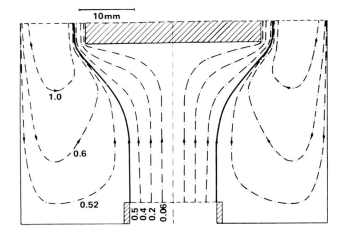

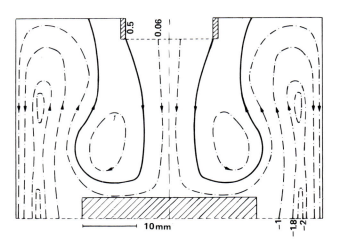

FIG. 5.13. Calculated flow lines for vertical pancake reactor with gas flow directed (a) downward, and (b) upward. (From Wahl, 1977.)

5.4. Interface Kinetics

The basic premise of crystal growth is that the growth velocity is the difference of two thermally activated processes—the arrival and departure rates of atoms at the surface,

$$v = R_A - R_D \tag{5.15}$$

where $R = R° \exp(-Q/kT)$ which yields

$$v = R_A° [\exp(-Q/kT)] \left[1 - \exp\left(\frac{-L\Delta T}{kT_E T} \right) \right] \tag{5.16}$$

where the preexponential factor R_A^2 equals the interatomic distance times the atomic vibration frequency, Q is associated alternately wth the activation energy for diffusion (Wilson, 1900) or the temperature dependence of viscosity (Frenkel, 1932), L is the latent enthalpy of the solid to vapor transition, i.e., the difference between the activation energies for the arrival and departure processes, and ΔT is the difference between the actual temperature and the equilibrium temperature and hence represents the driving force for the reaction. For small values of ΔT the growth rate should be a linear function of ΔT.

The surface kinetically limited growth rate calculated as above is orders of magnitude higher than experimentally observed. This can be interpreted to indicate that not all surface sites are actually growth sites. On low index faces, where a deposited atom could form only a single bond with the crystal, growth is thought to occur *not* by simply attaching atoms to the solid surface at random but by physical adsorption and diffusion along the surface to a site such as a step where several bonds can be formed to the solid. The formation of such steps may occur by formation of two-dimensional nuclei on the surface. Calculations of the value of ΔT needed for this nucleation process are also much larger than the actual degree of supersaturation necessary to sustain crystal growth (Jackson, 1975).

Burton *et al.* (1951) proposed that an inexhaustible supply of steps could be produced by screw dislocations. Such sources of steps have been observed in many systems but, even including these sources, calculations still predict values of supersaturation considerably larger than actually needed to sustain crystal growth.

The growth rate versus supersaturation behavior for two-dimensional nucleation and screw dislocation mechanisms can be dealt with very effectively using computer simulation of the crystal growth process. Gilmer (1977) has performed such simulations using the Ising model and Monte Carlo calculation techniques. The two-dimensional nucleation calculations indicate that even for very high values of supersaturation and rapid surface diffusion, the simulated growth rate is still much lower than that predicted by the Wilson–Frenkel theory. Crystal growth at low values of supersaturation is found to occur only when spiral growth at screw dislocations is included. The simulations also treat the effect of impurities which result in increased growth rate.

Jackson (1958) proposed a model for resolving the disparity between simple theories for the generation of surface steps and the actual low values of supersaturation required for crystal growth. He proposed that thermal roughening of the atomically smooth low index crystal faces occurs above a critical temperature T_c. This would give a built-in supply of steps which would account for the low values of supersaturation required for crystal growth. Jackson (1958) developed a model where the change in free energy of an initially plane interface due to the addition of molecules at random was calculated. The key parameter affecting the surface roughness is the ratio of the binding energy at the interface to the thermal energy kT. This is most conveniently expressed as $\Delta S/R$, the entropy difference between the two phases divided by the gas constant.

For growth of crystals where $\Delta S/R$ is small, for example the growth of metals from the melt where $\Delta S/R$ (1000°C) ≈ 1, the surface is very rough and surface nucleation is no hindrance to crystal growth. Thus one degree or less of supercooling is sufficient for rapid crystal growth. For the growth of Si or Ge from the melt, $\Delta S/R$ is about 3 and only a few degrees of supercooling are required to initiate growth. For growth of Si or Ge from the vapor $\Delta S/R$ is ~ 10, hence ~ 10 degrees of supercooling would be required.

Generally the steps that occur at the solid interface to transfer an atom from the vapor phase into the solid phase may be illustrated as

$$A(v) \xrightleftharpoons[\text{Desorption}]{\substack{\text{Physical}\\\text{adsorption}}} A^* \xrightarrow[\substack{\text{Attachment}\\\text{at step}}]{\substack{\text{Surface}\\\text{diffusion}}} A^s + B^* \xrightarrow[\text{products}]{\substack{\text{Desorption}\\\text{of}}} B(v) \qquad (5.17)$$

The reactant molecule A is physically adsorbed onto a random site on the solid surface. The adsorbed molecule diffuses around on the surface until it either desorbs or reaches a site where more than one bond can be formed, for example at a step. It then probably diffuses along the step until it reaches a kink where it can form an additional bond. There it reacts chemically to be incorporated into the solid releasing one or more product molecules, designated B* in the diagram. These then diffuse along the surface until desorbed.

Kinetic aspects of vapor phase epitaxial growth of Si and other semiconductors have not been totally defined but have been studied using the following techniques:

1. Specific surface studies in high vacuum apparatus equipped with some or all of the following instruments: mass spectrometer, Auger electron spectrometer, LEED, HEED. The vapor phase contacts the well-defined surface admitting the molecular species of interest into the reactor either as a gas or as a molecular beam from a high temperature Knudsen cell source.
2. Macroscopic kinetic studies of growth rate versus temperature, gas phase composition, substrate orientation, etc.

Two examples will be described here. First, the use of the UHV technique for Si growth from SiH_4 will be described in some detail. Then the second technique as applied to GaAs will be reviewed.

5.4.1. Si GROWTH FROM SiH$_4$

The growth of Si using SiH_4 is a process for which the thermodynamic analysis is rather simple. SiH_4 is unstable at high temperatures and thus decomposes completely on the hot substrate. The limitations to the rate of this process are purely kinetic, due to either mass transport or reaction rates at the surface. The most definitive studies of the kinetics of this process are those by Henderson and Helm (1972) and Farrow (1974). Earlier work was complicated by the contamination of the substrate surface with O and especially C. This caused the initial growth to occur by three-dimensional nucleation. Proper cleaning of the surface *in situ* has eliminated those problems.

The Henderson and Helm (1972) experiments were carried out in an UHV apparatus using HEED and Auger spectroscopy to monitor the cleanliness of the surface before growth. A mass spectrograph was used to monitor the adsorption of SiH_4 admitted into the sealed vacuum chamber and desorption of products during the reaction. The growth rate is proportional to the reaction rate which they found to be first order, i.e.,

$$R = K_1 p_{SiH_4} \qquad (5.18)$$

The activation energy for SiH_4 decomposition was found to be 20 ± 5 kcal mole^{-1}. Their data led to the conclusion that most adsorbed SiH_4 molecules desorb before decomposing to form a Si adatom. Even the fraction of Si adatoms eventually incorporated in the solid was found to be as low as 0.32 at 823°C and 0.15 at 987°C. They interpreted this to mean that the supersaturation necessary to achieve two-dimensional nucleation had not been reached, since at this point all Si adatoms would be incorporated into the layer. Assuming the steps to be due to dislocation sources, Henderson and Helm (1972) interpreted the activation energy for the exponential temperature dependence of the condensation coefficient (the fraction of adatoms incorporated into the solid) to be the sum of the enthalpies of desorption and surface diffusion, i.e.

$$\Delta G_{des} - \Delta G_{sd} = 26 \text{ kcal mole}^{-1} \tag{5.19}$$

Assuming $\Delta G_{des} = 50$ kcal mole^{-1}, the energy of a simple Si–Si bond yields a value for the activation energy for surface diffusion, ΔG_{sd}, of 24 kcal mole^{-1}.

Farrow (1974) reached similar conclusions about the growth process using a slightly different technique. SiH_4 was admitted to the vacuum system continuously to maintain a steady state growth reaction. The products were analyzed using a mass spectrometer. Typical results shown in Fig. 5.14 indicate that for temperatures above about 600° C the pyrolysis of SiH_4 occurs by the reaction

$$SiH_4 = Si + 2H_2 \tag{5.20}$$

in agreement with Henderson and Helm's result that the reaction is first order.

Knowing the growth rate R and the rate of production of free silicon atoms per unit area of specimen surface α, Farrow was able to calculate the condensation coefficient for Si, σ

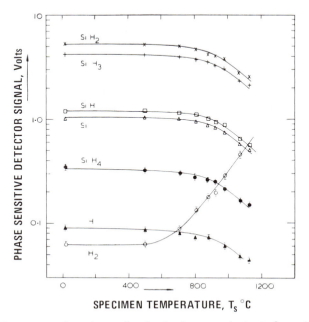

FIG. 5.14. Temperature dependence of molecular beam products at $p^{\circ}_{SiH_4} = 10^{-4}$ torr. (From Farrow, 1974.)

$(2.32 \times 10^{-3} \, R/\alpha)$. The values of R, σ, and α, including Henderson and Helm's values of σ, are plotted in Fig. 5.15. Using the analysis described above results in a value of ΔG_{sd} of $36 \, \text{kcal mole}^{-1}$ for Farrow's data.

In conventional SiH_4 systems, where $p_{H_2} \approx 1 \, \text{atm}$, the H_2 has been found to competitively adsorb on the growing Si surface, thus reducing the growth rate and affecting the growth morphology. The low temperature limit for the growth of good quality Si films is found to be reduced by using an inert gas such as He or Ar rather than H_2 as the carrier gas (Richman and Arlett, 1969; Chiang and Looney, 1973; Seto, 1975). In such systems the surface reaction is apparently the same (Seto, 1975) as reported by Henderson and Helm (1972) and Farrow (1974). For poly Si growth Seto (1975) found the activation energy for the growth process to be $11.9 \, \text{kcal mole}^{-1}$ in approximate agreement with the value of $10 \, \text{kcal mole}^{-1}$ reported by Farrow (1974). Duchamin et al. (1978) studied the growth kinetics as a function of total pressure in a low pressure CVD system using H_2 as the carrier gas. They found that for H_2 pressures in the range of 10–50 torr the reaction rate was inversely proportional to the square root of hydrogen pressure (at constant p_{SiH_4}). This behavior was taken to indicate that the adsorption of H_2, which takes two surface sites, competes with SiH_4 for sites. At higher pressures the growth rate was found to be proportional to $p_{H_2}^{-1}$ and to be nearly independent of temperature. This indicates the growth rate to be controlled by mass transport in agreement with results presented in section 2.2.

Farrow (1974) also studied the effect of As and B on the kinetics of the pyrolysis of SiH_4. As was found to significantly retard the reaction while B had the opposite effect. Two possible mechanisms for the impurity effect were suggested. The As might be thought of as blocking key sites on the Si surface which are active in inducing pyrolysis. Alternatively As and SiH_4, both of which are present at rather low concentrations, could be attracted on the surface with the presence of As in some manner retarding the pyrolysis. On the other hand, one would have to postulate that in some manner B would enhance the pyrolysis rate.

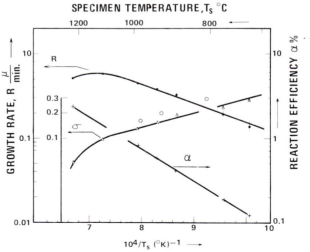

FIG. 5.15. Temperature dependence of silicon growth rate R, surface reaction efficiency α and silicon condensation coefficient σ for $p_{SiH_4} = 10^{-1} \, \text{torr}$. (From Farrow, 1974.) Henderson and Helm's (1972) data for σ are also included ($\bigcirc$).

5.4.2. GaAs GROWTH KINETICS

The growth processes occurring for GaAs, and III V compounds and alloys in general, are much more complex than those for Si. The systems using chloride transport of the group III element(s) as shown in Fig. 5.2d are most widely used. These systems are of necessity hot walled because HCl or $AsCl_3$ must be reacted with hot Ga, In, or Al to form volatile compounds, and because the reactions are exothermic and hence are suppressed on the hot walls upstream from the cooler substrate. For both the $AsCl_3$ and HCl systems the overall chemical reaction occurring at the substrate may be written

$$GaCl(g) + \tfrac{1}{2}As_2(g) + \tfrac{1}{2}H_2(g) = GaAs(s) + HCl(g). \tag{5.21}$$

The kinetics of the growth process have generally been studied by examining deposition or growth rate versus growth parameters such as substrate temperature and orientation and gas phase composition. Some of the most thorough work has been that of Shaw (1968b, 1970, 1975a) using an electrobalance for continuous monitoring of deposition kinetics. The results using elemental As and HCl to transport the Ga for the growth of GaAs are a good example of the characteristics of the kinetically limited growth regime as contrasted to the thermodynamic plus mass transport limited regime. The deposition rate is plotted versus reciprocal temperature in Fig. 5.16 for substrates with several low index orientations. Each curve has a maximum with growth rate decreasing at both high and low temperatures. Thermodynamically, the growth process is exothermic; thus the reaction efficiency should decrease as temperature is increased over the entire temperature range

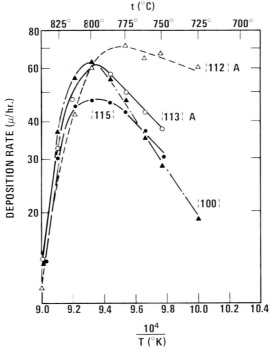

FIG. 5.16. Deposition rate versus temperature for GaAs of various orientations. (After Shaw, 1975).

with perhaps a flattening at low temperatures. Indeed, the high temperature slope of the experimental curves in Fig. 5.16 is approximately equal to that predicted from thermodynamics. The independence of growth rate on orientation is also characteristic of the thermodynamic mass transport limited growth regime. The Sedgwick (1964) model may be used to describe the data in this temperature range with $\beta \approx 0.1$ (Shaw, 1970).

In the low temperature range the data have three characteristics indicative of surface kinetically limited growth. (1) Growth rate is dependent on substrate orientation. (2) A plot of log growth rate versus reciprocal temperature has a negative slope. (3) Growth rate is independent of total flow rate (Shaw, 1968b, 1975a). The slope in the low temperature regime of Fig. 5.16 for each orientation represents the activation energy for the surface reaction limiting the growth rate. Typical numbers are $\sim 50 \, \text{kcal mole}^{-1}$ (Shaw, 1975a).

The more detailed information related to the orientation dependence of growth rate shown in Fig. 5.17 is useful for understanding more about the kinetic process limiting the surface reaction rate. The data are for individual growth experiments on oriented substrates (Shaw, 1968a, b) and for a single growth run using a hemispherical substrate (Hollan and Schiller, 1972). Lobes of low growth rate are observed on most low index faces (110), (111)B, and (100), but the maximum rate is observed near (111)A. However, the relative growth rates on (111)A and (111)B are reported to be very dependent on vapor phase stoichiometry (Shaw, 1975a). One might suppose the deposition kinetics to be limited by the factors described above, namely intrinsic surface roughness, step generation, adsorption onto the surface, surface mobility, or the kinetics of the surface chemical reaction presumably occurring at a step.

For the zinc blende structure the various low index surfaces are shown in Fig. 5.18. The (100) surface atoms form two bonds to the remainder of the crystal with two dangling bonds. This might be considered a "rough" surface in the Jackson sense of not requiring generation of steps for atoms to be attached to the surface. The (100) and both (111)A and

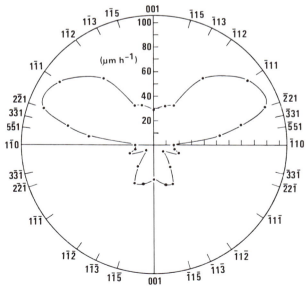

FIG. 5.17. Polar diagram of deposition rate versus crystallographic orientation. (After Shaw, 1968b.)

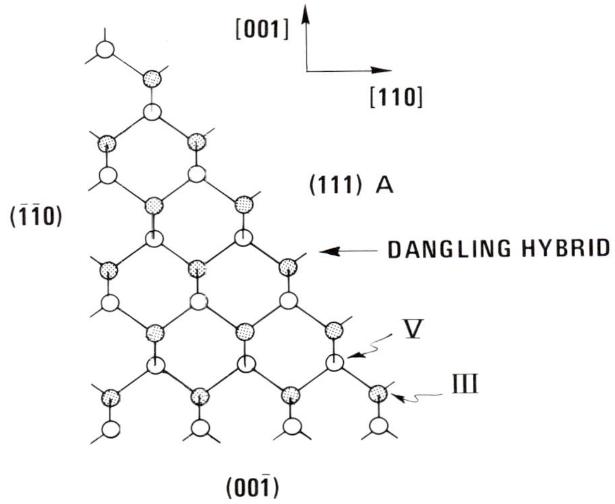

FIG. 5.18. Atomic configurations on (100), (110), and (111)A zinc blende surfaces.

(111)B surface atoms form only one bond to the bulk. The effect of surface roughness may explain the high growth rates for non-close-packed faces but does not explain well the relative rates for the low index faces when the growth rates typically follow the pattern (111)A > (100) > (110) > (111)B.

Very little is known about the degree of supersaturation required to initiate crystal growth in these systems. Since the bulk of the vapor phase is highly supersaturated it may be that at the surface the supersaturation is sufficient for two-dimensional nucleation. The thermal roughening is probably small since $\Delta S/R$ would be expected to be ~ 10 in these systems. No evidence of spiral growth patterns suggestive of screw dislocation sources of growth steps has been reported.

The most easily interpreted results on GaAs surface kinetics are obtained by molecular beam techniques. Here chopped Ga and As_2 beams are focused onto clean GaAs substrates and adsorption and desorption kinetics are studied using a mass spectrometer. The results are relatively simple. On (111) surfaces Arthur (1966) found that both Ga and As_4 surface mobilities remain high to temperatures well below the substrate temperatures used for conventional chloride crystal growth processes discussed here. Arthur (1968) also found the growth rate followed simple kinetics, being a linear function of the Ga flux when $J_{As_2} > J_{Ga}$. He also found that the As incorporation to take place by the reaction (Arthur, 1974)

$$As_s(g) + 2V_s \rightleftarrows As_2(g) \rightleftarrows 2As(s) \tag{5.22}$$

where V_s is a vacant As site on the surface.

The kinetics of the chloride growth process are considerably more complex. As shown in Fig. 5.19, in the low temperature regime an increase in GaCl partial pressure actually results in a decrease of the growth rate. HCl has been observed to produce a similar effect. The kinetics were found to follow the empirical relationship

$$R \propto p_{As}^{0.22} p_{GaCl}^{-0.59}. \tag{5.23}$$

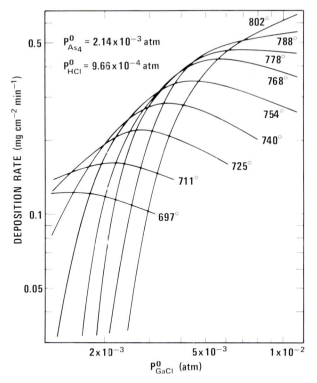

FIG. 5.19. GaAs deposition rate versus input partial pressure of GaCl at various substrate temperatures. (After Shaw, 1975.)

Shaw (1975a) proposed a model to explain this phenomenon where GaCl, HCl, and As all compete for the same surface sites. The model may be illustrated by the following simplified surface diagram; for the (100) surfaces:

Cadoret and Cadoret (1975) proposed a more elaborate and detailed model for the surface reaction mechanisms:

$$\tfrac{1}{2}H_2(g) \rightleftarrows H^* \tag{5.24a}$$

$$\tfrac{1}{4}As_4(g) \rightleftarrows \tfrac{1}{2}As_2(g) + V_s \rightleftarrows \tfrac{1}{2}As_2^* \rightleftarrows As^* \tag{5.24b}$$

$$H^* + As^* + GaCl(g) \rightleftarrows (As\text{–}Ga\text{–}Cl)^* + H^* \rightleftarrows (As\text{–}Ga\text{–}Cl\text{–}H)^* \\ \rightleftarrows GaAs(s) + HCl(g). \tag{5.24c}$$

The calculated dependence of growth rate on temperature and GaCl partial pressure appears to agree with the experimental results. Mizumo and Watanabe (1975) examined the effect of H_2 on the reaction kinetics of (100) GaAs using the $AsCl_3$ transport system. They found the growth rate to be proportional to p_{H_2}. On this basis they proposed the following model for the surface reaction mechanism:

$$H_2(g) + GaCl(g) \rightleftarrows H_2(g) + GaCl^* \rightleftarrows GaH_2Cl^* \qquad (5.25a)$$

$$\tfrac{1}{2}As_4(g) \rightleftarrows As_2^* \rightleftarrows 2As^* \qquad (5.25b)$$

$$GaH_2Cl^* + As^* \rightleftarrows GaAs(s) + HCl(g) + \tfrac{1}{2}H_2(g). \qquad (5.25c)$$

They contend that a surface species GaH_2Cl might be expected based on the stability of chlorosilanes.

Clearly the surface reactions occurring for the growth of GaAs by chloride transport are very complex and at this time not completely understood. However, it seems reasonable to conclude that the kinetics are controlled by surface reactions rather than by adsorption, step generation, or the other classical crystal growth processes.

The organometallic transport system is in some ways simpler than the chloride systems. It resembles the SiH_4 system in that it is carried out in a cold wall reactor and no thermodynamic equilibrium is involved. The reactions

$$Ga(CH_3)_3(g) + AsH_3(g) \rightarrow GaAs(s) + 3CH_4(g) \qquad (5.26)$$

or

$$Ga(C_2H_5)_3(g) + AsH_3(g) \rightarrow GaAs(s) + 3C_2H_6(g) \qquad (5.27)$$

for trimethyl Ga and triethyl Ga respectively as the Ga source material should go essentially to completion under all experimental conditions. The absence of Cl precludes the etching reaction and simplifies the kinetics by eliminating the possibility of adsorption of chloride species on the surface. The ratio of As to Ga in the input gas stream is >1 for the most favorable growth conditions, and the growth rate is found to be linearly proportional to the Ga flow rate (Manasevit and Simpson, 1969; Inoue and Asahi, 1972; Ito et al., 1973; Bass, 1975). Using TMG (eqn. 5.26), the growth rate was found by Inoue and Asahi (1972), Ito et al. (1973), and Bass (1975) to be nearly independent of temperature as shown in Fig. 5.20. Here the data are plotted normalized by the thermodynamic equilibrium growth rate, i.e., using the Sedgwick (1964) model $\beta(=R/\eta F_T)$ is plotted versus reciprocal temperature. Also included is a simple calculation of β vs $1/T$ taking β to be proportional to $\sqrt{D(T)\tau(T)}$ where $D(T)$ is the typical diffusion coefficient, $D \approx 9.5 \times 10^{-5} T^{3/2}$, and $\tau(T)$ is the time taken for the gas to flow over the substrate. The temperature dependence and magnitude of growth rate are typical of the behavior expected for mass transport limited growth kinetics.

Using TEG (eqn. 5.27) as the Ga source results in much lower growth rates (Seki et al., 1975; Stringfellow and Hall, 1978a). This low growth rate is not due to thermodynamics, since this reaction should go nearly to completion at equilibrium. Neither is it due to mass transport where only small differences in the diffusion coefficients of $Ga(CH_3)_3$ and $Ga(C_2H_5)_3$ in H_2 would be expected. The effect must be related to slow surface kinetics using TEG. Two processes occurring on the surface might be candidates to explain this effect. Since the growth rate is proportional to Ga transport to the surface, the growth rate in this case must be dominated by the residence time of $Ga(C_2H_5)_3$ on the surface before incorporation. The slow growth rate for $Ga(C_2H_5)_3$ relative to $Ga(CH_3)_3$ could be related to either a shorter surface lifetime before desorption or more likely a steric hindrance of

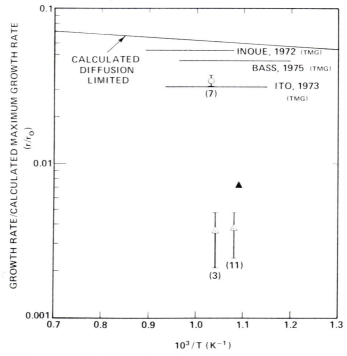

FIG. 5.20. Temperature dependence of growth rate (normalized to calculated maximum). Data for growth using TMG are from Inoue *et al.* (1972), Bass (1975), and Ito *et al.* (1973). Data for growth using TEG are from Seki *et al.* (1975) (▲) and Stringfellow and Hall (1978a) (△). Stringfellow and Hall (1978a) also reported data for TEG growth with the addition of HCl (○). (After Stringfellow, 1978b.)

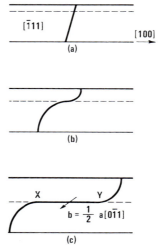

FIG. 5.21. The generation of a length *X Y* of misfit dislocation by mismatch induced glide of an inclined dislocation. (After Matthews, 1975.)

attachment at a step due to the larger ethyl radicals (Stringfellow and Hall, 1978a). The growth rate can be increased by adding HCl to the system as shown in Fig. 5.21. The HCl has the effect of "cleaving" the $Ga-C_2H_5$ bond, substituting Cl for the ethyl radical, and removing the hindrance to rapid surface kinetics. The growth rate again becomes mass transport limited (Stringfellow and Hall, 1978a).

The kinetics of GaAs growth using TMG have been studied by Inoue and Asahi (1972), Petzke *et al.* (1974), and Schlyer and Ring (1976, 1977). At temperatures below 625°C Inoue and Asahi (1972) found the growth rate to decrease with decreasing temperature, indicating that in this temperature range growth becomes limited by surface kinetics. Petzke *et al.* (1974) interpret their results to indicate that the surface reaction is first order with an activation energy of $13 \, kcal \, mole^{-1}$. Schlyer and Ring (1976, 1977) investigated the decomposition kinetics of TMG in the presence of AsH_3 on a GaAs surface in the temperature range 200–260°C. They obtained an activation energy of $13 \, kcal \, mole^{-1}$, which indicates the mechanism may be the same as that limiting the growth rate at higher temperatures. They propose a series of surface reaction mechanisms by which the growth process occurs:

$$AsH_3(g) + Ga(CH_3)_3(g) \rightarrow (CH_3)_2 \, GaAsH_2 + CH_4(g) \tag{5.28a}$$

$$(CH_3)_2 \, GaAsH_2 \rightarrow (CH_3) \, GaAsH + CH_4(g) \tag{5.28b}$$

$$(CH_3) \, GaAsH \rightarrow GaAs(s) + CH_4(g). \tag{5.28c}$$

Schlyer and Ring point out that these reactions may have important consequences for growth of GaAs. Care must be taken that the CH_4 molecules are desorbed from the surface before they are buried by further deposition, which would likely produce a high concentration of C in the layer. C is probably an acceptor in GaAs but is also possibly a component of nonradiative complexes (Stringfellow and Hall, 1979). Schlyer and Ring (1977) propose that a high ratio of AsH_3 to TMG in the input gas stream would remove CH_4 from the surface and reduce this problem. This may explain the well-established dependence of conductivity type of undoped GaAs layers on AsH_3/TMG ratio in the gas stream. High values of this ratio produce *n*-type layers while low values result in *p*-type layers (Ito *et al.*, 1973; Seki *et al.*, 1975).

5.5. Defect Generation

PVD at low temperatures and high growth rates, conditions used from most applications, results in polycrystalline or in the extreme case amorphous films. This would occur even for growth on single crystalline substrates because the surface atoms lack the time to find their lowest energy sites before other atoms cover them. In fundamentally oriented research heteroepitaxial growth of single crystalline islands at low deposition rates has been studied. TEM observations of these islands has yielded much of our understanding of defect generation during epitaxial growth (Matthews, 1975).

In semiconductors, crystalline defects such as grain boundaries, stacking faults, twins, or dislocations are known to have deleterious effects on many device properties. They have been shown to act as nonradiative recombination centers in many materials (Stringfellow *et al.*, 1974; Ettenberg, 1974). Thus they act to reduce the minority carrier lifetime and quantum efficiency in LEDs (Stringfellow *et al.*, 1974; Werkhoven *et al.*,

1977). Similarly they act as generation centers to reduce the performance of transistors, solar cells, detectors, and other devices. In addition, crystalline defects introduce problems in fabrication of devices, for example by acting as diffusion pipes (Ravi, 1976), and have been shown to play a role in the degradation of devices, particularly LEDs and injections lasers (Petroff and Hartman, 1974; Petroff et al., 1976). For these reasons considerable effort has been devoted to the study and elimination of crystalline defects in semiconductor materials.

5.5.1. SOURCES OF DEFECTS

The sources of defects occurring in epitaxial layers may be divided into five categories:

(1) Propagation of defects from the substrate into the epitaxial layer. This is the major source of dislocations in homoepitaxial layers. Grain boundaries, stacking faults, and twins are not generally found in substrates. In Si, GaAs, and GaP dislocation free substrates may be obtained (de Kock, 1973; Roksnoer et al., 1977). However, even for these "dislocation free" substrates, loops which intersect the surface result in the propagation of dislocation pairs into the layer (Stringfellow et al., 1974; Petroff et al., 1976; Werkhoven et al., 1977; Vink et al., 1978).

(2) Stacking faults due to improper substrate preparation. In many VPE layers stacking faults forming a closed figure (triangle for (111), square for (100)) are observed on the top surface. These are well established to propagate from the initial interface. Booker and Stickler (1962) studied stacking faults in (111) Si homoepitaxial layers and concluded that the triangle stacking fault patterns originated from nuclei having stacking faults at the interface with the substrate. A model was developed showing how coalescence of these nuclei with surrounding normal nuclei would produce the triangular stacking fault arrays. Continuing growth would cause the triangles formed at the intersection with the top surface to become progressively larger.

(3) Formation of precipitates or dislocation loops caused by a supersaturation of impurities, dopants, or native defects during cooling. These effects have been widely studied in semiconductor systems (Petroff et al., 1976; Roksnoer et al., 1976) but will not be dealt with here because they are not intrinsically associated with the growth process. They are more related to reactions occurring in the solid after growth due to intentional or accidental doping.

(4) Formation of low angle grain boundaries and twins due to rotation of islands. For heteroepitaxial growth where the lattice parameter of the epitaxial layer differs from that of the substrate there is some ambiguity in the exact orientation of small nuclei. Matthews (1972) has formulated a rule of thumb that "the relaxation of elastic misfit strain leads to a variation in the orientation of crystal planes (in radians) that is approximately equal to the misfit f". The misfit f is defined as

$$f = (a_s - a_0)/\bar{a}, \qquad (5.29)$$

where a_0 and a_s are the stress free lattice parameters of the layer and substrate respectively. Islands may also nucleate in several twinned orientations. When such misoriented islands meet and coalesce, defects such as low angle grain boundaries (arrays of dislocations to accommodate the slight misorientation between nuclei) or twins may result. Some such defects may anneal out by dislocation motion during the growth process to form a single, perfect island, but often the defects propagate into the film. The reader is referred to the review article by Stowell (1975) for a more detailed description of these phenomena.

(5) Formation of dislocations due to the lattice parameter mismatch between substrate and epitaxial layer. This is a very interesting area where a great deal of progress has been made in the last 10 years, with practical semiconductor systems being understood in terms of models and fundamental experiments developed for metal systems. A milestone occurred in 1949 with the publication of a theoretical paper by Frank and van der Merwe (1949) which predicted that any epitaxial layer with a lattice parameter mismatch (Δa_0) with the substrate of $\lesssim 12\%$ would initially grow pseudomorphically, i.e., for very thin layers the deposit would be elastically strained to have the same interatomic spacing as the substrate. Thus, the interface would be coherent. For very thin layers, the elastic strain energy is lower than the energy of the array of misfit dislocations needed to relieve the strain. With increasing thickness the elastic energy would increase, eventually becoming greater than the energy of the misfit dislocation array. Ideally, at this point dislocations would be generated to relieve a fraction of the misfit. As the layer thickness increased this fraction would increase until at infinite thickness the elastic strain would be totally eliminated. Since the mid 1960s experimental evidence has accumulated, based on TEM observations of heteroepitaxial growth of metal films such as Ni/Cu, Pd/Au, Pt/Au, and others, that the Frank and van der Merwe predictions are basically correct (Matthews, 1975). However, in practice the amount of elastic strain relaxed may be less than anticipated because of lack of dislocation sources, dislocation immobility due to interaction between dislocations, and other hindrances to the development of the equilibrium misfit dislocation network. This work has recently been reviewed in detail (Matthews, 1975). Matthews suggested several unconventional generation mechanisms for misfit dislocations. Two mechanisms are illustrated in Figs. 5.21 and 5.22. The first involves the bending of an edge dislocation propagating from the substrate into the interface plane where it acts as a mismatch dislocation. The second involves nucleation of a dislocation loop at the surface and propagation of the loop into the interface plane by the elastic stress in the layer. The loop propagation may occur by either dislocation glide or climb.

The critical thickness at which misfit dislocations are formed was derived by Jesser and Kuhlmann-Wilsdorf (1967):

$$h_c = \frac{G_s b^2}{2\pi a_s |f| (1 + 2v)(G_0 + G_s)} \ln \left[\frac{(G_0 + G_s)G(1 - v)}{2\pi G_0 G_s |f|} \right], \tag{5.30}$$

where G, G_0, and G_s are the shear moduli of the "interface", overgrowth, and substrate, b is the magnitude of the Burgers vector, and v is Poisson's ratio. Blanc (1978) has derived a simplified form having more obvious physical significance by letting $G = G_0 = G_s$ and ignoring the weaker logarithmic term:

$$h_c \approx \frac{b}{2|f|}. \tag{5.31}$$

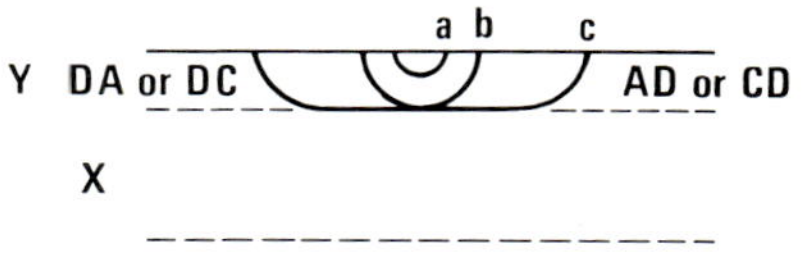

FIG. 5.22. Generation of a misfit dislocation by the nucleation (at the surface) and growth of a dislocation loop. (After Matthews, 1975.)

This can be interpreted to mean that the layer will be pseudomorphic until the accumulated misfit $(h_c \cdot |f|)$ exceeds about half of the unit cell dimension ($\sim b$).

In the remainder of this section defect generation in specific systems will be considered in more detail as examples of the concepts discussed above.

5.5.2. Si/Si:B

Perhaps the simplest semiconductor system for which defect generation due to lattice parameter mismatch has been studied is the SiH_4 VPE growth of undoped Si on B doped Si substrates. The B, being smaller, results in a reduction in lattice parameter; hence this system offers the opportunity to study defect generation at interfaces with small, controlled amounts of mismatch. Sugita *et al.* (1969a) studied 20 μm thick epilayers grown on 200 μm thick B doped substrates. Their results, summarized in Table 5.2, indicate that the qualitative features of the Frank and van der Merwe model are consistent with experimental observations, although the predicted values of misfit dislocation density are higher than experimental values. The critical thickness was explored in a series of experiments where the film thickness was varied with f held constant at 9×10^{-5} (Sugita and Tamura, 1969b). The results plotted in Fig. 5.23 show a critical film thickness above which misfit dislocations are generated. The value of h_c increases for thinner substrates as

TABLE 5.2. *Summary of Dislocation Generation in* Si/Si:B *(Data from Sugita, Tamura, and Sugawara, 1969a)*

f	Experimental Dislocation Density (cm^{-1})	Calculated Dislocation Density (cm^{-1})
3×10^{-5}	~ 1	$(f_c \sim 1.2 \times 10^{-5})$
$1.1–1.6 \times 10^{-4}$	$200–300$	3.4×10^3
2.0×10^{-4}	770	5.3×10^3
3.0×10^{-4}	$>5 \times 10^3$	8.0×10^3

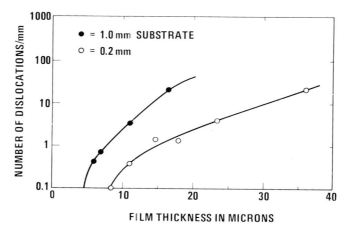

FIG. 5.23. Average misfit dislocation density versus film thickness for undoped Si grown on B doped Si substrates. (After Sugita and Tamura, 1969b.)

expected. For comparison, the calculated value of the critical thickness for an infinitely thick substrate is $\sim 3\,\mu$m.

These results agree semi-quantitatively with theory, but factors other than those considered here must be important in some cases. This is indicated by the failure of similar experiments performed at RCA to reproduce these results (Blanc, 1978).

5.5.3. III V ALLOY SYSTEMS

Defect generation in systems with larger values of Δa_0 may be illustrated by considering the heteroepitaxial growth of III V alloys such as $GaAs_{1-x}P_x$ and $Ga_{1-x}In_xP$ on GaAs substrates. $GaAs_{.6}P_{.4}$ is the material generally used for commercial red LEDs. The value of f for this alloy on the GaAs substrate is 1.5×10^{-2}. This is so large that $GaAs_{.6}P_{.4}$

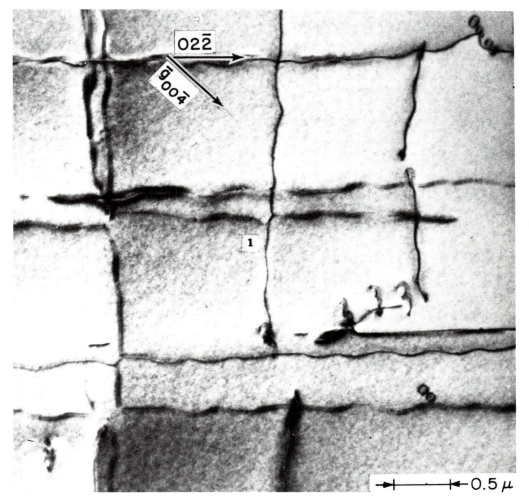

FIG. 5.24. Dislocation network in compositionally graded layer of $GaAs_{1-x}P_x$ grown on a GaAs substrate with (100) orientation. (After Abrahams *et al.*, 1969.)

grown directly on GaAs produces a high density of mismatch dislocations which interact to force some of the dislocations from the plant of the interface into a direction such that they propagate up into the layer itself (Stringfellow and Greene, 1969; Abrahams *et al.,* 1969). In order to reduce the density of these "inclined" dislocations, which deleteriously affect the photon generation efficiency of the material, a gradual increase in x from 0 at the interface to 0.4 after the growth of several microns of material was instituted. A TEM photograph of $GaAs_{1-x}P_x$ in this graded region is shown in Fig. 5.24. The segmented misfit dislocations are seen as well as inclined dislocations which propagate out of the plane of the photograph and appear as zigzag lines. Figure 5.25 shows how the density of these inclined dislocations decreases as the misfit dislocations are spread out by grading the composition increasingly more gradually.

In later studies on the growth of $Ga_{1-x}In_xP$ on GaAs, Olsen *et al.* (1975) discovered that an abrupt, small step in composition is beneficial because it bends the inclined dislocations into the growth plane, as indicated by Fig. 5.21, where they apparently terminate at the edge of the crystal causing a reduction in inclined dislocation density. Using this technique dislocation densities as low as $10^3 \, cm^{-2}$ have been obtained as compared to $\sim 10^5 \, cm^{-2}$ in crystals grown with a continuous grading (Stringfellow and Greene, 1969; Abrahams *et al.,* 1969).

The performances of most III V devices such as LEDs, solar cells, and photon detectors are severely degraded by the presence of dislocations. Thus, in recent years research has focused on alloys lattice matched to their substrates. $Al_xGa_{1-x}As$ is especially important because the covalent radii of Al and Ga are nearly equal, thus an alloy of any composition can be grown nearly lattice matched to the GaAs substrate. Furthermore, abrupt changes in composition necessary for modern devices such as solid state injection lasers (Kressel, 1977; Casey and Panish, 1978) and high performance solar cells (Woodall and Hovel, 1977) can be made without introducing undesirable dislocations at the interface.

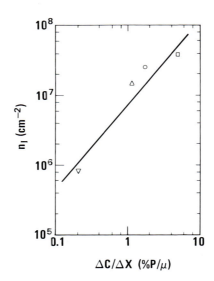

FIG. 5.25. Variation of mean values of inclined dislocation density with the rate of composition change during grading of $GaAs_{1-x}P_x$ layers on GaAs substrates. (After Abrahams *et al.,* 1969.)

Quaternary alloys such as $Ga_{1-x}In_xAs_{1-y}P_y$ are also being developed so that by changing x and y simultaneously the band gap and refractive index can be varied without affecting the lattice parameter (Sugiyama *et al.*, 1977; Moon, Chap. 11 of this book).

5.5.4. Si ON SAPPHIRE

The advantages of fabricating transistors in thin Si layers grown on insulating substrates have been recognized for some years. Such devices have lower parasitic capacitance, which leads to higher frequency operation and lower power dissipation (Cullen, 1978). The major problem blocking progress in this field has been the large number of defects generated due to the lattice parameter mismatch between Si and insulating substrate. For various reasons, including availability and freedom from harmful impurity contamination of the Si, single crystalline sapphire has become the standard insulating substrate. Single crystalline (100) oriented Si is found to grow on $(1\bar{1}02)$ sapphire. The sapphire is rhombohedral so the lattice parameter mismatch with Si is anisotropic. In Fig. 5.26 the Al atoms on the $(1\bar{1}02)$ sapphire surface are shown schematically with (100) Si atoms superimposed. The difference in interplanar spacing between the Al atoms and the unstressed Si atoms is 4.2% in the $(1\bar{1}01)$ direction and 12.5% in the $(1\bar{1}20)$ direction. Of course, there is no grading possible in this system, and the abrupt change in lattice parameter produces a high density of defects in the Si.

The nucleation and growth processes have been studied in detail using TEM techniques by Abrahams and coworkers (1976a, b) and Ham *et al.* (1977). The growth of Si islands has been followed from dimensions of $<100\,\text{Å}$ to coalescence into a continuous film. Problems related to defect generation are observed from the very beginning of the growth

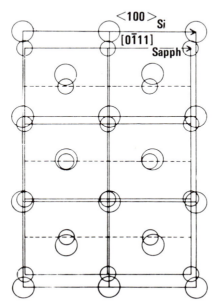

FIG. 5.26. Superposition of Si (100) surface atoms and Al_2O_3 $(01\bar{1}2)$ surface Al atoms. (After Larsson, 1966.)

process. Because of the large lattice parameter mismatch, individual islands are seen to contain many mismatch dislocations (Abrahams *et al.*, 1976a). In addition to angular rotations of $\sim 3°$, which would be expected as discussed in section 5.5.1, islands are observed having both (100) and (110) orientations. Twin orientations for each are also observed. When the islands coalesce this leads to the formation of microtwins, tilt boundaries, and stacking faults. The defect density is very large at the interface but annihilation of defects results in a rapidly decreasing defect density away from the

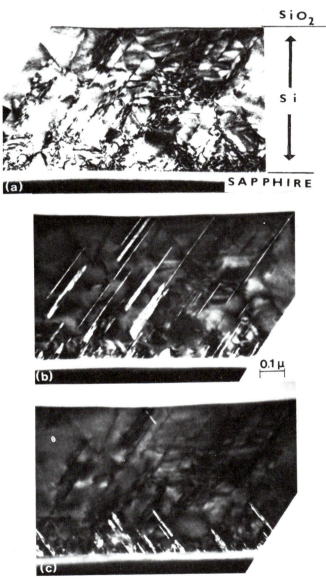

FIG. 5.27. Cross-sectional TEM micrograph of silicon on sapphire layer $\sim 0.6\,\mu$m thick. (After Ham *et al.*, 1977): (a) many beam bright field, (b) and (c) dark field images.

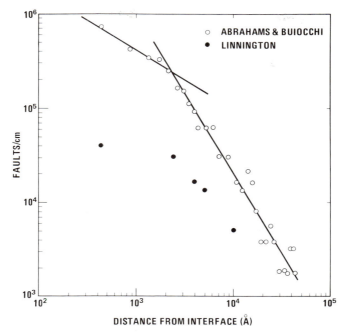

FIG. 5.28. Fault density in SOS layer versus distance from the interface with experimental data from Linnington (1971) and Abrahams and Buiocchi (1975). (After Abrahams and Buiocchi, 1975.)

interface. A cross-sectional TEM micrograph of a 0.6 μm thick Si layer is shown in Fig. 5.27. The defect density has clearly dropped abruptly after $\sim 0.2 \,\mu$m. In Fig. 5.28 the defect density plotted versus distance from the interface for a thicker film shows quantitatively the exponential decrease in dislocation density with distance from the interface.

Even in this imperfect Si, MOS transistors can be fabricated which have reasonable properties. This is due to the fact that enhancement mode transistors depend on transport of majority carriers in the material near the top surface, which has a relatively low defect density. The defects do act as generation centers in the drain–source depletion region and thus add to the leakage current (McGrievy, 1977). This limits the application of these SOS devices. To date they are not useful, for example, in dynamic RAM memory circuits. It would also be difficult to make bipolar transistors in SOS because the dislocations result in very poor minority carrier lifetimes (Kranzer, 1974).

III V compounds and alloys have been grown on insulating substrates such as sapphire (Wang, 1978) with relatively poor results since most devices fabricated in III Vs such as light-emitting diodes and photon detectors are minority carrier devices which cannot tolerate high dislocation densities.

5.6 Conclusions

This chapter is meant to be the basis of an understanding of some fundamental underlying aspects of vapor phase growth, particularly of single crystalline epitaxial layers. In the brief space available the emphasis has been placed on concepts rather than on extensive descriptions of practical processes. Examples of practical systems have been used to illustrate the concepts, and applications have been treated only very briefly. The topics of thermodynamics, mass transport, surface kinetic processes, and defect generation have been treated in the most detail. Other important topics such as point defects, both native defects and impurities and dopants, have not been treated, again because of lack of space.

It is clear that the fundamental aspects of vapor phase growth processes are not totally understood. Unfortunately, at the current time too little effort is being devoted to this area in contrast with the widespread efforts to develop vapor phase growth systems for use in production.

References

ABRAHAMS, M. S., WEISBERG, L. R., BUIOCCHI, C. J., and BLANC, J. (1969) *Material Science* **4**, 223.
ABRAHAMS, M. S. and BUIOCCHI, C. J. (1975) *Appl. Phys. Lett.* **27**, 324.
ABRAHAMS, M. S., BUIOCCHI, C. J., CORBOY, J. F., and CULLEN, G. W. (1976a) *Appl. Phys. Lett.* **28**, 275.
ABRAHAMS, M. S., BUIOCCHI, C. J., SMITH, R. T., CORBOY, J. F., BLANC, J., and CULLEN, G. W. (1976b) *J. Appl. Phys.* **47**, 5134.
ALEXANDER, E. G. (1967) *J. Electrochem. Soc.* **114**, 65c.
ARTHUR, J. R. (1966) *J. Appl. Phys.* **37**, 3057.
ARTHUR, J. R. (1968) *J. Appl. Phys.* **39**, 4030.
ARTHUR, J. R. (1974) *Surf. Sci.* **43**, 449.
BAN, V. S. (1971) *J. Electrochem. Soc.* **118**, 1473.
BAN, V. S. and ETTENBERG, M. (1973) *J. Phys. Chem. Solids* **34**, 1119.
BAN, V. S. and GILBERT, S. L. (1975a) *J. Electrochem. Soc.* **122**, 1382.
BAN, V. S. and GILBERT, S. L. (1975b) *J. Cryst. Growth* **31**, 284.
BAN, V. S. (1978) *J. Electrochem. Soc.* **125**, 217.
BASS, S. J. (1975) *J. Cryst. Growth* **31**, 172.
BERKMAN, S., BAN, V. S., and GOLDSMITH, N. (1978) in *Heteroepitaxial Semiconductors for Electronic Devices* (Cullen, G. W. and Wang, C. C., eds.), Springer-Verlag, New York.
BLANC, J. (1978) in *Heteroepitaxial Semiconductors for Electronic Devices* (Cullen, G. W. and Wang, C. C., eds.), Springer-Verlag, New York.
BLOEM, J. (1975) *J. Cryst. Growth* **31**, 256
BOOKER, G. B. and STICKLER, R. (1962) *J. Appl. Phys.* **33**, 3281.
BRADSHAW, S. E. (1966) *Int. J. Electronics* **21**, 205.
BRODSKY, M. H. (1978) *Thin Solid Films* **50**, 57.
BUNSHAH, R. F. (1974) *J. Vac. Sci. Technol.* **11**, 633.
BUNSHAH, R. F. (1976) in *New Trends in Materials Processing*, Am. Soc. Metals, Metals Park, Ohio, p. 363.
BURD, J. W. (unpublished results).
BURTON, J. A., PRIM, R. C., and SLICHTER, W. P. (1953) *J. Chem. Phys.* **21**, 1987.
BURTON, W. K., CABRERRA, N., and FRANK, F. C. (1951) *Phil. Trans. Roy. Soc.* (*London*) **234A**, 299.
BYLANDER, E. G. (1962) *J. Electrochem. Soc.* **109**, 1171.
CADORET, R. and CADORET, M. (1975) *J. Cryst. Growth* **31**, 142.
CARLSON, D. E. (1977) *IEEE Trans. on Electron Devices* **ED–24**, 449.
CASEY, H. C., JR., and PANISH, M. B. (1978) *Heterostructure Lasers*, Academic Press, New York.
CHIANG, Y. S. and LOONEY, G. W. (1973) *J. Electrochem. Soc.* **120**, 550.
CHO, A. Y. (1977) *Jpn. J. Appl. Phys.* **S16-1**, 435.
CONRAD, R. W., HOYT, P. L., and MARTIN, D. D. (1967) *J. Electrochem. Soc.* **114**, 164.
CULLEN, G. W. (1978) in *Heterostructure Semiconductors for Electronic Devices* (Cullen, G. W. and Wang, C. C., eds.), Springer-Verlag, New York.
CURTIS, B. J. and DISMUKES, J. P. (1972) *J. Cryst. Growth* **17**, 128.
DUCHAMIN, M. J. P., BONNET, M. M., and KOELSCH, M. F. (1978) *J. Electrochem. Soc.* **125**, 637.
DUPUIS, R. D. and DAPKUS, P. D. (1977) *Appl. Phys. Lett.* **31**, 839.
ETTENBERG, M. (1974) *J. Appl. Phys.* **45**, 901.
EVERSTEYN, F. C., SEVERIN, P. J. W., VAN DER BREKEL, C. H. J., and PECK, H. L. (1970) *J. Electrochem. Soc.* **119**, 925.
FARROW, R. F. C. (1974) *J. Electrochem. Soc.* **121**, 899.

FRANCOMB, M. H. (1975) in *Epitaxial Growth*, Part A (Matthews, J. W., ed.), Academic Press, New York.

FRANK, F. C. and VAN DER MERWE (1949) *Proc. Roy. Soc. (London)* **A198**, 216.

FRENKEL, J. (1932) *Physik Sowjet Union* **1**, 498.

GILMER, G. H. (1977) *J. Cryst. Growth* **42**, 3.

GOSSARD, A. C., PETROFF, P. M., WEIGMAN, W., DINGLE, R., and SAVAGE, A. (1976) *Appl. Phys. Lett.* **29**, 323.

HAM, W. E., ABRAHAMS, M. S., BUIOCCHI, C. J., and BLANC, J. (1977) *J. Electrochem. Soc.* **124**, 634.

HENDERSON, R. C. and HELM, R. F. (1972) *Surf. Sci.* **30**, 310.

HOLLAN, L. and SCHILLER, C. (1972) *J. Cryst. Growth* **13/14**, 325.

HUNT, L. P. and SIRTL, E. (1972) *J. Electrochem. Soc.* **119**, 1741.

ILEGEMS, M. and PEARSON, G. L. (1969) in *Proc. 1968 Symposium on GaAs*, The Institute of Physics and the Physical Society, London, p. 3.

INOUE, M. and ASAHI, K. (1972) *Jpn. J. Appl. Phys.* **11**, 919.

ITO, S., SHINOHARA, T., and SEKI, Y. (1973) *J. Electrochem. Soc.* **120**, 1419.

JACKSON, K. A. (1975) in *Treatise on Solid State Chemistry*, Vol. 5 (Hannay, N. B., ed.), Plenum Press, New 174.

JACKSON, K. A. (1975) in *Treatise on Solid State Chemistry*, Vol. 5 (Hannay, N. B., ed.), Plenum Press, New York, p. 233.

JESSER, W. P. and KUHLMANN-WILSDORF, D. (1967) *Phys. Stat. Solidi* **19**, 95.

JONES, M. E. (1969) in *Reactivity of Solids* (Mitchell, J. W., Devries, R. C., Roberts, R. W., and Cannon, P., eds.), Wiley, New York.

KAJIYAMA, K. (1976) *J. Electrochem. Soc.* **123**, 423.

DE KOCK, A. J. R. (1973) *Philips Res. Rept.* **S-1**.

KRANZER, D. (1974) *Appl. Phys. Lett.* **25**, 103.

KRESSEL, H. and BUTLER, J. K. (1977) *Semiconductor Lasers and Heterojunction LEDs*, Academic Press, New York.

LARSSEN, P. D. (1966) *Acta Cryst.* **20**, 599.

LINDQUIST, P. F. (unpublished results).

LINNINGTON, P. F. (1971) *Proc. 25th Ann. Mts. EMAG*, Inst. of Phys., Cambridge, England, p. 182.

MANABE, J. and GEJYO, T. (1971) *Jpn. J. Appl. Phys.* **10**, 1466.

MANASEVIT, H. M. and SIMPSON. W. I. (1969) *J. Electrochem. Soc.* **116**, 1725.

MANKE, C. W. and DONAGHEY, L. F. (1977) *J. Electrochem. Soc.* **124**, 561.

MATTHEWS, J. W. (1972) *Surf. Sci.* **31**, 241.

MATTHEWS, J. W. (1975) in *Epitaxial Growth*, Part B (Matthews, J. W., ed.), Academic Press, New York, p. 560.

McGRIEVY, D. J. (1977) *IEEE Trans. on Electron Devices* **ED–24**, 730.

MIZUMO, O. and WATANABE, H. (1975) *J. Cryst. Growth* **30**, 240.

MULLIN, J. B. and HURLE, D. T. J. (1973) *J. Luminescence* **7**, 176.

NAGAI, H., SHIBATA, T., and OKAMOTO, H. (1971) *Jpn. J. Appl. Phys.* **10**, 1337.

OLSEN, G. H., ABRAHAMS, M. S., BUIOCCHI, C. J., and ZAMEROWSKI, T. J. (1975) *J. Appl. Phys.* **46**, 1643.

PANISH, M. B. and ILEGEMS, M. (1972) in *Progress in Solid State Chemistry*, Vol. 7 (Reiss, H. and McCaldin, J. O., eds.), Pergamon Press, New York, p. 39.

PETROFF, P. M. and HARTMAN, R. L. (1974) *J. Appl. Phys.* **45**, 3899.

PETROFF, P. M., LORIMOR, O. G., and RALSTON, J. M. (1976) *J. Appl. Phys.* **47**, 1583.

PETROFF, P. M. and DE KOCK, A. J. R. (1976) *J. Cryst. Growth* **35**, 4.

PETZKE, W. H., GOTTSCHALCH, V., and BUTLER, E. (1974) *Krist. Tech.* **9**, 763.

PHILLIPS, J. C. and VAN VECHTEN, J. A. (1970) *Phys. Rev.* **B2**, 2147.

PHILLIPS, J. C. (1973) *Bands and Bonds in Semiconductors*, Academic Press, New York.

VAN DER PUTTE, P., GILING, L. J., and BLOEM, J. (1975) *J. Cryst. Growth* **31**, 249.

RAVI, K. V. (1976) in *Characterization of Epitaxial Semiconductor Films* (Kressel, H., ed.), Elsevier, New York.

REED, T. B. and LAFLEUR, W. J. (1972) *J. Cryst. Growth* **17**, 123.

RICHMAN, D. and ARLETT, R. H. (1969) *J. Electrochem. Soc.* **116**, 872.

ROKSNOER, P. J., BARTELS, W. J., and BULLE, C. W. T. (1976) *J. Cryst. Growth* **35**, 245.

ROKSNOER, P. J., HUIJBREGTS, J. M. P. L., VAN DER WIJGERT, W. M., and DE KOCK, A. J. R. (1977) *J. Cryst. Growth* **40**, 6.

ROSLER, R. S. (1977) *Solid State Technology*, April, **63**.

SCHLYER, D. J. and RING, M. A. (1976) *J. Organometallic Chem.* **114**, 9.

SCHLYER, D. J. and RING, M. A. (1977) *J. Electrochem. Soc.* **124**, 569.

SEDGWICK, T. O. (1964) *J. Electrochem. Soc.* **11**, 1381.

SEKI, Y. and MINAGAWA, S. (1972) *Jpn. J. Appl. Phys.* **11**, 850.

SEKI, Y., TANNA, K., IIDA, K., and ICHEKI, E. (1975) *J. Electrochem. Soc.* **112**, 1108.

SHAW, D. W. (1968a) in *Proc. 1968 Symp. on GaAs*, Inst. Phys. Soc., London, p. 50.

SHAW, D. W. (1968b) *J. Electrochem. Soc.* **115**, 405.

SHAW, D. W. (1970) *J. Electrochem. Soc.* **117**, 683.

Shaw, D. W. (1975a) *J. Cryst. Growth* **31**, 130.

Shaw, D. W. (1975b) in *Treatise on Solid State Chemistry*, Vol. 5 (Hannay, N. B., ed.), Plenum Press, New York, p. 283.

Sirtl, E., Hunt. L. P. and Sawyer, D. H. (1974) *J. Electrochem. Soc.* **121**, 919.

Smith, J. E. and Sedgwick, T. O. (1977) *Thin Solid Films* **40**, 1.

Steinmaier, W. (1963) *Philips Res. Rept.* **18**, 75.

Stowell, M. J. (1975) in *Epitaxial Growth*: Part B (Matthews, J. W., ed.), Academic Press, New York, p. 437.

Stringfellow, G. B. and Greene, P. E. (1969a) *J. Phys. Chem. Solids* **33**, 665.

Stringfellow, G. B. and Greene, P. E. (1969b) *J. Appl. Phys.* **40**, 502.

Sringfellow, G. B., Lindquist, P. F., Cass, T. R., and Burmeister, R. A. (1974) *J. Electron. Mat.* **3**, 497.

Stringfellow, G. B. (1974) *J. Cryst. Growth* **27**, 21.

Stringfellow, G. B., Weiner, M. E., and Burmeister, R. A. (1975) *J. Electron. Mat.* **4**, 363.

Stringfellow, G. B. and Hall, H. T. Jr. (1978a) *J. Cryst. Growth* **43**, 47.

Stringfellow, G. B. (1978b) *Ann. Rev. Mat. Sci.* **8**, 73.

Stringfellow, G. B. and Hall, H. T. Jr. (1979) *J. Electron. Mat.*

Suawara, H. (1972) *J. Electrochem. Soc.* **119**, 1740.

Sugita, K., Tamura, M., and Sugawara, K. (1969a) *J. Appl. Phys.* **40**, 3089.

Sugita, K. and Tamura, M. (1969b) *J. Vac. Sci. Technol.* **6**, 585.

Sugiyama, K., Kojima, H., Endo, H., and Shibata, M. (1977) *Jpn. J. Appl. Phys.* **16**, 2197.

Takahashi, R., Koga, Y., and Sugawara, K. (1972) *J. Electrochem. Soc.* **119**, 1406.

Theuerer, H. C. (1961) *J. Electrochem. Soc.* **108**, 649.

Thornton, J. A. (1977) *Ann. Rev. Mat. Sci.* **7**, 239.

Tietjen, J. J., Maruska, H. P., and Clough, R. B. (1969) *J. Electrochem. Soc.* **116**, 492.

Vink, A. T., Werkhoven, C. T., and von Opdorp, C. (1978) in *Semiconductor Characterization Techniques* (Barnes, P. A. and Rozgonyi, G. A., eds.), The Electrochemical Society, New Jersey.

Wahl, G. (1977) *Thin Solid Films* **40**, 13.

Wang, C. C. (1978) in *Heteroepitaxial Semiconductors for Electronic Devices* (Cullen. G. W. and Wang, C. C., eds.), Springer-Verlag, New York.

Werkhoven, C., von Opdorp, C., and Vink, A. T. (1977) *Inst. Phys. Conf. Ser.* **33a**, 317.

Werkhoven, C., Hengst, J. H. T., and Bartels, W. J. (1977) *J. Cryst. Growth* **42**, 632.

Weiner, M. A. (unpublished results).

Wilson, H. A. (1900) *Phil. Mag.* **50**, 238.

Woodall, J. M. and Hovel, H. J. (1977) *Appl. Phys. Lett.* **30**, 492.

CHAPTER 6

MBE—Molecular Beam Epitaxial Evaporative Growth

R. Z. BACHRACH

Xerox Palo Alto Research Center, 3333 Coyote Hill Road, Palo Alto, CA 94304, USA

6.1. Introduction

Molecular Beam Epitaxy (MBE)[1-3] is a crystal growth technique whereby constituents of the resultant solid are simultaneously evaporated from separate sources onto the growth surface. MBE has had a long evolution from general evaporative deposition techniques to its position today as an epitaxial growth method which can control mechanical, electrical, and optical properties. The evolution has been driven by the desire to use evaporative techniques to grow "device quality" crystals. Underpinning the evolution of evaporative methods into MBE has been the use of *in situ* diagnostic techniques which have allowed the epitaxy process to be decomposed and problems isolated. Thus as MBE has evolved, the application of surface analysis tools has had an instrumental impact on the success of the evolution. Today it is possible to grow with MBE a wide variety of semiconductor crystals in device structures as lasers, diodes, or microwave transistors which are competitive with devices fabricated from material grown by other methods. The attributes of MBE which have stimulated its development are listed in Table 6.1. Many of these characteristics are difficult to achieve with other epitaxial growth methods. Thus MBE creates the possibility for growing composite crystalline structures for device applications which have a number of unique aspects. These are perhaps best exemplified by integrated optics structures which are now being explored (see section 6.7).

TABLE 6.1. *Attributes of MBE*

1. Growth rates $1-10\,\mu/hr$
2. Low growth temperatures
3. Thin epitaxial crystals $<10-100\,\mu$
4. Dimensional control with large area uniformity
5. Atomically smooth interfaces
6. Hyper-abrupt composition changes
7. Hyper-abrupt doping profiles
8. Pattern growth through masks
9. Pattern growth with focused ionized beams
10. Selective area growth defined by electron beam writing

MBE is one of several methods presented in this book which can grow a variety of homo- or heteroepitaxial crystals. One can consider in general the sequence LPE (Chapter 11), VPE (Chapter 5), and MBE where the primary parameter is the ambient pressure over the substrate. MBE is performed at the $P \approx 0$ end of the growth phase diagram and with temperatures significantly lower than used in LPE or VPE. Figures 5.1 and 5.3 in Stringfellow's chapter on Vapor Phase Growth further discuss these relationships.

Table 6.2 lists examples of semiconductors grown by MBE and some references. In addition to these references, several reviews are also available.[2,3] As will become clear below, different crystal systems require individual considerations.

Because of the specialized nature of MBE, a discussion of MBE related apparatus is presented in section 6.2. For fuller detail on some aspects, the reader is referred to Chapter 4 on Environment for Crystal Growth by J. S. Shah, and a recent article by Luscher and Collins.[51]

A fairly good understanding of crystal growth with MBE exists today; however, many parameters and aspects remain to be explored more deeply. For example, the achievable growth rates as a function of crystal "quality" are not established. Recent evidence would indicate that the optimum growth temperatures are also not firmly established. Many of the questions that remain relate to the interplay of surface structures with doping phenomena. Interdiffusional processes are also receiving more attention. These latter questions will directly impact on the types of devices and their dimensions that can be grown with MBE.

TABLE 6.2. *Representative Semiconductors Grown by MBE*

IV	III–V	IV–VI	II–VI
Silicon (4, 5)	Binary	Binary	Binary
Germanium (6–8)	GaAs (9–16)	PbTe (34–36)	ZnTe (47–48)
	GaP (11, 12, 17–19)		
	InP (20–23)	PbS (35–38)	ZnSe (46, 48, 49)
	AlAs (24–27)	PbSe (34, 35, 39, 40)	CdTe (47)
	Ternary	SnTe (35)	CdS (50)
	GaAlAs (11, 14, 28–30)	Ternary	Ternary
	GaAsP (9, 31, 17, 18)	PbSnTe (36, 41–44)	ZnSeTe (46)
	InGaAs (32)	PbSnSe (39, 45)	
	GaSbAs (32, 33)	PbSSe (38)	

6.1.1. EVAPORATIVE METHODS OTHER THAN MBE

Other evaporative deposition methods besides MBE exist that can be used for epitaxial growth. The hot wall technique has recently been described in a review article by A. Lopez-Otero.[52] The hot wall technique is useful for growing materials with high vapor pressures at low temperatures. Thus in these systems, the evaporants are not pumped off after the first impingement on the substrate. Rather a hot wall containment vessel within the vacuum chamber makes it more likely for condensation on the substrate than escape.

Certain types of closed tube vapor transport techniques are also evaporative in nature. These are discussed in Chapter 5 by Stringfellow in the context of VPE.

6.2. Apparatus and Instrumentation

6.2.1. GENERAL

MBE is often referred to as a highly instrumented technique. This has occurred in the evolution from the technique of free evaporative deposition because of the complexity of the overall process. The initial failures to achieve deposition of semiconductor material with the desired qualities required that experiments be performed to understand the overall process including determination of evaporative species, adsorption processes, and resulting surface structures at the growth temperature. The research apparatus usually employed multiple instrumentation.

Three techniques contributed significantly to the progress of MBE. These are quadruple residual gas analysis, high energy electron diffraction analysis,[53, 4] and Auger electron spectroscopy.[55, 6] Most research systems now include some aspects of these techniques, although growth is possible without them.

The utilization of quadruple mass analyzers allowed the effective diagnosis of the gas ambient as well as the investigation of source composition and desorption phenomena. Auger spectroscopy is very powerful for examining surface cleanliness and composition. High energy electron diffraction, which was developed in the 1930s, has had an important impact on the development of epitaxy since it allows crystallinity to be determined at growth initiation. HEED and then LEED in turn have shown the complexity of surface structures that exist on semiconductor surfaces and in co-ordination with Auger the dependence of the structures on composition.

The discussion will first describe vacuum system requirements, then growth-related equipment, and, finally, equipment for *in situ* analysis.

6.2.2. VACUUM SYSTEMS

The use of UHV for MBE resulted from an attempt to minimize the contamination resulting from unintentional adsorption processes. The sticking coefficient for a particular species relative to the growth rate is an important parameter determining impurity incorporation. In most cases, adsorption to the metals before they have reacted is more important than interaction with the semiconductor formed. The contaminants arise from several sources: the ambient gas background; reacted gases on the sources; other hot filaments; or unintentional evaporants from the sources. The result of these effects can range from disruption of epitaxy to poor control of doping, to poor control of minority carrier lifetime.

The vacuum technology used in MBE has typically been introduced as it has become available. The emphasis has been on removing or minimizing the static and dynamic ambient background. Today it is both accessible and common place to configure primarily stainless steel MBE systems capable of 5×10^{-12}–5×10^{-11} base pressure. Systems for the IV–VI's still often use glass bell jar type vacuum enclosures.

Examples of modern multifunction systems are shown pictorially in Fig. 6.1,[57] with an associated interlock fixture in Fig. 6.2, and in Figs. 6.3[58] and 6.4.[51] These systems use a variety of vacuum pumping techniques[59] including: (i) carbon vane, (ii) sorption, (iii) titanium sublimation, (iv) liquid nitrogen cryopanels, (v) ion pump, and (vi) He cryo pump. The first two are used for initial roughing while the latter are particularly effective

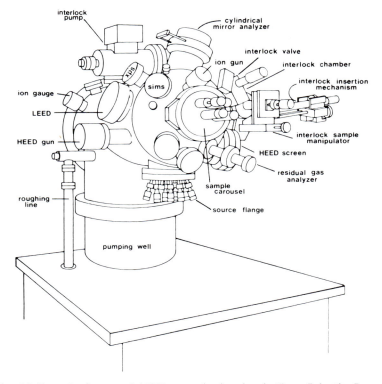

FIG. 6.1. Example of a research MBE system developed at the Xerox Palo Alto Research Center, and built by Physical Electronics Inc. The system incorporates a high capacity UHV pump system, a work area, and an interlock preparation chamber. The various elements are shown on the figure and described in the text.[57]

in combinations for obtaining ultra high vacuum (UHV). Modern design turbo molecular pumps and diffusion pumps can be used in some cases.

A wide variety of systems which have been individually configured are in use today and complete systems are currently available from Physical Electronics, Riber, VG and Varian among others. These systems, as those shown in Figs. 6.1, 6.3, and 6.4, are typically multichambered with a mechanism for passing the sample between them.

Figure 6.1 shows a system used in our laboratory which was designed for research studies into the physics of film growth, surfaces, interfaces, and devices. The system incorporates a wide range of capabilities for studying the molecular beam epitaxy films *in situ* during and after growth. The analytical tools provided for include HEED, LEED, scanning Auger, XPS, UPS, and SIMS. Residual gas analysis and evaporation monitoring can be performed in several configurations. Ion milling can be utilized both for surface cleaning and depth profiling. A tee-shaped 18 in. vacuum system incorporating an integral 500 l/sec ion pump, titanium sublimation, and water and cryo-cooled surfaces effectively delivers high pumping speed to the source–growth region. The growth region is defined by a cryo-cooled shroud which confines the evaporants. Seven evaporator positions with a 7 cm source to substrate distance are individually water isolated and shuttered. Three side ports give access for additional evaporators or instrumentation for during growth measurements. A substrate carousel incorporates four 700°C heaters and a Faraday cup. A precision *xyz* manipulator allows accurate resettable sample alignment. A port on the manipulator allows direct access to the evaporation position. Vacuum performance of the system has been excellent, and a base pressure of $\leq 4 \times 10^{-11}$ torr is readily achieved.

The vacuum interlock and transfer mechanism was designed subsequent to the design of the primary system. It allows removable substrate holders to be inserted into the main chamber while maintaining the system base pressure. The transfer substrate holder and the carousel receiver are shown in Fig. 6.2. The holder is physically passed and locked in this design so that during growth the interlock valve is closed.

The receiver and holder are made from molybdenum. The heater sits in the well with the thermocouple passing through the middle and spot welded to the spring probe which contacts the substrate holder when the two are mated. This provides reproducible temperature measurements.

The interlock chamber incorporates associated processing so that intermediate steps can be incorporated. Thus before growth, the substrate can be thermally outgassed and sputter cleaned in the interlock. Following growth, metal evaporation can be performed for contacts prior to removal from the vacuum system.

Figure 6.3 shows a commercial MBE system currently available from Physical Electronics Inc. which was principally specified by J. R. Arthur. The system consists of a growth chamber which is cryopumped, an ion and titanium sublimation pumped analysis chamber, and a sample entry chamber. At this time, the sample sits on a single probe which accesses all chambers. The probe transport when inserted fills the interchamber aperture

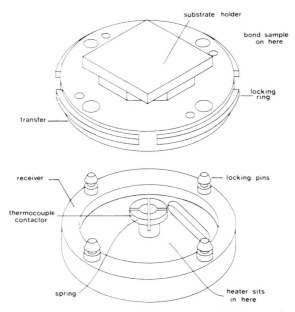

FIG. 6.2. Transfer mechanism used in MBE system in Fig. 6.1. The molybdenum substrate holder is physically inserted and locked to the receiver. The heater and thermocouple remain intact so that reproducibility is maintained. An objective of this design was to provide a definitive and stable orientational transfer. This is established by the four pins.

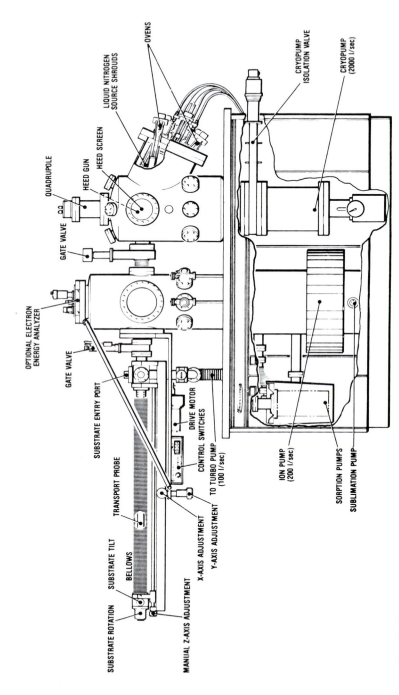

FIG. 6.3. A commercial modular MBE system developed by Physical Electronics Inc. A single probe accesses an interlock chamber, an analysis chamber, and the growth chamber. The growth chamber is pumped by a closed cycle helium cryopump. (Courtesy J. R. Arthur, Physical Electronics Inc.)

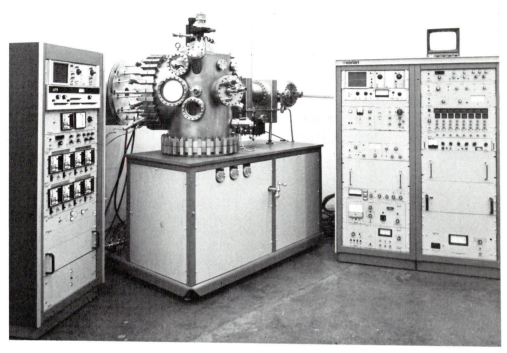

Fig. 6.4. A commercial MBE system and supporting electronics developed by Varian Inc. This system incorporates an interlock and a growth-analysis chamber. The source flange on the left in this case incorporates larger sources. (Courtesy P. Luscher, Varian Inc.)

so that they are effectively isolated. This configuration is similar to one used by Smith.[44] The primary pumping chambers in this system are 12 in. in diameter.

Figure 6.4 shows a commercial MBE system currently available from Varian. This system uses a primary 18 in. diameter growth and analysis chamber and transfer chamber. As in the first system described, the substrate is physically transferred onto another manipulator. Both of the latter two systems use sources with the substrate in the vertical so that flaking does not return into the sources. This can be a problem with near vertical evaporation systems such as in Fig. 6.1 and special precautions need to be taken.

6.2.3. EVAPORATION SOURCES

Evaporation source design is a key element in effective MBE systems. The two most commonly used heating methods are resistive ovens and electron beam.

Figure 6.5 shows a schematic of an effusion oven system as is commonly used in III–V MBE growth with one evaporator per element.[61] These sources are not true effusion type ovens but provide many of the control aspects. They also provide larger flux and more easily controlled beam uniformity at the substrate.

The material is typically contained in a pyrolitic boron nitride crucible. For the highly reactive Ga, Al, or In, this is essential while for other materials it is convenient. The

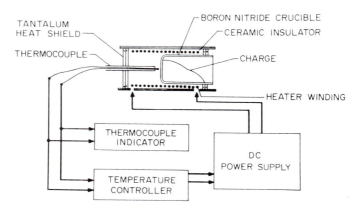

FIG. 6.5. Schematic of a typical MBE evaporation source in use today with the associated circuitry. A pyrolitic boron nitride crucible is heated by a well-insulated oven. The thermometer is closely coupled and with a proportional controller maintains a stable temperature.[50]

crucible sits in a heater winding which is well shielded by multiple layers of thin tantalum foil. The foil is typically dimpled for spacing and conductive isolation. A thermocouple sits in close proximity to the crucible and, with a proportional controller, regulates the temperature. A variety of thermocouples are useful, but in general they need to be isolated from contact with Ga, Al, or In. Often used thermocouples are chromel–alumel; platinum–platinum rhodium; and tungsten–rhenium.

For many of the evaporants, the temperature is high enough that radiative coupling equilibrates the source. For low temperature sources such as arsenic (where the material is also typically lumps) the difficulty of establishing conductive equilibrium effects the design strategy.

With effective design the primary power loss from the sources will be radiative from the evaporation aperture. Thus 30 W is sufficient to reach 1000°C with a typical crucible aperture of 1 cm. Lateral coupling between the sources is not strong, but it is typically useful to remove the excess heat by surrounding the sources with a water cooled shroud as shown in Fig. 6.6.

Figure 6.6 shows the multiple port evaporation shroud used in the system of Fig. 6.1. Each of seven sources is isolated by the inner water shroud as well as being individually shuttered. The water shroud and growth region is surrounded by a liquid nitrogen shroud. Pumping is through the inner annulus and side holes not shown.

Figure 6.7 shows the angular distribution obtained from sources such as shown in Fig. 6.5 and used in source shrouds such as in Fig. 6.6.[51] The intensity as a function of angle is parameterized by the collimation tube length l to crucible diameter r. The flux from a source is discussed further in section 6.3. The data in Fig. 6.7 can be used to estimate the evaporation uniformity for a given source substrate configuration.

For Si MBE, electron beam evaporation sources are required to achieve the desired $1-2\,\mu$/hr growth rates. Figure 6.8 shows an example of an evaporator used by Ohta.[5] The major innovation here is the incorporation of the silicon collecting rings. High purity silicon can be used for the source and more conductive silicon for the collector. The collecting rings create a return path to ground for secondary electrons without introducing unwanted contamination.

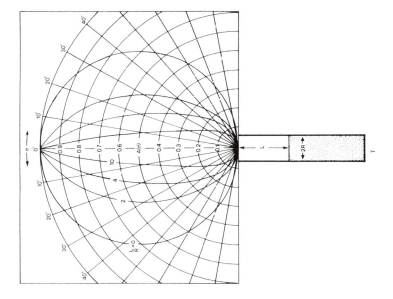

FIG. 6.7. Evaporation angular distribution function on a polar plot parameterized as a function the source diameter r to collimation length l ratio. $1/r = 0$ corresponds to a full crucible. (Luscher and Collins.[51])

FIG. 6.6. Example of the evaporation housing used in the MBE system shown in Fig. 6.1. The sources are isolated by a water-cooled shroud and then surrounded by a liquid nitrogen reservoir. Each source is individually shuttered.

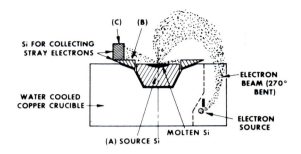

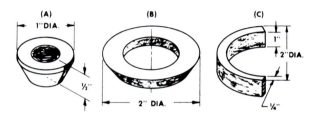

FIG. 6.8. Silicon electron beam evaporator source developed by Ohta. An important part is the silicon collecting rings for secondary electrons which eliminates contamination. (Ohta.[5])

6.2.4. SUBSTRATE HOLDERS AND HEATERS AND SAMPLE MANIPULATORS

The specific design of substrate holders and heaters depends on the particular crystals to be grown and the substrate temperatures required. For the III–V's, heater arrangements such as shown in Fig. 6.2 suffice.[57] The substrate is typically held to the molybdenum block with indium which is well behaved in the temperature range to 700°C.

For silicon, much higher temperatures are required. Figure 6.8 shows the scheme used by Ohta where resistive substrate heating was used to achieve temperatures to 1100°C. In this case, one contact is mechanically floating so that the crystal is not strained.

Direct radiative heating has also been used. In this approach an oven is held directly behind the substrate. One design uses a graphite meander hot plate.[62] In these designs, one has to consider the possible effect of evaporation onto the substrate. With the large temperature gradient across the sample, heater material can diffuse through onto the growth surface. Radiative heating is also typically used for the IV–VI's since they also must be held in a strain free manner.

Typically the substrate holder needs to incorporate thermometry which can monitor or control the surface temperature. Because of the high thermal conductivity of molybdenum, measuring the holder block temperature near the substrate suffices. For silicon, it is typically necessary to use some form of optical pyrometer.

Determining the actual surface temperature accurately is difficult. For example in the configuration shown in Fig. 6.9 typical of that used for III–Vs, the optical power incident on the substrate from the multiple source heaters can amount to 100 W.

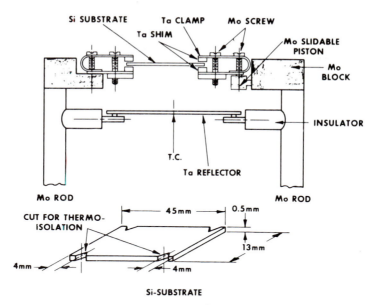

FIG. 6.9. Silicon substrate holding scheme using radiant heating. A back reflector helps minimize the heat loss. The clamp on the right is free to minimize straining the substrate. (Ohta.[5])

6.2.5. INSTRUMENTATION

A full treatment of instrumentation is beyond the scope of this chapter, but briefly discussing the most important techniques is useful.

6.2.5.1. *Gas Pressure and Composition*

Knowledge of the ambient gas pressure is important. The Bayard–Alpert ionization gauge is the standard type in use in UHV systems and covers a pressure range from approximately 10^{-4} to 5×10^{-11} torr. The primary disadvantage of this type of gauge is that the ionization of some gas species significantly enhances their surface reactivity. Thus some care is required, particularly where control of low level oxides are important. (This is true of hot filaments in general.) Standard gauges and controllers are commercially available. Another type of gauge which avoids the ionization problem and has a wider range is the trigger gauge. This gauge is useful in some circumstances.

The quadruple electrostatic residual gas analyzer is useful for determining gas composition in the range 1–300 AMU. Good units typically have 1/2 AMU resolution and 10^{-12} torr partial pressure sensitivity.

The RGA can monitor simultaneous source fluxes directly, while an ionization gauge can be used to monitor the sources individually. Both require that the ionizer be inserted in the beam for accurate results. In some cases, the emission from the RGA can interfere with an ion gauge.

6.2.5.2. *Deposition Monitors*

Standard deposition monitoring techniques can be useful for calibrating sources, but specific problems arise. Arsenic alone, for example, does not stick well while gallium tends to destroy the crystal metallization. The electron induced atomic or molecular fluorescence technique is useful but difficult to calibrate absolutely unless a quartz crystal monitor can be used as a calibration standard. Smith has used back to back heated quartz crystal pairs to monitor deposition of simultaneous fluxes of Ga and As or Zn and Te.[44] This approach avoids some of the problems discussed above.

6.2.5.3. *Thermometry*

Accurate thermometry is the key to controlling MBE. Typically one needs to reproducibly control the sources and substrate each to ± 0.2 K. The most common thermocouples in use are chromel–alumel; $Pt-6\%Rh-Pt-30\%Rh$; and $W-5\%Re-W-26\%Re$. Each of these has certain advantages or disadvantages. In all cases, it is necessary to arrange the wires so that gallium, which readily amalgamates, is not deposited. This is particularly so for the heated wires.

Positioning the thermocouples so that their readings are representative of the surface temperatures is a difficult task. The considerations change depending on the temperature as one goes from a radiative to conductive coupling. Optical pyrometry is useful in some circumstances, but often physically difficult to accomplish with the necessary accuracy.

6.2.5.4. *HEED—High Energy Electron Diffraction*

High energy electron diffraction has played an important role in studies of epitaxy. Figure 6.10 shows a typical schematical arrangement. A high energy electron gun illuminates the growth substrate at a shallow angle ($\approx 3°$) and the diffracted beams are

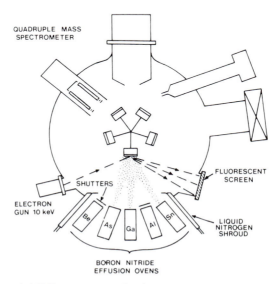

FIG. 6.10. Schematic MBE arrangement showing evaporation ovens and HEED geometry.

detected by a fluorescent screen. Because the gun to sample distance can be large, HEED can be performed during growth. Pashley has a good review of this subject.[63]

Figure 6.11 shows some examples of HEED patterns and micrographs of the associated surfaces.[15] Full analysis of surface structure requires that the substrate be rotatable around the azimuthal axis with electron polar angle fixed. For growth diagnostics, however, a fixed arrangement can be used. One can align substrates with respect to a cleavage side. The presence or absence of the HEED pattern is then an indication that the epitaxy is proceeding. One can also observe changes in surface structures depending on the impinging surface fluxes. For example with GaAs, one can distinguish between an As-rich and Ga-rich regime.

A smooth surface such as shown in Fig. 6.11b gives a bar like diffraction pattern whereas rough surfaces result in spotted patterns. Beeby has recently given a theoretical description of HEED which supersedes some of the historical explanations.[64]

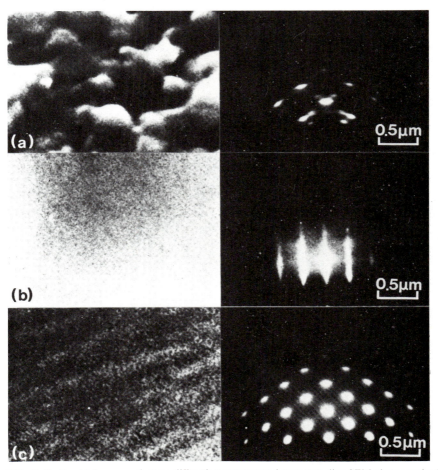

FIG. 6.11. Reflection high-energy electron diffraction patterns and corresponding SEM photographs. The substrate temperature was (a) 430°C, (b) 480°C, and (c) 580°C. The thickness of the layers was 1000 Å on a GaAs (100) semi-insulating Cr-doped substrate. This is an example of how HEED pattern appearance relates to surface morphology.

6.2.5.5. *Auger Electron Spectroscopy*

The addition of functional Auger electron spectroscopy added an important element to the success of MBE development. Extensive reviews of Auger spectroscopy exist.[55, 56] Briefly, a focused high energy ($\geq 2\,keV$) electron beam incident on the sample ejects core electrons. The de-excitation process, which fills the core hole, includes a process whereby another electron with an energy characteristic of the transition is emitted. This Auger electron is energy analyzed and detected and spectra such as shown in Fig. 6.12 obtained. Figure 6.12 shows the derivative of the current versus electron energy for a GaAs surface at various stages of cleaning. This is discussed more in section 6.4. Important aspects are that

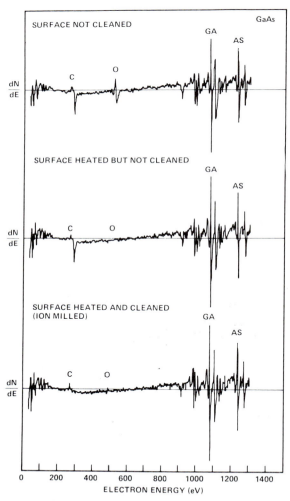

FIG. 6.12. Auger derivative spectra showing the cleaning stages of GaAs. The upper spectrum shows the surface after polishing and insertion from ambient. The middle spectrum shows the surface after heating to 550°C which removes the oxides. Removal of the carbon shown in the bottom spectrum utilized argon ion bombardment. As discussed in the text, step three can be bypassed if the surface is properly polished.

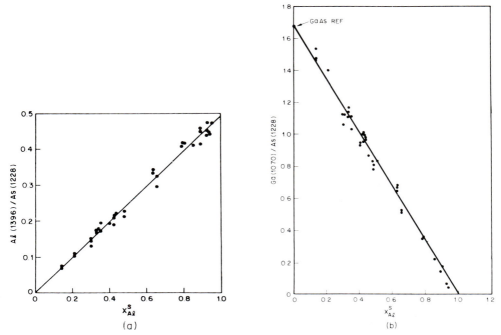

FIG. 6.13. (a) Aluminum to arsenic Auger peak ratio. (b) Gallium to arsenic Auger peak ratio versus the aluminum concentration for $Ga_xAl_{1-x}As$.[68]

the peaks can be associated with particular elements and that the intensities can be related to concentrations. For example, the data shown in Fig. 6.13a, b were obtained by Arthur[65] and they show the ratios of the Ga/As and Al/As peaks versus Al concentration in $Ga_{1-x}Al_xAs$. Thus one can quantitatively determine the composition. In general, significant effort is required to establish Auger to an accurate quantitative level. Scanning Auger can also be used to examine both area distribution of elements and, in conjunction with sputter etching, depth distributions.

For well characterized growth systems it is now often common to dispense with using Auger on a routine basis.

6.2.5.6. *Other Surface Analysis*

A variety of other surface analysis tools can be included in an MBE system for *in situ* studies. These are in general not essential for the growth process, but they make use of MBE's capability to prepare exposed atomically clean surfaces for study. These tools include:

(a) SIMS: secondary ion mass spectrometry. This uses a modified mass analyzer in conjunction with an ion etching source to measure composition levels which are too low to detect by Auger. The sensitivity limit in special cases is $\leq 10^{17}\,cm^{-3}$.

(b) XPS and UPS. Photoemission with properly selected excitation energy can provide a highly surface sensitive measurement. Valence band spectroscopy at low energies and core level spectroscopy at higher energy can give significant surface chemical and electronic information. Photoemission experiments are just beginning on *in situ* grown MBE. In the future, one can anticipate systems being configured to also utilize synchrotron radiation sources as excitation.

(c) LEED. Low energy electron diffraction is a powerful technique for studying surface structure. LEED is typically done with 20–500 eV electrons incident normally on the sample. Because of the close coupled geometry, LEED cannot be done during growth. Thus although one directly can obtain the surface structure, it is most useful for after growth analysis.

6.3. MBE—Crystal Growth

Several aspects distinguish MBE from other growth techniques besides the background vacuum ambient. The principal one is the fact that the substrate and material sources are held at different temperatures. In the original attempts to grow binary semiconductors, this came to be termed by Gunther as the three-temperature method.[1] From an evolutionary point of view, the exploration of growth with multiple sources is the progenitor of MBE growth. The advances of Arthur and of Cho were most responsible for establishing MBE as a viable materials technology.[2]

We will use below several growth compound systems as examples. The considerations for these are somewhat different because of the basic phase diagram and structural differences. For the most part in this chapter, the GaAs and GaAlAs system will be used as the specific example. The basic MBE system for a binary compound consists of evaporating elements M and N:

$$
\begin{array}{cccc}
\mathrm{M} & \mathrm{N_n} & \mathrm{MN} & \\
| & + \quad | & \rightarrow \quad | & \quad (6.1) \\
T_1 & T_2 & T_3 &
\end{array}
$$

to form the compound MN. The subscript n indicates a molecular species in the gas phase. The source temperatures T_i result in the various fluxes. In general, the major vapor species over the compound semiconductors MN are MN, M, and N_2. For some systems discussed below, an MN source is preferable or M and N are evaporated from separate crucibles held at the same temperature in a common oven and the flux ratio determined by adjusting the source aperture areas A, in eqn. (6.2). Simple extension of relation (6.1) applies to ternary, quaternary, or doped systems. Crucible evaporation sources which meet the effusion cell criteria produce a flux density along the normal to the cell

$$
F \simeq 1.11 \times 10^{22} \, [AP(T)]/[D^2(mT)^{1/2}] \quad \text{molecules/cm}^2/\text{sec} \quad (6.2)
$$

where $P(T)$ is the equilibrium pressure at the cell temperature T, D is the distance to the substrate, m the atomic mass of effusing species. In practice, the sources such as shown in Fig. 6.5 are more convenient and the equation will provide a lower bound estimate. Figure 6.7 shows the angular distribution from such a source.

For real systems as typically used $A \sim 0.8$–$1 \, \text{cm}^2$, $D \sim 7$–$12 \, \text{cm}$ and $P(T) \sim 10^{-2}$–10^{-3} torr in order to achieve a flux at the substrate of 10^{15}–10^{16} molecules/cm²-sec and a growth rate of 1–10 monolayers/sec.

General elemental evaporation data are available in the tabulations of Honig and Kramer.[66] These can provide a general guide. Gallium, for example, can be characterized by[51]

$$P(T) = 10^{[-(11021.9/T) + 7 \cdot \log T - 15.42]} \tag{6.3}$$

from these data and then F evaluated.

Pressure versus temperature diagrams provide one with the surface evaporation rate for a free surface as a function of temperature. Smith has collected the data[67] shown in Fig. 6.14 which represents the temperature at which the equilibrium vapor pressure equals 10^{-6} torr or the surface loses approximately a monolayer/sec. MBE growth will typically occur most optimally below this temperature.

The general aim of epitaxial growth is to produce controlled stoichiometry crystals. Thus one needs to understand the thermodynamics of the grown crystal as well as the deposition process. The principle concepts are introduced in section 6.31, but the reader is referred to the references for expanded details.

6.3.1. PHASE EQUILIBRIA AND STOICHIOMETRY—GaAs

The Ga–As phase diagram has been determined by a number of investigators,[68] whose similar results showed the 50–50 congruently melting compound with a melting point

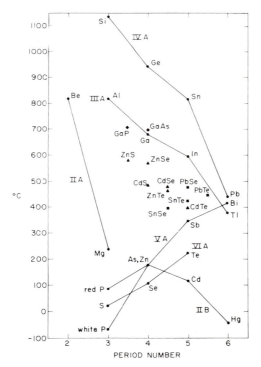

FIG. 6.14. Temperature at which evaporation rates of the binary semiconductors and their constituent elements reach 10^{-6} torr (approximately 1 monolayer/sec or 1 μm/hr evaporation rate).[67]

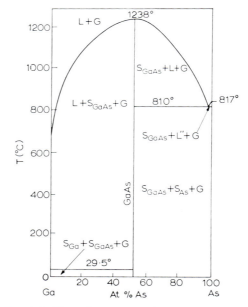

FIG. 6.15. Phase diagram $(T–x)$ for Ga–As systems.

considerably above that of either component. Thurmond has also constructed the liquidus curve which shows a eutectic of composition 2.4 at./%Ga at a temperature of 810°C.[69] The m.p. of arsenic is 817°C and of GaAs 1238°C. The complete phase diagram ($T–x$ projection) is shown in Fig. 6.15. The narrow stoichiometry range is not shown.

The liquidus curve provides the basis for the evaluation of equilibrium pressure measurements, and it was computed by Thurmond as follows. Vieland has shown that the liquidus curve for a pure compound in equilibrium with a regular solution is given in terms of an interaction energy α

$$\alpha = -(RT)/[2(0.5 - x)^2]\,[\ln 4x(1 - x) + (\Delta S^F)/R(T_M/T - 1)] \tag{6.4}$$

where x is the atom fraction of arsenic in the saturated liquid phase, ΔS^F is the entropy of fusion per mole of compound, assumed temperature independent, and T_M is the melting point. For a regular solution, α is independent of temperature and composition (in this context regular solutions are those for which the heat of mixing is proportional to the atom fractions of the two components, and the excess entropy of mixing is ideal). Using available data it was found that α is not constant, i.e., the liquid phases of this system are not regular, but are given by

$$\alpha = 5160 - 9.16T \tag{6.5}$$

using a ΔS^F value of 16.64 e.u. mole^{-1}. The liquidus curve corresponding to α given by eqn. (6.5) can then be plotted.

Having established the liquidus curve, the following reactions must be considered in order to calculate the pressure–temperature curves along the liquidus. Standard enthalpies are as quoted by Arthur:[70]

$$GaAs_{(s)} \rightarrow Ga_{(s)} + 1/2As_2(g), \qquad \Delta H_{298°} = 44.9 \pm 0.5 \, kcal \qquad (6.6)$$

$$GaAs_{(s)} \rightarrow Ga + 1/4As_4(g), \qquad \Delta H_{298°} = 29.4 + 0.7 \, kcal \qquad (6.7)$$

$$2As_2(g) \rightarrow As_4(g), \qquad \Delta H_{298°} = -62.5 + 1.5 \, kcal \qquad (6.8)$$

$$GaAs_{(s)} \rightarrow Ga(g) + As(g), \qquad \Delta H_{298°} = 155 + 2.0 \, kcal \qquad (6.9)$$

Several investigators have determined the equilibrium pressures of the species As_2, As_4, and Ga over GaAs, and the most reliable data are probably due to Arthur.[70] In common with other workers he used a Knudsen cell technique, but ensured that the mass spectrometer sampled only material at the cell temperature. Using these data, the activity coefficients for gallium and arsenic along the binary liquidus can be calculated. The activity coefficient of arsenic in gallium is given by

$$\gamma_{As} = 1/X_{As}[P_{As_4}/P^0_{As_4}]^{1/4} \qquad (6.10)$$

where X_{As} is the atom fraction of arsenic, P_{As_4} is the partial pressure of As_4 above saturated solution, and $P^0_{As_4}$ is the pressure over pure arsenic. γ_{Ga} may be calculated from the γ_{As} curve, since for the decomposition of GaAs to form As_4 and a liquid phase (eqn. 6.7),

$$K_2 = X_{Ga}\gamma_{Ga}(P^0_{As_4})^{1/4} \qquad (6.11)$$

Therefore from eqn. (6.10)

$$X_{Ga}\gamma_{Ga} = K_2/[\gamma_{As}X_{As}(P^0_{As_4})^{1/4}] \qquad (6.12)$$

$K_2 = (P_{As_4})^{1/4}$ at low temperatures since $\gamma_{Ga}X_{Ga} \rightarrow 1$ in dilute solution and at high temperatures can be calculated from free energy and heat of vaporization data. From the activity data, the vapor pressure data and Thurmond's[69] values for $P^0_{As_4}$, Arthur[70] calculated the vapor pressures of As, As_2, As_4, and Ga along the complete GaAs liquidus. His results are shown in Fig. 6.16 with m.p. pressures of $P_{As_4} = 0.648$ atm and $P_{As_2} = 0.328$ atm. At a certain temperature the vapor phase will have the same composition as the solid phase, i.e. when $P_{Ga} = 2P_{As_2} + 4P_{As_4}$, and at this temperature GaAs will evaporate congruently, while at higher temperatures it will decompose. Using Arthur's data, this temperature is 910 K. Thus below 910 K there must be a crystal composition within the solidus curve which evaporates congruently.

6.3.2. STOICHIOMETRY

Stoichiometry will be discussed for bulk GaAs in this section and then in the context of MBE doping in section 6.6.2.

The stoichiometric behavior of a compound MN can be described in terms of the T–x projection of the equilibrium phase diagram. The generalized form is shown in Fig. 6.17a for the excess metal fraction x plotted versus temperature T.[67] The single solid phase MN is stable over some range below the melting point of MN. A second phase appears in the form of an M or N precipitate beyond the edge of this region (the solidus boundary) so that the solidus boundary represents the solubility limit of point defects. The width of the single-phase existence region for most compound semiconductors is very small or of the order of 10^{-3} to 10^{-6}. The creation of point defects in the lattice which effect

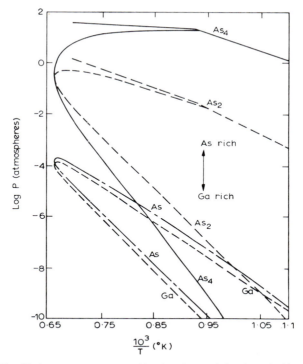

FIG. 6.16. Equilibrium vapor pressures of As, As$_2$, As$_4$, and Ga along the binary liquidus. The straight lines are vapor pressures of As$_2$ and As$_4$ over pure liquid and solid As.[70]

stoichiometry can be described by an activation energy E_a. The solubility limit of the defects generally increases exponentially with T or

$$Vm \sim e^{-Ea/kT} \tag{6.13}$$

in the region well away from the melting point. This retrograde solubility is often a source of difficulty in crystal growth. If the solidus is crossed upon cooling a nonstoichiometric crystal, microprecipitates may be formed in the bulk.

A vapor phase in equilibrium with the condensed matter exists at any point on the T–x diagram and the equilibrium partial pressures of various vapor species will be functions of T and x. Since nonstoichiometric MN is effectively a dilute solid solution of excess M or N, the equilibrium partial pressure of N$_2$ would be expected to increase with concentration of dissolved N as the N-rich solidus boundary is approached. The increase in pressure from boundary to boundary of the existence region is typically many orders of magnitude. Figure 6.17b illustrates this in generalized form for a given T shown by the dashed line in Fig. 6.17. The second condensed phase begins to form when the solidus boundaries are reached and the partial pressures are then fixed. This is in accordance with Gibbs's phase rule that the degrees of freedom equals the components minus the phases plus two. For two components and with two condensed phases plus the vapor phase, only one degree of freedom is left. Thus once T is specified, so are all the partial pressures as well as x for each phase. Within the single-phase region, however, there is a second degree of freedom so that

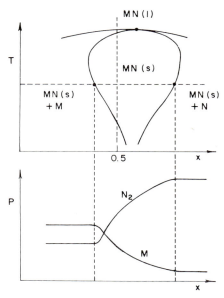

FIG. 6.17. General form of phase equilibria for nonstoichiometric binary semiconductors $M_{1-x}N_x$: (a) is the T–x projection, and (b) is the P–x projection.[67]

x may be determined by setting one of the partial pressures. This last fact is most significant for MBE. Even though MBE cannot be considered an equilibrium process, the composition of the grown film may be driven toward the N-rich or M-rich boundaries by controlling the N_2/M ratio of molecular beam fluxes. By consideration of data such as in Figs. 6.14 and 6.16, and using data from eqn. (6.2) and ref. 67, one finds in general that the impinging fluxes equivalent partial pressure at the surface far exceeds the equilibrium vapor pressure at the substrate temperature. MBE typically occurs under supersaturated conditions. The equilibrium phase diagrams also apply to MBE to the extent that the solidus boundaries are encountered in the form of defect concentration saturation or of precipitates on the growth surface if N_2/M is varied over too wide a range. Aspects of this effect are discussed in section 6.6.2.

The available variable one can adjust in MBE growth to effect the composition x of the grown film are the substrate temperature and the fluxes of the evaporants onto the substrate. The substrate temperature is usually constrained within a range of $\sim 100°C$ bounded by re-evaporation of the film on the high end and by loss of crystal quality on the lower end. For a fixed substrate temperature, one can adjust N_2/M in the beams to avoid precipitates and to regulate x within the single-phase region to optimize the electronic properties of the film.

From purely thermodynamic considerations any compound must have an existence region of finite width at temperatures $> 0\,K$.[71] The width of the region is determined by the curvature of the Gibbs free energy–composition curve, the Gibbs free energy of formation of the compound, and the mutual solubility energies of the components. Thus in principle it is possible for the stoichiometric composition of a compound to be outside its existence region at $T > 0\,K$.

A considerable body of experimental evidence shows high concentrations of point defects in GaAs. From lattice parameter measurements Potts and Pearson[72] observed concentrations of 10^{19} cm^{-3}. More recently, Willoughby et al.[73] have carried out similar measurements and concluded that concentrations of arsenic interstitials of up to 2×10^{18} cm^{-3} exist in as-grown material. They consider that annealing in vacuum leads to the formation of arsenic monovacancies, at a quenched-in concentration of 4×10^{17} cm^{-3}. These quenched-in defects anneal out in two stages: a rapid first stage, by recombination of Frenkel pairs from last interstitial migration, and a slower second stage by the diffusion of arsenic monovacancies.

A theoretical analysis using standard thermodynamic methods has been carried out by Logan and Hurle[74] who have derived expressions for the equilibrium concentration and have derived expressions for the equilibrium concentration, as a function of As$_2$ pressure, of arsenic monovacancies, gallium monovacancies, gallium divacancies, and charged versions of these defects. Equilibrium constants were estimated and matched with experimental data where possible. Defect concentrations of 10^{18}–10^{19} cm^{-3} in melt grown material were obtained in this way. The values obtained for the equilibrium constants were used to calculate the minimum practical deviation from stoichiometry and also the width of the existence region of solid GaAs as a function of temperature. Their results are shown in Fig. 6.18. Note the highly expanded scale compared with Fig. 6.14. GaAs MBE is typically performed in the 825–928 K temperature range so that the stoichiometry range is very narrow.

6.3.3. DEPOSITION

Before discussing MBE processes in detail in sections 6.4 through 6.7, general deposition considerations will be introduced.[1, 3, 52] The number of molecules deposited

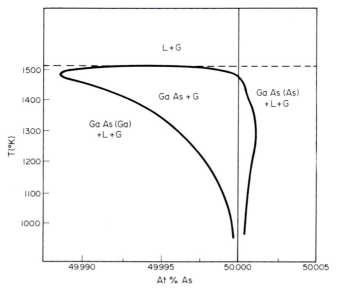

FIG. 6.18. The calculated existence region for GaAs on an expanded scale.[74]

on a substrate can be expressed without establishing the details of the growth process. The rate of deposition for an element per unit area is from eqn. (6.2):

$$R_d = \alpha\{\,[F_e(T_e)]/[(2\pi M_e k T_e)^{1/2}] - [F_s(T_s)]/[(2\pi M_s k T_s)^{1/2}]\,\} \qquad (6.14)$$

where α is a condensation coefficient,[1,3,52] e refers to the evaporant, and s to the substrate. Note the respective fluxes are at the substrate surface. In general $R_d > 0$ only if the incident molecular rate R_e exceeds some critical value $R_c(T)$, which is a function of the type and condition of the substrate and its temperature. In the ideal case, as might happen for homoepitaxial growth, no supersaturation is necessary to start the deposition. In most practical cases, and particularly for heteroepitaxial growth, it is necessary that the supersaturation F/F_e be greater than unity to start the growth. The adsorbed molecules need to overcome an energy barrier in order to combine and form a stable nucleus which can subsequently grow. The value of this energy barrier increases with decreasing supersaturation and is theoretically infinite for $F/F_e = 1$.

Similar consideration applies to binary or ternary compounds. When the vapor phase in equilibrium with a compound MN is mainly composed of AB molecules, the deposition process is basically similar to that for a single element. This is the case for many for the IV–VI compounds. Most binary compounds dissociate upon evaporation so that the vapor phase consists of two separate components M and N. When these two elements have comparable vapor pressures, dissociation does not present a difficult problem for the epitaxial growth of a compound from a single source. This is because the interaction of the components with each other on the substrate surface can lead, under favorable growth conditions, to the formation of MN molecules and therefore to the eventual growth of the compound.[1] However, when one constituent is much more volatile than the other, condensation of the stoichiometric compound becomes almost impossible, and separate sources containing the elements A and B have to be employed.

Whatever the situation, the density N_{MN} of adsorbed MN molecules will be proportional to the product $N_A N_B$ of the densities of adsorbed atoms M and N and a mean diffusion coefficient D, i.e.,

$$N_{MN} = \text{const.} \times N_M N_N D. \qquad (6.15)$$

The density $N_j (J = M, N)$ of adsorbed particles is given by

$$N_j = R_i^j \tau_j \qquad (6.16)$$

where R_i^j and τ_j are the incident rate and the mean stay time (before being desorbed) respectively of the particles j. τ_j is given by[75]

$$\tau_j = (1/v_j)\exp(E_{des}^j/KT) \qquad (6.17)$$

where v_j and E_{des}^j are the surface vibrational frequency and the desorption energy respectively of the particles j. We observe that the density N_{MN} should also be proportional to the product of the incident fluxes $R_i^M R_i^N$.

A simple way of describing the deposition process for a binary compound is shown in Fig. 6.19 where we plot, following Gunther,[1] the total deposition rate R_d^{M+N} of M and N as a function of the incident rate of one of the components R_i^A. The incident rate of the other component R_i^N and the substrate temperature T are kept constant. We assume that $R_e^{MN} < R_i^M < R_c^N$ so that no condensation of the pure component N alone takes place such that condensation of the compound MN will be possible when R_i^M becomes sufficiently

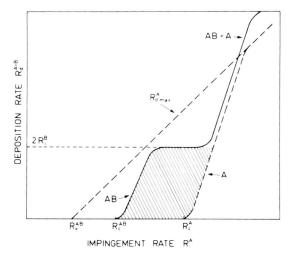

FIG. 6.19. The deposition rate of a compound AB as a function of the impingement rate of component A for fixed impingement rate of component B and constant temperature.

large. R_e^{MN} corresponds to the vapor pressure in the gas phase of the compound MN in equilibrium with its condensed phase at temperature T. For $R_i^M < R_e^{MN}$, fewer MN molecules will be formed than are necessary to keep the equilibrium state corresponding to the substrate temperature. However, at a critical incidence rate $R_i^M = R_c^{MN} > R_e^{MN}$, which again depends on the nucleation mechanism and on the type and condition of the surface, sufficient molecules MN are formed to start nucleation and growth. With increasing R_i^M the deposition rate R_d^{M+N} increases rapidly to a maximum value $R_{d\,max}^{M+N}$ $= 2R_i^N$. At this stage almost all the impinging particles of component B react with the impinging particles of component M to form the compound MN. The maximum value $R_{d\,max}^{M+N}$ is limited by the fixed value R_i^N so that a further increase in R_i^A does not result in an increase in R_d^{M+N} (the excess M particles are re-evaporated). At a value $R_i^M > R_c^M$ condensation of noninteracting fluxes takes place, resulting in a layer rich in the component M.

The deposition process of a binary compound can also be described by plotting R_d^{M+N} in terms of the substrate temperature T. For some fixed values of the incident rates R_i^M and R_i^N, the formation of the compound MN on the surface of the substrate can only take place below a certain critical temperature T_c^{MN} until at temperatures $T < T_c^M$ and $T < T_c^N$ additional condensation of the pure components M and N takes place. In Fig. 6.20 the composition of the condensed layers as a function of R_i^M and R_i^N is shown for a given substrate temperature T. The region of existence of the compound AB is limited by the lines $R_i^M = R_c^M R^{MN}$, the lines $R_i^N = R_c^N R_c^{MN}$, and the curve $R_i^M R_i^N = $ const. The curve is connected with the probability of interaction between M and N on the substrate surface. It is clear from Fig. 6.20 that the growth of a compound from separate components in the vapor phase is possible provided that the growth conditions are right. Important information on the establishment of the appropriate growth conditions is obtained from the phase diagrams of the compounds discussed previously.

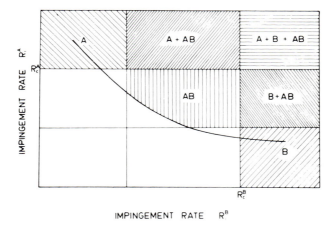

FIG. 6.20. The composition of the material condensed on a substrate as a function of the impingement rates of components A and B for constant temperature T.

6.4. Substrate Preparation

The MBE process typically starts with a crystalline substrate inserted from room ambient after some preparation. The substrate typically is covered with a contamination layer consisting of carbonaceous compounds and oxides.[2,77,78] Two examples for discussion are GaAs and Si. In both these cases, the final polishing preparation is often terminated in such a way that an oxide film is left. Surface studies have clarified the preparation necessary to initiate epitaxial growth. Substrate surface preparation is a critical part of successful MBE. Surface damage, which is beyond the scope of this review, can yield poor epitaxy.

In the case of GaAs, several different cleaning regimes exist which depend on how well carbon is excluded from the GaAs–oxide interface upon termination of polishing. In the worst case for GaAs, heat treatment in vacuum can remove the oxide, but ion sputtering is required to remove the carbonaceous component. This is exemplified in Fig. 6.12, which shows Auger derivative spectra at three stages of cleaning. Oxygen desorption becomes appreciable above 500°C, and the oxide film is removed within several minutes. These cleaning procedures leave the surface potentially in a variety of surface reconstructions which are discussed in the next section. Studies have shown that contamination must be removed to 0.1 monolayer or less in order for the epitaxy to proceed well. Once a clean surface is achieved, annealing can restore crystallinity. Typically, however, the resulting surface will not be stoichiometric.

If the surface is prepared properly in terminating the polishing, the need for argon sputtering can be avoided. The heat treatment required is as high as 650° to remove all the carbon and oxygen. Stoichiometry is restored with the initial arsenic flux prior to initiating growth.

Similar procedures work for silicon, but silicon is more reactive. The silicon surface is particularly sensitive to carbon contamination, presumably through the formation of silicon carbide. The carbonaceous contamination can disrupt epitaxy. Silicon MBE, which produces "device quality" layers, seems to be best achieved in helium cryo-pumped vacuum systems, but turbo-pumped systems have also been successfully employed. Such systems achieve lower partial pressures of carbon containing molecules such as CO, CO_2, CH_4, etc. Silicon cleaning is typically more difficult and temperatures in the $1200°C$ range are used. Argon ion sputtering can remove these layers, but damage is introduced which must be annealed prior to initiation of epitaxy. Thermal cleaning appears to be the preferred method for silicon to date. Bean *et al.* have investigated the damage associated with argon sputtering using helium backscattering.[79] They find that although the surface damage can be annealed away for a wide variety of conditions as observed with *in situ* techniques such as HEED and AES, other techniques which are more sensitive show that the extent and concentrations of defects depend strongly on substrate temperature. Cleaning with a cold substrate and then annealing leaves the best substrate from which to initiate epitaxy if argon sputtering is used.

6.5. Surface Structures

Surface structures of semiconductors have been under active investigation for many years.[80] The free surfaces tend to undergo both structural reconstruction and relaxation. These effects depend on the surface composition in the case of GaAs or other III–V's. Overlayers add further complication.

LEED patterns can provide surface structures in the vicinity of room temperature.[81] HEED is able to provide information at the growth temperature. The importance of surface structures will appear repeatedly in the following discussions of effects related to surface states, doping effects, and overlayers. At the outset therefore the structural aspects have to be elucidated. Many of these surface structures represent phases with varying surface composition while some represent reconstructions without compositional change. In addition to the structure, positional relaxations can occur which maintain the bulk periodicity. These effects can be exemplified by discussion of the epitaxy of Si, GaAs, and ZnSe or from covalent to relatively ionic crystals.

Figure 6.21 shows schematically the side and top views of the zincblende structure (100), (110), and (111) surfaces. These low index faces are typically used for epitaxy. The (100) surface is used for GaAs laser related work because two parallel (110) planes can be used to cleave mirror faces. Examining these figures is instructive since one can readily see the origin of the reconstructions that occur and the plausible way they are related to stoichiometry.

The complexity of these surface structures is exemplified by a study of the MBE of ZnSe on GaAs.[82] These two semiconductors nominally lattice match. However, in the growth on a GaAs (100) surface, ZnSe forms in a C (2×2) and even for a thick film retains this surface structure. Silicon (100) typically exhibits a 2×1 surface reconstruction, and the (111) face is quite complex.[83]

The following discussion will focus on GaAs since much of the MBE work has involved this III–V semiconductor. Figure 6.21 represents the ideal surface termination. Referring to the side view, the distances between the outermost layers are $1/4a_c$, $1/[2(2^{1/2})]a_c$, and $(3^{1/2}/4)a_c$ respectively for the (100), (110), (111) faces. For GaAs at room temperature, $a_c = 5.657\,Å$. The (110) face is now known to relax in such a way that the arsenic moves out by $0.2\,Å$ and the gallium moves in by $0.45\,Å$.[84] The displacement of $0.65\,Å$ is comparable to the $0.82\,Å$ interplanar separation for the ideal (111) face. All of the real faces then

ZINCBLENDE

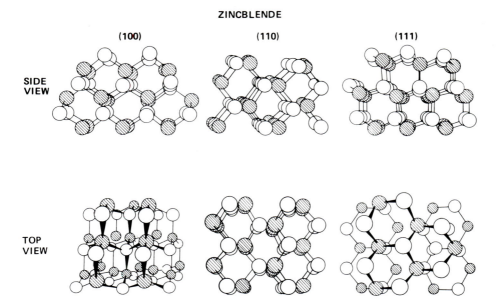

FIG. 6.21. Zincblende (100), (110), and (111) ideal surface models showing side and top views. The (110) surface exhibits a (1 × 1) reconstruction where the arsenic moves out of the plane and the gallium into the plane. The (100) and (111) surfaces exhibit a variety of stoichiometry related reconstructions discussed in the text. (After Duke.[80])

display a layer of atoms where one or the other species is predominantly exposed. In the case of the (100) and (111) faces, this leads to a variety of reconstruction which are determined by the site occupancy.

Studies of GaAs by Cho showed that two principal regimes occur on the (100)[85] and (111)[86] faces and the transition between them is driven by the surface stoichiometry. This result was established conclusively in a study by Arthur in which structural determination was coordinated with Auger.[87] The details of these surface structures have been extended by a number of investigators.[88–90] The issue arises since stoichiometric changes in surface position can also result in reconstructions. This in fact happens on the silicon surface.

Cho showed using HEED that the (111) GaAs face had two phases, a 2×2 reconstruction and a $19^{1/2} \times 19^{1/2} R23.4°$ structure which depended on the relative As to Ga flux incident on the surface. These are termed Ga-stabilized and As-stabilized.[84] Ranke and Jacobi investigated the compositional dependence of these structures.[91] Auger signals for differently reconstructed (111) surfaces where obtained and related to the composition.

Cho found similar structural results for the GaAs (100) surface.[85] The (100) surface, however, shows a large number of phases. Arthur[2,87] and later Joyce and Foxon[92] investigated the compositional dependence of the surface reconstructions using flash desorption measurements. Arthur showed that substantially different As$_2$ mass desorption curves as a function of temperature were obtained for the As-stabilized and

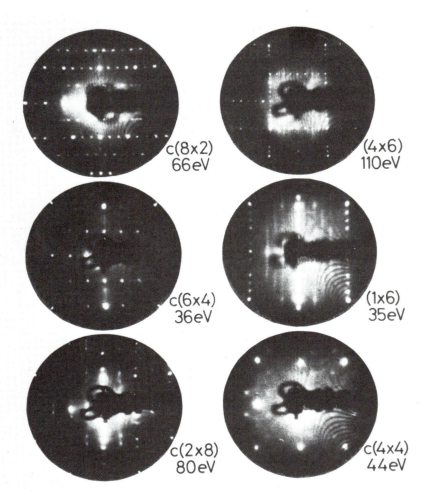

FIG. 6.22. LEED photographs showing some of the different room temperature GaAs (100) surface reconstructions that can be observed as a function of surface composition. The relative As/Ga ratio was measured with Auger spectroscopy. A (1 × 1) surface not shown can also be prepared.[93]

Ga-stabilized structures. A peak associated with excess arsenic being driven off was observed and the structure converted to Ga-stabilized.

Figure 6.22 shows LEED photographs obtained by Drathen *et al.* which exhibit the range of structures.[93] The phase boundaries of these structures have been studied by Massies *et al.* who obtained similar LEED data.[94] The fractional arsenic coverage associated with these phases is shown in Table 6.3. At the growth temperatures typically used (550–600°C) the dominant phases are the C(8×2) Ga and the C(2 × 8) As which converts at approximately 1/2 monolayer arsenic excess to deficit. Figure 6.23 shows the phase ranges established by Massies *et al.*[94]

TABLE 6.3. *Vapor Pressure Versus T*

	10^{-2}	10^{-3}	$Tm°C$
Arsenic	584	544	817
Gallium	1320	1205	30
Aluminum	1495	1370	660
Indium	1215	1105	157
Tin	1510	1380	232
Beryllium	1480	1360	1278
Silicon	1905	1750	1410

6.6. Adsorption and Desorption

MBE deposition proceeds via adsorption processes on the crystalline substrate. The primary adsorbates are the elements of which the epitaxial semiconductor is composed, secondly the evaporated dopants, and, thirdly, the unintentional contaminants. The eventual growth of the crystal on the cleaned substrate proceeds in stages which can include physisorption, chemisorption, and chemical reaction. At the temperatures at which semiconductor epitaxy occurs, physisorption binding is weak. If physisorption is the main sticking process, then the surface residence time will be short and no deposition will occur. Chemisorption and chemical reaction are the more important processes. The details of how these proceed depend on the surface structures which have already been introduced. In addition, the nucleation and growth can generate or propagate structural defects.

The primary surface structures discussed in the previous section markedly affect the surface interaction with adsorbates. Three examples of epitaxy can be drawn from GaAs

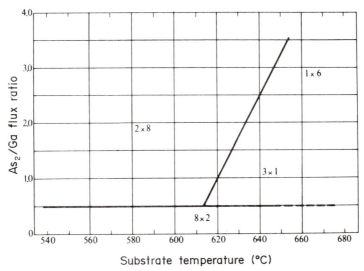

FIG. 6.23. Phase ranges of surface reconstruction on GaAs (100) in the vicinity of the growth temperature as a function of As$_2$/Ga flux ratio.[94]

substrates and epitaxy of GaAs,[90] ZnSe,[82] and Al[95, 96] respectively. The interaction of the elements involved is very different. This depends also on the molecular species in the gas phase. Gallium and aluminum evaporate monatomically, As evaporates typically as As_2 and As_4,[97] Zn as Zn_2[98] and Se as Se_2, Se_4, and Se_8.

The discussion needs to be divided into the primary composition elements and the intentional and unintentional dopants.

6.6.1. INTERACTION OF As, Ga, AND Al ON GaAs

Growth of GaAs or GaAlAs on GaAs is controlled by the gallium and aluminum arrival rate at epitaxial growth temperatures since arsenic does not stick. The kinetic processes involved with gallium and arsenic interaction have been reviewed by Joyce and Foxon.[92, 99] The relevant details are discussed here. Similar studies although less detailed have been carried out for the II–VI's by Smith.[46]

The initial issue is how the elemental flux interacts with an isolated surface and then secondly how the interaction is modified in the presence of concurrent fluxes.[1] Figure 6.24 shows as an example the desorption of arsenic from a GaAs (100) surface as a function of surface temperature in the presence of an As_2 flux of 10^{13} mol cm^{-2} sec^{-1}. The source in this case was GaAs. The As_4 then results from an association reaction on the surface.

The desorption energies for As_2 and As_4 have been determined to be 0.38 eV and 0.58 eV respectively. This compares with gallium, which has a desorption energy of 2.48 eV for the GaAs (111) face and is strongly chemisorbed. The gallium is mobile, however, and has a surface diffusion range of 200 Å at the growth temperature.[2] The origin of the chemisorption is seen from photoemission studies of gallium deposited on the (110) GaAs

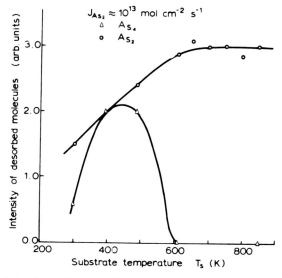

FIG. 6.24. Relative desorption rates of As_2 and As_4 for an incident As_2 flux as a function of substrate temperature.[97]

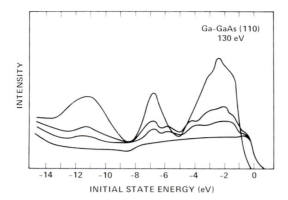

FIG. 6.25. Evolution of the valence band photoemission spectra going from cleaved GaAs (110) to a thick gallium metal overlayer. The intermediate exposures represent one-half and one monolayer of gallium on GaAs. The new peaks that arise at -4.2 eV and -5.8 eV reflect bonding across the interface.[100]

face at room temperature.[100] Figure 6.25 shows valence band density of states of GaAs as a function of coverage. In the low coverage range, interface states are observable which indicate that bonding is occurring across the interface.

Aluminum with GaAs is an interesting case because one can form epitaxial AlAs, GaAlAs ternary alloy, or one can epitaxially grow aluminum metal.[95, 96] The work of Dingle *et al.* has shown that the interdiffusion coefficient of AlAs in GaAs is very small so that one can grow alternating layers of GaAs and GaAlAs which are stable to 900°C.[103, 104] In the case of aluminum on GaAs, however, the work of Bachrach *et al.* has shown that an exchange reaction takes place and an interfacial layer of AlAs forms between GaAs and Al.[100, 103] This is an example of how the reactions can differ in the presence of concurrent fluxes.

The exchange reaction is shown by the Al2p and Ga3d core level photoemission spectra presented in Fig. 6.26 as a function of aluminum coverage. The initial deposit of 0.53Å Al onto room temperature GaAs (110) was almost completely converted into AlAs. This is seen by the 0.7 eV chemically shifted aluminum peak to higher binding energy. Gallium, on the other hand, shows a corresponding shift to lower binding energy indicating metallic gallium. The fractional peak intensities are consistent with a complete monolayer of AlAs being formed. The difference between the chemisorption case and the chemical exchange reaction is exemplified in Fig. 6.27. The heat of reaction which favors this exchange also results in GaAs diffusion into the overlayer. The presence of arsenic during growth would presumably suppress the exchange reaction and allow abrupt heterojunctions to be grown.[101, 102]

6.6.2. DOPING AND STOICHIOMETRY IN MBE

Doping effects with MBE for groups IV, III–V, II–VI, and IV–VI semiconductors are complex. Many of the dopants which are useful in other forms of epitaxy do not behave well with MBE. The actual behavior is strongly dependent on substrate temperature and

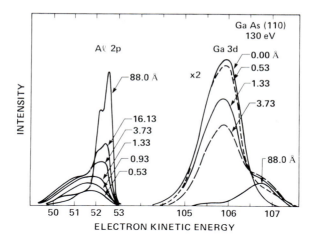

FIG. 6.26. Aluminum 2p and gallium 3d core level photoemission spectra as a function of aluminum coverage. The chemical shift indicating an exchange reaction takes place creating an interfacial layer of AlAs and releasing gallium metal into the overlayer. The aluminum coverage sequence is 0.53, 0.93, 1.33, 1.73, 3.73, 16.13, and 88Å.[104]

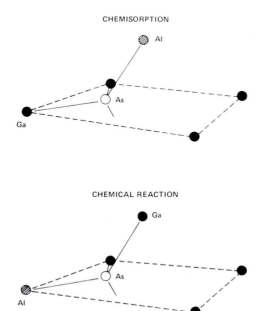

FIG. 6.27. Schematic of the GaAs (110) unit cell depicting difference for deposition with chemisorption and chemical reaction.

the surface reconstruction involved with growth. The doping regime is typically $< 10^{19}\,\mathrm{cm}^{-3}$. Above $10^{20}\,\mathrm{cm}^{-3}$ it becomes an alloy regime.

Commonly used dopants in GaAs are Sn, Ge, and Si for n-type[105] and Mn and Be for p-type.[106, 107] In silicon, antimony is used for p-type and gallium for n-type. Aluminum does not behave well in silicon for profile control because of the strong interdiffusion.[5, 108] The IV–VI's can be doped by control of stoichiometry.[34, 67]

Ilegems has provided a general organization to impurity behavior in GaAs grown by MBE[109] as shown in Fig. 6.28. Figure 6.28 plots the ratio of P_{min}/P_{eq} as a function of substrate reciprocal temperature. P_{eq} is the impurity partial pressure in equilibrium with the doped solid; and P_{min} is the impurity flux needed to achieve the desired impurity concentration in the grown solid during MBE growth if every incident impurity atom is incorporated. To first order, P_{min}/P_{eq} is independent of impurity concentration, and for values greater than 0 the incorporation is arrival rate controlled while for values less than

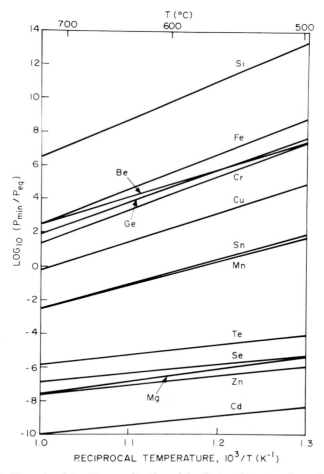

FIG. 6.28. The ratio of P_{min}/P_{eq} as a function of the GaAs substrate reciprocal temperature. P_{eq} is the desired impurity partial pressure in the grown solid and P_{min} is the minimum impurity flux needed to achieve the impurity incorporation.

TABLE 6.4. *Impurity Incorporation in GaAs by MBE*[a]

Element	Donor or Acceptor	Maximum Carrier Concentration (cm^{-3})	Sticking Coefficient	Reference
Si	D	$\sim 5 \times 10^{18}$	1	g
Sn[b]	D	$\sim 10^{19}$	1	h
Te	D	$\sim 10^{19}$	0.5–1	i
Ge[c]	D or A	$\sim 5 \times 10^{18}$	1	j
Be	A	$> 10^{19}$	1	k
C[d]	A	—	—	l
Cd	Not incorporated	—	Very low	m
Mg	A	10^{16}–10^{17}	$\sim 10^{-5}$	n
Mn[e]	A	$\sim 10^{18}$	—	o
Zn[f]	Not incorporated	—	Very low	p

[a] At the usual MBE substrate temperature range of 500–600°C.

[b] Doping profile depends on substrate temperature.

[c] Conductivity type depends on the arsenic to gallium ratio in the beams. For Ga-rich conditions, layer is *p*-type, while for As-rich conditions, layer is *n*-type.

[d] Carbon concentration increases with the ratio of arsenic to gallium in the beams.

[e] For As-rich growth conditions. Less manganese is incorporated when the arsenic to gallium beam intensity ratio is decreased.

[f] Zinc has been incorporated with an ionized-zinc beam (see ref. p).

[g] A. Y. Cho and I. Hayashi, *Metall. Trans.* **2**, 777 (1971).

[h] A. Y. Cho, *J. Appl. Phys.* **46**, 1733 (1975).

[i] J. R. Arthur, *Surf. Sci.* **43**, 449 (1974).

[j] A. Y. Cho and I. Hayashi, *J. Appl. Phys.* **42**, 4422 (1971).

[k] M. Illegems, *J. Appl. Phys.* **48**, 1278 (1977).

[l] M. Illegems and R. Dingle, *Gallium Arsenide Related Compounds: 1974 Symp. Proc.*, p. 1, Inst. of Phys., London, 1975.

[m] J. R. Arthur, private communication.

[n] A. Y. Cho and M. B. Panish, *J. Appl. Phys.* **43**, 5118 (1972).

[o] J. R. Arthur, *Surf. Sci.* **38**, 394 (1973).

[p] M. Naganuma and K. Takahashi, *Appl. Phys. Lett.* **27**, 342 (1975).

zero the incorporation depends primarily upon equilibrium vapor pressure of the impurity. Impurities in the first category will be expected to be easily incorporated and to be effective dopants for MBE GaAs.

Table 6.4 summarizes the observed behavior of a number of impurity elements in GaAs grown by MBE. Comparison of Table 6.4 and Fig. 6.28 shows that the qualitative approach is reasonably predictive. Tellurium is an exception because of the formation of a stable surface phase.[109]

Beryllium is a particularly well behaved *p*-type dopant in GaAs and will be discussed in more detail. Figure 6.29 shows the induced carrier concentration versus reciprocal source temperature, for beryllium in GaAs, one of the better controlled *p*-type dopants. The actual quality of the crystal obtained is dependent on the surface reconstruction which is exemplified with photoluminescence measurements in Fig. 6.30. The crystal grown under Ga-rich conditions is considerably more efficient. This result is consistent with studies of deep traps with capacitative spectroscopy by Lang *et al.*[110] Figure 6.31 shows an example of such a result, in this case for an *n*-type sample. In these studies, a variety of unidentified levels appear. The number and concentration, however, are substantially larger when

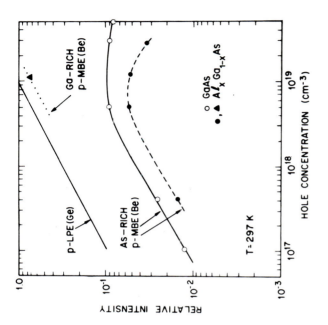

FIG. 6.30. Relative photoluminescence intensity for beryllium doped GaAs (○) and $Al_{0.17}Ga_{0.7}As$ (●) layers grown under As-rich conditions and for $Al_{0.17}Ga_{0.83}As$ (△) layer grown under Ga-rich conditions. The results for Ge-doped LPE GaAs are shown as a reference.[107]

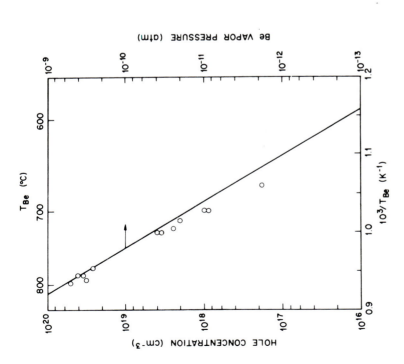

FIG. 6.29. Hole concentration versus beryllium effusion temperature. The average growth rate was 1.4 μm/hr with a cell substrate distance of ~ 5 cm. The solid line represents the beryllium equilibrium vapor pressure curve.[107]

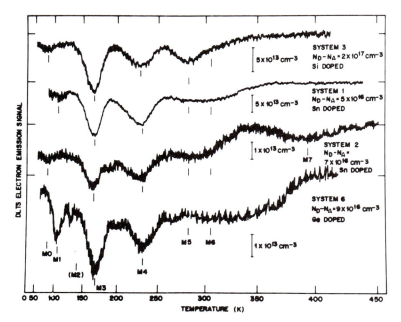

FIG. 6.31. DLTS capacitance spectra of electron traps in n-GaAs. The two lower curves are for samples grown under As-rich conditions and the upper curve was for a sample grown under Ga-rich conditions. Each of the labeled peaks represents a different impurity or defect related center.[110]

growth occurs under As-stabilized conditions. Better understanding of the origin and incorporation of these defects will be an important area of study in the next few years.

Figure 6.32 gives another example of doping behavior in MBE which is different than other growth techniques. Figure 6.32 characterizes the $n-p$ transition as a function of As_2/Ga flux ratio for fixed germanium flux.[67] For $As_2/Ga < 0.7$, germanium is p-type with a mobility equal to good LPE material indicating that all the germanium is entering arsenic sites. The transition to n-germanium is very abrupt, although the mobility only reaches half the LPE value, indicating strong compensation. The amphoteric behavior of germanium in MBE is not found in LPE. Silicon, however, behaves oppositely and is amphoteric in LPE and not in MBE.

The kinetic and interactive aspect of the doping problem is further exemplified by the result of Cho for magnesium doping.[117] Figure 6.33 shows the increase in "effective" magnesium sticking coefficient deduced from the electrically active magnesium and corresponding increase in doping level as a function of the aluminum mole fraction. Joyce and Foxon have shown that in fact magnesium has unity sticking coefficient but that on GaAs the magnesium diffuses away from the surface and sits at electrically inactive interstitial sites.[99] The co-evaporation of aluminum modifies the tendency for magnesium to sit interstitially, but the mechanism has not been established.

An opposite form of behavior occurs for tin which has a tendency to accumulate at the surface.[105, 112] Tin doping has been well studied and results in a uniform doping level in GaAs.[113] In the range 5×10^{16}–5×10^{17} carriers cm^{-3} the compensation level was

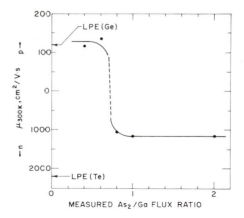

FIG. 6.32. Doping behavior of germanium in GaAs as a function of As_2/Ga impingement flux ratio during growth: 560°C growth on GaAs (100), 2×10^{18} Ge/cm^3.[67]

Molecular beam epitaxy

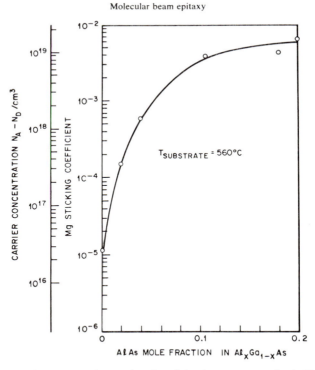

FIG. 6.33. Carrier concentration as a function of aluminum concentration in $Al_x Ga_{1-x} As$. The electrically active magnesium concentration is a strong function of aluminum concentration. In GaAs, the magnesium which also has unity sticking coefficient does not sit at an electrically active substitutional site.[38]

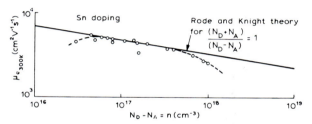

FIG. 6.34. Electron mobility as a function of net donor concentration ($N_D - N_A$) for Sn-doped films. Deviation from the theoretical line is an indication of the level of compensation.[32]

low.[99] This was established by comparing the mobility as a function of carrier concentration with the calculation of Rode and Knight[114] as is shown in Fig. 6.34.

Electrical doping with beryllium p-type GaAs can result in good mobilities as shown in Fig. 6.35. Trapping in AlGaAs, however, results in significantly poorer mobility than achieved by LPE growth. Understanding the problems of growing high quality AlGaAs with MBE is a remaining area of research.

Two other examples of doping are the interaction of zinc[98] with GaAs and the group VI elements which in the bulk form neutral donors. Oxygen, sulphur, selenium, and tellurium when evaporated typically evolve as molecular species which at the growth temperature interact weakly with the surface. This low sticking coefficient leads to low doping. These effects are strongly temperature dependent. Studies of the oxidation of GaAs are protypical of these doping phenomena and are discussed in the next section.

Zinc is one of the primary p-type dopants in LPE grown GaAs, but for MBE, the sticking coefficient is so low that no appreciable doping can be achieved with a thermal evaporation source. By using an ionized source, electrically active zinc can be

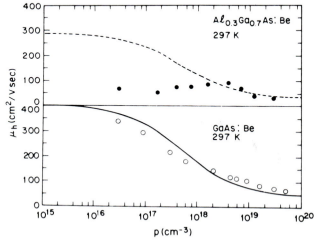

FIG. 6.35. Room-temperature hole mobilities versus net hole concentration for Be-doped GaAs and $Al_{0.3}Ga_{0.7}As$ layers. The solid line represents the highest mobilities measured in GaAs grown by other techniques and the dashed line represents the corresponding estimate for $Al_{0.3}Ga_{0.7}As$.[107]

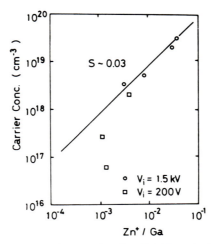

FIG. 6.36. Hole carrier concentration as a function Zn^+/Ga arrival rates. The zinc ion accelerating voltages were 1.5 kV (○) and 200 V (□). The solid line corresponds to an effective sticking coefficient of 0.03. Without the ionizer, the sticking coefficient is ~0.[49]

incorporated.[115] This is shown in Fig. 6.36 where a sticking coefficient of 0.03 is deduced. The incorporation has been explained as an ion implantation result. The actual mechanism, however, has not been clarified since thermally evaporated zinc is presumably composed of Zn_2 while the ionization would probably break up these dimers. A similar approach has been used for nitrogen doping in GaAs.[116]

6.6.3. OXIDATION

A discussion of oxidation is relevant because the interactions are typical of the doping reactions for other group VI elements in GaAs grown by MBE. The group VI elements are not useful for dopants in GaAs. Sulphur, selenium, and tellurium have a weak interaction with GaAs and result in few incorporated impurities at the growth temperature. This results from the competing desorption reaction.

Oxygen in GaAs forms a deep substitutional donor with low solubility and on the surface at high concentration can form a native oxide. Studies of oxidation of semiconductors prepared *in situ* with MBE provide access to other than the cleavage faces. Examples of this type of work are studies of the GaAs (100), (111),[88] and (110)[117] faces, and the ZnSe (100) face.[82] A related problem is the oxidation of $Ga^{(100)}$ and Al metal.[118,119] For example, a monolayer coverage of oxygen on GaAs requires a 10^6 Langmuir exposure while ~10^2 L on aluminum and 10^3 L on gallium provide a saturated coverage. In the MBE growth environment, the oxidation of the metals prior to reaction into the semiconductors is the more likely step if oxygen or oxygen-containing molecules are present. When an oxygen beam is present during the growth of GaAlAs, semi-insulating material can be grown.[120]

With reference to Fig. 6.21, one sees that the unreconstructed polar (100) and (111) faces have surfaces with a specific outermost atom. The (110) surface as discussed earlier

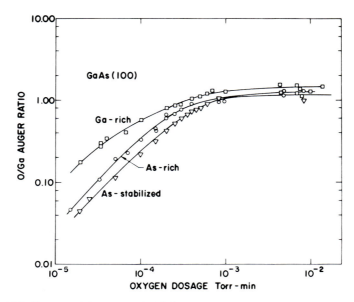

FIG. 6.37. Oxygen uptake measured with Auger spectroscopy; room temperature oxygen exposure curves are for Ga-rich (4 × 6), As-stabilized c(2 × 8), and As-rich (1 × 1) surface structures. (10^{-3} torr min = 6 × 10^4 L.)[21]

undergoes a relaxation whereby the arsenic moves out and the gallium moves in so that the three faces are structurally not all that different, so the compositional effects might be more important.

The initial oxidation of GaAs has been studied in detail both on MBE grown GaAs[88,91,117] and on *in situ* cleaved GaAs (110).[121] This oxide depends on whether the exposure is to molecular oxygen or to excited oxygen. Compositionally only very thin oxides are achieved in vacuum without plasma activation. These thin oxides seem to differ compositionally from native oxides established from aqueous solution,[122] but this discussion is beyond the scope of this article. The interaction of excited oxygen may also relate to the higher sticking coefficient of zinc evaporated with an ionizing source.

Figures 6.37 and 6.38 show oxygen uptake data respectively for the GaAs (100)[88] and (111)[91] faces as a function of oxygen dose. In both cases, the Ga-rich surface adsorbs oxygen at a much faster rate. These can be compared with the uptake curves for Al (100)[118] and Ga[100] metal in Figs. 6.39 and 6.40.

Figure 6.37 shows oxygen uptake curves measured with Auger spectroscopy as the ratio of the oxygen to gallium signals. These exposures are made statically with a maximum pressure of 10^{-4} torr (note that 10^{-4} torr min = 6 × 10^3 Langmuirs). Initially a Ga-rich surface chemisorbs oxygen at a much faster rate.

A similar result is obtained for the (111) face as seen in Fig. 6.38. (Note that 1 torr sec = 10^6 L.) The as-grown surfaces are substantially less reactive than the (100) face which is comparable to the argon bombarded surfaces. Presumably this is related to bombardment related defects.

Pianetta *et al.* have found for *in situ* cleaved GaAs (110) that saturation requires close to 10^{12} L of ground-state oxygen.[2] Using excited oxygen, i.e., oxygen exposed to hot

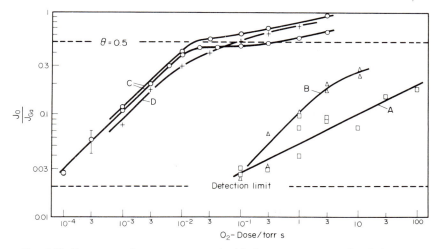

FIG. 6.38. Oxygen uptake curves measured with Auger spectroscopy for GaAs (111) surfaces. (A) As-stabilized (2×2); (B) Ga-stabilized $(19^{1/2} \times 19^{1/2})$; (C) argon bombarded and annealed for 10 min at 500°C (two separate runs); (D) argon bombarded only (1 torr sec. $= 10^6$ L.)[24]

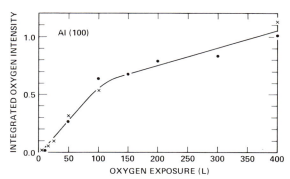

FIG. 6.39. Oxygen uptake curve measured with valence band photoemission spectroscopy for the oxygen 2p resonance on Al (100).[52]

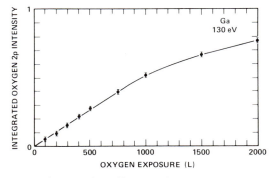

FIG. 6.40. Oxygen uptake curve for gallium metal evaporated on GaAs measured with spectroscopy of the oxygen 2p resonance.[36]

filaments or ionizing elements, can result in substantially increased reactivity. This aspect is not characterized in the experiments discussed above. The effect arises both because the excited oxygen dissociates more readily and also has a greater capability to break surface back bonds. Pianetta also found for the (110) surface with UPS that the oxygen chemisorbed to arsenic. This result led to some controversy since other ELS evidence suggested that oxygen chemisorbed to gallium.

The evidence for gallium involvement on the (111) and (100) faces is particularly strong, but in these cases there is some evidence that molecular oxygen chemisorbs. Evidence for the (110) face suggests in this case the chemisorption is dissociative. Desorption measurements typically show a gallium oxide evolution;[123] however, the question of surface mobility needs to be investigated more. If the oxygen becomes mobile, it has a greater ability to break a gallium bond than an arsenic bond. Consistent with this, the sticking coefficient also decreases with increasing arsenic content. Charge transfer certainly occurs from the surface arsenic, but the structure and occurrence of special sites is more important for the adsorption of oxygen than the surface concentration of arsenic. The calculation of Mele and Joannopoulos[124] has clarified many of the issues relating to the interpretation of UPS and ELS studies. Their result for the (110) surface indicates that oxygen adsorption can produce spectral changes in both arsenic and oxygen derived features. Their calculation indicates a preferential chemisorbed bond to the surface arsenic. More theoretical work is called for with respect to the polar faces.

Studies of the oxidation of GaAlAs are only beginning, but preliminary results indicate that the aluminum strongly chemisorbs oxygen.[104] This is to be expected based upon studies of the oxidation of aluminum metal. The question of the oxidation of GaAlAs is important to the growth of heterostructure laser devices and the quality of the GaAlAs–GaAs interface.

MBE has also been used to investigate the oxidation of ZnSe(100)[82] epitaxially grown on GaAs. GaAs and ZnSe lattice match well, but during growth the ZnSe surface exhibits a (2×2) reconstruction. When exposed to oxygen, the ZnSe does not show a gradual uptake similar to GaAs. Rather, above a critical pressure of 0.08 torr, an abrupt exchange reaction takes place. This seems to be accompanied by a loss of selenium from the surface.

6.7. Specific Materials and Specialized Structures

In this section specific results on some representative materials will be presented where successful epitaxy has been achieved with good electrical and optical properties. In addition, aspects of specialized structures and integrated optics will be introduced in the context of the particular material where the results have been achieved. These are mainly in the III–V's and IV–VI's, but progress is being made in silicon-related MBE. The discussion presented here is not exhaustive so the reader should refer to the references for more detail.

6.7.1. SILICON

Silicon MBE has been slower in reaching the desired level of material quality because of several difficulties. Most of these are related to the quality of the vacuum required. Silicon evaporates at very high temperatures (10^{-2} torr at 1632°C) so that electron beam evaporation is typically required to achieve useful deposition rates. Several problems accrue from the thermal and electrical loading associated with these sources. The heat load can cause other parts of the vacuum system to outgas; and the secondary electron flux can introduce unwanted impurities by striking other surfaces. Designs such as discussed in Fig. 6.8, which collect the secondaries into silicon, seem to be effective in this regard.[5]

The silicon epitaxial deposition temperatures are also quite high, 800–900°C, so that temperature control of an exposed surface is required. Silicon cleaning as discussed in section 6.4 also is best performed by thermal desorption and must raise the substrate to approximately 1000°C. The clean silicon surface is quite reactive. The biggest problem is carbide formation, which can disrupt the epitaxy or result in inclusions. Thus silicon MBE requires more stringent vacuum than many of the other semiconductors. The availability of closed cycle helium cryopumps is making the necessary vacuum requirements more accessible.

Doping experiments have shown that silicon can be reliably doped n- or p-type with MBE growth.[62,108,125,126] Doping superlattices have been demonstrated.

Silicon-related MBE work has included demonstrating doping superlattice and with germanium compositional superlattices.[127] Silicon diode structures have been characterized and silicon-MBE layers are reported to have been used to fabricate Schottky diodes for an x-band mixer, and high frequency bipolar transistors although the details are not available.[62] Millimeter-wave PIN switching diodes have also been fabricated.[128]

6.7.2. GaAs–GaAlAs AND OTHER III–V'S

By far the most successful utilization of MBE has been for the growth of GaAs and GaAlAs. This has been extended to other III–V's in many cases to achieve desired specific results which require different material properties such as smaller bandgaps, etc.

In addition to GaAs, significant work has been performed on InP and this is reviewed by Farrow.[21]

Some aspects of this section have been covered in the general discussion and here some specific details will be covered to point the interested reader to the appropriate literature. This section of necessity covers a broad area so the discussions will be brief. All the properties of MBE described in Table 6.1 have been achieved or demonstrated with GaAs or GaAlAs.[2,3,129] From the point of view of bulk binary or ternary properties, the electronic and optical properties such as doping range, mobility, diffusion length, and internal scattering can be controlled as well as with other growth techniques.

A wide variety of devices of varying complexity have been fabricated.[129,130] These start with metal–semiconductor Schottky diodes where aluminum metal is epitaxially grown on GaAs (100).[96] High quality p–n homo- and heterojunction diodes can be fabricated. Microwave devices such as hyperabrupt varacter, mixer and IMPATT diodes and FET transistors have been successfully achieved. GaAs–GaAlAs lasers have been successfully fabricated which have thresholds and lifetimes of lasers comparable to lasers grown by other techniques. Tsang in particular has demonstrated lasers with thresholds as low as have been grown by LPE. MBE holds the promise of allowing other more complicated integrated optic structures to be achieved.

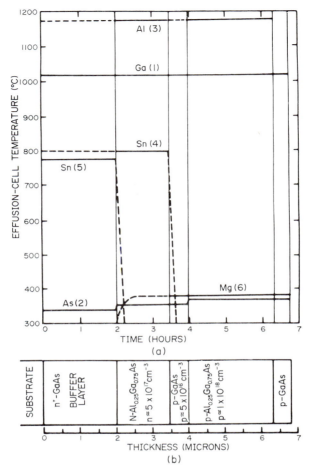

GIG. 6.41. (a) Temperature–time cycles used to grow MBE five-layer double heterostructure lasers. Solid lines indicate shutter was opened while dotted lines indicate shutter was closed. Substrate temperature was 625°C. (b) Schematic representation of the layers grown by the temperature–time cycles in (a).[29]

Figure 6.41 shows an example of the growth time–temperature sequences for a double heterostructure laser.[2] Figure 6.42 shows a scanning electron micrograph of such a structure which in this case was incorporated in an integrated fashion with a tapered coupler to a passive AlGaAs waveguide.[130] The tapered coupler was grown by inserting a knife-edge in front of the substrate during growth of the GaAs active layer.

This is an example of the use of masks to selectively define the growth areas on the substrate. Figure 6.43 shows, for example, the growth of GaAs and GaAlAs areas on a substrate with the use of a shadow mask.[131] This results because of the collimated aspect of MBE growth. Figure 6.44 further exemplifies this where it is shown how the collimation of MBE can be used to provide channel-isolated growth.[132,133]

Figure 6.44 shows an etched channel and the resulting layer growths. The channel overhang provides a mask which isolates the growth on the central island.

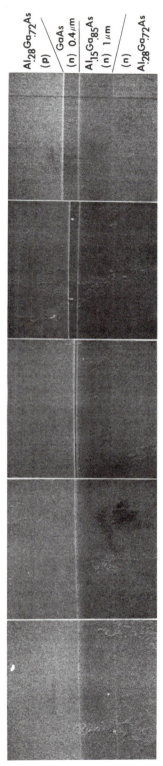

$Al_{.28}Ga_{.72}As$ (p)

GaAs (n) 0.4 μm

$Al_{.15}Ga_{.85}As$ (n) 1 μm

(n) $Al_{.28}Ga_{.72}As$

FIG. 6.42. Scanning electron micrograph showing an integrated optic structure grown by MBE. A laser, tapered coupler and waveguide were fabricated during the growth.

GaAs Epilayer

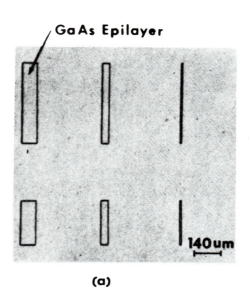

(a)

Ga$_{0.7}$Al$_{0.3}$As Epilayer

GaAs Epilayer

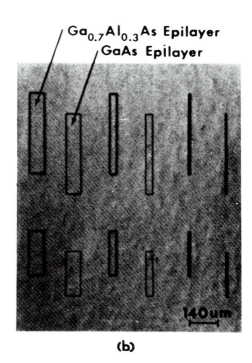

(b)

FIG. 6.43. Nomarski optical micrographs of single level (a) and two level (b) depositions of GaAs (Al$_{0.3}$Ga$_{0.7}$As) patterns. For (b) the same mask (shifted in position) was reused for the second deposition after etching off the polycrystalline growth (Tsang and Illegems[131]).

The low growth temperature and collimation of MBE also allows crystal growth on specifically prepared substrates such as seen in Fig. 6.45. A grating was argon ion etched into the substrate and then a layer of AlGaAs grown.[134] Gratings such as these are used in distributed feedback lasers or in grating optical couplers. In the case of lasers, they provide feedback without the use of end mirrors.[135]

MBE has also allowed the growth of ultrathin layers in a controlled fashion.[136] Perhaps the ultimate demonstration is that of Gossard *et al.*, where the control was established such that alternating monolayers of –Ga–As–Al–As–Ga– were grown with finite lateral coherence lengths or domain sizes.[137] From there one can extend to superlattice composites such as –GaAs$_n$–GaAlAs$_m$– with the number of atomic layers n, m adjusted for desired properties. Detailed studies have shown that the introduction of these artificial periodicities can result in quantum size dependent electronic properties.[138]

Figure 6.46 gives an example of a multilayer waveguide grown by MBE.[139] The alternate light and dark lines represent different compositional layers. Note the fine interlayer spacing.

Annealing experiments on GaAs–GaAlAs structures such as these have shown them to be very stable to temperatures in the neighborhood of 900°C. From the specific results which show an abrupt disordering transition, very low interdiffusional coefficients are deduced.[137, 140]

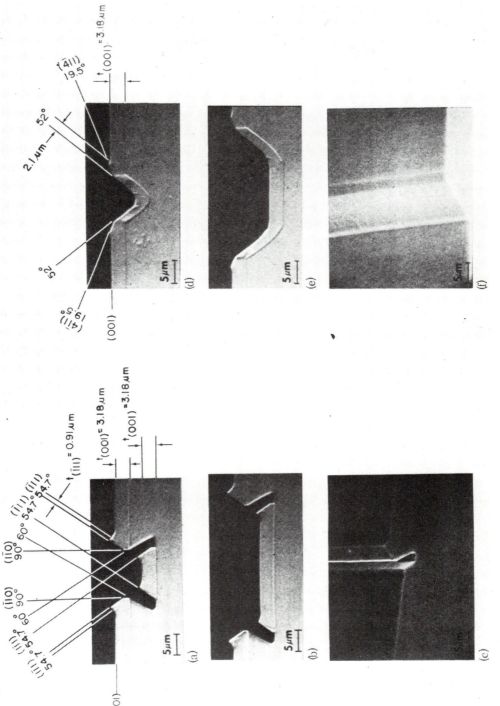

Fig. 6.44. Examples of growth in channels for providing isolated structures.

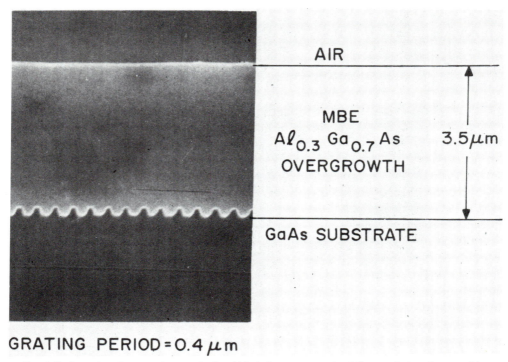

GRATING PERIOD = 0.4 μm

FIG. 6.45. Scanning electron micrograph of an AlGsAs layer grown epitaxially on a GaAs substrate with a periodic corrugation created by ion-beam milling (Illegems[134]).

As an example of new electronic properties achieved with these superlattices, Dingle *et al.* have combined modulation doping with superlattice formation to achieve substantial mobility enhancement within the GaAs layers.[141]

6.7.3. IV–VI'S: PbSnTe

$Pb_{1-x}Sn_{1-x}Te$ is a good example of a IV–VI material grown by MBE which has advanced to device quality development.[34, 142] Evaporative techniques are particularly useful for IV–VI semiconductors because they have convenient vapor pressures and sublime predominantly as diatomic molecules. The bandgap of $Pb_{1-x}Sn_xTe$ is extremely sensitive to the composition so that even with evaporative techniques, care is needed to deposit the desired pseudobinary alloy.

For example, Northrup[143] has shown that the gas phase in equilibrium with $Pb_{1-x}Sn_xTe$ is enriched approximately 30% in SnTe over the range of compositions ($x \leq 0.2$) that are suitable for IR detectors. A $Pb_{1-x}Sn_xTe$ source becomes depleted of SnTe with use and the band gap epitaxial of the layer becomes larger. This effect can be avoided by using an open crucible rather than a Knudsen cell and by using a homogeneous source material in a solid rather than a powdered or granular form. Separate evaporative

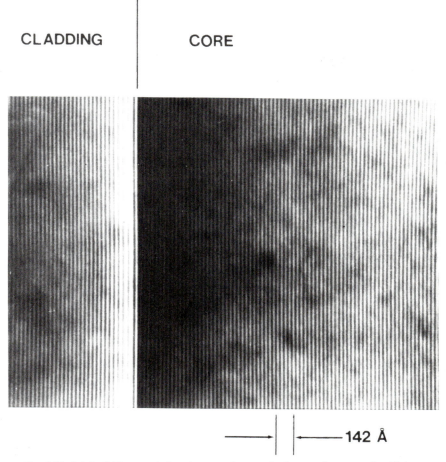

CLADDING   **CORE**

——————| |——————**142 Å**

FIG. 6.46. Bright-field transmission electron micrograph of a monolayer sample with two distinct regions. The marked interface separates a core region on the right with 1.0 monolayer of AlAs alternating with 9.3 monolayers of GaAs and the cladding region on the left with 3.2 monolayers of AlAs alternating with 7.4 monolayers of GaAs. In both cases, the total periodicity is approximately 10 layers since each monolayer is 2.83 in. thick; the spacing of the individual lines is 28.3 Å. (Merz, Gossard, and Wiegmann[139].)

sources can also be used for the IV–VI's. To obtain the necessary compositional control, typically a single oven is used and the relative flux is adjusted by varying the effusion cell apertures.[34] A variety of vacuum systems have been used for the IV–VI's including ion pumped glass bell jar type. Radiant heating is typically used so that the substrates can be held in a strain free manner.

Much of the good device work has been done on IV–VI substrates because of the good lattice match which reduces strain related defects. The growth of IV–VI layers on insulating substrates has a substantial literature[3, 144] which is mostly devoted to growth on alkali halide substrates. Occasional use has been made of other materials, such as CaF_2 and mica. The application of such semiconducting IV–VI films to p–n junction device technology demands a more critical approach to crystalline perfection than has been customary in the literature on epitaxial growth. Determination of the degree of epitaxy in IV–VI layers has been mostly by diffraction techniques that are insensitive to the spread in orientation that is associated with a small-grained mosaic structure. An early study[145] of PbS on NaCl, for example, demonstrated the presence of a mosaic structure with grain sizes of the order of 1000 Å. Similar results were obtained later with PbTe on NaCl,[146] which was found to have about 5×10^{10} dislocations/cm^2 that were mostly arranged to form grain boundaries with about 2000 Å spacing and about 2° misorientation.

Despite the early indications of a mosaic structure in epitaxial IV–VI semiconductors on alkali halide substrates, much of the subsequent work ignored the crystal perfection of the layers and, consequently, has little relevance to p–n junction device applications. The situation resembles that in the early stages of development of epitaxial III–V semiconductors, such as GaAs. Conventional diffraction techniques can be inadequate for the assessment of the crystal perfection of the IV–VI semiconductors. Such problems of characterization appear to be widespread with epitaxial semiconductors.

Among the substrates used for IV–VI MBE presented in Table 6.5, BaF$_2$ and SrF$_2$ have been particularly successful. Because they are transparent in the infrared, they can reduce optical losses. This is useful in a variety of device contexts.

IV–VI devices grown by MBE include infrared sensitive photodiodes, infrared lasers, and infrared waveguides. The lasers include distributed feedback lasers such as exemplified in Fig. 6.43. This makes use of the low growth temperature and collimation that can be achieved with MBE.

The optoelectronic device results achieved with MBE grown layers demonstrate that MBE grown devices are competitive with devices fabricated from bulk crystals. The MBE devices have technological advantages and performance features that have not yet been demonstrated in bulk-crystal devices.

TABLE 6.5. *Properties of the Lead Chalcogenides and Some Substrates*

Compound	Lattice Const. (Å) at 300 K	Thermal Expansion Coefficient $(K^{-1} \times 10^6)$ near 300 K	Cleavage	P(torr) at 700 K
PbS	5.94	20		
PbSe	6.12	19		
PbTe	6.46	20		
NaCl	5.64	39	(100)	4.5×10^{-7}
NaBr	5.96	42	(100)	3.7×10^{-6}
NaI	6.46	45	(100)	4.0×10^{-5}
KCl	6.29	37	(100)	2.2×10^{-6}
KBr	6.59	38	(100)	1.1×10^{-5}
KI	7.05	40	(100)	3.9×10^{-5}
CaF$_2$	5.40	19	(111)	$\sim 6 \times 10^{-20}$
SrF$_2$	5.80	18	(111)	$\sim 1 \times 10^{-20}$
BaF$_2$	6.20	18	(111)	$\sim 3 \times 10^{-17}$

6.7.4. ZnSe AND OTHER II–VI'S

Extensive literature exists on the evaporative deposition of II–VI semiconductors,[3,21,147] but most of this would not be characterized as MBE and little of the work has resulted in epitaxial layers useful for devices.

Among the more recent II–VI MBE studies, D. L. Smith has grown a large number of epitaxial II–VI semiconductors that show good morphology and optical properties.[46] Little electrical characterization has been done on such layers. Some luminescence studies have been performed,[147] but these are typically done at such large excitation densities that it is difficult to relate them to the crystalline or impurity perfection of the layers.

Because ZnSe lattice matches to GaAs, one can anticipate that this will be an interesting system for further investigation by MBE techniques. Significant scientific advances are required in II–VI materials technology, however, before electronic device related achievements are made similar to those discussed in the preceding sections.

References

1. K. G. GUNTHER, Interfacial and condensation processes occurring with multicomponent vapors, in *The Use of Thin Films in Physical Investigations* (J. C. Andersen, ed.), Academic Press (1966), p. 213.
2. A. Y. CHO and J. R. ARTHUR, *Progress in Solid State Chemistry* **10**, 157 (1975). This work contains extensive references and a bibliographic review.
3. L. L. CHANG and R. LUDEKE, *Epitaxial Growth*, Part A (J. W. Matthews, ed.), Academic Press (1975), p. 37; and L. ESAKI and L. L. CHANG, *Chemistry and Physics of Solid Surfaces* (R. Vanselow and S. Y. Tong, eds.), CRC Press (1977), p. 111; R. F. C. FARROW, *1976 Crystal Growth and Materials* (E. Kaldes and H. J. Scheel, eds.), North-Holland (1977), p. 238.
4. E. KASPER, H. J. HERZOG, and H. KIBBEL, *Appl. Phys.* **8**, 199 (1975).
5. Y. OHTA, *J. Electrochem. Soc.* **124**, 1795 (1977).
6. L. I. DATSENKO, A. N. GUREEV, N. F. KOROTKEVICH, N. N. SOLDATENKO, and YU. A. TKHORIK, *Thin Solid Films* **7**, 117 (1971).
7. V. N. VASILEVSAKAY, N. N. SOLDATENKO, and YU. A. TKORIK, *Thin Solid Films* **7**, 127 (1971).
8. K. ITO and K. TAKAHASHI, *Jpn. J. Appl. Phys.* **7**, 821 (1968).
9. J. R. ARTHUR and J. J. LePORE, *J. Vac. Sci. Technol.* **6**, 545 (1969).
10. A. Y. CHO, *J. Appl. Phys.* **41**, 1780 (1970).
11. A. Y. CHO, M. B. PANISH, and I. HAYASHI, *Proc. 3rd Int. Symp. GaAs and Related Compounds*, Aachen, Germany (5–7 Oct. 1970), Inst. Phys. Phys. Soc. Conf., Ser. No. 9.
12. A. Y. CHO and Y. S. CHEN, *Solid State Commun.* **8**, 377 (1970).
13. A. Y. CHO, *J. Vac. Sci. Technol.* **8**, S31 (1971).
14. L. L. CHANG, L. ESAKI, W. E. HOWARD, R. LUDEKE, and G. SCHYUL, *J. Vac. Sci. Technol.* **10**, 655 (1973).
15. S. GONDA, Y. MATSUSHIMA, Y. MAKITA, and S. MUKAI, *Jpn. J. Appl. Phys.* **14**, 935 (1975).
16. A. Y. CHO, *J. Appl. Phys.* **47**, 2841 (1976).
17. K. TATEISHI, M. NAGANUMA, and K. TAKAHASHI, *Jpn. J. Appl. Phys.* **15**, 785 (1976).
18. Y. MATSUSHIMA and S. GONDA, *Jpn. J. Appl. Phys.* **15**, 2093 (1976).
19. A. Y. CHO, *J. Appl. Phys.* **41**, 782 (1970).
20. R. F. C. FARROW, *J. Phys. D: Appl. Phys.* **7**, L121 (1974).
21. R. F. C. FARROW, in *1976 Crystal Growth and Materials* (E. Kaldes and H. J. Scheel, eds.), North-Holland, New York (1977); *J. Phys. D: Appl. Phys.* **8**, L87 (1975).
22. Y. MATSUSHIMA, Y. HIROFUJI, S. GONDA, S. MUKAI, and M. KIMATA, *Jpn. J. Appl. Phys.* **15**, 2321 (1976).
23. J. H. McFEE, B. I. MILLER, and K. J. BACHMANN, *J. Electrochem. Soc.* **124**, 259 (1977).
24. L. L. CHANG, A. SEGMULLER, and L. ESAKI, *Appl. Phys. Lett.* **28**, 39 (1976).
25. L. L. CHANG and A. KOMA, *Appl. Phys. Lett.* **29**, 138 (1976).
26. A. C. GOSSARD, P. M. PETROFF, W. WIEGMANN, R. DINGLE, and A. SAVAGE, *Appl. Phys. Lett.* **29**, 323 (1976).
27. J. L. MERZ, A. S. BARKER, JR., and A. C. GOSSARD, *Appl. Phys. Lett.* **31**, 117 (1977).
28. A. Y. CHO, *Appl. Phys. Lett.* **19**, 467 (1971).
29. A. Y. CHO and H. C. CASEY, *J. Appl. Phys.* **45**, 1258 (1974).
30. A. Y. CHO, R. W. DIXON, H. C. CASEY, JR., and R. L. HARTMAN, *Appl. Phys. Lett.* **28**, 501 (1976).
31. S. GONDA and Y. MATSUSHIMA, *J. Appl. Phys.* **47**, 4198 (1976).

32. H. Sakaki, L. L. Chang, R. Ludeke, C. Chang, G. A. Sai-Halasz, and L. Esaki, *Appl. Phys. Lett.* **31**, 211 (1977).
33. A. Y. Cho, H. C. Casey, Jr., and P. W. Foy, *Appl. Phys. Lett.* **30**, 397 (1977).
34. H. Hollaway and J. N. Walpole, *Progress in Crystal Growth and Characterization*, **2**, 1 (1979).
35. J. N. Zemel, J. D. Jensen, and R. B. Schoolar, *Phys. Rev.* **140A**, 330 (1965).
36. H. Holloway, *J. Nonmetals* **1**, 1347 (1973).
37. R. B. Schoolar and J. N. Zemel, *J. Appl. Phys.* **35**, 1848 (1964).
38. G. F. McLane and K. J. ..., *J. Electron. Mater.* **4**, 465 (1975).
39. D. K. Hohnke and S. W. Kaiser, *J. Appl. Phys.* **45**, 892 (1974).
40. R. B. Schoolar and J. R. Lowney, *J. Vac. Sci. Technol.* **8**, 224 (1971).
41. T. O. Farinre and J. N. Zemel, *J. Vac. Sci. Technol.* **7**, 121 (1970).
42. H. Holloway, E. M. Logthetis, and E. Wilkes, *J. Appl. Phys.* **41**, 3543 (1970).
43. R. F. Bis, J. R. Dixon, and J. R. Lowney, *J. Vac. Sci. Technol.* **9**, 226 (1972).
44. D. L. Smith and V. Y. Pickhardt, *J. Electron. Mater.* **5**, 247 (1976).
45. D. K. Hohnke, H. Holloway, K. F. Yeung, and M. Hurley, *Appl. Phys. Lett.* **29**, 98 (1976).
46. D. L. Smith and V. Y. Pickhardt, *J. Appl. Phys.* **46**, 2366 (1975).
47. T. Yao, S. Amano, Y. Makita, and S. Maekawa, *Jpn. J. Appl. Phys.* **15**, 1001 (1976).
48. T. Yao, Y. Miyoshi, Y. Makita, and S. Maekawa, *Jpn. J. Appl. Phys.* **16**, 369 (1977).
49. J. T. Calow, D. L. Kirk, and S. J. T. Owen, *Thin Solid Films* **9**, 409 (1972).
50. G. Shimaoka, *Thin Solid Films* **7**, 405 (1971).
51. P. E. Luscher and D. M. Collins, *Progress in Crystal Growth and Characterization*, (1979).
52. A. Lopez-Otero, *Thin Solid Films* **49**, 3 (1978).
53. D. W. Pashley, *Epitaxial Growth*, Part A (J. W. Mathews, ed.), Academic Press, New York (1975).
54. K. Ploog and A. Fischer, *Applied Physics* **13**, 111 (1977).
55. C. C. Chang, in *Characterization of Solid Surfaces* (P. F. Kane and G. B. Larabee, eds.), Plenum Press, New York (1974).
56. J. R. Arthur, *J. Vac. Sci. Tech.* **10**, 136 (1973).
57. R. Z. Bachrach, R. D. Burnham, D. R. Scifres, and R. S. Bauer, *Materials Research Society Symposium on MBE, Cambridge, Mass., November 1976.*
58. J. Arthur, Physical Electronics Inc.
59. S. Dushman, *Scientific Foundations of Vacuum Technique*, John Willis & Sons, New York (1966).
60. L. Holland, W. Steckelmacher, and J. Yarwood, *Vacuum Manual*, Ex F. N. Spon, London (1974).
61. H. C. Casey, Jr., and M. B. Panish, *Heterostructure Lasers*; Part B, Academic Press, New York (1978).
62. U. Konig, H. Kibbel, and E. Kasper, *J. Vac. Sci. Tech.* (1979).
63. D. W. Pashley, *Epitaxial Growth*, Part A (J. W. Mathews, ed.), Academic Press, New York (1975), ch. 1.
64. S. Holloway and J. L. Beeby, *J. Phys. C· Solid State Phys.* **11**, L247 (1978).
65. J. R. Arthur and J. J. LePore, *J. Vac. Sci. Technol.* **14**, 979 (1977).
66. R. E. Honig and D. A. Kramer, *RCA Rev.* **30**, 285 (1969).
67. D. L. Smith, *Progress in Crystal Growth and Characterization*; the discussion presented here is based on this paper.
68. W. Koster and B. Thomas, *J. Metallic* **46**, 293 (1955); J. van der Boomgaard and K. Schol, *Philips Res. Rep.* **12**, 127 (1957).
69. C. D. Thurmond, *J. Phys. Chem. Solids* **26**, 785 (1965).
70. J. R. Arthur, *J. Phys. Chem. Solids* **28**, 2257 (1967).
71. W. Albers and C. Haas, *Philips Techn. Rev.* **30**, 82 (1969).
72. H. R. Potts and G. L. Pearson, *J. Appl. Phys.* **37**, 2098 (1966).
73. A. F. W. Willoughby, C. M. H. Driscoll, and B. A. Bellamy, *J. Mat. Sci.* **6**, 1389 (1971).
74. R. M. Logan and D. T. J. Hurle, *J. Phys. Chem. Solids* **32**, 1739 (1971).
75. B. Lewis, *Thin Solid Films* **7**, 179 (1971).
76. J. R. Arthur and J. J. LePore, *J. Vac. Sci. Technol.* **6**, 545 (1969).
77. K. Ploog and A. Fischer, *Appl. Phys.* **13**, 111 (1977).
78. C. E. C. Wood and B. A. Joyce, *J. Appl. Phys.* **49**, 4854 (1979).
79. J. C. Bean, G. E. Becker, P. M. Petroff, and T. E. Seidel, *J. Appl. Physics* **48**, 907 (1977).
80. C. B. Duke, *Critical Rev. Solid State and Material Sciences* **8** (1978).
81. J. B. Pendry, *Low Energy Electron Diffraction*, Academic Press, New York (1974).
82. R. Ludeke, *Solid State Comm.* **24**, 725 (1977).
83. J. E. Rowe and H. Ibach, *Phys. Rev. Lett.* **32**, 421 (1974).
84. D. J. Chadi, *Phys. Rev. Lett.* 1062 (1978), and *Phys. Rev.* **B19**, 2074 (1979).
85. A. Y. Cho, *J. Appl. Phys.* **42**, 2074 (1971).
86. A. Y. Cho, *J. Appl. Phys.* **41**, 2730 (1970).
87. J. R. Arthur, *Surf. Sci.* **43**, 449 (1974).
88. R. Ludeke and A. Koma, *CRC Critical Rev. Solid State Sci.* **5**, 259 (1975).

89. K. Jacobi, G. Steinert, and W. Ranke, *Surf. Sci.* **57**, 571 (1976).
90. J. Massies, P. Etienne, and N. J. Linh, *Thompson-CSF Technical Rev.* **8**, March (1976).
91. W. Ranke and K. Jacobi, *Surf. Sci.* **63**, 33 (1977).
92. B. A. Joyce and C. T. Foxon, *J. Cryst. Growth* **13**, 122 (1975).
93. P. Drathen, W. Ranke, and K. Jacobi, *Surf. Sci.* **77**, L162 (1978).
94. J. Massies, P. Devoldere, and N. T. Linh, *J. Vac. Sci. Technol.* **15**, Aug/Sept. (1978).
95. R. Ludeke, L. L. Chang, and L. Esaki, *Appl. Phys. Lett.* **23**, 201 (1973).
96. A. Y. Cho and P. D. Dernier, *J. Appl. Phys.* **49**, 3328 (1978).
97. C. T. Foxon and B. A. Joyce, *Surf. Sci.* **64**, 293 (1977).
98. J. R. Arthur, *Surf. Sci.* **38**, 394 (1973).
99. B. A. Joyce and C. T. Foxon, *Jpn. J. Appl. Phys.* **16**, 17 (1976).
100. R. Z. Bachrach and A. Bianconi, *J. Vac. Sci. Technol.* **15**, 525 (1978).
101. R. Dingle, *Advances in Solid State Physics* (H. J. Queisser, ed.), Pergamon Press, Oxford (1975), p. 21.
102. A. C. Gossard, *Thin Solid Films* **57**, 3 (1979).
103. R. Z. Bachrach, *J. Vac. Sci. Technol.* **15**, Aug/Sept. (1978).
104. R. Z. Bachrach, R. S. Bauer, J. C. McMenamin, and A. Bianconi, *Proc. 14th Int. Conf. on Phys. of Semiconductors, Edinburgh, Scotland, 1978.*
105. A. Y. Cho, *J. Appl. Phys.* **46**, 1733 (1975).
106. M. Illegems, R. Dingle, and L. W. Rupp, Jr., *J. Appl. Phys.* **46**, 3059 (1975).
107. M. Illegems, *J. Appl. Phys.* **48**, 1278 (1977).
108. G. E. Becker and J. C. Bean, *J. Appl. Phys.* **48**, 3395 (1977).
109. M. Illegems, unpublished and ref. 61.
110. D. V. Lang, A. Y. Cho, A. C. Gossard, M. Illegems, and W. Wiegmann, *J. Appl. Phys.* **47**, 2558 (1976).
111. A. Y. Cho and M. B. Panish, *J. Appl. Phys.* **43**, 5118 (1972).
112. K. Ploog and A. Fischer, *J. Vac. Sci. Technol.* **15**, 255 (1978).
113. G. Abstreiter, E. Bauser, A. Fischer, and K. Ploog, *Appl. Phys.* (1978).
114. D. L. Rode and S. Knight, *Phys. Rev.* **B3**, 2534 (1971).
115. N. Matsunaya, M. Naganuma, and K. Takahashi, *Jpn. J. Appl. Phys.* **16**, 443 (1976).
116. Y. Matsusima, S. Gonda, Y. Makita, and S. Mukai, *J. Cryst. Growth* **43**, 281 (1978).
117. R. Ludeke, *Solid State Comm.* **21**, 815 (1977).
118. R. Z. Bachrach, S. A. Flodstrom, R. S. Bauer, S. B. M. Hagstrom, and D. J. Chadi, *J. Vac. Sci. Technol.* **15**, 488 (1978).
119. S. A. Flodstrom, R. Z. Bachrach, R. S. Bauer, and S. B. M. Hagstrom, *Phys. Rev. Lett.* (1978).
120. H. C. Casey, Jr., A. Y. Cho, and E. H. Nicollian, *Appl. Phys. Lett.* **32**, 678 (1978).
121. P. Pianetta, I. Lindau, M. Garner, and W. E. Spicer, *Phys. Rev.* B, **37**, 1166 (1976).
122. C. C. Chang, R. P. H. Chang, and S. P. Murarka, *J. Electrochem. Soc.* **128**, 481 (1978); and B. Schwartz, F. Ermanis, and M. Bradstad, *J. Electrochem. Soc.* **123**, 1089 (1976).
123. J. R. Arthur, *J. Appl. Phys.* **38**, 4024 (1967).
124. E. J. Mele and J. D. Joannopoulos, *Phys. Rev.* B (1978), to be published; E. J. Mele and J. D. Joannopoulos, *Phys. Rev. Lett.* **40**, 341 (1978).
125. J. C. Bean, *Appl. Phys. Lett.* **33**, 654 (1978).
126. Y. Ohta, *J. Electrochem. Soc.* (1980), to be published.
127. E. Kasper and W. Pabst, *Thin Solid Films* **37**, L5 (1976).
128. Y. Ohta, W. L. Buchanan, and O. G. Petersen, *International Electron Devices Meeting Technical Digest, Washington DC 1977*, p. 375.
129. A. Y. Cho, *J. Vac. Sci. Technol.* Mar./Apr. (1979).
130. A. T. Cho, *1976 International Conference on Solid State Devices, Tokyo, Japan* (1976), *Jpn. J. Appl. Phys.* **16**, 435 (1975).
131. W. T. Tsang and M. Illegems, *Appl. Phys. Lett.* **31**, 301 (1977).
132. W. T. Tsang and A. Y. Cho, *Appl. Phys. Lett.* **30**, 293 (1977).
133. S. Nagata, T. Tanaka, and M. Fukai, *Appl. Phys. Lett.* **30**, 293 (1977).
134. M. Illegems, H. C. Casey, Jr., S. Somekh, and M. B. Panish, *J. Cryst. Growth* **31**, 158 (1975).
135. R. D. Burnham and D. R. Scifres, *Progress in Crystal Growth and Characterization*, 2 (1979).
136. A substantial literature exists on superlattices so that the references here are only representative.
137. A. C. Gossard, P. M. Petroff, W. Wiegmann, R. Dingle, and A. Savage, *Appl. Phys. Lett.* **29**, 232 (1976); and A. C. Gossard, *Thin Solid Films* (1979).
138. P. M. Petroff, *J. Vac. Sci. Technol.* **14**, 973 (1977).
139. J. L. Merz, A. C. Gossard, and W. Wiegmann, *Appl. Phys. Lett.* **629** (1977).
140. R. Dingle, *Festkorperprobleme XV (Advances in Solid State Physics)*, Pergamon–Vieweg, Braunschweig (1975), p. 21.
141. R. Dingle, H. L. Stormer, A. C. Gossard, and W. Wiegmann, *Appl. Phys. Lett.* **33**, 665 (1978).
142. The discussion in this section follows that given recently by Hollaway and Walpole in ref. 34.

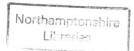

143. D. A. NORTHRUP, *J. Electrochem. Soc.* **118**, 1365 (1971).
144. J. N. ZEMEL, in *Solid State Surface Science* (M. Green, ed.), Dikkar, New York (1969).
145. J. N. ZEMEL, J. D. JENSEN, and R. B. SCHOOLAR, *Phys. Rev.* **140A**, 330 (1965).
146. J. H. MYERS, R. H. MORRIS, and R. J. DECK, *J. Appl. Phys.* **42**, 5578 (1971).
147. D. B. HOLT, *Thin Solid Films* **24**, 1 (1974).
148. T. YAO, Y. MAKITA, and S. MAEKAWA, *Jpn. Appl. Phys.* **16**, 451 (1977).

CHAPTER 7

Crystal Pulling

C. D. BRANDLE

Union Carbide Corporation, San Diego, CA 92123, USA

7.1. Introduction

Of the many crystal growth methods in use today, one method which can produce crystals weighing from several grams to many kilograms is the crystal pulling technique. Many variations of this technique exist throughout the literature and these will be discussed separately at the end of the chapter. However, prior to such a discussion, an understanding of the basic principles, the advantages, and limitations of this technique are in order.

All crystal pulling processes are based upon a technique developed by Czochralski.[1] This technique and its various modifications have become the dominant process used in industry today for the production of semiconductor and oxide single crystals. Materials which are routinely grown today using this technique include silicon, sapphire (Al_2O_3), GaP, GaAs, InP, $Gd_3Ga_5O_{12}$, $Nd:Y_3Al_5O_{12}$, germanium, and $LiNbO_3$ to name only a few. All of these materials find their use in the electronics industry, primarily as substrates for electronic devices. Because this technique is a relatively fast crystal growth process, it also finds wide application in the laboratory for the synthesis in single crystal form of many new materials.

The basic process is rather simple and is shown schematically in Fig. 7.1. The material to be "grown" is placed in a suitable container or crucible. The crucible is then heated either

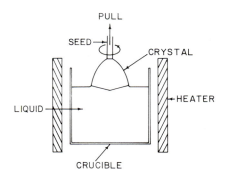

FIG. 7.1. Basic elements of a crystal pulling system.

275

by resistance or induction heating until the charge is melted. The temperature of the molten charge is then adjusted so that the center of the liquid is at its freezing point. A seed crystal is dipped into the liquid and the crystal growth or "pulling" process begins by slowly withdrawing the seed. With proper temperature control of the liquid, crystallization on the seed crystal can be started as the seed is withdrawn from the liquid. Further adjustments of the liquid temperature during the "pulling" process provide control of the crystal diameter. When the desired crystal length has been reached, the crystal is quickly raised from the liquid surface or the liquid temperature is slowly increased to reduce the diameter. When the crystal is free from the liquid, the temperature is lowered to room temperature and the crystal withdrawn from the growth apparatus.

7.2. Material Considerations

When considering crystal growth using a crystal pulling technique, several requirements and restrictions are placed upon the materials which can be used or grown. Therefore, one must consider the following questions:

(1) Is the material to be grown compatible with the restrictions placed upon it by the crystal pulling technique?
(2) Are crucible materials available which meet certain requirements?
(3) Is a heat source available consistent with the crucible selection and growth environment?

Each of these points is discussed in the following sections.

7.2.1. LIQUID MATERIAL LIMITATIONS

For a material to be considered as a possible candidate in a crystal pulling technique, several restrictions are placed upon the material. These are briefly summarized below:

(1) The material should have a congruent melting point, i.e. the material should not decompose upon or before melting.
(2) It should have a relatively low vapor pressure.
(3) It should not have any first order solid–solid phase transitions or reconstructive phase transitions.
(4) There must be a crucible material which is nonreactive with the material above its melting point.

In any crystal growth system one is always dealing with phase equilibrium between a solid and a liquid or, as discussed elsewhere in this book, a solid and a gas. Crystal pulling deals with the equilibrium that exists between a solid and its liquid and as such it is imperative that these be understood so that the results of a crystal pulling experiment can be interpreted correctly.

Consider Fig. 7.2, which is typical for many binary oxide compounds, e.g. $LiNbO_3$,[2] $LiTaO_3$,[3] and $Gd_3Ga_5O_{12}$.[4, 5] The stoichiometric composition shown in Fig. 7.2 is AB; however, the figure also shows that some solid solution of A exists in AB and that the congruent melting point is a compound (A^+B) which is rich in A. If one then prepared the compound AB, melted it, and grew a crystal from that liquid, the composition of the

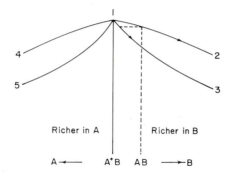

FIG. 7.2. Typical phase diagram in the region of a binary oxide compound showing nonstoichiometric congruent point.

resulting crystal would follow the solidus curve 1–3 and the liquid would follow the curve 1–2. Thus, the crystal would have a varying composition and properties throughout its length. To grow a crystal with uniform properties, a solid with the composition A^+B would have to be prepared and used for growth. Such behavior as outlined above is quite common in semiconductor solid solutions and binary oxide compounds.

A second consideration as already mentioned is that the compound has a low vapor pressure and that it vaporizes congruently. Once again incongruent vaporization of the liquid leads to an enrichment of either A or B in an AB liquid, and thus a nonstoichiometric melt, creating the same type of problems as discussed above. This behavior is quite common in semiconductors such as GaP, where the phosphorus vapor pressure is extremely large leading to the loss of phosphorus from the liquid. Specialized techniques as discussed in section 7.6 have been developed to control this problem.

The third requirement is that no first order, solid–solid phase transitions exist. Numerous examples of this behavior exist in the literature. Consider the case of CuCl. Cuprous chloride exists in two polymorphs. The low temperature form has the wurtzite structure whereas the high temperature form has the cubic zinc blende structure. The transition temperature is about 5°C below the melting point. Any attempt to grow this material by standard crystal pulling techniques results in a polycrystalline mass instead of a single crystal. Thus, to grow this material an alternate method such as traveling solvent must be employed which keeps the growing crystal below the transition point.

The last requirement is based on the chemical reactivity of the material to be grown. Since all crystal pulling techniques require a crucible to contain the liquid, any reaction between the liquid and crucible would contaminate the liquid. Furthermore, for the crystal material being considered for growth, a crucible material must exist which has a melting point sufficiently high to ensure safe containment of the liquid. These points and others are discussed in the section on crucible selection.

7.2.2. CRUCIBLE SELECTION

Once it has been determined that the material can be grown by a crystal pulling technique, the first equipment selection which must be made is the choice of a crucible and its size. Selection of the crucible material is based upon (1) compatibility with the melt, (2) melting point of the crucible material versus that of the compound, (3) type of heating, (4) chemical stability, and (5) mechanical properties.

For semiconductor materials, i.e., silicon, germanium, GaAs, GaP, etc., the most commonly used crucible material is fused silica. It provides the most suitable compromise between chemical stability, cost, ease of fabrication, and the other factors mentioned above. Usually, these crucibles are used only once and discarded because upon cooling, expansion of the freezing liquid causes the crucible to fracture.

For the oxide crystals (Al_2O_3, $LiNbO_3$, $Gd_3Ga_5O_{12}$, $LiTaO_3$, etc.), which generally have higher melting points, refractory metals or the noble metals are used. Once again the use of platinum, iridium, or in special cases tungsten or molybdenum, provides the most reasonable compromise between chemical stability, ease of fabrication, and mechanical strength. Generally in oxide crystal growth, the crucible is cleaned and reused repeatedly because of the initial expense. Some of the more commonly used materials are listed in Table 7.1 along with their useful range.

The second factor which must be considered is the size and shape of the crucible, for these dimensions can strongly affect the results of any crystal growth experiment. In any heated system the primary source for fluid motion within the crucible is due to liquid density changes resulting from temperature gradients. This effect is always present except in space (see Chapter 15). Carruthers[6] has examined the effects of crucible geometry and orientation on convective flow and has shown that the type of fluid flow present is determined by aspect ratio of the crucible and the thermal and viscous properties of the liquid.

The final factor which must be considered is the crystal size which can be grown from a crucible of given dimensions. A general rule of thumb, as discussed in section 7.3.1, is the final crystal diameter should be approximately 50% of the crucible diameter. When this value is exceeded, growth becomes more difficult for several reasons:

(1) Small irregularities in the crucible diameter can produce large changes in the growth rate.
(2) The rotating crystal has a much stronger influence on the fluid flow and stability regions for different flow types become increasingly narrow giving rise to abrupt flow transitions and therefore large temperature fluctuations.[7]

TABLE 7.1. *Common Crucibles*

Material	Maximum Operating Temperature (°C)	Melting Point (°C)	Use
Platinum	1400	1773	1, 2
Iridium	2150	2452	1, 2
Molybdenum	2300	2620	1, 2
Tungsten	2800	3370	1, 2
Carbon	3000	—	1, 2
Silica	1550	1700	1
Alumina	1800	2050	1

1 = resistance; 2 = r.f. heating.

7.2.3. HEAT SOURCES

In all crystal pulling techniques, the material from which the crystal is to be grown must be kept in the molten state. Although various unique power systems have been developed,[8–10] today two common sources are used. These are resistance heating and induction heating.

Resistance heated furnaces are usually limited to the lower temperature ranges (less than 1500°C) and can have one of several types of elements (Table 7.2). Because of temperature and ambient gas limitations, resistance heating usually finds use in the growth of semiconductor materials and a few of the low melting oxides, e.g. $LiNbO_3$, $Bi_{12}SiO_{20}$, and TeO_2. Higher temperatures require either special elements such as platinum–rhodium alloys or a hard vacuum operation to protect refractory metal elements from oxidation.

A general advantage of resistance heated furnaces over other types is greater electrical efficiency and therefore reduced operation costs. Also power input into the furnace, and therefore temperature control of the furnace, can be easily accomplished without the need for elaborate control equipment.

The second source of heating is r.f. or inductive heating and is generally used for the higher melting oxides such as sapphire (Al_2O_3) and $Gd_3Ga_5O_{12}$ (GGG); however, it too can be used for the growth of semiconductor materials. Usually r.f. generators operate in the frequency range of 250–500 kHz and a wide power range (20–100 kW). For general laboratory use, a 20 or 25 kW r.f. generator is suitable and provides the capability of growing crystals up to about 3 cm in diameter.

The use of an r.f. generator places an additional constraint on the crucible in that it must be conductive since r.f. heating induces a current flow in the crucible. The depth of current flow or penetration of the r.f. field into the crucible walls is determined by the frequency and is given by

$$D = \frac{3570\sqrt{\rho}}{\sqrt{(f\mu)}} \text{ cm}, \tag{7.1}$$

where ρ is the resistivity in ohms, μ is the permeability, f is the frequency in Hz, and D is the penetration or "skin depth". Table 7.1 lists commonly used r.f. crucibles and their maximum operating range. Of those materials listed in Table 7.1, platinum is the only one which has to be operated far below its melting point because of large changes in its resistivity with temperature as one approaches its melting point. This causes localized "hot spots" to develop in the crucible wall which can result in crucible failure.

TABLE 7.2. *Resistance Elements*

Material	Useful Temperature (°C)	Atmosphere
Nichrome	≤1200°C	Air
SiC	≤1500°C	Air
Pt–50 %Rh	≤1800°C	Air
Carbon	≤3000°C	Neutral, reducing
Molybdenum	≤2400°C	Vacuum, neutral, or reducing
Tungsten	≤3000°C	Vacuum, neutral, or reducing

Because such crucible defects as small cracks, impurities in grain boundaries, and variations in wall thickness can cause localized resistivity changes, the physical condition of the crucible is much more important in r.f. heating than in resistance heating. One method of reducing these effects is to use lower frequencies and thereby increase the "skin depth". This reduces the effect on resistivity of crucible imperfections and therefore provides more uniform heating of the crucible. Usually lower frequencies are supplied by solid state generators (30–100 kHz) or motor generators (~ 10 kHz).

Another factor which must be considered in r.f. heating is load matching. Improper matching of the loaded work coil to the tank circuit can cause an enormous loss of efficiency. Furthermore, as one decreases the frequency, the matching becomes more critical and coil/crucible setup less versatile. Thus in choosing a r.f. heating unit, one must select a unit which provides the best compromise between heating uniformity and versatility.

7.2.4. FURNACE CONSTRUCTION

The furnace which is used for crystal pulling can vary from the very simple, e.g., a resistance wound heating element, to one which is extremely complex, because of thermal and chemical constraints placed upon it by the crystal. These furnaces can also be divided into two types. The first type is for oxide crystal growth and generally is composed of ceramic and noble metal parts, whereas the second type is for semiconductor growth and usually is composed of graphite and fused silica parts.

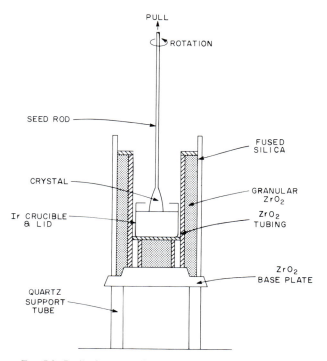

FIG. 7.3. Radio frequency furnace used to grow oxide compounds.

A typical furnace used for oxide crystal pulling is shown in Fig. 7.3. Mounted on a ceramic base are the outer sleeve (usually fused silica), loose granular insulation followed by an inner ceramic tube, usually ZrO_2 or Al_2O_3. The choice of ZrO_2 or Al_2O_3 depends only on the operating temperature. A second base or pedestal is placed inside this inner tube to position the crucible within the furnace. A r.f. coil outside the silica sleeve then completes the assembly. A bell jar or some other enclosure is then placed around the furnace to provide atmosphere control. Using this type of furnace and an iridium crucible, it is possible to reach and maintain for several weeks a temperature in excess of 2000°C. As shown in the figure, only the seed and crystal are rotated.

In contrast to the above furnace, a typical r.f. semiconductor furnace is shown in Fig. 7.4. An outer sleeve of fused quartz is placed between the r.f. work coil and the carbon susceptor. The susceptor has been machined so that a fused quartz crucible will fit inside. A thermocouple well for temperature control is machined in the center of the carbon susceptor. This entire assembly can be raised or lowered and rotated within the work coil of the r.f. generator. For production semiconductor growth, the r.f. coil has been replaced with a resistance element, carbon, in a "picket fence" design. As already mentioned, this type of heating is more efficient than r.f. heating.

7.3. Crystal Growth

The successful growth of a crystal using a crystal pulling technique is dependent upon the interaction of many variables to produce the desired results. The "art" of growing crystals by the pulling technique is the manipulation of the pulling rate, rotation rate, thermal geometry, and atmosphere to minimize crystal imperfections. In this section, methods of altering the thermal geometry are discussed as well as various effects on crystal perfection due to the thermal growth environment. This is done to give not exact solutions for various crystal pulling problems but only guidelines for their solution.

7.3.1. GROWTH RATE

First one must realize that the pull rate, i.e., the rate of seed withdrawal, is not the growth rate of the crystal. Witt and Gatos[11,12] have shown that the instantaneous growth rate of

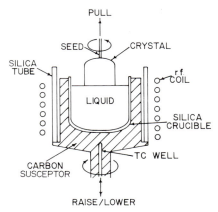

FIG. 7.4. Radio frequency semiconductor furnace.

any crystal is subject to wide variations and that these growth fluctuations are due to temperature oscillations in the liquid or rotation about an axis other than the thermal axis. The actual growth rate is therefore the time average of the instantaneous growth rate and must include the liquid level drop.

The actual time averaged growth rate can be calculated in the following manner using a mass balance. Referring to Fig. 7.5 and assuming a circular crucible cross-section, the amount of liquid crystallized per unit time is then give by

$$G_l = \pi R^2 L \rho_l, \tag{7.2}$$

where R is the radius of the crucible, L is the height of liquid crystallized/unit time, and ρ_l is the average liquid density.

A similar analysis for the crystal yields

$$G_c = A_c h \rho_c, \tag{7.3}$$

where A_c is the surface area of the crystal interface, h is the height of the crystal grown/unit time, and ρ_c is the solid density of the crystal.

Furthermore, from Fig. 7.5 it is obvious that $h = P + L$, i.e., the total length of crystal grown is the sum of the pull rate P and the liquid level drop L. If one now assumes a crystal with a circular cross-section and a flat interface, eqn. (7.3) can be written as

$$G_c = \pi r_c^2 h \rho_c = \pi r_c^2 (P + L), \tag{7.4}$$

where r_c is the radius of the crystal.

Since the amount of material solidified must equal that lost by the liquid, eqns. (7.2) and (7.4) can be equated, thus yielding for L,

$$L = r_c^2 P / (R^2 \kappa - r_c^2), \tag{7.5}$$

where $\kappa = \rho_l / \rho_c$ and the linear growth rate h is then

$$h = P \frac{R^2 \kappa}{R^2 \kappa - r_c^2}. \tag{7.6}$$

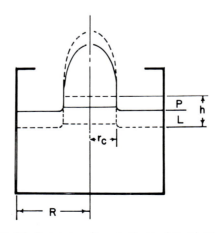

Fig. 7.5. Parameters for growth rate determination.

Division by R^2 then yields

$$h = P \frac{\kappa}{\kappa - D^2}, \qquad (7.7)$$

where $D = r_c/R$.

Thus, one can see that the actual time averaged growth rate of the pulled crystal is dependent upon the ratio of the liquid to solid density κ and the ratio of the crystal radius (or diameter) to that of the crucible radius (or diameter) D. Rearrangement of eqn. (7.7) then yields

$$\frac{h}{P} = \frac{\kappa}{\kappa - D^2}. \qquad (7.8)$$

This equation is plotted for various values of κ as shown in Fig. 7.6. This figure can then serve as a useful guide for estimating actual time averaged growth rates in any crystal pulling system. Furthermore, it is evident from the shape of the curves in Fig. 7.6 that for $D \geq 0.5$, small changes in D can result in large changes in h/P, i.e., the crystal growth rate. Thus, the general "rule of thumb" that the ratio of the crystal diameter to the crucible diameter should never exceed 0.5–0.6 so that problems associated with growth rate fluctuations are minimized. An example of the types of problems encountered when this value is exceeded is given in Brandle.[7] Growth rates for the pulling technique tend to fall into two categories, oxides (0.1–1.2 cm/h) and semiconductors (2.5–8 cm/h).

One further point should be made. In oxide crystal growth the crucible is usually stationary and most values of κ are less than unity whereas in semiconductor growth the

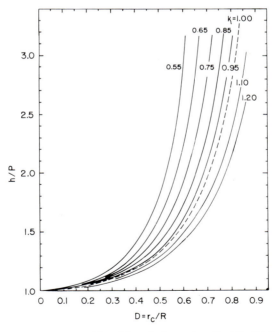

FIG. 7.6. Growth rate versus crystal/crucible ratio for various liquid/solid density ratios.

crucible is raised during growth to compensate for the liquid level drop; thus the pull rate P is equal to the crystal growth rate. Also, all κ values are greater than unity. A κ value greater than unity causes unique control problems with semiconductors which does not exist for oxides and will be discussed under the control section of this chapter.

7.3.2. THERMAL GRADIENTS

Once a power supply and heat source have been selected along with the crucible material, the placement of the crucible within the "hot zone" of the furnace can result in success or failure. One of the prerequisites for good control of the growth operation is a suitable thermal gradient. This gradient can be established in one of several ways and usually a combination of several is used by the crystal grower.

One of the easiest ways to establish a known thermal gradient is with resistance wound furnaces with split elements. In these furnaces very precise thermal gradients can be established and maintained over long periods of time. Because of this factor, these furnaces are usually used for crystal growth processes which require extended time periods such as vapor growth or directional freezing as discussed in other chapters.

One convenient way of classifying thermal gradients with respect to the crystal pulling process is either as external or internal. An external thermal gradient is defined as one which is outside the crystal or crystal growing medium whereas an internal gradient is within the crystal or crystal growing medium. For example, the vertical gradient in an afterheater would be an external gradient and that present in the liquid or crystal is an internal gradient.

It is not completely possible to separate the internal from the external gradients; however, some degree of independence can be achieved by selection of the method of heating. Most resistance heated furnaces tend to have a long hot zone since heat is supplied not only to the crucible but to the surrounding furnace parts as well. This leads to very uniform heating of the crucible and therefore low liquid thermal gradients. Small changes of the position of the crucible within a resistance furnace have little effect on the internal or liquid gradient since all gradients are determined by that produced from the resistive elements. Zupp et al.[13] have used this fact to investigate the effect of temperature oscillations on growth striations.

Because of the nature of r.f. heating, heat is supplied only to the crucible walls and any heating of the furnace bottom and walls is by conduction or radiation from the hot crucible. This results in much higher internal and external thermal gradients. It also gives a certain degree of independence which cannot be achieved using a resistance furnace. This independence has been shown by Brandle and Miller[14] in the growth of $LiTaO_3$. By changing the position of the crucible with respect to the work coil, they were able to alter the vertical liquid gradient without appreciably changing the external vertical and radial gradients. Their results are shown in Fig. 7.7. They also found that this liquid gradient greatly effected the quality of the resultant crystal. Thus crucible placement, particularly in a r.f. furnace system, can significantly alter the growth environment of the crystal.

A way of changing or modifying external thermal gradients is by the use of afterheaters. These can be classified as active or passive. An active afterheater acts as its own heat source and can be a resistance element or a r.f. element. In either case it is used to alter the thermal gradients in the growth furnace above the liquid although some change in the internal gradient is unavoidable.

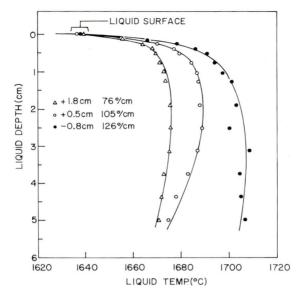

FIG. 7.7. Measured liquid temperature profiles in molten LiTaO$_3$ for various crucible/coil positions.[14]

The passive afterheater relies solely on the insulating properties of its components to alter the external thermal gradient and is usually composed of ceramic tubes and baffles. Several common arrangements of each of these afterheaters are shown in Fig. 7.8 along with typical thermal gradients they produce.

The selection of a particular type of afterheater depends upon the thermal properties and characteristics of the crystal being grown. Afterheaters are generally used to reduce the strain associated with dislocation formation, control nondestructive phase changes, or reduce thermal stresses in the growing crystal.

7.3.3. THERMAL EFFECTS

Rotation of either the crystal or the crucible or both is designed to accomplish one primary objective—to smooth out asymmetric temperature profiles present in the growth furnace during the crystal pulling process. Regardless of how carefully a furnace is built, there always exists a small or sometimes large deviation of the thermal axis of symmetry from the mechanical axis of symmetry. This deviation of the two axes causes growth irregularities which manifest themselves as wandering of the growth axis, excessive facet formation, poor diameter control, or striations.

The misalignment of the thermal axis of symmetry with the mechanical axis of symmetry is most obvious in the liquid. Rotation of the growing crystal tends to stir the liquid thereby forcing the two axes to approach each other. Additional mixing of the liquid can be accomplished by rotating the crucible. For most semiconductor materials, both crystal and crucible rotation are used, while in oxide growth only crystal rotation is used. This is because the furnaces associated with oxide growth are generally more complex than in semiconductor growth.

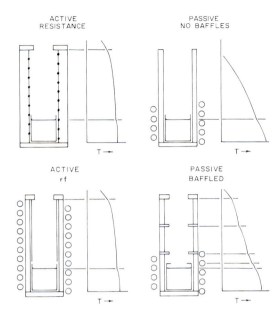

FIG. 7.8. Different types of afterheaters and thermal gradients produced.

As already discussed, one of the concerns in a crystal pulling system is that of thermal gradients. As a direct result of the liquid thermal gradients, density gradients are generated and bulk fluid flow results within the crucible. Fluid flow can be classified into two broad areas: (1) natural convection wherein the driving forces for fluid motion are density gradients created by local temperature differences, and (2) forced convection wherein the driving forces for the fluid flow are mechanical in origin and usually due to crystal/crucible rotation. In all crystal pulling systems, either one or both of these convective forces exist and it is the interaction of these forces which can determine the growth and perfection of the crystal. Carruthers[15] and Carruthers and Nassau[16] first looked at the fluid flow in a crucible which was driven by forced convection, i.e., rotation of the crucible and/or the crystal. Their results showed the existence of cellular flow within the liquid. These cells, called Taylor–Proudman cells, are two-dimensional with respect to the axis of the rotating fluid. Furthermore, their work showed that the size and number of cells formed is dependent upon the relative rotation rates of the crucible to that of the crystal.

Recently, simulation experiments of the crystal pulling process by Shiroki,[17] Brandle,[18] Langlois and Shir,[19] and Kobayaski and Arizumi,[20] in which both natural and forced convection were present, have shown a rather complicated flow within the crucible but still cellular in structure. Figure 7.9 is an example of the fluid flow observed in a series of simulation experiments. These simulation experiments agree quite well with actual observed fluid flow on the surface of oxide melts by Takagi et al.[21] and Whiffin et al.[22] Fluid flow in oxide melts is much easier to observe than in semiconductor melts due to differences in the thermal properties of the liquid. Therefore, most direct observations of liquid flow during the crystal pulling process have been in oxide systems.

In addition to the bulk fluid flow observed during growth, sinusoidal temperature variations associated with unstable thermal convection can occur. Carruthers[6] has

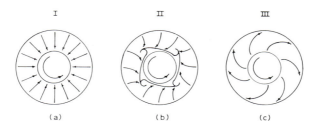

FIG. 7.9. Simulated fluid flow in crystal pulling system: (a) slow rotation, (b) moderate rotation, and (c) fast rotation.[18]

examined several models of the origins of these temperature fluctuations. The results sown that these instabilities depend on a critical Rayleigh number and basic thermal properties of the fluid and system. It is therefore essential that in the design of a crystal pulling system the factors affecting the fluid flow in the crucible be considered so that the system does not produce unstable flow conditions as discussed by Carruthers.[6]

The height of the fluid within the crucible can also have pronounced effects on thermal oscillations. Whiffin and Brice[23] have pointed out that the thermal stability of the fluid can be increased considerably by positioning baffles within the liquid. The most stabilizing effect resulted when the baffle was positioned between $h = d_l/3$ and $h = 2d_l/3$ (d_l = liquid depth and h = baffle depth).

This same effect is observed in normal crystal growth runs of $Gd_3Ga_5O_{12}$. Typically, 85 % of the liquid is crystallized, and thus near the end of the growth the ratio of the liquid depth to crucible diameter decreases considerably. Those sections of crystals grown with this low melt level show considerably fewer striations than the top of the crystal, indicating smaller temperature oscillations in the melt during this period of growth (see section 7.3.4).

7.3.4. GROWTH STRIATIONS

As a result of the fluid flow and the temperature variations associated with it, the growing crystal interface is subjected to a constantly changing temperature. This results in growth rate changes, which can be a factor of ten different from the time-averaged growth rate discussed earlier. These rapid changes in growth rate manifest themselves as striations in the crystal and are due to local variations in the chemical composition.[24] Striations are particularly evident in doped crystals or crystals which exhibit some degree of nonstoichiometry, e.g., $Gd_3Ga_5O_{12}$ and $LiTaO_3$ (Fig. 7.10).

Growth striations have two sources.[25] The first is a nonsymmetric temperature distribution about the rotations axis. As the crystal rotates it "sees" different temperatures and therefore experiences different growth rates. In some cases where considerable asymmetry exists, actual remelting of areas of the growth interface can occur. A second source for growth striations can be the sinusoidal temperature variations associated with unstable thermal convection as discussed by Carruthers.[6]

Growth striations can be observed by numerous techniques.[26-28] Among these is the use of an etch which is sensitive to small variations in chemical composition. Such etches have been reported for both semiconductor[29] and oxide crystals.[30] The study of growth

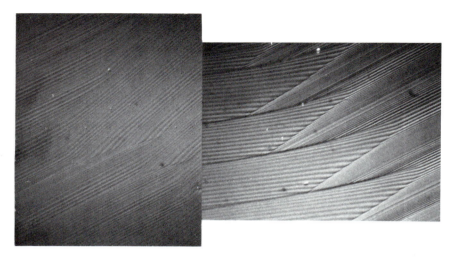

FIG. 7.10. Growth striations in $Gd_3Ga_5O_{12}$ showing remelt regions near facet of crystal.
(Courtesy D. C. Miller.)

striations within a crystal can provide a detailed look at the growth history of the crystal and provide clues to the sources of imperfections.[31]

Although growth striations can serve as an excellent tool for the study of the actual growth process,[31] they do affect the quality of the resulting crystal. Because growth striations are variations in chemical composition, small changes in the lattice parameter of the crystal result. These small lattice parameter variations produce localized strain within the crystal which is not relieved during fabrication. In addition to lattice parameter changes, growth striations can produce variations in the electrical resistivity and refractive index caused by the periodic chemical variation throughout the crystal.

Striations can be considerably reduced in magnitude by suppressing thermal oscillations in the liquid either by reducing the temperature gradients[13] or by using baffles in the melt.[23] Reducing the temperature gradient makes crystal growth more difficult and promotes facet formation while baffles reduce the usable volume of melt. An alternate approach to the use of baffles is to use a crucible with a height to diameter ratio (H_c/C_c) of less than one—ideally using a value of $\frac{1}{2}$ or $\frac{1}{3}$. This produces crystals with fewer striations but as in the low gradient technique, growth is more difficult. A similar approach is to use a normal crucible $H_c/D_c = 1$, but to have a liquid height of only $\frac{1}{2}H_c$. Crystals grown under these conditions have fewer striations but they are much more difficult to control. Because the strain associated with growth striations is normal to the striations, another approach would be to grow the crystal so that the striations would be parallel to the fabricated surface. The angle of intersection between the striations and the fabricated surface would then be very small and little, if any, strain would be transmitted to the finished device.

7.4. Solid Solutions and Impurities

Many of the industrially important single crystals grown today by the pulling technique are truly solid solutions of several compounds. Many of the semiconducting materials grown have dopant ions added to produce some desirable property. Similarly, all single crystal laser materials have dopant or impurity ions added which in the host crystal become the active ion. Therefore, the control of the concentration level and distribution of the dopant through the length of the crystal is extremely important.

Consider Fig. 7.11 which represents a continuous solid solution between A and B. If one starts with a liquid whose composition is represented by point 1, then the solid crystallized from that liquid would have a composition given by point 2. As the growth process continues, the composition of the liquid would follow curve 1–3 and that of the solid, curve 2–4. Thus, it can be seen that the solid does not have the same composition of the liquid and furthermore, that it varies from one point in the crystal to the next. A measure of the separation of the solidus from the liquidus curves can be expressed by a single coefficient k called the distribution or segregation coefficient which is the ratio of the solid concentration C_s to the liquid concentration C_l. The more k differs from unity, the greater the separation and the more difficult the crystal pulling process becomes.

At equilibrium, the concentration in the solid is given by

$$c_s = kc_0(1 - g)^{k-1}, \tag{7.9}$$

where c_0 is the initial concentration of the dopant ion and g is the fraction of melt that has been crystallized.

Many complications can arise since k as given above corresponds to a system in complete equilibrium, which certainly is not the case during crystal growth. For those interested in a detailed discussion of these effects, it is suggested that the cited references be consulted.[32–34]

If one assumes that k is independent of concentration and that k can be replaced by $k_{effective}$ which includes the effects of growth rate dependence and diffusion layer dependence, then eqn. (7.9) can be used to determine a value for an effective distribution coefficient in an actual growth system. This distribution coefficient can then be used as a guide for predicting concentration variations through the length of the crystal, shifts in lattice parameter due to concentration changes, and shifts in stoichiometry due to melt composition changes.

An example of the change in distribution with concentration is given by Brandle and Valentino[35] for rare-earth gallium garnet solid solutions. In several garnet systems it has been shown that the distribution coefficient can vary from less than unity to greater than

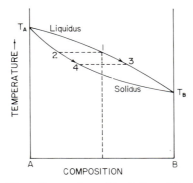

FIG. 7.11. Phase diagram for solid solution of A and B.

unity depending on the initial ion concentration in the melt.[36, 37] It has also been shown in the garnet systems that the distribution coefficient of a given ion at the same concentration can be related solely to the lattice parameter of the host garnet.[38]

Thus, using results obtained from several crystal growth experiments, sufficient data can be obtained to make reasonable estimates of k. In general, the lower the value of k the slower the growth rate and the more critical the rotation rate of the crystal. This phenomenon, called constitutional supercooling, is a direct result of the crystal pulling process (growth rate) exceeding the mass transfer rate (diffusion rate) in the liquid at the growing crystal interface. For any impurity or dopant with $k \neq 1$ there is a concentration gradient between the solid and liquid and between the growth interface and the bulk liquid as shown in Fig. 7.12. This concentration gradient introduces a local change in the melting point of the solution near the growth interface. If the temperature gradient in the liquid is too low, crystallization will take place in front of the growing crystal interface resulting in poor crystal quality. Since the distribution coefficient is closely tied to stirring of the liquid, the selection of the crucible shape (H_c/D_c ratio) and positioning of the crucible within the work coil for r.f. heating can aid in the prevention of constitutional supercooling.

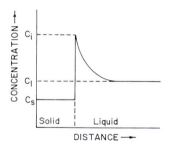

FIG. 7.12. Concentration gradients produced in the liquid due to $k \neq 1$.

7.5. Growth Control

Control of the crystal pulling process can be considered as two separate processes. First, the problem of providing a stable temperature within the furnace must be considered and, second, that of providing a suitable feedback system for detecting changes in the crystal diameter due to temperature changes. The interaction of these two control mechanisms can then provide complete control of the crystal pulling process.

7.5.1. TEMPERATURE CONTROL

The traditional method of temperature control of a furnace is through the use of thermocouples. For resistance heated furnaces, this form of temperature sensor is usually quite adequate. However, for the case of a r.f. heated furnace, the power source imposes several limitations on the type of sensor which one can use. Usually the operating temperature and/or environment is too severe for conventional thermocouples and r.f. interference due to the generator prevents reliable readings. Therefore, alternate methods of generating a control signal are required. This signal can be derived from the oscillator voltage or any other suitable voltage or current which changes with generator power. This in turn means that the temperature of the melt is *not* the controlled variable but rather the *power* delivered to the load, i.e. crucible, is controlled. This method of control assumes that the temperature of the liquid is proportional to the power input and is constant for a constant power input.

Adequate control of the crystal pulling process can be obtained in this manner using a diameter feedback network to control power as discussed in the next section. However, changes in the load characteristics and therefore frequency can cause a temperature increase at the same power input. This situation usually arises when a crucible leaks some of its charge into the furnace insulation resulting in coupling of the power to the furnace instead of the crucible.

7.5.2. DIAMETER CONTROL

Until the early 1970s most crystal growth was controlled by open loop systems, i.e., the growth of a crystal with a uniform diameter required a skilled operator to make adjustments in the furnace power. As the size of the crystals became larger and the number of crystal pulling stations increased, the need for automation became apparent. Today, automatic diameter control in both the laboratory and industry is common practice. These techniques are based on either a weighing system or an optical system.

The method of control using the weighing technique is based upon the assumption that the diameter of the crystal will remain constant for a mass growth rate. This assumption implies that the interface shape and buoyant forces are also constant throughout the entire crystal pulling operation. The errors introduced by these assumptions are generally small and in the case of oxide crystals can be neglected. However, for the case of semiconductor crystals, where the density of the solid is less than that of the liquid, severe control problems result because of buoyancy effects.

The first type of weighing control is based on the melt weighing technique. The weight of the crystal is not measured directly but determined through the loss of weight of the melt. The usual sensor is an electronic balance on which the entire furnace and therefore the crucible with melt rests. During the growth process, the rate of weight loss from the balance is monitored and from this loss a voltage is generated proportional to the weight loss. The weight signal can be processed in several ways to furnish the final signal to control the furnace power, either analog[39−41] or digital.[42,43] An example of analog melt weighing, diameter control loop, is shown in Fig. 7.13.

The melt weighing technique has several advantages:

(1) The output of the balance is not subject to periodic variations due to the rotation of the growing crystal.
(2) It does not require modifications of an existing puller.
(3) There is no lever arm associated with current electronic balances and therefore little motion of the furnace assembly occurs during growth.
(4) The entire assembly can be housed inside the growth chamber.

Among its disadvantages are:

(1) Except for a unique crucible position, the balance output is sensitive to power changes due to levitation of the crucible when a r.f. generator is used. These changes require a second compensating levitation circuit.

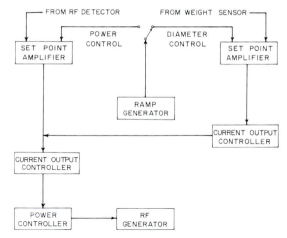

FIG. 7.13. Analog control loop for both power and diameter control.[39]

(2) As the size of the crystal increases, so must the furnace weight and therefore sensitivity must be sacrificed to accommodate the large furnace weight.

An alternate weighing technique is to weigh the crystal. The main advantages of the crystal weighing technique are:

(1) The balance capacity can be closely matched to that of the final weight.
(2) There are no levitation forces to consider.
(3) The furnace assembly does not have to be "free-floating".

Disadvantages include:

(1) Noise in the input signal due to the crystal rotation.
(2) Puller modifications.
(3) Depending upon the exact weighing system, possible lateral motion of the crystal during growth.

As already mentioned, in the case of semiconductor crystals, where $\rho_{solid} < \rho_{liquid}$, the assumptions mentioned above are no longer valid and a more sophisticated signal processing technique must be used in order to achieve good, uniform diameter control. For details concerning this particular situation, it is suggested that the reader refer to the papers by Bardsley et al.[44,45]

Regardless of which weighing technique (crucible or crystal) is used, the signal generated from the weight sensing device is compared to that produced from a ramp generator. The ramp-generated signal represents the crystal weight signal for the ideal case. Thus, the error signal which goes to the controller is the difference between the ideal case (ramp signal) and the actual case (weight generated signal). A typical weight versus time curve A is shown in Fig. 7.14. Because the final weight of the crystal has been fixed as a function of time, any error in weight must be compensated by an equal and opposite error further into the growth. In this manner, one error generates an equal and opposite additional error.

If, however, one takes the derivative of the weight signal[39] with respect to time, this problem can be eliminated since this new signal now represents the slope of curve A in Fig.

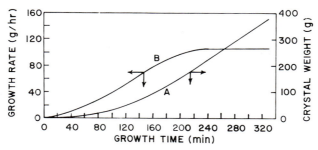

FIG. 7.14. Crystal weight as a function of time and its derivative.[39]

7.14 and is shown as curve *B* in Fig. 7.14. In this way the growth rate and not the mass of the crystal becomes the control variable, and one removes the absolute dependence of the crystal weight on time. The primary disadvantage of this control technique is the low noise level required in the input signal prior to the differentiation.

As is obvious from the above discussion, each weighing system has its advantages and disadvantages. In general, for the growth of crystals weighing less than 2000 g, the melt weighing technique is the easier and more efficient to use. For larger diameter crystals in the 2500–10,000 g range, the advantage of weighing only the crystal becomes significant. This does not imply that large, massive crystals cannot be grown using a melt weighing technique, but the other factors in their growth must be considered to determine the appropriate control system to use.

A second method for crystal diameter control is based on optical techniques and truly measures the diameter of the crystal. In this system optical sensors detect the position of the solid–liquid boundary during growth[46, 47] and make changes in the generator power to maintain some predetermined diameter. This type of control has the unique advantage that the system is truly sensing diameter changes in the crystal and not some other variable and the sensitivity of the control system is independent of diameter or mass growth rate. Its main disadvantages are:

(1) Compensation must be made in either the crucible position or "view angle" to adjust for the drop in liquid level during growth.
(2) Materials which produce vapors that can condense on the optical parts can distort the image and produce erroneous output signals.

Despite these disadvantages, optical systems for diameter control were first used in the semiconductor industry because the geometry of the growth furnace and crucible allowed easy adjustment for the liquid level drop during growth.

As mentioned above, both analog and digital methods for diameter control have been developed. Where the control algorithm is rather simple and the number of growth furnaces small, analog control provides the best method of controlling the crystal diameter. However, as the control algorithm becomes more complex and the number of growth furnaces increases, the use of digital control becomes justified.

7.6. Special Techniques

In this section, specialized techniques, which have been developed to overcome some of the material limitations placed on the crystal pulling process, are discussed. In addition, several techniques, which produce shaped single crystals via a "pulling" process, will be introduced. However, because of the importance of silicon in the world today it is felt that a special section briefly describing that growth process be included.

7.6.1. SILICON GROWTH

Today the most commercially successful single crystal pulling operation is the growth of silicon single crystals for the semiconductor industry with crystal diameters in the range 75–125 mm. Figure 7.15 gives an example of currently produced silicon single crystal. Although considerable effort has gone into development of the detailed growth techniques, the basic process remains unchanged from that described earlier in this chapter. The main effort in the growth of silicon single crystals has been to control impurities and imperfections.

FIG. 7.15. Typical commercial silicon crystal (4 in. in diameter and weighing 50 lb). (Courtesy Monsanto Co., St. Peters, MO.)

Silicon, because of its melting point, reacts slowly with the crucible material, vitreous silica, to form SiO. The oxygen from the crucible has an appreciable solubility in molten silicon and therefore acts as an impurity in the final silicon crystal. In addition to oxygen, other impurities in the crucible can be dissolved into the molten silicon and affect its electrical properties. These impurity effects can be minimized through atmosphere control and proper stirring of the melt.

Crystal lattice imperfections such as dislocations have been controlled and eliminated by using a technique first reported by Dash[48] and illustrated in Fig. 7.16. This technique requires the reduction in diameter of the seed followed by growth at the reduced diameter before enlarging the crystal to the final diameter. In this manner dislocations have an opportunity to grow out of the crystal toward the surface prior to the bulk growth.

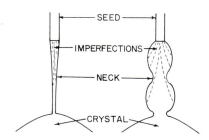

FIG. 7.16. Elimination of crystal imperfections using the technique developed by Dash.[48]

7.6.2. LIQUID ENCAPSULATION CZOCHRALSKI (LEC)

This technique[49] is a crystal pulling technique which has been developed to overcome one of the main material limitations of crystal pulling, namely that the material to be grown should have a relatively low vapor pressure. It finds wide use today in the growth of III–V semiconductor compounds such as GaP, GaAs, and InP.

In all of these compounds, the dissociation pressure of phosphorus or arsenic at the melting point is greater than 1 atmosphere, the actual vapor pressure depending upon the compound. The apparatus is shown in Fig. 7.17. In this case the crystal pulling chamber is a pressure chamber with a r.f. heating coil arranged inside. Heating is accomplished through the use of a carbon susceptor. The charge is placed in a silica crucible along with the B_2O_3 encapsulant. This material is chosen as an encapsulant because of its chemical stability, density less than that of the melt, transparency, and low melting point. As the temperature of the charge is increased, the B_2O_3 melts coating the charge. Once the melting point of the charge has been reached, the B_2O_3 forms a liquid layer above the molten charge. As long as the pressure inside the chamber is greater than the dissociation pressure of the charge, free phosphorus or arsenic is not released. Thus, the molten B_2O_3 acts as a liquid seal to contain the volatile component. This has the added advantage that the external balancing pressure can be produced by an inert gas such as N_2 and that the walls of the growth chamber can be at room temperature without having to worry about vapor condensation and reaction.

The actual growth process is monitored through the use of a closed circuit television system while power adjustments for diameter control can be made either manually or

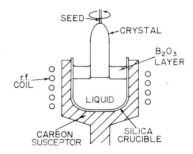

FIG. 7.17. LEC crystal pulling arrangement.

using an automatic system as described earlier. These systems áre usually of the weighing type rather than the optical type because some of the volatile component is usually lost during the initial heatup and impairs visibility.

Although this technique overcomes a primary deficiency in crystal pulling and extends the usefulness of the process to other materials, it also introduces several additional problems. The major source of problems is the addition of a second liquid layer above the first. If the encapsulant is too thick, it acts as a thermal blanket thereby making growth more difficult. If it is too thin, it fails to hold in the volatile component creating melt stoichiometry problems. Since the crystal must be pulled through the encapsulating layer, there is the danger of improper draining of the encapsulant from the crystal thus causing cracking during cooldown.

The third problem associated with the encapsulant is that of contamination. There is a tendency for small particles of the melt to become trapped within the encapsulating layer impairing visibility. Conversely, some of the encapsulant can be dissolved into the melt and act as an impurity ion. Despite these problems, GaP in weight ranges of 2–5 kg can be grown using this technique and it has become a commercial production process.

7.6.3. FLUX PULLING

As mentioned at the beginning of the chapter, another limitation placed on the crystal pulling technique is that the compound to be grown must have a congruent melting point and show no reconstructive phase transitions between its melting point and room temperature. A technique which has been devised to overcome this limitation is called flux pulling and consists of pulling a crystal as in the conventional technique from a liquid which contains the desired compound in solution. In this case crystallization results from a change in the solubility of the compound with temperature.

This technique can be further subdivided into two classes: (1) homogeneous, and (2) heterogeneous. The homogeneous case is illustrated in Fig. 7.18 and consists of adding one of the end members of the compound to depress the melting point of the liquid so that an undesirable phase transition is avoided. Again referring to Fig. 7.18 it is necessary to produce the compound A_2B which undergoes a peritectic decomposition at the temperature T_1. The addition of pure A to A_2B would produce a solution A_2^+B which has a melting point T_2, i.e., lower than the decomposition temperature. Thus by seeding the liquid with an A_2B seed combined with the simultaneous pulling and cooling of the liquid, it is possible to grow an A_2B crystal. Examples of this method can be found for

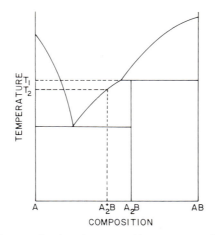

FIG. 7.18. Phase diagram showing the compositions necessary for growth using the homogeneous flux technique.

GaP[50] and numerous oxides, e.g. BaTiO$_3$.[51] This approach allows crystals to be grown without the introduction of foreign compounds which could be incorporated into the growing crystal.

The heterogeneous case is usually used where the melting point of the desired crystal is too high for the available crucible materials. If this approach is to be used, one usually seeks a solvent which has a high solubility to allow as concentrated solution as practical. A recent example of the application of this technique is the growth of BeO from CaO–BeO solutions.[52] This system showed BeO solubilities from 50 to 80 % and produced crystals several centimeters in length.

In either of these flux pulling techniques it must be remembered that the crystal growth process is dependent upon the mass transfer of the compound from the solvent to the crystal interface. Therefore, stirring of the liquid, both mechanically and thermally, is extremely important for good growth, and careful attention should be paid to ensure a homogeneous solution and high thermal gradients.

7.6.4. SHAPED GROWTH

Any crystal grown by one of the previously described techniques will tend to have a circular cross-section. For fabrication into substrates, the resulting crystal must be ground into a cylinder, sliced into blanks, and finally polished. Thus, the fabrication of the as-grown crystal into a final product requires considerable time and expense. Despite these drawbacks, all commercial production of substrates relies on the above mentioned process.

To overcome these drawbacks, two processes have been developed which allow for the pulling of shaped crystals. The first of these processes is called the EFG (edge-defined, film-fed growth) process and has been highly successful in producing shaped Al$_2$O$_3$ crystals.[53, 54] A second process, which has been applied to semiconductor materials, produces long, thin ribbon crystals and is called the dendritic web process.[55–57] Both of these techniques are variations of the basic crystal pulling process described earlier.

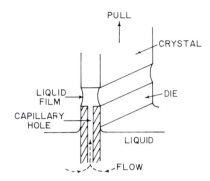

FIG. 7.19. Crystal and die arrangement in the EFG technique. Note the liquid covers the entire die surface.

7.6.4.1. *Edge-defined, Film-fed Growth*

A schematic drawing of the apparatus is shown in Fig. 7.19. The crystal is grown from a thin film of liquid on top of a suitable die surface. The shape of the film and therefore of the crystal is determined by the external shape of the die. Replenishment of the thin liquid film during crystallization is dependent upon capillary action. The capillaries are designed so that the liquid will rise to the top of the die regardless of the liquid depth within the crucible. This allows almost complete utilization of the melt and essentially "decouples" the melt from consideration in the crystal growth process. Thus, the liquid from which growth takes place is sandwiched between the die and the growing crystal and is "held" in that position by its surface tension.

For Al_2O_3 growth, the die and crucible are usually made from molybdenum. Unlike the more conventional crystal pulling techniques, growth rates are extremely high being in the range 1–5 cm/min as compared to 0.01–0.05 cm/min for conventional crystal pulling.

More recently, this technique has been used to produce silicon ribbon for photovoltaic devices. For this case, carbon dies are used[58,59] since the technique depends upon the liquid wetting the surface of the die and quartz (fused silica) is not sufficiently wet by molten silicon to provide enough capillary action. The use of carbon as the die material does introduce the problem of contamination via the reaction between the die and the molten silicon. The reaction product, SiC, is then trapped by the growing ribbon crystal and can act as sites for defect propagation within the ribbon. Furthermore the erosion of the die by the silicon limits the amount of material which can be produced. Despite these problems, EFG silicon holds great promise for producing single crystal ribbon for photovoltaic applications at a low cost.

7.6.4.2. *Web Growth*

A second form of shaped growth is dendritic web growth which is most suitable for semiconductor systems. The growth geometry is shown schematically in Fig. 7.20. Liquid silicon is contained within a silica crucible heated by a carbon susceptor. A slotted lid is placed over the crucible to form the proper thermal geometry in the liquid. A dendrite seed is then brought into contact with the liquid and the temperature of the liquid adjusted as in

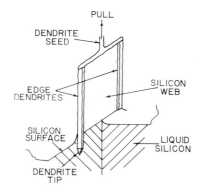

FIG. 7.20. Crystal growth of silicon using the web growth technique. Note that the dendrites extend into and below the liquid surface.

the normal crystal pulling process. Growth is initiated by lowering the temperature of the liquid several degrees so that supercooling of the liquid occurs in the vicinity of the seed. The seed initially grows in a lateral direction until thermal equilibrium is established. Dendrites then grow down into the silicon liquid. Pulling of this structure produces a web of liquid silicon supported by the side dendrites. This liquid web then crystallizes into a well defined and orientated single crystal ribbon of silicon. Growth rates are comparable to those achieved in the EFG process.

This technique has the advantage that no foreign material such as the die in the EFG process need be introduced into the liquid. Thus, contamination of the liquid is kept to a minimum. One problem with web growth is associated with the stability of the dendrite structure during growth. Droplets of liquid can be trapped within the dendrite and if they freeze at the wrong point during growth they can cause distortion of the web or in severe cases cause a polycrystalline web to form. Additional modifications of the thermal geometry from cylindrical to oblong and allowing for "continuous feeding" during growth have improved the economics to the point where this technique can be considered as a source for photovoltaic silicon.

References

1. J. CZOCHRALSKI, *Z. Phys. Chem.* **92** (1918) 219.
2. R. L. BYER, J. F. YOUNG, and R. S. FIEGELSON, *J. Appl. Phys.* **41** (1970) 2320.
3. S. MUJAZAUA and H. IWASAKI, *J. Cryst. Growth* **10** (1971) 276.
4. S. GELLER, G. P. ESPINOSA, L. D. FULLMER, and P. B. CRANDALL, *Mat. Res. Bull.* **7** (1972) 1219.
5. C. D. BRANDLE and R. L. BARNS, *J. Cryst. Growth* **26** (1974) 169.
6. J. R. CARRUTHERS, *J. Cryst. Growth* **32** (1976) 13.
7. C. D. BRANDLE, *J. Appl. Phys.* **49** (1978) 1855.
8. D. B. GASSON and B. COCKAYNE, *J. Mat. Sci.* **5** (1970) 100.
9. W. G. FIELD and R. W. WAGNER, *J. Cryst. Growth* **314** (1968) 799.
10. T. B. REED and E. R. POLLARD, *J. Cryst. Growth* **2** (1968) 243.
11. A. F. WITT and H. C. GATOS, *J. Electrochem. Soc.* **113** (1966) 808.
12. A. F. WITT and H. C. GATOS, *J. Electrochem. Soc.* **115** (1968) 70.
13. R. R. ZUPP, J. W. NIELSEN, and P. V. VITORIO, *J. Cryst. Growth* **5** (1969) 269.
14. C. D. BRANDLE and D. C. MILLER, *J. Cryst. Growth* **24/25** (1974) 432.
15. J. R. CARRUTHERS, *J. Electrochem. Soc. Solid State Sci.* **114** (1967) 959.
16. J. R. CARRUTHERS and K. NASSAU, *J. Appl. Phys.* **39** (11) (1968) 5205.

17. K. Shiroki, *J. Cryst. Growth* **40** (1977) 129.
18. C. D. Brandle, *J. Cryst. Growth* **42** (1977) 400.
19. W. E. Langlois and C. C. Shir, *Computer Methods in Appl. Mech. Engng* **12** (1977) 145.
20. N. Kobayaski and T. Arizumi, *J. Cryst. Growth* **30** (1975) 177.
21. K. Takagi, T. Fukazawa, and M. Ishü, *J. Cryst. Growth* **32** (1976) 89.
22. P. A. C. Whiffin, T. M. Bruton, and J. C. Brice, *J. Cryst. Growth* **32** (1976) 205.
23. P. A. C. Whiffin and J. C. Brice, *J. Cryst. Growth* **10** (1971) 91.
24. R. F. Belt and J. P. Moss, *Mat. Res. Bull.* **8** (1973) 1197
25. K. Morizane, A. F. Witt, and H. C. Gatos, *J. Electrochem. Soc.* **114** (1967) 738.
26. G. H. Schurtte in *Direct Observations of Imperfections in Crystals* (J. B. Newkirk and J. H. Wernick, eds.), Interscience, New York (1962) 497.
27. A. F. Witt, M. Lichtensteiger, and H. C. Gatos, *J. Electrochem. Soc.* **120** (1973) 119.
28. W. T. Stacy, *J. Cryst. Growth* **24/25** (1974) 137.
29. B. E. Warren, *X-ray Diffraction*, Addison-Wesley (1969).
30. D. C. Miller, *J. Electrochem. Soc.* **120** (1973) 678.
31. D. C. Miller, A. J. Valentino, and L. K. Shick, *J. Cryst. Growth* **44** (1978) 121.
32. W. G. Pfann, *Zone Melting*, 2nd edn., Wiley, New York (1966).
33. J. A. Burton, R. C. Prim, and W. D. Slichter, *J. Chem. Phys.* **21** (1953).
34. J. C. Brice, *The Growth of Crystals from Liquids*, North-Holland, Amsterdam (1973).
35. C. D. Brandle and A. J. Valentino, *J. Cryst. Growth* **12** (1972) 3.
36. K. Chow, G. A. Keig, and A. M. Hawley, *J. Cryst. Growth* **23** (1974) 58.
37. D. Mateika, J. Herrnring, R. Rath, and Ch. Rusche, *J. Cryst. Growth* **30** (1975) 311.
38. C. D. Brandle and R. L. Barnes, *J. Cryst. Growth* **20** (1973) 1.
39. A. J. Valentino and C. D. Brandle, *J. Cryst. Growth* **26** (1974) 1.
40. T. Kyle and G. Zydzik, *Mat. Res. Bull.* **8** (1973) 443.
41. W. Bardsley, G. W. Green, C. H. Holliday, and D. T. J. Hurle, *J. Cryst. Growth* **16** (1972) 277.
42. D. F. O'Kane, V. Sadagapan, and E. A. Giess, *J. Electrochem. Soc.* **120** (1973) 1272.
43. A. E. Zinnes, B. E. Nevis, and C. D. Brandle, *J. Cryst. Growth* **19** (1973) 187.
44. W. Bardsley, D. T. J. Hurle, and G. C. Joyce, *J. Cryst. Growth* **40** (1977) 13.
45. W. Bardsley, D. T. J. Hurle, G. C. Joyce, and G. C. Wilson, *J. Cryst. Growth* **40** (1977) 21.
46. K. J. Gärtner, K. F. Rittinghaus, and A. Seiger, *J. Cryst. Growth* **13/14** (1972) 619.
47. D. F. O'Kane, T. W. Kwap, L. Gulitz, and A. L. Berdnourtz, *J. Cryst. Growth* **13/14** (1972) 624.
48. W. C. Dash, *J. Appl. Phys.* **30** (1959) 459.
49. J. B. Mullin, R. J. Heritage, C. H. Holiday, and B. W. Straughan, *J. Cryst. Growth* **3/4** (1968) 281.
50. A. R. von Neida, L. J. Oster, and J. W. Nielsen, *J. Cryst. Growth* **13/14** (1972) 647.
51. A. von Hippel, Tech. Rept. 178, Laboratory of Insulation Research, MIT, March 1963, 44.
52. R. C. Linares, presented at *Fourth American Conference on Crystal Growth, Washington DC, 1978.*
53. H. E. LaBelle, Jr., *Mat. Res. Bull.* **6** (1971) 581.
54. B. Chalmers, *J. Cryst. Growth* **13/14** (1972) 84.
55. S. N. Dermatis and J. W. Faust, Jr., *IEEE Commun. Electron.* **65** (1963) 94.
56. D. L. Barrett, E. H. Myers, D. R. Hamilton, and A. I. Bennett, *J. Electrochem. Soc.* **118** (1971) 952.
57. R. G. Seidensticker. *J. Cryst. Growth* **39** (1977) 17.
58. T. Surek, B. Chalmers, and A. I. Mlavsky, *J. Cryst. Growth* **42** (1977) 453.
59. K. V. Ravi, *J. Cryst. Growth* **39** (1977) 1.

CHAPTER 8

Zone Refining and Its Applications

J. S. SHAH

University of The Andes, Merida, Venezuela †

8.1. Introduction

Zone refining and its application in the last 30 years has had a profound influence on the course of solid state devices in industry, and indeed, many advances in different branches of the solid state science.

The main reason for this impact is that zone refining—a simple process—is capable of producing a variety of organic and inorganic materials of extreme high purity. The technique, when it was first applied to purify silicon and germanium, generated a new industry of transistors. Since then it has been an important part of the technology of fabrication of all kinds of solid state devices. The applications of zone refining are so diverse that literally thousands of materials, elements, inorganic compounds, and organic compounds have been successfully purified and/or converted into single crystals and utilized for fundamental studies or used for applications.

8.1.1. BRIEF DESCRIPTION

Zone melting is a portmanteau title given to a large family of techniques which have in common the following feature: "A liquid zone is created by melting a small amount of material in a relatively large or long solid charge or ingot. It is then made to traverse through a part or the whole of the charge."

Zone-melting techniques basically enable one to manipulate distribution of soluble impurities or phases through a solid. They can therefore be used in many ways. For example:

Uniform doping of a known impurity. From one end of the ingot to the other. This can be done by passing a zone through a solid in both forward and reverse directions many times. The process is then known as zone levelling.

Controlled discontinuities in impurity distribution. For fabricating p–n junctions one requires a discontinuous band of impurities. This can be achieved by manipulating the conditions of the molten zone.

Impurity removal. The "removal" of unwanted impurities in a solid can be achieved by repeated passes of a molten zone or several zones in one direction. This process is called the course of solid state devices in industry and, indeed, many advances in different

† On leave from University of Bristol, H. H. Wills Physics Laboratory, Bristol, UK.

zone refining. It shifts the impurities towards one end of the ingot to give very pure material at the other end. Because of its widespread application the process of zone refining is the most important of zone techniques. It can be confused with the process of fractional fusion or fractional crystallization (Goodman, 1954). The difference between the two methods is in the mode of melting. In fractional fusion one repeatedly melts the *whole* of the fractions. This method was, in fact, used by Curie to isolate radium. In zone refining one always, repeatedly, melts a small part of the ingot.

Zone-melting techniques in addition to impurity redistribution may be employed to produce single crystals with some control on inclusion structural defects such as dislocations.

8.1.2. BRIEF HISTORY

Before examining the mechanism for redistribution of impurities it may be interesting and useful to recapitulate, in brief, the history of zone melting.

The birth of zone melting and the discovery of its power of impurity manipulation should be attributed to the paper published by Pfann (1952) in July 1952. In this paper, entitled "Principles of zone melting", Pfann described zone melting and zone refining processes and presented a simple theory to show the impurity distributions which may be achieved by the above processes. Later Pfann and Olsen (1953) described their initial experiments of repeated motion of a molten zone through a rod of germanium and a dramatic degree of purification achieved thereby.

It is of interest to record that the phenomena involving solid–liquid equilibrium, on which zone melting processes are based, were understood long before the techniques were born. It is even more intriguing to note that at least two distinguished scientists had utilized a passage of a molten zone through rods of metals without apparently realizing the true power of the technique. Kapitza (1928) used a traversing zone for moving a high temperature gradient to grow single crystals of bismuth. His concern was to obtain single crystals of bismuth, in which the perfect cleavage plane of the crystal was perpendicular to its boule axis, in order to study magnetoresistance of the metal. Andrade and Roscoe (1937) described a travelling furnace to melt a short section of cadmium and lead wires. They were growing single crystals of the metals to study plastic deformation. It must be stated that from their papers it is clear that the above authors were acutely aware of the influence of impurities on the properties they were observing.

Pfann (1967) himself recalls that he used a molten zone back in 1939 to produce lead–antimony alloy crystals of uniform composition. Then the idea did not strike him as particularly remarkable. It was not until after the Second World War, when he was trying to obtain pure germanium for transistors, that he thought again and realized the full potential of the technique for the removal of impurities.

Since 1952 literally hundreds of papers have been published on application, instrumentation, and theory of zone melting. During the period Pfann has written numerous original and significant publications which contribute substantially to the growth of the field. He has also reviewed the subject of zone melting during the above period. Zone melting has been generously surveyed and reviewed by a number of other people, and many books have been published on the subject. For the benefit of the reader the above reading matter is fully listed in the references.

8.2. Theoretical Aspects of Zone Melting

For intelligent applications of zone-melting processes it is absolutely vital to understand the mechanism responsible for the impurity redistribution in processes involving a solid–liquid interface. The key parameters, which need to be defined before any mathematical treatment can be introduced, are the distribution coefficients. These are dealt with in section 8.2.1. For the benefit of the new entrants in the field the simple cases of normal-freezing and single-pass zone melting are fully derived together with a review of more complicated cases in sections 8.2.2 and 8.2.3. The literature on multipass distribution and the ultimate distribution is also reviewed. The mathematics and description of continuous zone refining, however, are not included. The mathematical theory of the thin alloy zone techniques, etc., is not presented either, but the literature on the above techniques is briefly reviewed in section 8.6.

8.2.1. THE DISTRIBUTION COEFFICIENTS

8.2.1.1. *The Equilibrium Distribution Coefficient*

The equilibrium distribution coefficient was first defined by Pfann with the aid of a phase diagram of a binary system with a solute (a soluble impurity) and a solvent (a host material) as components. Figure 8.1a and b schematically represent portions of such diagrams near the melting points of solvents. The equilibrium distribution k_0 is defined as the ratio of the concentration of the solute in the solid C_S to that in the liquid C_L when the solid and liquid phases are in equilibrium, i.e.

$$k_0 = \frac{C_S}{C_L}. \tag{8.1}$$

From Fig. 8.1 it can be clearly seen that C_S and C_L are the values of solute concentration given by the intersection of a tie line (at a temperature T) with the solidus and the liquidus.

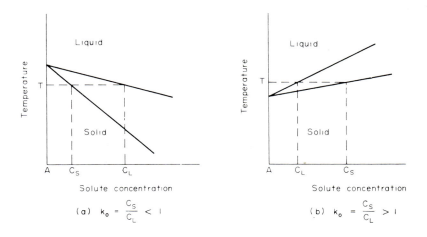

FIG. 8.1. Schematic solvent–solute phase diagrams: (a) $k_0 < 1$, (b) $k_0 > 1$.

The distribution coefficient as defined by eqn. (8.1) actually represents the ratio of the increment in thermodynamic potential due to the interaction and entropy of mixing of the solute with the solvent in the solid and liquid phases. It is therefore similar to other definitions of distribution coefficients between two phases. This fact can be demonstrated by the use of the well-known concept of the activity of the solute defined by Lewis (1907a, The chemical potentials of the solute in the solid and liquid phases are, therefore,

$$\mu = \mu_0 - RT\ln a, \tag{8.2}$$

where μ_0 is a constant, R is the gas constant, and a is the activity of the solute.

The chemical potential of the solute in the solid and liquid phases are, therefore,

$$\mu_S = \mu_0^S - RT\ln a_S$$

$$\mu_L = \mu_0^L - RT\ln a_L.$$

At equilibrium $\mu_S = \mu_L$;

therefore $\dfrac{a_S}{a_L}$ is a constant.

The definition of the activity prescribes that as the concentration of the solute in the solution tends to zero the activity tends to be equal to the concentration. That is, for dilute solutions one may write

$$\frac{a_S}{a_L} = \frac{C_S}{C_L}. \tag{8.3}$$

From eqn. (8.3) it may be further inferred that the equilibrium distribution coefficient is generally a function of the concentration of the solute except for very dilute solutions.

It may be pointed out referring to Fig. 8.1a and b again that if the addition of the solute lowers the melting point (Fig. 8.1a), then $k_0 < 1$ and, conversely, if the melting point is raised by the addition, then $k_0 > 1$.

It is evident from the foregoing that in principle k_0 can be calculated from the phase diagram data. If, however, the accurate phase diagram is not available at the desired concentration (this is often the case for very dilute solutions) then *for dilute solutions* k_0 may be deduced by plotting $\ln k_0$ against $1/T$, where T is the absolute temperature. It was demonstrated by Thurmond and Struthers (1953) that for dilute Ge–Cu, Ge–Sb, and Ge–Si alloys $\ln k_0$ varies linearly with $1/T$.

For very dilute solutions, k_0 may be estimated by the thermodynamic method of Hayes and Chipman (1939). If just a solidus concentration or a liquidus concentration is known at the required temperature k_0 can be calculated from the following equation:

$$N(t) = \frac{\Delta H\, \Delta T}{RT^2(1 - k_0)}, \tag{8.4}$$

where $N(t)$ is the mole fraction of the solute in the liquid, ΔH is the heat of fusion of the solvent, ΔT is the difference in freezing point between the pure solvent and the solution, R is the gas constant, and T is the temperature.

As the name implies, the equilibrium coefficient is applicable to equilibrium solidification processes in which the temperature gradient across the solid–liquid system is negligible and the rate of solidification very small so that impurity gradients are wiped out by diffusion processes. In zone-melting and other practical solid–melt techniques, solidification does not take place in equilibrium conditions. There are often large temperature and concentration gradients present, solidification rates are high, and the mixing of the solute in the liquid imperfect. It is therefore desirable to define two more distribution coefficients.

8.2.1.2. *Interface Distribution Coefficient k**

This is defined as the ratio of the solute concentration in the solid to that in the liquid, actually at the interface.

It is necessary to define the interface distribution coefficient separately because in practice, with only imperfect mixing in the liquid solution and $k_0 < 1$, the rate of rejection of the solute with advancing solidification is much higher than the rate of diffusion of the solute in the liquid. Figure 8.2 shows a schematic solute concentration normally encountered near the solid–liquid interface for systems with $k_0 < 1$.[†] We may therefore write[‡]

$$k^* = C_S/C_{L(0)}. \tag{8.5}$$

Mullin and Hulme (1960) and Carruthers and Nassau (1968) experimentally showed that in general the interface coefficient k^* is different from the equilibrium coefficient k_0. It has been suggested by Brice (1965), Jindal and Tiller (1968), and Baralis (1968) that k^* should be a function of the velocity of solidification (or growth) since the rates of the exchange reactions between the solid and the liquid at the interface (i.e. the rates of the chemical processes taking place) is dependent upon the velocity of growth. Krumnacker and Lange (1969) have measured k^* directly and, indeed, have found it to vary with the growth

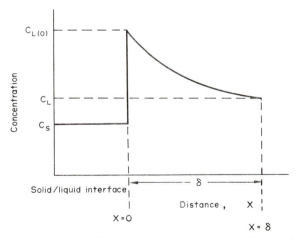

FIG. 8.2. Concentration profile and the diffusion layer at a solid–liquid interface.

[†] The argument also applies for systems with $k_0 > 1$ for which an enhanced depletion of the solute occurs in the vicinity of the solid–liquid interface.

[‡] Normally the theoretical treatments are carried out in the melt and the concentrations in the solids are deduced by using distribution coefficients. In defining various distribution coefficients therefore a single symbol C_S is used for denoting concentration in the solids.

velocity. Brice (1971) has fitted the interpolated Krumnacker and Lange data for zinc dissolved in tin to the following equation derived on a model in which the solute incorporation in the solid is assumed to be by means of exchange reactions across solid–liquid interface:

$$k^* = k_0 + \frac{f(1 - k^*)}{\beta V}, \tag{8.6}$$

where k_0 is the equilibrium distribution coefficient, f is the velocity of growth, β is the dissociation coefficient of the observed solute molecule at the interface; it describes molecules leaving the interface and incorporating into the liquid, and V is the molecular velocity of diffusion of the solute.

It is to be noted that the equilibrium distribution coefficient in the above model can be written as

$$k_0 = \frac{\alpha}{\beta}, \tag{8.7}$$

where α is the sticking coefficient of the solute molecule and describes the entrance of the solute in the crystal.

8.2.1.3. Effective Distribution Coefficient (k_{eff} or k)

Because interface coefficient, at non-zero growth velocities, is different to the equilibrium segregation coefficient, the overall incorporation of solute into the solid in practical crystal growth methods cannot be described by the equilibrium segregation coefficients. One, however, can define the effective distribution coefficient k to do this. The functional relationship between k and k^* may be written as

$$k = k^* F, \tag{8.8}$$

where F is a function of such parameters as growth rate, the degree of mixing (or the degree of stirring), diffusion of the solute in the liquid, and the concentration of the solute.

8.2.1.4. Burton, Prim, and Slichter Theory for k

The manner in which k is defined above prescribes that it ought to be used in all the zone-melting mathematics. It is therefore important to know how to calculate it from the equilibrium coefficient k_0. The BPS (Burton, Prim, and Slichter, 1953) derivation of k was based on the Nernst (1904) postulate that the chemical change at the interface is instantaneous and so the overall rate of the solid–solute–liquid reaction is mainly dependent on diffusion-governed transport processes. The equation they solved was a one-dimensional steady-state diffusion equation expressing conservation of the solute molecular across the interface, namely

$$D\frac{d^2C}{dx} - v_x\frac{dC}{dx} = 0, \tag{8.9}$$

where D is the diffusion coefficient of the solute, C is the concentration of the solute, and v_x is the fluid velocity of liquid in the x-direction. $v_x = f + \omega$, where f is the solidification velocity and ω is the normal fluid velocity. The boundary conditions applied to the eqn. (8.8) may be better understood from Fig. 8.2. The interface is at $x = 0$ and the positive x direction is extended into the melt.

$$C = C_{L(0)} \quad \text{at} \quad x = 0, \qquad \text{(a)}$$
$$C = C_L \quad \text{at} \quad x = \delta. \qquad \text{(b)}$$

The condition (b) means that beyond the distance δ the concentration in the melt due to fluid flow is uniformly equal to C_L. The layer of the thickness δ at the interface is generally referred to as the diffusion layer. Within the diffusion layer

$$v_x = f, \quad \text{i.e.} \quad \omega = 0. \qquad \text{(c)}$$

Equation (8.9) is very similar to the heat flow equation. Its solution with the above boundary conditions yields†

$$k = \frac{k_0}{k_0 + (1 - k_0)e^{-f\delta/D}}. \qquad (8.10)$$

It is interesting to note that the above value of k_{eff}, based on the boundary layer analysis can be further employed to study morphological stability of the solid–liquid interface in the presence of convection; see, for instance, Hurle (1961, 1969). It turns out that δ is a function of the diffusion coefficient kinematic viscosity of the liquid and conditions of the fluid flow. The dependence of δ on the growth velocity can be better seen in the eqns. (8.10a) and (8.10b) which are obtained by rearranging eqn. (8.10), namely

$$\ln\left(\frac{1}{k} - 1\right) = \ln\left(\frac{1}{k_0} - 1\right) - \frac{f\delta}{D} \quad \text{for} \quad k < 1, \qquad (8.10a)$$

$$\ln\left(1 - \frac{1}{k}\right) = \ln\left(1 - \frac{1}{k_0}\right) - \frac{f\delta}{D} \quad \text{for} \quad k > 1. \qquad (8.10b)$$

Experimentally therefore the quantity δ/D can be obtained by measuring k in crystals grown at different velocities with identical stirring conditions and plotting $\ln(1/k - 1)$ or $\ln(1 - 1/k)$ against the growth velocity f. It should be a straight line and its slope is $(-\delta/D)$. It will be shown, however, in the discussion following eqn. (8.14) that if the experimental system incorporates an interface instability such as melting back then $(-\delta/D)$ measured in the above fashion may involve large errors. In general, the dependence of δ on f is weak. However, it varies strongly with the conditions of stirring. For example $\delta \approx 10^3$ cm for vigorous stirring and $\delta \approx 10^{-1}$ cm for negligible stirring. Theoretical solutions of δ for conditions in zone melting are difficult and have not been found satisfactory.

The diffusion layer is also characterized by the actual impurity distribution profile within it. Tiller *et al.* (1953) have obtained the following equation, assuming planar interface, constant growth velocity, and no stirring:

$$C_{L(x)} = C_{L(0)} \exp\left(-\frac{f}{D}x\right) + C_0, \qquad (8.11)$$

† Strictly speaking, in eqn. (8.9) k_0 should be replaced by k^*.

where $C_{L(x)}$ is the concentration of the solute in the liquid at a distance x from the interface, C_0 is the mean initial concentration, $C_{L(0)}$ is the concentration at the interface. In steady state $C_{L(0)} = 1 - k^*/k^*C_0$, f is the growth velocity, and D is the diffusion coefficient of the solute in the liquid.

It is well known that in a melt crystal growth system the solid–liquid interface may be subjected to temperature fluctuations either due to inherent hydrodynamic effects associated with the convective motion of the melt or due to external variations of the heat source. Hurle *et al.* (1968) have shown that in the presence of perturbations of the thermal field the average interface distribution coefficient $\bar{k}^*$ (averaged over a cycle of the perturbation) is in general different to the interface distribution coefficient in the absence of the fluctuations, i.e.

$$\bar{k}^* \neq k^*. \tag{8.12}$$

The change in the effective distribution coefficient in the presence of the perturbation is given by

$$\frac{\delta k}{k} = \frac{\bar{k}^* - k^*}{k^*}. \tag{8.13}$$

Hurle and Jakeman (1969) have obtained the expression for eqn. (8.13), namely†

$$\frac{\delta k}{k} = \frac{1 - k^*}{\Omega} \times \mathrm{Im}\left\{\frac{k^* + i\Omega}{\sqrt{(1 + 4i\Omega)}\coth\left[\frac{1}{2}\Delta\sqrt{(1 + 4i\Omega)}\right] + 2k^* - 1}\right\} \ddagger \tag{8.14}$$

where

$$i = \sqrt{(-1)}, \quad \Omega = \frac{\omega D}{f^2(0)}, \quad \Delta = \frac{f(0)\delta}{D}, \quad \delta = \left(\frac{D}{v}\right)^{1/3}\delta m$$

in which ω is the temporal frequency of the perturbation, D is the diffusion coefficient of the solute in the liquid, f_0 is the mean growth velocity, δ is the thickness of the diffusion layer, v is the kinematic viscosity of the melt, and δm§ is the momentum boundary layer.

The above authors (Hurle and Jakeman, 1969) have shown the validity of eqn. (8.14) by applying it to the experiments of Barthel and Eichler (1967) in which the effective distribution coefficient of tungsten in electron beam zone melted molybdenum was measured as a function of the length of the molten zone.

It was mentioned earlier that the experimental determination of δ/D with the help of eqns. (8.10a) and (8.10b) will lead to errors if temperature fluctuations, and therefore interface instabilities, are present in the experimental system. For the presence of low frequency fluctuations one may write

$$\frac{\delta k}{k} = \eta(1 - k_0)f(0)\delta/D, \tag{8.15}$$

where η is effectively a parameter which represents the error in δ/D. This is because the experimentally measured distribution coefficient is actually the average effective distribution coefficient $\bar{k}$ and

$$\bar{k} = k(1 + \eta) = k_0\{1 + (1 + \eta)(1 - k_0)f(0)\delta/D\}. \tag{8.16}$$

† In eqn. (8.14) k^* may be replaced by k_0.
‡ Im { } is the imaginary part of the bracketed function.
§ For the definition of δm and discussion of its relation to δ, see Levich (1962).

It is now evident that the slope of the plot obtained from eqn. (8.10a) or (8.10b) is $(1 + \eta)\delta/D$ and not δ/D.

One may summarize the foregoing discussion on the distribution coefficients in a steady state by the following qualified equations:

$$k_0 = \frac{C_S}{C_L}, \qquad f = 0 \qquad 2.\mathrm{I}$$

$$k^* = \frac{C_S}{C_{L(0)}}, \qquad f = f \qquad 2.\mathrm{II}$$

$$k = \frac{C_S}{C_L}, \qquad f = f \qquad 2.\mathrm{III}.$$

At equilibrium therefore

$$k = k^* = k_0 \qquad 2.\mathrm{IV}.$$

For very high growth rates (i.e. as $f \to \infty$)[†]

$$\left.\begin{array}{l} k^* \to 1 \\ \text{and} \quad k \to 1 \end{array}\right\} \quad \text{as} \quad f \to \infty \quad 2.\mathrm{V}.$$

Finally, it should be pointed out that eqn. (8.10) and the above exposition does not explain the experimental variations with orientation of growing crystals; for example, the facet effect first described in detail by Mullin and Hulme (1960) and more recently by Abe (1974). They can be explained if one accepts that k^*, and therefore k, is influenced by the orientation of the growing face, i.e. the exchange rates of solute atoms between the solid and liquid vary with orientation of the growing facet.

Hall (1953), Trainor and Bartlett (1961), and Holmes (1963) independently discussed segregation on the phenomenon of sheet growth of a monolayer and the effect of the motion of growth steps on the impurity incorporation. The concentration of impurity increases with the velocity of growth steps and decreases with an increase in the rate at which the impurity diffuses out of the growth steps into the melt. It may well be that higher impurity concentration in the faceted regions is due to the higher velocity of growth steps. Abe (1974) has recently estimated from experimental conditions of (111) faceted growth in silicon that a relatively large difference in supercooling ($\sim 9°\mathrm{C}$) occurs in the faceted region, relative to the rest of the interface. Such a difference in supercooling is sufficient to generate faster motion of growth steps in the faceted region.

8.2.2. SOLUTE DISTRIBUTION IN NORMAL FREEZING PROCESSES

The processes of solidification in which the whole of a charge is melted initially and then gradually solidified unidirectionally are generally referred to as normal freezing processes. The two basic normal freezing methods popularly used for growing crystals are:

The Bridgman process. In this method, due to Bridgman (1925), a crucible containing a molten charge is moved slowly, relative to a stationary temperature gradient for unidirectional solidification.

[†] Experimental studies on "splat" cooling by Duwez (1965) confirm that as $f \to \infty$, $k \to 1$.

The Stöber process. This method was used by Stöber (1925). Here a temperature gradient is moved through a crucible containing a molten charge.

It is important for the reader to be familiar with expressions for impurity distribution in normal freezing processes. At any moment a normal freezing process may be schematically represented by Fig. 8.3, where g is the fraction solidified. The following assumptions are made for the derivation of the first expression.†

(1) Diffusion of the solute, under consideration, is negligible in the solid, i.e.

$$x^2 \gg D_S \cdot t, \tag{8.17}$$

where x is a length of the fraction solidified in time t and D_S is the diffusion coefficient of the solute in the solid.

(2) The effective distribution coefficient k is constant.

(3) The density change of the solution during freezing is zero.

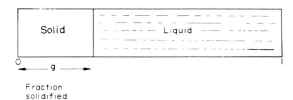

Fraction
solidified

FIG. 8.3. A schematic representation of a normal freeze process.

Then the solute concentration (e.g. solute atoms/unit volume) in the solid immediately behind the interface is given by

$$C_S = -\frac{ds}{dg}, \tag{8.18}$$

where C_S is the solute concentration in the solid, s is the amount of solute in the liquid, and g is the solidified fraction of the original volume of unity.

Recalling eqn. (8.1) and the definition of the diffusion coefficient we have

$$C_S = kC_L$$

where C_L is the concentration of the solute in the liquid. But

$$C_L = \frac{s}{1-g},$$

therefore

$$C_S = \frac{ks}{1-g}. \tag{8.19}$$

Substituting C_S in eqn. (8.18) and integrating we get

$$\int_{s_0}^{s} \frac{ds}{s} = \int_{0}^{g} -\frac{k}{1-g}\,dg,$$

† The assumptions made here are in fact valid for (1) systems with complete mixing (i.e. C_L uniform) and (2) steady-state systems with partial mixing where the growth velocity and the thickness δ of the diffusion layer (see section 8.2.1.4) are constant.

where s_0 is the total amount of the solute and therefore the amount, initially, in the liquid when $g = 0$, i.e.

$$s = s_0(1 - g)^k.$$

Therefore eqn. (8.18) can be rewritten as

$$C_S = -\frac{ds}{dg} = ks_0(1 - g)^{k-1},$$

because the original volume was unity,

$$s_0 = C_0,$$

where C_0 is the original concentration.

Therefore
$$C_S = kC_0(1 - g)^{k-1}. \tag{8.20}$$

Equation (8.20) can be rewritten in the logarithmic form:

$$\ln\left(\frac{C_S}{C_0}\right) = \ln k + (k - 1)\ln(1 - g). \tag{8.20a}$$

It can now be used for a simple experimental determination of k. The logarithmic plot of C_S/C_0 against $(1 - g)$ is a straight line and its slope and intercept at $g = 0$ will provide a measure of k. An experimental determination of the solute concentration C_S as a function of g can be made by solidifying a cylinder of material in a normal freeze manner. The experimental logarithmic plot thus obtained may be compared with those calculated from (8.20a). For a fuller account of the various methods of determination based on eqn. (8.20) the reader is referred to Peyzulayev *et al.* (1965).

Distribution curves of normal freezing computed from eqn. (8.20) are presented in Fig. 8.4 for some typical values of k.

For a system containing a volatile solute, in addition to the solute exchange between the solid and the liquid. An exchange reaction between the gaseous and the liquid phases has to be accounted for. The above case has been considered by van den Boomgaard (1955). The concentration in the liquid is dependent on the reaction rates of the liquid and gaseous phases and therefore the quantity dC_L/dt is proportional to the reaction constant k_R, the surface area of the melt A, and the reciprocal of the melt volume V. The actual expression for dC_L/dt is given by

$$\frac{dC_L}{dt} = \frac{k_R}{h}(C_{eq} - C_L) \tag{8.21}$$

where $h = V/A$, i.e. volume/surface ratio of the melt, and C_{eq} is the concentration of the solute in the liquid in equilibrium with the gas phase.

By substituting the value of C_L from (8.21), for the instant when the fraction g of the volume is solidified, in an equation similar to eqn. (8.18) one may obtain

$$\frac{dC_S}{dg} + \left(\frac{k - 1}{1 - g} + \frac{k_R}{hf}\right)C_S = \frac{k_R}{hf}C_{eq}, \tag{8.22}$$

where $g = ft$. Integral solution of eqn. (8.22) can be evaluated if k_R, h, and f are known. For $k_R = 0$ the solution reduces to that given by eqn. (8.20).

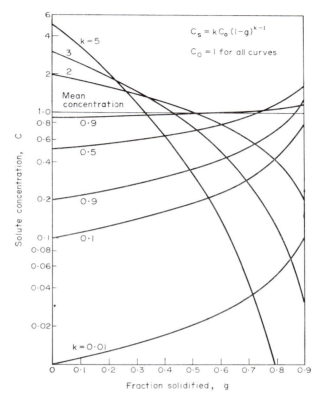

FIG. 8.4. Computed solute distribution curves for normal freezing. (From *Zone Melting*, second edition, by William G. Pfann. Copyright © 1958, 1966, John Wiley and Sons Inc., by permission of John Wiley and Sons Inc.)

For non-volatile solute–solvent systems, if the growth velocity f and/or thickness δ of the diffusion layer vary with g, the fraction solidified, then k is not a constant. The variations of k may be evaluated from the solute conservation condition, namely

$$\left[\frac{C_S(g)}{k(g)} - C_0 \right](1 - g) = \int_0^g [C_0 - C_S(g)]\,dg. \tag{8.23}$$

From eqn. (8.23)

$$k(g) = \frac{C_S(g)}{\left\{ C_0 + \dfrac{\left[\displaystyle\int_0^g \{C_0 - C_S(g)\}\,dg \right]}{(1 - g)} \right\}}. \tag{8.24}$$

It may be recalled from section 8.2.1.4 that k must lie in the limits $k_0 \leqslant k \leqslant 1$. Therefore the constants in eqn. (8.24) should be so chosen that the value of k given by eqn. (8.24) obeys the above-mentioned constraints. Johnston and Tiller (1962) have demonstrated in the case of Pb–Sn system the possibility of programming growth conditions, in accordance with eqns. (8.23) and (8.24) to obtain C_S/C_0 as a linear and cosine function of the distance along the length of a bar ingot.

For rapid freezing rates for which no effective stirring occurs (i.e. convective stirring is assumed to be negligible) the solute distribution is governed by the diffusion of the solute in the liquid. The rate of change of concentration in the liquid is given by

$$\frac{\mathrm{d}C_L}{\mathrm{d}t} = D\frac{\mathrm{d}^2C_L}{\mathrm{d}x^2} + f\frac{\mathrm{d}C_L}{\mathrm{d}x}. \tag{8.25}$$

Here $D(\mathrm{d}^2C_L/\mathrm{d}x^2)$ is the rate of change of the solute concentration due to diffusion and $f(\mathrm{d}C_L/\mathrm{d}x)$ is the rate of change of the solute concentration due to the solidification velocity.

Appropriate boundary conditions applicable to eqn. (8.25) are:

(1) $C_L = C_0$ at $x = \infty$ and $t > 0$ and at $t = 0$ for $x > 0$.
(2) $f(1 - k^*)C_L = -D(\mathrm{d}C_L/\mathrm{d}x)$ at $x = 0$ (the flux condition at the interface).

An approximate solution to eqn. (8.25), as given by Tiller *et al.* (1953), is eqn. (8.11). Then C_S is given by

$$C_S = C_0\left[(1 - k^*)\left\{1 - \exp\left(-\frac{f \cdot k^*}{D} \cdot x\right)\right\} + k^*\right]. \tag{8.26}$$

It is argued by Pohl (1954a, b) that eqn. (8.28) is only valid for the case when $k^* \approx 0$. For a more rigorous solution and further discussion the reader is referred to Tiller *et al.* (1953), Hulme (1955), and Memelink (1956).

In the summary of "no mixing" cases it may be stated that for situations with growth velocity $f = $ constant and planar interface, $k \to 1$ in steady state. For mixing freezing processes with variable growth rates and/or plane interface the effective distribution coefficient departs from the unity.

8.2.3. SOLUTE DISTRIBUTION IN ZONE MELTING

To discuss the solute distribution in zone melting the following variables need to be specified:

(1) The zone length l.
(2) Length of the charge L.
(3) The initial concentration of the solute. (If it is uniform throughout the ingot, it is denoted by a constant C_0.)
(4) The traverse velocity of the zone or the zone velocity f.

In addition to the above, the knowledge of the conditions of mixing and vapour phase data are essential.

8.2.3.1. *Single-pass Distribution*

The distribution of the solute in the ingot from one end to the other is first evaluated for a simple case defined by the assumptions:

(1) The distribution coefficient k is a constant.
(2) The zone length l is constant.

(3) The initial concentration C_0 of the solute is uniform throughout the ingot.

(4) The densities of the liquid and the solid under consideration are the same.

(5) Diffusion of the solute in the solid is negligible as defined by eqn. (8.17).

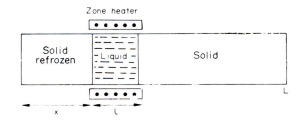

FIG. 8.5. A schematic representation of a zone melting process.

Figure 8.5 is a schematic diagram of a non-continuous zone-refining process. Let the direction of the zone traverse be in the direction $x = 0$ to $x = L$. As the liquid zone advances along the ingot it leaves behind the portion of the ingot resolidified. The equation for solute transfer due to the solidification of the charge of an incremental volume dx as the zone advances may be formulated from the following argument. If C_L is the concentration of the solute in the liquid then the amount of solute leaving the zone due to solidification is $kC_L\,dx$. The solute entering the zone due to the melting of the charge of the volume dx is $C_0\,dx$. Therefore, the net change in the total amount of solute (5) in the liquid is

$$ds = (C_0 - kC_L)\,dx. \tag{8.27}$$

Assuming the cross-sectional area of the ingot and the zone to be unity, the concentration in the liquid zone is

$$C_L = \frac{s}{l}, \tag{8.28}$$

where s is the amount of the solute in the zone at a distance x.

Therefore eqn. (8.27) becomes

$$ds = \left(C_0 - \frac{ks}{l}\right)dx \tag{8.29}$$

or

$$\frac{ds}{dx} + \frac{k}{l}s = C_0. \tag{8.30}$$

Solution of eqn. (8.30) is

$$\int_s^{s_0} s\exp\left(\frac{k}{l}\right)x = C_0\int_0^x \exp\left(\frac{k}{l}x\right)dx$$

or

$$s\exp\left(\frac{k}{l}x\right) - s_0 = \frac{C_0 l}{k}\left\{\exp\left(\frac{k}{l}x\right) - 1\right\}. \tag{8.31}$$

Here s_0 is the amount of the solute in the zone at $x = 0$, i.e.

$$s_0 = C_0 l. \tag{8.32}$$

Substituting from eqn. (8.32) and rearranging eqn. (8.31),

$$s = \left\{ C_0 l + \frac{C_0 l}{k} \left[\exp\left(\frac{k}{l}x - 1\right) \right] \exp\left(-\frac{k}{l}x\right) \right\}. \tag{8.33}$$

The solute concentration C_S in the solid at any x is given by

$$C_S = \frac{k \cdot s}{l}. \tag{8.34}$$

Therefore from eqns. (8.33) and (8.34)

$$\frac{C_S}{C_0} = \left\{ 1 - (1 - k)\exp\left(-\frac{k}{l}x\right) \right\}. \tag{8.35}$$

In Fig. 8.6, which is a schematic profile of the solute distribution ($k < 1$) for a single pass, eqn. (8.35) represents the distribution in the region marked *I*. It embraces the portion of the ingot from $x = 0$ to $x = L - l$, where L is the total length of the ingot. Region II in the figure is the last liquid zone to freeze out. It is actually normally frozen and therefore the distribution in this region is given by the normal freeze, eqn. (8.20).

Figure 8.7 shows curves calculated from eqn. (8.35) for some typical values of k and the total ingot length $L = 10l$. It is of interest to compare a typical normal freeze curve to a typical single-pass curve. These are superimposed in Fig. 8.8. The curves for $k = 0.1$ in Fig. 8.8 show that purification obtained by a single zone pass is worse than that by one normal freezing step.

If the initial solute distribution in the charge is not a constant but a function of the distance along the charge (as it would be if the charge were prepared by any melt technique

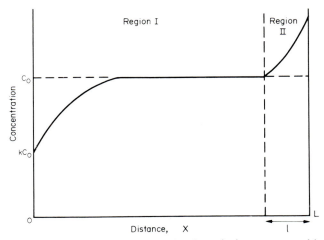

FIG. 8.6. A schematic solute distribution for a single-pass zone melting.

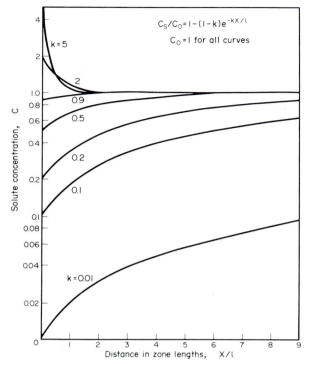

FIG. 8.7. Computed single-pass solute distribution curves. (From *Zone Melting*, second edition, by William G. Pfann. Copyright © 1958, 1966, John Wiley and Sons Inc., by permission of John Wiley and Sons Inc.)

except zone levelling), a general solute conservation equation of obtaining a concentration $C_S(x_1)$ at any point x_1 is given by

$$\left\{ \frac{C_S(x_1)}{k} - \int_{x_1}^{l} C_0(x)\,dx \right\} = \int_0^{x_1} [C_0(x) - C_S(x)]\,dx. \tag{8.36}$$

The above equation may be applied to two simple cases of zone levelling described by Pfann (1952):

(1) Simple zone melting with a starting charge: in this case the starting charge has an initial concentration C_0/k for the first zone length and C_0 in the rest of the ingot, i.e.

$$C_0(x) = \frac{C_0}{k} \quad \text{for} \quad 0 \leqslant x \leqslant l \quad \text{and} \quad C_0(x) = C_0 \quad \text{for} \quad x > l.$$

(2) Starting charge into pure solvent: this case is suitable for solutes with $k < 0.1$. Here the solute is contained in the first zone length of the charge and the remainder is the pure solvent initially, i.e. $C_0(x) = C_0$ for $0 \leqslant x \leqslant 1$ and $C_0(x) = 0$ for $x > 1$.

In both cases the solution of the equation yields

$$\frac{C_S(x_1)}{C_0} = k\,e^{-kx_1/l} \tag{8.37}$$

except for the last zone length.

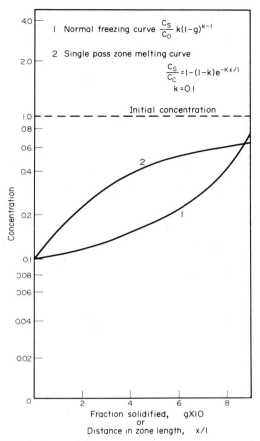

FIG. 8.8. Typical normal-freeze and single-pass curves—superimposed.

In volatile solute systems, single-pass zone distribution would be dependent upon (1) the reaction between the liquid and vapour, (2) segregation of the solute at the trailing solid–melt interface of the zone, and (3) the dissolution of the solute at the leading melt–solid interface. Under idealized conditions (i.e. when the assumptions listed in the beginning of section 8.2.3.1 hold and, in addition, when the gaseous phase does not change the composition of the solid) the equation for the solute distribution for $0 \leqslant x \leqslant L - l$ was formulated by van den Boomgaard (1955),

$$\frac{\mathrm{d}C_S}{\mathrm{d}x} + \left(\frac{k}{l} + \frac{k_R}{hf}\right)C_t = \frac{k_R}{hf}C_{\mathrm{eq}} + \frac{k}{l}C_0, \tag{8.38}$$

where k_R is the reaction constant for the liquid and solid phases, $h = V/A$, i.e. volume–surface ratio of the melt, and C_{eq} is the concentration of the solute, in the liquid, in equilibrium with the gaseous phase. The solution of the eqn. (8.38) is given by†

$$C_S = kC_0 \exp\left[-\left(\frac{k}{l} + \frac{k_R}{fh}\right)x\right] + \frac{k_R l C_{\mathrm{eq}} + kfhC_0}{k_R l + kfh}\left\{1 - \exp\left[-\left(\frac{k}{l} + \frac{k_R}{fh}\right)x\right]\right\}. \tag{8.39}$$

† It is assumed that the zone at $x = 0$ starts to move just as the reaction with the gaseous phase begins.

By comparing eqn. (8.35) with eqn. (8.39) one can estimate the influence of the atmosphere due to volatile solute. It can be shown that for an infinite charge the solute concentration will level off at a value $C_{S(t)}$ where

$$C_{S(t)} = \frac{k_R l C_{eq} + kfhC_0}{k_r l + kfh}.$$ (8.40)

8.2.3.2. *Multipass Distributions*

It was pointed out earlier that the real power for purification of the zone refining is realized when repeated unidirectional passes of a molten zone are executed on an ingot. Qualitatively the purification ability of the multipass processes may be understood by referring to Fig. 8.6. On successive passes the concentration in the beginning of the region I is lowered as the molten zone, in each pass, accumulates the solute ($k < 1$) leaving behind a more pure solid. The impurity contained in the zone begins to be piled up in front of region II as the front edge of the zone (the leading edge) "sees" a higher slope of the concentration curve of the region II than that in the previous pass. The process of piling up may be alternatively described as being reflected one zone length during each successive pass with diminishing intensity.

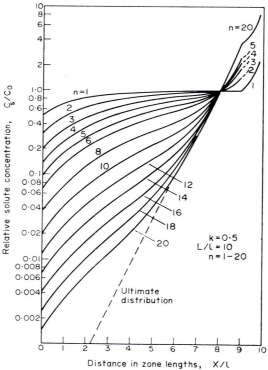

FIG. 8.9. Computed multipass distribution curves for $k = 0.5$, $L/l = 10$ and $n = 1$–20. (From *Zone Melting*, second edition, by William G. Pfann. Copyright © 1958, 1966, John Wiley and Sons Inc., by permission of John Wiley and Sons Inc.)

It may be guessed from the above discussion that the mathematical description of the multipass operations is tedious and difficult, involving considerable computations. The basic approach of all the mathematical methods reported is based on solving a differential or integral equation relating the change in solute concentration of the moving zone to the fluxes of solute leaving and entering the zone (i.e. the principle of solute conservation is applied under appropriate conditions). For a simple case as defined by the assumptions in the beginning of the section 8.2.3.1, both Lord (1953) and Reiss (1954) independently derived the so-called differential–difference concentration equation:

$$\frac{l}{k} \, dC_n(x) = [C_{(n-1)}(x+1) - C_n(x)] \, dx, \tag{8.41}$$

where $C_n(x)$ is the solute concentration freezing out at a distance x from the starting end of the ingot in the nth pass.

A series of special solutions for eqn. (8.41) or for an equivalent integral equation have been reported:

(1) For values of k near unity by Lord (1953).
(2) For a semi-infinite bar by Reiss (1954) and Aleksandrov *et al.* (1956).

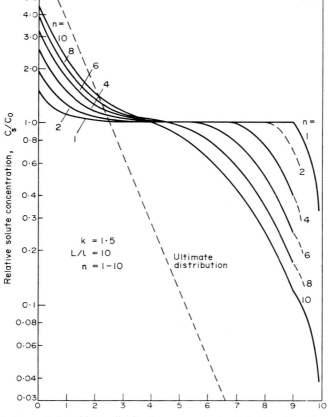

FIG. 8.10. Computed multipass distribution curves for $k = 1.5$, $L/l = 10$ and $n = 1$–10. (From *Zone Melting*, second edition, by William G. Pfann. Copyright © 1958, 1966, John Wiley and Sons Inc., by permission of John Wiley and Sons Inc.)

(3) For an infinite bar by Reiss and Helfand (1961) and Kirgintsev (1963).

(4) For a toroid by Reiss and Helfand (1961).

(5) For a finite ingot but small number of passes by Braun and Marshall (1957).

(6) For a finite ingot and any number of passes by Kirgintsev et al. (1963a) and Helfand and Kornegay (1967).

(7) For two-component systems by Kirgintsev et al. (1963b) (see section 8.2.3.5).

In addition, a computational technique devised by Hemming (first referred to by Lord (1953)) has been extensively used for calculating zone-refining curves for ingots with arbitrary initial distribution of the solute. Briefly an ingot in this method is divided into a number of cells with cell width ≈ 0.1–0.2 zone length. The zone is allowed to move in one cell jumps so that the material in the vacated cell is solidified segregating the solute. At the end of the ingot where normal freezing commences the technique takes into account the decreasing volume of the zone. The numerical method used was the so-called trapezoidal rule for integration. Burris et al. (1955) calculated some curves by the above method. A collection of the curves also appears in Pfann (1966). Here two examples of the computed curves for $k = 0.52$ and $k = 1.5$ are shown in Figs. 8.9 and 8.10.

Analogue computing machines for deriving solute distribution in zone refining have been utilized. An electric analogue, using charging and discharging capacitors, and a mechanical analogue have been reported by Bertein (1958a, b). Hydraulic or liquid analogues are described by Pfann (1966). Additionally, Mason (1961) has used an analogue simulation of a zone melting in a small diameter tungsten rod heated by electron bombardment to determine optimum power input in relation to the rod diameter, heating rate, and the zone speed. He did not evaluate concentration–distance profiles.

8.2.3.3. *Ultimate Distributions*

It is evident that the purification is achieved by zoning due to a non-equilibrium solute concentration distribution between the solid and the liquid zone. Therefore, the maximum purification or ultimate distribution is reached when the distribution reaches a steady-state equilibrium. In the steady state the forward flux of the solute due to segregation at solidification equals the backward flux due to mixing in the zone, so that the driving force for the solute is reduced to zero.

A simple expression for the ultimate distribution was derived by Pfann (1952) as follows:

Let $C_u(x)$ be the ultimate distribution of the solute. If at any stage the molten zone leaves behind the concentration $C_u(x)$ at a point x then the concentration $C_L(x)$ in the zone is given by

$$kC_L(x) = C_u(x), \qquad (8.42)$$

where k is the distribution coefficient. But for the unit cross-sectional area of the zone $C_L(x)$ can also be expressed by

$$C_L(x) = \frac{1}{l} \int_x^{x-1} C_u(x)\,dx, \qquad (8.43)$$

where l is the zone length.

Substituting $C_L(x)$ from eqn. (8.42) and eqn. (8.43),

$$C_u(x) = \frac{k}{l} \int_x^{x-1} C_u(x) \, dx. \tag{8.44}$$

The solution of eqn. (8.44) is of a simple exponential form

$$C_u(x) = A \, e^{Bx}, \tag{8.45}$$

where the constants A and B may be calculated from

$$k = \frac{Bl}{e^{Bl} - 1} \tag{8.46}$$

and

$$A = \frac{C_0 BL}{e^{BL} - 1}, \tag{8.47}$$

where C_0 is the mean solute concentration per unit volume and L is the total length of the ingot. Equation (8.47) does not hold for the last zone length where normal freezing takes place. It also cannot account for the enhanced solute pile up or the back reflection of the solute in front of the last zone length.

More rigorous expressions for the ultimate distributions under different boundary conditions have been derived by Birman (1955), Davies (1958a), Volchok (1962), Velicky (1964), Davies (1964), and Helfand and Kornegay (1967). The results show a general agreement with curves computed by the Hemming method. The calculations of Davies (1958a) were found to be in reasonable agreement with the experimental curves for gallium in germanium (Davies, 1958b).

8.2.3.4. Cropping and Ultimate Distributions

An ultimate distribution is a steady-state distribution of the solute. Therefore, to accomplish further purification, a non-equilibrium in the solute equilibrium has to be introduced. This may be achieved by cropping of the impure end (terminal zone end of the ingot for $k < 1$). Then further zone refining can yield a still lower ultimate impurity concentration in the leading end of the ingot. It is, however, important for establishing non-equilibrium solute distribution, to allow the cropped length of the ingot to fill up the original length. Neglecting the back reflection of the impurity from the terminal zone an expression to estimate the improvement in the purification by the above procedure may be derived as follows:

For an original ingot length L the first ultimate distribution, before cropping, is given by the eqn. (8.45), namely,

$$C_1 = A_1 \, e^{Bx},$$

where

$$A_1 = \frac{C_{01} BL}{e^{Bl} - 1}, \tag{8.48}$$

and C_{01} is the original mean concentration. For the cropped ingot length L' the end ultimate distribution (assuming that the cropped ingot fills the original length L) is

$$C_2 = A_2 e^{Bx}. \tag{8.49}$$

The constant B in eqns. (8.48) and (8.49) is the same as it is independent of the zone length. The constant A_2 is

$$A_2 = C_{02} BL(e^{BL} - 1), \tag{8.50}$$

where C_{02} is the mean concentration after cropping; it is given by

$$C_{02} = \frac{1}{L'} \int_0^{L'} A_1 e^{Bx} \, dx = \frac{A_1(e^{BL'} - 1)}{BL'}. \tag{8.51}$$

Substituting C_0 from eqn. (8.50) in eqn. (8.51) and dividing by eqn. (8.48),

$$\frac{A_2}{A_1} = \frac{BL(e^{BL} - 1)^{-1}}{BL'(e^{BL'} - 1)^{-1}}. \tag{8.52}$$

It is pointed out by Pfann (1966) for $k = 0.58 L/l = 10$ and $L' = \frac{1}{2}L$, $A_2/A_1 \approx 0.014$, i.e. a substantial improvement can be obtained, the ratio for constant L' decreasing with increasing k and increasing L. It should, however, be remembered that a similar improvement in the ultimate purification may be obtained without cropping and using the longer ingot lengths in the first place.

8.2.3.5. *Zone Melting, Variable k, and Phase Relationships*

In the single, multipass, and ultimate distributions calculations cited in sections 8.2.3.2, 8.2.3.3, and 8.2.3.4, k, the effective distribution coefficient, was assumed to be constant. It may be recalled from eqn. (8.8) that in general the distribution coefficient is a function of

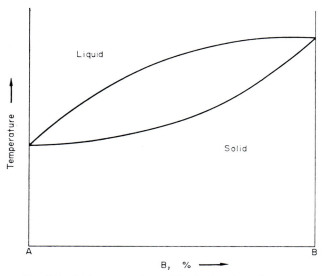

FIG. 8.11. A binary complete solid–solution phase diagram.

growth, rate, orientation, and concentration. For a fixed growth velocity and the orientation of growing front it is possible to express k as a function of concentration of the solute in the liquid

$$k = k_0 = k(C_L) = k(0)\,[1 + AC_L], \tag{8.53}$$

where A is a constant, $k(C_L)$ the distribution coefficient at concentration C_L in the liquid, and $k(0)$ is the distribution coefficient at zero concentration.

Romanenko (1960) has obtained the following expression (together with one for the normal freezing) for a single-pass distribution under the assumption that the length of the liquid zone is independent of the solute concentration (i.e. the temperature gradient at the edges of the zone is very sharp):

$$\frac{C_S(x)}{C_0} = 1 - [1 - k(0)]\exp\left[-k(0)\frac{x}{l}\left\{1 + \frac{2AC_0}{k(0)}\right\}\right]$$

$$+ 2\left[\frac{AC_0}{k(0)}\right]\exp\left[-k(0)\frac{x}{l}\left\{1 + \frac{2AC_0}{k(0)}\right\}\right]$$

$$- \left[\frac{AC_0}{k(0)}\right][2 - k(0)]^2 \exp\left[-k(0)\frac{x}{l}\left\{1 + \frac{2AC_0}{k(0)}\right\}\right]$$

$$+ \left[\frac{2AC_0}{k(0)}\right][1 - k(0)]^2 \exp\left[-2k(0)\frac{x}{l}\left\{1 + \frac{2AC_0}{k(0)}\right\}\right]. \tag{8.54}$$

If the terms in AC_0 are neglected eqn. (8.54) reduces to the Pfann equation (8.35).

For very dilute solutions the value of A may be derived with the help of expressions of Thurmond and Struthers (1953).

For large concentrations, the variability of k may be determined from the relevant T–x type phase diagram. A typical phase diagram for a simple binary solid solution is schematically shown in Fig. 8.11 and qualitatively the variation of k with composition (or concentration) of one component in the other is immediately apparent.

The above case of the two-component system $A_X B_{(1-X)}$, where X and $1 - X$ (i.e. the molar fractions of the components A and B) are comparable, is considered by Kirgintsev et al. (1963b). The differential equation solved by them was

$$\frac{dX_n(z)}{dz} = \frac{1}{\lambda}[1 - (1 - \lambda)X_n(z)]^2\,[X_{n-1}(z + 1) - X_n(z)], \tag{8.55}$$

where $X_n(z)$ is the molar fraction of the component A at a distance x after the nth pass of the zone $z = x/l$ where l is the length of the zone and λ is a constant of the system defined by the equation:

$$\frac{C_{B(S)}}{C_{B(L)}} = \lambda\frac{C_{A(S)}}{C_{A(L)}} \tag{8.56}$$

in which $C_{A(S)}$ and $C_{B(S)}$ are the concentrations of AB in the solid and $C_{A(L)}$, $C_{B(L)}$ are the concentrations in the liquid. The boundary condition applied was

$$X_n(0) = \frac{\displaystyle\int_0^1 X_{n-1}(z)\,dz}{\lambda + (1 - \lambda)\displaystyle\int_0^1 X_{n-1}(z)\,dz}. \tag{8.57}$$

A similar equation for a normally frozen part of the ingot was solved and thus the curves of distribution for a large and infinite number of passes were computed for different values of λ. The significant conclusion of the calculations is that the zone melting can be used for separating components in solution in comparable amounts.

Zone melting of a semiconducting compound of type AB with negligible vapour has been treated by van den Boomgaard (1956). In this instance the phase diagram, shown in Fig. 8.12, has a maximum, but not necessarily at the stoichiometric composition AB. The expression for the deviation from the stoichiometric composition as a function of distance along the ingot was obtained by defining two distribution coefficients k and k', involving the concentration of the "vacancies" of A and B in the solid and liquid phases. For the case $k = k'$, the solution for the distribution of a component in excess of the stoichiometric composition is given by the Pfann equation. For the cases where k and k' differ by small amounts, it is possible to express the distribution by an equation similar to (8.35) by using a coefficient k_Δ which is the weighted average of k and k'. The exact value of excess concentration of the component in the leading end of the ingot, i.e. at $x = 0$, however, is *not* given by

$$C(x)_{x=0} = k_\Delta \cdot \Delta C_0 \qquad (8.58)$$

(where ΔC_0 is the initial and constant excess), but by a more complicated equation.

For a simple case of a binary eutectic system, forming no solid solution and having a phase diagram of the type shown in Fig. 8.13 is given by Vetter (1960). He showed that a single-pass distribution can be described by dividing the ingot into three distinct regions as shown in Fig. 8.14.

(I) $x = 0$ to $x = x_e$ Contains the pure component A.

(II) $x = x_e$ to $x = L - l$ Contains initial composition C_0 of A.

(III) $x = L - l$ to $x = L$ Contains the eutectic mixture.

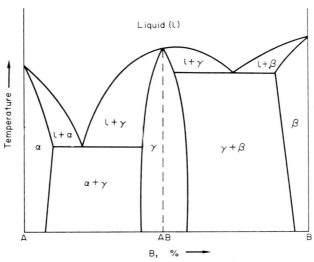

FIG. 8.12. A two-component phase diagram containing a compound.

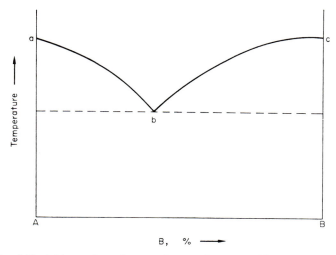

FIG. 8.13. A binary phase diagram of a eutectic system with no solid solution.

For maximum separation of the component A the intermediate region of initial composition should reduce to zero,

i.e.
$$x_e = L - l = \frac{l(C_e - C_0)}{C_0 - C_A} \qquad (8.59)$$

or
$$\frac{L}{l} = \frac{C_e - C_0}{C_0 - C_A} + 1, \qquad (8.60)$$

where C_0 is the initial concentration of A, C_e is the concentration of A in the eutectic mixture, C_A is the concentration of pure A, and L and l are the usual total ingot length and the length of the zone.

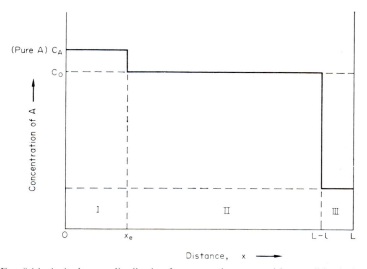

FIG. 8.14. A single-pass distribution for a eutectic system with no solid solution.

A typical phase diagram of a two-component eutectic system, with limited solid solubility is shown in Fig. 8.15. After a single pass, again the distribution of *A* is typified by three distinct regions similar to Fig. 8.14. In the first region a phase α will be solidified as the continuous solid solution exists in the solidus region *ab*. In the intermediate region the frozen solid has initial concentration of *B* (say *X*). This, however, is a mixture of α dendrites and the eutectic. In the third region the concentration of *A* is that of the eutectic.

Multipass zone refining tends to shift all the eutectic composition in the end region. In practice it is often possible to produce a "pseudo-eutectic microstructure" of the solid by zone refining for non-eutectic volume ratio of the phases. It is, however, necessary to use extremely pure components and take due care in adjusting growing conditions for growing highly perfect lamellar or rod-like eutectic structures.

The zone melting phenomenon in a peritectic system was first described by Goodman (1954). A phase diagram of a binary peritectic system is shown in Fig. 8.16. Here the formation of a solid phase β is possible due to the following reaction between the phases

$$\alpha + \text{liquid} \leftrightharpoons \beta.$$

It is well known that the peritectic reaction such as above is slow because it primarily depends upon the rates of diffusion of *A* atoms from the α phase to the β phase and of *B* atoms from liquid to the β phase. Furthermore, because the diffusion rates decrease rapidly with temperature the reaction rate falls as well. Therefore, the zone melting of a uniform composition ingot first freezes out phase α in accordance with the solid solution region *a, b*. As the concentration in the zone reaches point *c* the peritectic reaction will begin and due to the presence of a sharp temperature fall at the zone/solid interface a thin layer (the reaction layer) is formed within which the peritectic reaction is confined. After the formation of the reaction layer the zone leaves behind the phase β following the composition along the solidus *ce*. A typical distribution of a single pass in such a system is schematically shown in Fig. 8.17. The most notable feature of the distribution profile is

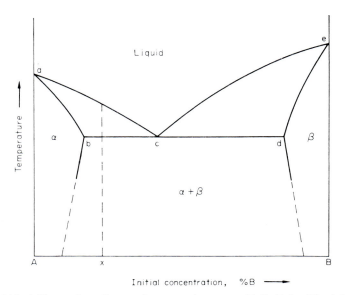

FIG. 8.15. A binary phase diagram for a eutectic system with limited solid solubility.

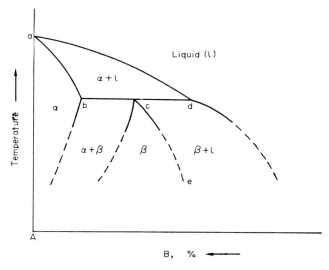

Fɪɢ. 8.16. A binary phase diagram of a peritectic system in the vicinity of the peritectic point.

that it exhibits a relatively sharp discontinuity in concentration at a certain distance x, where the concentration jumps from the value b to c. This "peritectic jump" occurs precisely where the peritectic reaction layer is formed. After the jump, the concentration is dictated by the formation of β, i.e. the solidus of the phase β. The effect of further zone passes is to shift the jump boundary further along the ingot. Thus by zone melting it should be possible to prepare homogeneous compositions from a to b or c to e. It is, however, not possible to prepare and separate compositions from b to d. Preparation of the relatively pure ternary semiconducting compounds such as $Cd\text{–}In_2\text{–}Te_4$ and $Zn\text{–}Cd\text{–}Sn\text{–}As_2$ (which has a peritectic phase diagram) were prepared by Mason and Cook (1961) and Shah (1967) respectively.

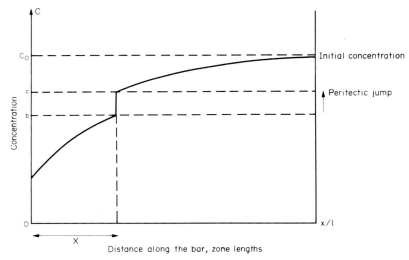

Fɪɢ. 8.17. A typical single-pass distribution in a peritectic system.

In principle it is possible to utilize zone melting with more complex systems. For instance Shah and Pamplin (1969) have reported preparation of a number of compositions in the quaternary system Zn–Cd–Sn–As. It becomes increasingly difficult, however, to make intelligent use of zone melting with complex systems due to the lack of precise phase-diagram data. A general rule for zone melting operations with regard to different mixed-crystal systems is that there is a marked tendency for low melting components to segregate in the direction of the zone movement while high melting components do the reverse.

Deliberate use of an additional component for impurity removal or separation may have the following advantages. It may lower the solidification temperature so that the compounds which are unstable at their melting points or just melt at a very high temperature may be prepared at a lower temperature, and the segregation coefficient of a desired component may improve in the new system. The above advantages, however, may be accompanied by a higher risk of contamination. In any case, the accurate knowledge of the relevant phase diagram is necessary. A general rule for the incorporation of an additional component is that it must have zero or very low solid solubility in the pure phase to be separated.

8.3. Factors Affecting the Practice of Zone Melting

In the foregoing section the detailed theoretical treatments of the distribution of impurities were described. In this section the general rule for the choice of operational variables such as zone length/zone traverse velocity will be described. Additionally the phenomenon of matter transport often encountered with zone refining will be discussed.

8.3.1. THE ZONE LENGTH

The zone length l is one of the most important operational parameters as it appears in nearly all the equations of zone melting. It is often desirable to express the total ingot length L in terms of multiple m of the zone length, i.e.

$$m = \frac{L}{l}. \tag{8.61}$$

By expressing the distance in terms of the zone length the eqn. (8.45) may be rewritten as

$$C_u(Z) = A\,e^{BZ}, \tag{8.62}$$

where $Z = x/l$.

The constants A and B in eqn. (8.62) will be given by the equations

$$k = B(e^B - 1)^{-1} \tag{8.63}$$

and

$$A = Bm(e^{Bm} - 1)^{-1}. \tag{8.64}$$

The r.h.s. of eqns. (8.63) and (8.64) are in fact the Einstein functions. Therefore, the values of A and B can be calculated from the Einstein function tabulations by Sherman and Ewell

(1942) if k is known. A and B can be substituted in the logarithmic form of the equation (8.62), namely,

$$\ln C_u(Z) = \ln A + BZ \tag{8.65}$$

or

$$\log_{10} C_u(Z) = \log_{10} A + 0.4343 BZ. \tag{8.65a}$$

Now for $k < 1$ the lowest concentration is obtained at $Z = x/l = 0$, i.e.

$$\log C_u(Z) = \log A; \tag{8.66}$$

$\log A$ can be further approximated to

$$\log A \approx m \log k \tag{8.67}$$

because eqns. (8.63) and (8.64) are predominantly exponentials. The above equation now clearly reveals the advantage of using large m (for $k < 1$) or a narrow zone for the zone refining.

Figure 8.18 shows approximate distribution for a different number of passes in an ingot with mean initial concentration C_0 and $k < 1$. Here the ultimate distribution is the one achieved for $n = \infty$. Table 8.1 shows the lowest values of $\log C/C_0$ for different values of n. It is evident therein that the greater the value of m the greater number of zone passes are required to approach the ultimate distribution. After m number of passes, however, each successive additional pass produces less additional purification. In practice the values of m less than 5 are not very useful.

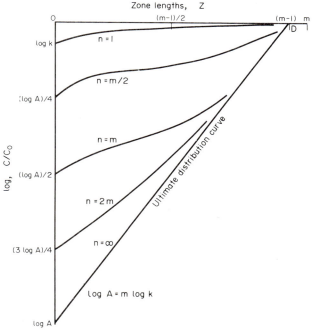

FIG. 8.18. Approximate distribution curves for a system with (1) $k < 1$; (2) the ingot length $L = m.l$; and (3) after a different number, n_1, of zone passes. (From *Zone Melting of Organic Compounds*, by E. F. G. Herington. Copyright © 1963 Blackwell Scientific Publications Ltd., by permission of Blackwell Scientific Publications Ltd.)

TABLE 8.1.

Number of Passes (n)	Ultimate Lowest Concentration ($\log C/C_0$)
∞	$\log A$
$2m$	$(3 \log A)/4$
$m (= L/l)$	$(\log A)/2$
$m/2$	$(\log A)/4$
1	$\log k$

It should be remembered that a useful interpretation of the term *zone length* was given by Pfann (1964) as the length of solid melted to form the zone and *not* the length of the molten zone. With this interpretation of the concentrations, the values of concentrations entered in the zone-melting equations are expressed in weight fraction or atom fractions and the difference in the density of the solid and liquid are accounted for.

8.3.2. ZONE TRAVERSE VELOCITY

The effective coefficient k of an impurity was shown to be the function of growth velocity in eqn. (8.10). It was also shown in section 8.2.1.4 that the thickness of the diffusion layer at the solid–liquid interface is a function of growth velocity. Therefore, the zone velocity is one of the determinant factors for the nature and shape of the solid–liquid interface. The detailed treatment of interface morphology and interface segregation is given by Delves (1975) in Chapter 3 of the first edition of this book. It will, however, suffice to say here that the onset of an interface stability such as constitutional supercooling is inherently dependent upon the zone traverse speed, the concentration of the solute, and the temperature gradient at the solid–liquid interface. It is, therefore, necessary for a homogeneous single-phase, single-crystal production to select the traverse velocity with due care. Generally speaking, slower growth rates are advantageous for producing more perfect crystals. On the other hand, very fast growing speeds may enable one to avoid interface instabilities simply because the very fast growth does not permit the diffusion processes to build an enhanced impurity layer at the interface.

For the purpose of maximum separation or purification in the shortest possible time, for a given k it is necessary to set the ratio n/f, where n is the number of passes and f is the zone traverse velocity, to its lowest possible value. It was in fact shown by Harrison and Tiller (1961) that a generalized operational rule for the above purpose is to set the growth speed so that

$$\frac{f\delta}{D} \approx 1. \tag{8.68}$$

8.3.3. TEMPERATURE GRADIENT AT THE SOLID–LIQUID INTERFACE

The temperature gradient at the solid–liquid interface is, as pointed out above, an important parameter governing the onset of interface instability. A correct combination of the imposed parameter gradient and traverse speeds is essential for avoidance of the microsegregation. The imposed temperature gradient, of course, depends upon the nature of the heat source, thermal conductivity of the ingot material, and other modes of heat losses. It should be noted that the magnitude of the temperature gradient would affect the zone length and the degree of mixing in the zone (see section 8.4.4).

8.3.4. THE DEGREE OF MIXING IN THE LIQUID

It was suggested in sections 8.2.1.3 and 8.2.1.4 that the effective distribution coefficient is a function of the degree of mixing in the liquid because the thickness of the diffusion layer in the vicinity of the solid/liquid interface depends upon the condition of fluid flow, the diffusion coefficient and the kinematic viscosity of the liquid. As a general rule if diffusion is the only mixing process then the width of the diffusion layer is comparable to the zone length. If convection currents are additionally available then δ is of the order of a millimetre or so. For liquid metals (which have kinematic viscosity of the order $2 \, \text{mm}^2 \, \text{s}^{-1}$) the effective width can be reduced to 0.1 mm by moderate stirring and to 0.01 mm by vigorous stirring. Therefore, it is seen from eqn. (8.68) that the degree of mixing will affect the zone traverse velocity for maximum purification. With a greater degree of mixing a higher value of zone velocity can be used.

8.3.5. MATTER TRANSPORT IN ZONE MELTING

Solute transport in zone refining is also accompanied by matter transport unless it is prevented. For instance, zone refining of an ingot in an open horizontal boat develops a taper after being zoned a few times. In fact the mechanism of matter transport is similar to that of solute transport. The driving force in the case of matter transport is actually provided by the difference in density between the solid and liquid phases (neglecting the surface tension and other forces). If the density of the liquid is higher than that of the solid, i.e. if volume contraction occurs on melting, then the matter is transported in the direction of the zone travel. Conversely, if expansion occurs on melting then the matter is transported in the opposite direction to that of the zone travel.

The physics of matter transport may be explained by referring to the Fig. 8.19a. Here the initial cross-section of the solid ingot has the height h_0. When a liquid zone of constant length l is created the height of the liquid, h, is given by

$$h = \alpha h_0, \tag{8.69}$$

where

$$\alpha = \frac{\rho_S}{\rho_L}, \tag{8.70}$$

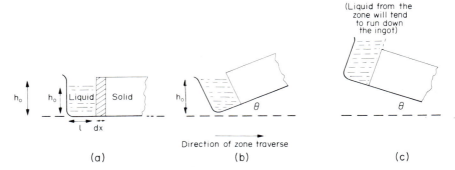

FIG. 8.19. (a) A schematic diagram showing how mass transport arises. (b) An experimental arrangement showing a critical angle of tilt to avoid mass transport in systems $\alpha < 1$. (c) An experimentally unattainable arrangement required for avoiding mass transport in systems with $\alpha > 1$.

where $\rho_S =$ the density of the solid and $\rho_L =$ the density of the liquid. If $\alpha < 1$ and the zone is made to traverse a distance dx, then the volume of the additional solid melted is given by $h_0\,dx$. At the same time the amount of solid frozen at the leading end is $h\,dx$.

Since the solid freezes at the level of the liquid, the change in the volume of the liquid zone dv due to the traverse dx is given by

$$dv = l\,dh = \alpha(h_0 - h)\,dx. \qquad (8.71)$$

Solution, for h, of eqn. (8.71), which is applicable in all the ingot but the last zone length, is of the form

$$\frac{h}{h_0} = 1 - (1 - \alpha)\exp\left(-\frac{\alpha x}{l}\right). \qquad (8.72)$$

In the last zone length the further increases in the height due to normal freezing is yielded by

$$\frac{h}{h_0'} = (1 - g)^{\alpha - 1}, \qquad (8.73)$$

where h_0' is the height of the liquid at the commencement of normal freezing and g is the fraction of the liquid frozen. It can be seen that eqns. (8.72) and (8.73) are of the same form as eqns. (8.35) and (8.20), respectively, for the solute transport, i.e. α is the matter transport equivalent of the distribution coefficient k. The ultimate shape of the ingot in terms of its cross-sectional height may be shown to be of the form

$$\frac{h}{h_0} = A_M \exp(B_M x), \qquad (8.74)$$

where the constants A_M and B_M are given by

$$\alpha = \frac{B_M l}{\exp(B_M l) - 1} \qquad (8.75)$$

and

$$A_M = \frac{B_M L}{\exp(B_M L) - 1},$$ (8.76)

where L is the total length of the ingot.

Following the simple treatment above, matter transport can be largely avoided by tilting the ingot. By adjusting the angle of tilt it is possible to make effective α to be unity. The critical angle of tilt, as shown in Fig. 8.19b, is such that the liquid surface at the trailing end (i.e. the surface in contact with freezing solid) is at the same height as that of the original solid before it was molten. Because the volume of the molten zone is the same before and after tilting the following equation holds

$$\alpha b h_0 l = \frac{l b [h_0 + (h_0 - l \tan \theta)]}{2},$$ (8.77)

where b is the cross-sectional width of the initial solid bar, i.e. critical angle θ is given by

$$\theta = \tan^{-1} \frac{2h_0(1 - \alpha)}{l}.$$ (8.78)

For $\alpha > 1$, unfortunately, the tilting in the opposite direction, as shown in Fig. 8.19c, which is required to bring effective α to unity, is not practical as the liquid would tend to run down the slope on to the solid, particularly if the surface tension of the liquid is low.

In such cases zone refined ingots of constant cross-section may be obtained by making the height of the solid at the beginning of the ingot h_0/α so that the height of the liquid in the zone when the solid is molten is h_0.

Matter transport also occurs in vertical arrangements of zone melting. It poses a problem of container cracking if the processing material is held in a tube container. For melts which contract on melting, however, the safe mode of operation is to move the zone upwards from the bottom of the container to the top. In the above operation as the zone freezes from the bottom it creates voids which travel with the zone upwards and finally bubble out at the top. Similarly, for melts expanding on melting, it is wise to move the zone downwards from the top to the bottom.

8.4. Design and Choice of Zoning Equipment

It is fairly obvious that a zoning apparatus must consist of:

(1) a means of producing liquid zone(s), i.e. heater(s) and/or cooler(s);
(2) a traverse mechanism for the transport of molten zone(s);
(3) a means of mounting or holding a charge.

In addition the apparatus may incorporate a means of stirring a liquid and maintaining constant or controlled ambient atmosphere and temperature conditions, particularly at the solid–liquid interface.

In discussing the design of the equipment it is useful to divide the zoning into two categories:

(1) zoning in a container;
(2) zoning without a container.

8.4.1. ZONING IN A CONTAINER

If a suitable container material for holding a charge is available, then the most convenient way may be to perform zone melting in a container. The basic physical and chemical criteria for the choice of a container material are presented by Shah (1974,(1981) elsewhere in this book.

The design of the container itself is one of the deciding factors in achieving a degree of control over zone length and zone spacing. For attaining a sharp zone boundary and a short zone it is necessary to have a container with a good lateral heat transfer, and at the same time poor longitudinal heat transfer. It is thus advantageous to have the walls of the container as thin as possible.

The cross-sectional shape of the container is often dictated by the minimum contamination requirement and therefore the cross-section is designed in such a way that the charge has a minimum surface in contact with the container and the ambient. Circular cross-sections in the vertical zoning and semicircular and rectangular cross-sections in the horizontal zone refining do fulfil the requirement for lesser contamination but the small contact area to volume ratio in the above geometries make them less favourable for production of short zones.

The cross-sectional design must also ensure that the container (particularly if the charge expands on solidifying) does not produce strain due to expansion or mass transfer and thus cause cracking of the container. For horizontal zone-melting, boats of semicircular, rectangular, or trapezoidal sections, with their ends sloping outwards at a small angle to the horizontal as shown in Fig. 8.20, are useful in this respect, as they allow the solid to expand without constraints.

Finally, it may be desirable to construct a cross-section so that the advantage of stirring the liquid is achieved by permitting the natural convection in the zone.

The longitudinal shape of the charge and the container is determined by the desirable total length of the ingot. The total length of the ingot can be increased in a confined space by using radial, spiral, and helical geometries which are described by Pfann (1966).

High melting-point materials or very reactive materials may be zone melted in a water cooled boat or a "cold hearth" made from a high thermal conductivity material such as copper. A thin layer of a noble metal such as iridium on the contact surface or just a highly polished inner surface of the boat removes the risk of the charge sticking to the boat. In fact the risk of contamination from a water-cooled boat is negligible as Berghezan and Bull-Simonsen (1961) and Bull-Simonsen (1962) in the course of melting various metals including Si, Fe, Nb, Ta, and W have estimated that the temperature of the crucible does not rise beyond 70°C. The cold hearth zone melting process may be used with a variety of heating methods, such as d.c. or a.c. arc, electron beam heating, and induction heating. Another melting method, popularly known as the "silver boat process" and developed by Sterling and Warren (1963), in which a water-cooled boat or basket of copper or silver tubing is used in conjunction with induction heating to levitate the molten charge, has

FIG. 8.20. A diagram of a boat with sloping ends to prevent strain due to expansion on solidification.

been very successfully employed in zone melting. Here again the risk of contamination from the crucible is negligible.

8.4.2. ZONE REFINING WITHOUT CONTAINERS

Zone-melting techniques in the above family are particularly useful for materials which are very effective solvents (or are reactive) at their melting points that they cannot be refined or retained at high purity at the slightest contact with other materials.

By far the most common and well-perfected method in this category is the "float-zone" melting (FZM). The method was developed by Keck and Golay (1953), Emeis (1954), and Theuerer (1956, 1962) for the preparation of high purity silicon. It is now used for a variety of materials such as refractory metals, alloys, and semiconducting compounds. In this method a molten zone is held in place between two vertical collinear solid rods (compacted) or otherwise) by its surface tension. The stability of the molten zone is obviously of paramount importance. It has been mathematically analysed by Heywang and Ziegler (1954) and Heywang (1956). From the condition of the stability that the inward pressure on the planar surface of the liquid due to surface tension should at all points be equal and opposite to the pressure exerted by the liquid head, it was shown that the maximum height of the liquid, lm, that can be supported in a floating zone technique is given by the expression

$$lm \approx 2.8 \sqrt{\frac{\gamma}{\rho g}}^{\dagger} \tag{8.79}$$

where γ is the surface tension of the liquid, ρ is the density of the liquid, and g is the gravitational acceleration.

This value for silicon is of the order of 1.5 cm. Figure 8.21 shows the stable zone length for a small range of rod diameters according to the Heywang analysis. It should be

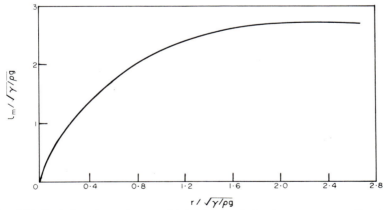

FIG. 8.21. Variation of a maximum stable height Cl_m of a floating zone versus radius r for cylindrical rods of equal diameter. (After Heywang, 1956, by permission of *Zeitschrift für Naturforschung.*)

† Clark Maxwell (1890) considered the problem of axisymmetric meniscus with a horizontal planar surface separating a heavier fluid from a lighter fluid below. He showed that the interface becomes unstable if the radius exceeds $3.8\sqrt{\gamma/\rho_d g}$, where ρ_d is the density difference between fluids.

particularly noted that the above treatment does not indicate any upper limit on the diameter of the rod, contrary to the results of an earlier treatment by Keck et al. (1953). In actual fact silicon rods with diameters of 80 mm are commercially grown by float-zone processes (Collins, 1977; Matlock, 1977). Recent work by Coriell et al. (1977) on liquid zones of cylindrical volume has confirmed the calculations of Heywang (1956).

Pfann et al. (1959) and Benson (1960) have shown that in large diameter pipes and plates (i.e. geometries with large width and small thickness) it is possible to establish stable wide zones with liquid height lh as large as the diameter. For the above geometries, Kramer et al. (1961), in agreement with Benson (1960), have theoretically shown the condition of zone stability in a circular disc to be

$$\left(\frac{3.83}{R}\right)^2 \geqslant \frac{\rho g}{\gamma}, \tag{8.80}$$

where R is the radius of the zone.

For rectangular geometry the condition is given by

$$\pi^2\left(\frac{4}{a^2} + \frac{1}{b^2}\right) \geqslant \frac{\rho g}{\gamma}, \tag{8.81}$$

where a is the width of the zone, b is its thickness, and $a \geqslant b$.

For processing of materials in the normal rod form, however, the suitability of utilizing float zone processes can be evaluated by the following general rules:

(1) surface tension of a liquid decreases with temperature;
(2) generally speaking the higher the melting point of the material the higher is its surface tension;
(3) the height of the stable zone is proportional to $\gamma^{1/2}$ in accordance with eqn. (8.79);
(4) the density of the liquid is compatible for a required zone length.

From the above rule it is inferred that floating zone techniques are suitable for metals with m.p. > 1200 K (particularly refractory group of metals), oxides, alkali halides, carbides, and other high melting-point materials. Low-melting-point materials, however, may be purified to a higher degree with the help of a float-zone technique. This was demonstrated by Otake and Matsuno (1971) in the case of bismuth.

Coriell et al. (1977) have investigated the stability of the liquid vapour interface for a liquid zone with and without rotation at a constant angular velocity. They showed that the shape of the liquid zone was governed by the ratio of the radii of the melting solid and the freezing solid R_m/R_f, the ratio of the zone length to the radius of the freezing solid L/R_f, and three other dimensionless parameters dependent on the volume of the liquid zone V, the density difference between the liquid and the vapour l, the angular velocity of rotation Ω, and gravitational acceleration g.

Perhaps the most interesting feature that came out in the above calculations is that for a growth of a constant diameter crystal (i.e. freezing solid) two stable zone shapes (Fig. 8.22) of different volumes can exist with all other parameters identical. The existence of two different zone shapes has been observed by Shah (unpublished results) in the electron beam float-zone growth of a number of materials including growth of a pale rose ruby (dilute solution of Cr_2O_3 in Al_2O_3-sapphire).

It must, however, be remembered that the actual shape and stability of the zone is greatly modified by the amount of impurity in the zone, the degree of stirring, and the

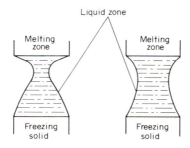

Fɪɢ. 8.22.

thermal conductivity of the solid rod because these parameters influence the curvature of the solid–liquid interface. Equation (8.79) and others therefore can only provide a guideline for the selection of the parameters discussed.

Recent experiments by Schwabe *et al.* (1978) on the floating zone melting of $NaNo_3$ have shown that growth stabilities can arise due to surface tension gradient induced by a concentration gradient of a surface active impurity in front of the growing surface. Their conclusion is that the surface tension driven flow in the zone is comparable to that due to natural convection.

Almost any form of heating may be employed. For example, carbon arc imaging by Kooy and Couwenberg (1962) and glow discharge by Trousil (1962); hollow cathode d.c. plasma by Class *et al.* (1966) and Class (1968); hollow anode discharge by Dugdale (1966) and Storey and Laudise (1970); r.f. plasma by Reed (1961) and Alford and Bauer (1966);[†] laser by Eickoff and Gurs (1969), Cockayne and Gasson (1970), and Savitski and Burkhanov (1978); and halogen lamps with elliptical reflectors by Okada *et al.* (1971), Takahashi *et al.* (1971), and Takai (1978). By far the most common modes of heating employed for FZM are the electron beam heating and r.f. heating (or induction heating).

8.4.2.1. *Electron Beam Heating*

The first electron beam float-zone melter (EBFZM) was constructed by Calverley *et al.* (1957). In EBFZM a molten zone is produced by bombarding a part of the vertically mounted material rod by electrons accelerated through a high voltage field. The earlier electron guns used in EBFZM were work-accelerated guns, i.e. they were two electrode affairs consisting of a heated filament as one electrode and the charge rod as the other. They were suitable for processing only high conductivity materials like metals and semiconductors. The electron current with these guns could not be held constant due to inherent instabilities. The cross-contamination between the filament and the charge material was a major problem in the earlier days. Neumann and Huggins (1962) devised a triode gun for processing insulating materials. Their gun included an additional helical tungsten wire grid to prevent negative electrostatic charge build-up on the sample, using secondary emission from the material being heated. For processing insulating (as well as conducting) materials modern electron guns are self-accelerating types, i.e. the accelerating voltage field in these guns is applied between the cathode and the anode. The

† Used in Verneuil process.

charge rod itself does not form a part of the electron gun assembly. A latest version of a self-accelerating gun designed by Kanaya *et al.* (1968) is schematically shown in Figs. 8.23A and 8.23B. Here the cathode is a tungsten wire in the form of a ring of radius c and the anode consists of two water-cooled copper rings forming a slit. Cross-contamination between the filament and the charge rod is avoided in the above assembly by the use of a biased grid which due to its electrostatic field deflects the electron beam emitted from the filament in the vertical plane and eventually focuses on the charge rod. Focusing of the beam in horizontal plane is necessary to attain higher power density on the charge rod. It is achieved in the above gun assembly by the magnetic field induced by the cathode current.

Power requirements for melting a zone in the material may be estimated by an empirical equation due to Belk (1959),

$$P = Ad + Bd^2,\tag{8.82}$$

where P is the power in watts, A is a constant proportional to T_m^4 where T_m is the melting temperature, d is the diameter of the rod, and B is a constant proportional to the thermal conductivity of the rod.

Relation between the power required to form a floating zone and its shape has been studied by Kobayashi (1978). He computationally solved the Laplace equation governing the temperature, both in the rod (unmolten solid) and the molten zone. It was shown that the temperature distribution can be characterized by five dimensionless parameters such as the ratio of thermal conductivity of the solid and the liquid, the ratio of the length of the solid and the zone length, the width of the heated region to the diameter of the crystal, and the non-dimensionalized power. By the above scheme it is possible to compute zone shapes and temperature distributions as a function of the power supplied to the zone.

A major source of instability in the EBFZM is the power fluctuation due to outgassing of the charge material. The released gases are ionized due to the high field and bombard the filament, raising its temperature and consequently its emission. Two basic methods to

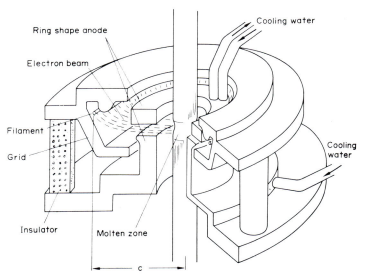

FIG. 8.23A. A self-accelerating electron gun: schematic construction.

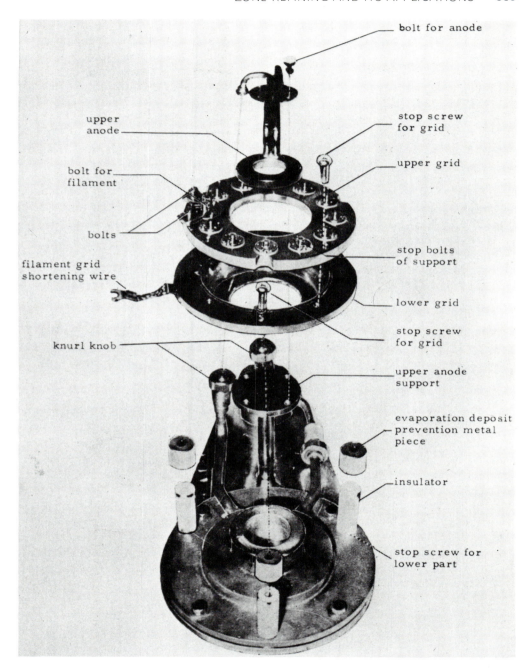

FIG. 8.23B. A self-accelerating electron gun: exploded view. (By permission of Japan Electron Optics Laboratories Ltd.)

control emission are used, namely (1) adjustment of the filament temperature, i.e. emission control, and (2) field control by the use of a constant current source. In the apparatus of Kanaya et al. (1968) the accelerating voltage was stabilized to one part in 10^3 by a feedback system. The beam current was similarly stabilized by a separate circuit.

Kamm (1977) has also described a system for emission control which uses a regulated d.c. power supply and operational amplifier control circuits to regulate the emission current and the filament power supply.

The subject of EBFZM is reviewed by Lawley (1962), Schadler (1963), and Lawley (1968). In conclusion the advantages and disadvantages of EBFMZ are listed here.

Advantages:

(1) Small melt volume.
(2) Well-defined thermal gradients.
(3) No crucible contamination.
(4) Probably the most efficient method of heating (see Donald, 1961).
(5) Additional purification achieved due to vacuum of impurity evaporation.

Disadvantages:

The most serious limitation is that the method can only be used in a vacuum, i.e. at pressure $< 1.33 \times 10^{-2}$ Pa. High vapour pressure or dissociating materials therefore cannot be processed by EBFZM.

8.4.2.2. *Induction Heating*

Induction heating in FZM may be applied at any positive pressure of inert, oxidizing or reducing atmosphere; it may also be applied in a vacuum. General considerations and details of h.f. heating are described by Shah (1981). Here the aspects pertinent to FZM are described.

In a FZM arrangement a h.f. coil surrounds a vertically mounted charge rod, either inside the work chamber envelope or outside it. Proximity of the coil to the charge rod is desirable for efficient coupling, but the environment inside the work chamber may prevent the placing of the work coil inside. Buehler (1957) has provided an equation for the selecting of the "optimum frequency" (i.e. a minimum frequency above which no appreciable power transfer is realized), namely

$$v_{\mathrm{opt}} = \frac{6.25\rho \times 10^9}{8\pi^2 \mu r_0^2} \tag{8.83}$$

where v_{opt} is the optimum frequency, as defined above in Hz, ρ is the resistivity of the charge rod in $\Omega\,\mathrm{m}^{-1}$, μ is the relative permeability of the material, and r_0 is the radius of the rod in mm. The expression for the power dissipated per mm^2 surface area of the rod is given by

$$p = \frac{H^2(\rho\mu v)^{1/2}\,10^{-2}}{8\pi}, \tag{8.84}$$

where v is the frequency of radiation used in Hz, p is the power in W, and H is the magnetic field strength in ampere-turns. It is given by $H = 4\pi NI$, where N is the number of turns of the r.f. coil and I is the current in A.

Frequency ranges normally used lie in the range 450 kHz to 5 MHz.[†] For producing short zones the power can be concentrated by using specially designed coils.

The following additional benefits may be gained by the use of induction heating:

(1) It provides automatic electromagnetic stirring in the molten zone, the extent of agitation depending upon the frequency used, the actual coil arrangement, and the length of the zone.

(2) As demonstrated by Oliver (1963) and Rutherford et al. (1965) levitation forces may be advantageously used to increase the stability of the molten zone.

(3) An automatic control of the zone diameter can be incorporated by detecting changes in "Q" or conductivity of the load as shown by Buehler (1957) and Warren (1962).

8.4.3. TRAVERSE MECHANISMS

The movement of a liquid zone along the length of a charge may be effected either by keeping the charge stationary and moving the heater producing a molten zone or by keeping the heater stationary and moving the charge. A large number n of passes through an ingot may be effected by three alternative methods, namely by passing n passes of a single zone heater, one pass through n zone heaters, or by a number of short reciprocating strokes (with fast return passes) with a smaller number h of equally spaced zone heaters $h \ll n$, the length of each stroke equalling the distance x between each interval. If L is the length of the ingot then the values H and d obey the following relationship:

$$H = Ld.$$

The number of reciprocating strokes N required to effect n passes are given by $N = n/H$.

Different kinds of drive mechanisms such as lead screw, chord and drum, cam, etc., may be used. The main requirements to be satisfied are the flexibility (e.g. variable speed) and constancy of the motion. If the apparatus is to be used for single crystal growth then the motion has to be smooth.[‡] For straight zone refining, however, the motion is required to avoid sudden freezing of layers of thickness >0.5 mm. For the growth of homogeneous alloys the rate of traverse needs to be in the region of a fraction of a mm hr^{-1}; for a general purpose zone-melting apparatus it is desirable to incorporate a variable speed drive.

Recently, an inexpensive and simple variable speed has become available using a stepping motor and a variable low frequency oscillator (10–300 Hz). A stepping motor is driven in steps by suitably shaped voltage pulses, hence the speed of the motor may be varied at will by changing the frequency of the incoming pulses. By transmitting this drive through a high reduction gear box to a lead screw it is possible to ensure a smooth motion. It is customary to incorporate a separate motor for fast reverse passes. With simple microswitches the traverse mechanism may be made automatic.

[†] Lower frequencies in the medium frequency (m.f.) band (i.e. 1–100 kHz) can be used for large charges.

[‡] It should be noted that Jindal (1972) found that the effective segregation coefficient of NaCl during freezing of the aqueous solution is affected by vibrations, i.e. it is found to increase with the increase in the frequency at a constant amplitude. It has been shown by Bachmann (1973) that the background "noise movement" in the traverse mechanism can be reduced to a level of 2 μm r.m.s.

8.4.4. DESIGN CONSIDERATIONS FOR "IDEAL ZONES"

The main desirable features of a molten zone for zone melting are:

(1) a constant zone length l;
(2) a stable solid–liquid interface;
(3) a small melt volume;
(4) well-defined thermal gradients.

A schematic temperature distribution for maintaining a molten zone with the above features in conventional zone melting (i.e. excluding thin alloy zone technique, etc., described in section 8.6) is characterized by two features, over the zone width l: (1) the temperature $> Tm$ (the melting temperature), and (2) there is a temperature gradient on both sides of the zone so that the temperature outside the zone width is $< Tm$. The actual shape of the temperature profile depends upon the power and the nature of the heat sources and heat sinks incorporated in the apparatus, the thermal conductivity of the charge ingot, and the solute content of the liquid.

The design of a zone-melting apparatus therefore should consider the heat characteristics of heat sources, the thermal losses inherent in the system, and additional heat sink incorporation to alter or fix the "imposed temperature gradient" at the zone boundaries. A convenient way of comparing various heat sources is by looking at their heat transfer intensities (HTI) defined by Reed (1966) as the capacity of the source for transferring heat per unit area. The steepness of the imposed temperature gradient primarily depends upon the placement of heat sources and heat sinks. It may be remembered that a high temperature gradient offers the following advantages:

(1) It helps minimizing fluctuations of zone width due to fluctuations of temperatures of the sink and the source.
(2) It reduces the influence of variations in longitudinal cooling by conduction along the charge due to changes in lengths of the ingot not directly heated.
(3) It reduces the possibility of incorporation of micro-inhomogeneities due to interface instability.

One significant difficulty encountered in horizontal zone melting, particularly for materials with low thermal conductivity (e.g. organic materials), is that the zone width tends to vary across the longitudinal cross-section of the zone volume parallel to the direction of traverse. This is because the natural convection current directions are limited in the horizontal geometry so that the hotter liquid tends to accumulate on the top of the container or on the surface of the zone and spreads by melting the adjacent solid (see Pfann, 1969). A constant zone length for the above materials was very simply obtained by Pfann (1969) by slowly rotating (0.5–25 r.p.m.) the charge during zoning. Fischer (1973) has, indeed, obtained a very high efficiency, in zone refining of aromatic hydrocarbons, by the use of intermittent rotation of the charge. The constancy of the zone width is thought to be restored due to angular flow pattern of the rotating liquid.

In the zone melting of organic compounds the liquid often tends to seep back under the resolidified material. Anderson (1969) claims to have solved this problem by subjecting the liquid to centrifugal force. Incidentally, it was also claimed that at least in the case of naphthalene a more rapid and effective zone refining is achieved due to the use of centrifugal force. For comparison of the various zone refining arrangements, for organic compounds, the reader is referred to Karl and Probst (1970) who have devised a standard critical test for fair assessment.

8.4.5. STIRRING

It has already been stated that effective distribution coefficient depends upon the thickness of the diffusion layer at the interface. One way to reduce the thickness of the layer is to stir the melt. Stirring also provides an improvement in the heat transfer and eradicates localized density difference arising from concentration differences and temperature differences.

Stirring may be provided by:

(1) Making natural convection processes more effective. A theoretical understanding of the convection process is discussed by Wilcox (1962).
(2) Forced convection.
(3) Mechanical stirring including vibration and pumping.
(4) Induced current stirring.
(5) Magnetic and electromagnetic stirring.

For detailed experimental arrangements several textbooks listed in the references may be consulted.

8.5. Modifications of Zone Refining

More and more complicated inorganic compounds and their alloys, requiring closer control in stoichiometry and other properties are processed by zone melting. More organic compounds continue to be zone refined for separation and purification. Innovations in employment of heating devices, some of which were mentioned in section 8.4.2, for melting high-melting-point materials, continue to be reported in literature. The most significant developments on this front over the past few years are the employment of laser radiation and plasma discharges. High-pressure float-zone melting is beginning to be employed for achieving higher purity of dissociating materials. For instance, Billingham *et al.* (1971) and Moldovanova *et al.* (1971) have obtained high purity VC (vanadium carbide) and GaP by high pressure FZM.

A significant trend in the modification of zone refining is that for achieving still higher purity levels it is combined with some specific chemical action for a removal of one or more species of trace impurities. Shah and Huntley (1970) zone-refined antimony in dry hydrogen atmosphere at 400 torr to achieve resistance ratio $\rho_{300\,\mathrm{K}}/\rho_{4.2}$ of around 6×10^3. Similarly, higher values of resistance ratios of platinum and niobium were obtained by Shah and Brookbanks (1972) and by Reed (1972) respectively. Recently Grytsiv *et al.* (1975), by using radioactively labelled impurities of calcium and mercury and the estimation of impurity scattering, showed that the concentration of the above atoms can be reduced in CdTe by zone refining saturated solution CdTe in Cd. The reduction was of the order of 10^3–10^4. Similarly, homogeneous crystals of alloy semiconductors such as $Zn_x Cd_{(1-x)} SnAs_2$ (Shah 1969) and $Hg_x C d_{(1-x)} Te$ were obtained by zone refining "off stoichiometric compositions" in sealed containers and using tellurium as solvent respectively. Travelling heater arrangement of the conventional zone refining when applied to a growth of a compound/alloys from solution can be often more advantageous than the travelling solvent *or* think zone alloy crystallization method discussed in section 8.6. For instance, Gillessen and von Münch (1973) found that for growing SiC from carbon solutions, travelling heater method offered constant growth conditions and permitted higher growth velocities (~ 0.3 mm hr^{-1}) than those achievable in crucible growth and the travelling solvent method. They attribute this to the fact that in a relatively thick solution zone generated by the travelling heater a temperature maximum within the zone and the total concentration drop between the source material and grown material takes places in two steps and thus fulfils stability criteria more easily.

Some experimental studies on floating zone have been performed in low gravity environment in conjunction with sky lab and other space programmes. Worth mentioning are the studies by Grodzka and Bannister (1970) and Reed *et al.* (1976) on surface assisted convection; Carruthers and Grasso (1971), Carruthers (1974), and Carruthers *et al.* (1976) on rotational and vibrational stabilities. It is interesting to note that the observed instabilities cannot be explained on existing theories. A concise review on the space crystal growth experiments is given by Carruthers (1977).

8.5.1. LIQUID ENCAPSULATION

The conventional zone-melting technique for zone melting of materials with volatile components was originally proposed by van den Boomgaard (1955). It used a completely enclosed system with a multi-temperature to maintain the vapour pressure of the volatile constituents in the system over or at the value of the equilibrium dissociation pressure over the melt. For high-melting-point materials with high vapour pressure constituent the zone-melting apparatus is difficult to fabricate. Not so long ago Mullin *et al.* (1965) proposed a simple technique of liquid encapsulation. It essentially uses an inert liquid seal to cover the melt in any melt growth apparatus. The loss of the volatile constituent is prevented simply by maintaining an inert gas pressure greater than the equilibrium dissociation vapour pressure of the volatile component. It is desirable that a liquid encapsulant should be less dense than the melt, be optically transparent, have a low viscosity at the melting point of the melt, and be chemically stable and non-reactive with the melt crucible and the environment. Also, the migration of the volatile constituent through it by diffusion and convection should be negligible and it should not be a solvent for the phase to be solidified or its constituent.

Although liquid encapsulation has been widely used in (Czochralski) growth of III–V compounds, notably GaAs and GaP, it can be used for zone refining. For instance, Swiggard (1967) and Shah (1967) utilized encapsulation in conventional zone-refining techniques. Hiscocks and Elliott (1969) have used it for float zone refining of Cd$_3$As$_2$.

Unhydrous B$_2$O$_3$ is the most popular encapsulant around and above 1000°C as it has all the desirable properties including low viscosity. It has, however, been successfully used below 800°C for the growth AgGaS$_2$ by Korczak and Staff (1974). Around these temperatures BaCl$_2$, CaCl$_2$, and CaCl$_2$ + KCl can also be used. At still lower temperatures, around 600°C Shah (1967) used a mixture of LiI + KI as an encapsulant for zone refining of CdSnAs$_2$. Organic acids have been used as encapsulant for growing bismuth crystals at around 150°C described by Mullin (1975).

8.5.2. MICROSCALE ZONE MELTING

The idea of microscale zone melting is very attractive because it offers the possibility of purifying and processing very small quantities of materials and the zone-refining time could be drastically reduced due to the use of a large number of zones in a small space and increase in zone speed due to more effective diffusive mixing.

The first microscale zone refiner containing three zone heaters was designed by Hesse and Schildknecht (1956). Since then, Handley and Herington (1956), Schildknecht and Mannl (1957), Mair *et al.* (1958), Ronald (1959), Schildknecht and Vetter (1959), and Schildknecht and Mass (1967) have designed microzone refiners that are capable of processing organic substances weighing from a few hundred micrograms (Schildknecht and Mass (1967)) to a few milligrams. The apparatuses contained a few tens of zones within a distance of a few centimetres.

In the sixties the microzone recrystallization was successfully applied to obtain thin films of semiconducting materials having electrical properties comparable with those of the bulk materials. The precursory idea to the thin film microzone recrystallization was presented by Leitz (1950) over 30 years ago. He proposed that recrystallization of a thin film could be accomplished by moving a molten zone established by radiation from a heated thin platinum wire. The major practical problem in the application of the idea was that the production of a well-defined molten zone was prevented by agglomeration due to surface tension. However, it appears that later Maserjian (1963) and Teede (1967a, b) and Clawson (1972) were able to produce large crystalline germanium and InSb films respectively by basically very similar techniques, though using different heat sources. The solution of the problem was arrived at by Wieder (1965) when he discovered that agglomeration could be prevented by the use of a thin "surface containment layer" of an oxide. It is believed by Billings (1969) and others that the prevention of the agglomeration is due to the reduction of surface tension forces due to the provision of an upper surface under which the liquid zone can flow. Fortunately, the use of a containment layer over a thin film may provide the following additional advantages, namely containment of vapours at the liquid–vapour interface, control of stoichiometries, and prevention of contamination. Wieder and Davis (1965) and Williamson (1969) produced a containment layer on InSb film by partially oxidizing an indium film deposited on the top. Teede (1967b), on the other hand, produced it (again on InSb) by directly growing In_2O_3. SiO has been used, as the containment layer, on II–VI compound films by Billings (1969).

An electron beam has been used as a heat source in microzoning by Gilbert *et al.* (1961), Maserjian (1963), Weinreich and Dermit (1963), Davis (1969), and Paureau (1972).

The familiar effect of mass transport in bulk zone melting has also been discovered and investigated by Beiziter *et al.* (1968).

8.5.3. DIRECT CURRENT EFFECTS AND ZONE MELTING

On the application of a d.c. current to a zone-melting system (both thick and thin zones) the kinetics in the molten zone is influenced by a variety of phenomena such as electrodiffusion, Peltier effect, Thomson effect, and the Joule effect. In practice these phenomena exert concurrent influences on the zone. Hurle *et al.* (1964b, 1967) have derived phenomenological equations with simplified assumptions for the zone kinetics taking into account the above effects. However, it is advantageous to describe experimental finding on electrodiffusion and thermoelectric effects separately.

Field-aided zone melting. Imposition of an electric field gradient on a conductive liquid solution zone causes ionic movement. Consequently, the changes in concentration of the solute at the freezing interface will occur. The effects of the phenomena have been studied by Angus *et al.* (1961), Pfann and Wagner (1962), Hay and Scala (1965a, b), Verhoeven (1965, 1966), and Grigorvev (1977). Briefly, the net result of the change of concentration at the interface due to ion migration may be expressed in a modified Burton–Prim–Slichter expression (i.e. eqn. (8.10)) for the effective distribution k:

$$k = \frac{1 + \dfrac{f'}{f}}{1 + \left[\dfrac{1}{k_0}\left(1 + \dfrac{f'}{f}\right) - 1\right]\exp\left[\left(-\dfrac{f\delta}{D}\right)\left(1 + \dfrac{f'}{f}\right)\right]}, \tag{8.85}$$

where f' is the velocity of the solute ions, towards the interface, due to the field, and all other symbols have the same meaning as in eqn. (8.10).

The value of f' is given by

$$f' = E\,\Delta\mu, \tag{8.86}$$

where E is the applied electric field and $\Delta\mu$ is the difference in ionic mobilities of the solute and the solvent.

The examination of eqns. (8.85) and (8.86) reveals that k may be varied in a range of values simply by varying the field, because the variations of the field affects the flux of the solute ions at the interface. The above method therefore can be extremely useful for removal of impurities which have $k \approx 1$ (in the absence of the field, i.e. when $E = 0$). Hay and Scala (1965a, b) in the case of tungsten and Wagner *et al.* (1966) for BiSn have experimentally demonstrated the usefulness of the field aides zone refining for the enhanced purification.

Hurle *et al.* (1967) have claimed that the interface stability in the presence of ionic transport is primarily dependent upon electrical resistivities of the solid and liquid phases.

Zone melting and thermoelectric effects. If a direct current is passed through a molten zone, heat is absorbed at one solid–liquid interface and evolved at the other, due to the Peltier effect. The phenomenon therefore would affect the velocity of the zone interfaces. It was independently proposed by Ioffe (1956) and Pfann *et al.* (1957) that the effect could be utilized for crystal growth of relatively pure semiconductors. Peltier coefficient with regard to zone melting is defined as +ve if the heat is absorbed at the interface when the direction of the current is from the solid to the liquid. It was shown by Ioffe (1956) that, taking into account the Joule heating (but neglecting the Thomson effect), the optimum value of the current for the production of a maximum change in the growth velocity is a function of the Peltier coefficient and the resistivities of the solid and liquid. Pfann *et al.* (1957) actually grew germanium crystals using the Peltier heat across an alloy zone. It was later argued by Hurle *et al.* (1964a) that to explain the discrepancies in the sign of Peltier coefficients, as deduced from the direction of the migration of the zone for various germanium/metal zones used by Pfann *et al.* (1957), the ionic transport should be taken into account. As stated before, they formulated a theory to take into account all thermoelectric effects and ionic migration. The relevance of the Thomson effect in the experiments of Pfann *et al.* (1957) was also pointed out by O'Connor (1960). Recently, Balloch and Dubiri (1978) have grown germanium ribbons of dimensions 25 × 1.1 × 0.6 mm by the use of Peltier current. Zone velocities of around 1.5, 0.21, and 0.16 mm min^{-1} were used.

8.5.3.1. *Solution Zone Refining*

This process is almost identical to the ordinary zone refining. The liquid zone, however, is a thick ($\sim$ cm) zone of a concentrated solution containing the phase desired to be refined as the solute. The process has also been christened, by Nicolau (1970), as the travelling heater zone refining (THZR).

It is apparent that a solution zone technique would normally have the disadvantages of very slow rates of traverse to avoid solvent entrapment, and the elimination of solvent at the end of each zone passage.

Nicolau (1970) has claimed to have overcome the above difficulties by using a two-temperature zone heater as shown in Fig. 8.24. The lower temperature following the solution zone in the above arrangement ensures that the crystallization at the freezing interface always occurs at the temperature above the melting point of the solvent. Thus the zone in this process is constantly agitated by forced convection. Due to the above temperature profile the solvent zone at the end of each zone passage can be easily decanted.

The requirements of the solvent used in the process are:

(1) The substance to be purified (the solute), and impurities contained in it, should be soluble in the solvent.
(2) The solubility of the solvent in the solid phase of the solute (and the impurities) should be negligible.

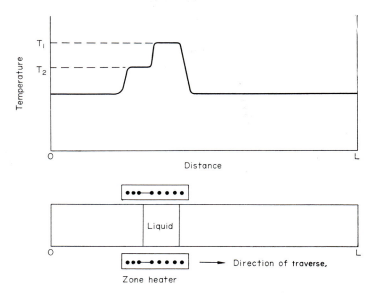

FIG. 8.24. A schematic representation of the solution zone refining (THZR) and the temperature distribution required therein.

Nicolau (1971) has described open-tube and closed-tube systems to purify alkali metal phosphates, sulphates, alums, etc. Water was used as the solvent and the traverse rates required for successful zone refining are quoted to be in the region 0.25 mm min^{-1} to 1 mm min^{-1}. Nicolau (1970) has also derived expressions for single-pass and multi-pass distributions in terms of the ratio impurity/solute.

The usefulness of the process for the purification of metals, semiconductors, and organic materials seems to be limited because it would be very difficult to find suitable solvents satisfying the requirements listed above. It is, however, possible that "solution zone refining" may show promise to purify materials which decompose on melting, have very high melting points, have very high vapour pressure at the melting point, and are highly reactive.

Provided suitable solvents are found the major advantage of the process seems to be that it would permit extension of zone purification of materials at temperatures much lower than their melting point.

8.6. Allied Techniques

8.6.1. THIN ALLOY ZONE TECHNIQUES

The term "thin alloy zone crystallization" (TAZC) was coined by Hurle *et al.* (1964a, 1967) as the family name of the series of techniques which show the following features:

(1) Crystallization of a solid phase occurs via diffusion through a liquid alloy zone.†
(2) At least one dimension of the zone is small, ∼25 μm.
(3) The zone forms a distinct phase and acts as a transport medium for the crystallizing phase.
(4) The driving force for the crystallization is provided by the free energy of the phase.

In the above techniques any zone geometry such as spherical, cylindrical, or lamellar (i.e. dot, wire, or a sheet) may be effected. The motion of the migration of the zone is provided by a gradient of thermodynamic potential across the zone. The zone velocity in most of the cases is proportional to the gradient of the potential. For a typical TAZC system the thermodynamic, or to be more precise the electrochemical, potential is a function of pressure P, temperature T, electrostatic potential, and the chemical potential (a function of the concentrations of the phase components). In practical situations the principal driving force is provided by a gradient of one or two variables while the others are held constant. Therefore, various TAZC processes are conveniently classified in Table 8.2 according to the principal driving forces they employ. The processes in the table are also labelled according to the phases they incorporate in the various stages of the crystal growth scheme, namely

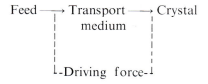

† The term "liquid alloy" includes references such as solution, dilute solution, fluxes, mixtures, etc.

TABLE 8.2

Driving Force	Name of the Technique	Original Proposers of the Technique
Temperature gradient (ΔT)	S–L–S process (1) Temperature gradient Zone melting (TGZM) or travelling solvent method (2) L–L–S process (3) V–L–S process	Pfann (1955) Mlavsky (1961) Delves (1965)
Electrochemical potential and electrostatic potential ($\Delta \bar{\mu}$ and $\Delta \phi$)	(1) Field aided zone melting (thick zone) (2) Field freezing (3) Peltier effect aided zone melting (4) Direct current induced TAZC. (Note: driving force is a result of thermoelectric effects and ionic migration)	Angus et al. (1961) Pfann and Wagner (1962) Ioffe (1956) and Pfann et al. (1957) Hurle et al. (1964a)
Pressure gradient (ΔP)	Pressure gradient TAZC S–L–S, L–L–S, V–L–S	Hurle et al. (1967)
Chemical potential gradient ($\Delta \mu$)	Metastable phase TAZC (1) V–L–S (2) L–L–S (3) S–L–S	Wagner and Ellis (1964) Hurle et al. (1964b) Hurle et al. (1964b)

For instance a melt growth process would be labelled as a S–L–S process because it employs solid feed medium, a liquid transport medium to obtain crystalline solid medium.

Wieder (1978) has recently described the use of electron beam for producing thin alloy zone of composition: 80% In and 20% Sb and m.p. 425°C for refining InSb thin layers.

8.6.1.1. *TGZM*

As stated before, in TGZM the movement of a thin zone is accomplished by a temperature gradient across the zone. In practice a temperature gradient $T_H T_c$ is applied across the charge of material A, which contains a sandwiched liquid zone containing a molten alloy A–B, as shown in Fig. 8.25. It should be noted that $T_m > T_H > T_c$, where T_m is the melting point of A. The motion of the zone may be understood in very simple terms by following the phase diagram of the system A–B shown in Fig. 8.25. The liquid zone ab initially tends to elongate, dissolving more A. The elongation of the zone first ceases, at the cooler interface, as it (the interface) "hits" the liquidus of AB at the point (C_a, T_a). Meanwhile, the dissolution of A at the hotter interface continues and is eventually stopped when the hotter interface reaches the liquidus at the point (C_b, T_b). Thus a concentration gradient of A is set across the zone which causes A to diffuse towards the cooler interface, where it crystallizes obeying the law of segregation, i.e. the concentration of B in the crystallized layer is given by kC_a. Dissolution–diffusion–solidification causes the zone to migrate, "along" the liquidus, towards the hot end of the charge.

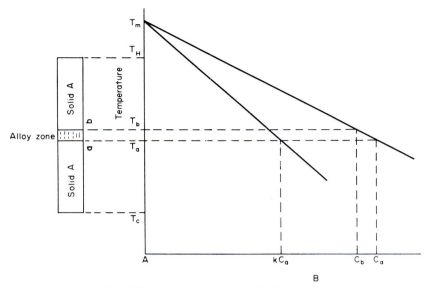

FIG. 8.25. A representation of a TGZM process.

Tiller (1963a, 1965) and Hurle *et al.* (1964a, 1967) have theoretically derived the expression for the migration velocity of the zone in TGZM process of a binary system. The zone velocity is found to be proportional to:

(1) the temperature gradient dT/dx;
(2) the reciprocal of the liquidus slope $(1/(dT/dC))$;
(3) the diffusion coefficient.

Lozovski (1971) has applied the Tiller (1963a) approximation to the case of a three-component TGZM. He showed that the structure of the formulae expressing the dependence of the zone velocity on the factors listed above remains the same. The zone velocity, however, does alter due to the presence of the third component.

The length of the zone in TGZM is time dependent and is influenced by the solubility of B in A, the volatility of B, and the slope of the liquidus dT/dC.

The distribution of B in A, in TGZM, depends upon the effective distribution coefficient of B and the temperature gradient across the charge. A smaller temperature gradient produces a more uniform distribution.

It is advocated by Hurle *et al.* (1964c, 1967) that the principal advantage of thin-zone techniques is that the gradient of constitutional supercooling is much smaller for a given velocity than that for the large liquid zone. Seidensticker (1966), however, applied a Mullins and Sekerka (1964) type analysis to TGZM and predicted the possibility of an instability at the dissolving interface analogous to the instability due to constitutional supercooling at the freezing interface. (The condition for the above instability is referred to as the constitutional superheating.) The instability was also shown to be affected by thermal conductivity of the solid and the liquid. Delves (1967) also produced a "dynamic theory of stability" for TGZM and confirmed the Seidensticker conclusions. He has also calculated the boundary between stable and unstable conditions. The general conclusions of the above workers may be summed up by saying that instabilities can be avoided by using thinner zones and small temperature gradients. It is, finally, pointed out that TGZM has been used in fabricating semiconducting junctions, joining, and single-crystal growing.

8.7. Conclusions

There is no better way of concluding a chapter on zone melting than to quote Pfann (1967), to whose discoveries we owe much gratitude for many a great advance in our scientific knowledge:

"I regard the conception and development of zone melting as an exciting scientific advance. And I cannot help being saddened to hear it occasionally referred to as simply a technical innovation that was mysteriously evoked by the need for transistor grade germanium and silicon. I regard zone melting as elegant both in its simplicity and its surprising complexity. I also regard it to this day as a wonderful adventure, filled with surprise and joy."

Acknowledgements

I wish to thank Brian Pamplin for reading the manuscript and various bodies for permission to reprint copyright illustrations.

References

ABE, T. (1974) *J. Cryst. Growth* **24/25**, 463.
ALEKSANDROV, B. N., BERKIN, B. L., LIFSHITS, I. M., and STEPANOVA, G. I. (1956) *Fiz. Metal i Metalloved* **2**, 105 (in Russian).
ALFORD, W. L. and BAUER, R. F. (1966) *Proc. ICCG, Boston*, p. 71.
ANDERSON, E. L. (1969) US Patent 3,428,437.
ANDRADE, E. N. DA C. and ROSCOE, R. (1937) *Proc. Phys. Soc.* **49**, 152.
ANGUS, J., RAGONE, D. V., and HUCKE, E. E. (1961) in *Physical Chemistry of Process Metallurgy*, Part II (St. Pierre, C. R., ed.), Interscience, New York, p. 833.
BACHMANN, K. J. (1973) *J. Cryst. Growth* **18**, 13.
BALLOCH, M. and DUBIRI, A. E. (1978) *J. Cryst. Growth* **43**, 277.
BARALIS, G. (1968) *J. Cryst. Growth* **3-4**, 627.
BARTHEL, J. and EICHLER, K. (1967) *Kristall und Technik* **2**, 205.
BEIZITER, L., VOVSI, A., and PATMALNEIKS (1968) *Lat. PSR Zinat. Akad. Vestis. Fiz. Tehen. Ser* (USSR) **3**, 42 (in Russian).
BELK, J. A. (1959) *J. Less-Common Metals* **1**, 50.
BENSON, K. E. (1960) *Metallurgical Soc. Conferences* **5**, 17.
BERGHEZAN, A. and BULL-SIMONSEN, E. (1961) *Trans. AIME* **221**, 1029.
BERTEIN, F. (1958a) *J. Phys. Radium* **19**, (12) supplement, 121A.
BERTEIN, F. (1958b) *J. Phys. Radium* **19**, (12) supplement, 182A.
BILLINGHAM, J., BELL, P. S., and LEWIS, W. H. (1971) *Proc. ICCG Marseille*, p. 693.
BILLINGS, A. R. (1969) *J. Vac. Sci. Technology* **6**, 757.
BIRMAN, J. L. (1955) *J. Appl. Phys.* **26**, 1195.
BRAUN, I. and MARSHALL, S. (1957) *Br. J. Appl. Phys.* **8**, 157.
BRICE, J. C. (1965) in *The Growth of Crystals from Melt*, North-Holland, p. 63.
BRICE, J. C. (1971) *J. Cryst. Growth* **10**, 205.
BRIDGMAN, P. W. (1925) *Proc. Am. Acad. Arts Sci.* **60**, 305.
BUEHLER, E. (1957) *Rev. Sci. Instr.* **28**, 453.
BULL-SIMONSEN, E. (1962) *J. Iron Steel Inst.* **200**, 193.
BURRIS, L. JR., STOCKMAN, C. H. and DILLON, I. G. (1955) *Trans. AIME* **203**, 1017.

BURTON, J. A., PRIM, R. C. and SLICHTER, W. P. (1953) *J. Chem. Phys.* **21**, 1987.
CALVERLEY, A., DAVIS, M. and LEVER, R. F. (1957) *J. Sci. Instr.* **34**, 142.
CARRUTHERS, J. R. (1974) *Proc. Intern. Colloq. Drops and Bubbles* (Collins, D. J., Preset, M. S., and Salfren, M. M., eds.), Cal. Tech./Jet Prop. Lab.
CARRUTHERS, J. R. and GRASSO, M. (1971) *Proc. ICCG Marseille*, p. 611.
CARRUTHERS, J. R. and NASSAU, K. (1968) *J. Appl. Phys.* **39**, 5205.
CARRUTHERS, J. R., GIBSON, E. G., KLETT, M. G., and FACEMIRE, B. R. (1976) *AIAA Tech. Publication*, No. 75–692.
CLARK MAXWELL, J. (1890) *Scientific Papers*, Vol. II, Dover, New York, p. 541.
CLASS, W. (1968) *Proc. ICCG Birmingham, UK*, p. 241.
CLASS, W., NESTER, H., and MURRARY, G. T. (1966) *Proc. ICCG Boston*, p. 75.
CLAWSON, A. R. (1972) *Thin Solid Films* **12**, 291.
COCKAYNE, B. and GASSON, D. B. (1970) *J. Materials Sci.* **5**, 837.
COLLINS, R. L. (1977) *J. Cryst. Growth* **42**, 490 (Proc. Vth ICCG, Cambridge, Mass.).
CORRIELL, S. R., HARDY, S. C., and CORDES, M. R. (1977) *J. Colloid Sci.* **60**, 126.
DAVIES, L. W. (1958a) *Phil. Mag.* **3**, 159.
DAVIES, L. W. (1958b) *Trans. AIME* **212**, 799.
DAVIES, L. W. (1964) *Solid State Electron.* **7**, 501.
DAVIS, N. M. (1969) *J. Vac. Sci. Technology* **6**, 768.
DELVES, R. T. (1965) *Br. J. Appl. Phys.* **16**, 343.
DELVES, R. T. (1967) *Phys. Stat. Sol.* **20**, 693.
DIMITROV, O. and FROIS, C. (1970) in *Physical Metallurgy* (Kahn, R. W., ed.), North-Holland.
DONALD, D. K. (1961) *Rev. Sci. Instr.* **32**, 811.
DUGDALE, R. A. (1966) *J. Materials Sci.* **1**, 160.
DUWEZ, P. (1965) in *Energetics in Metallurgical Phenomena*, vol. 1 (Mueller, M., ed.), Gordon & Breach, New York, p. 193.
EICKOFF, K. and GURS, K. (1969) *J. Cryst. Growth* **6**, 21.
EMEIS, R. (1954) *Z. Naturforsch.* **9a**, 67.
FISCHER, D. (1973) *Materials Res. Bull.* **8**, 385.
GILBERT, G. B., POEHLER, T. O., and MILLER, C. F. (1961) *J. Appl. Phys.* **32**, 1597.
GILLESSEN, K. and VON MÜNCH, W. (1973) *J. Cryst. Growth* **19**, 263.
GILMAN, J. J. (1963) *The Art and Science of Growing Crystals*, Wiley, New York.
GOODMAN, C. H. L. (1954) *Research* **7**, 168.
GRIGORVEV, V. D. (1977) *Russ. Metall. Izv. Akad. Nauk SSSR. Melt.* **5**, 112.
GRODZKA, P. G. and BANNISTER, T. C. (1970) *Proc. 4th Int. Conference Electron and Ion Beam Sci. and Technology, Los Angeles* (E. Chem. Soc.), p. 408.
GRYTSIV, V. N., FESH, R. N., NIKONYUK, E. S., PANCHUK, O. E., and SAVITSKII, A. V. (1975) *Inorg. Mater.* **11**, 1507.
HALL, R. N. (1953) *J. Chem. Phys.* **57**, 836.
HANDLEY, R. and HERRINGTON, E. F. G. (1956) *Chemical Ind.* p. 304.
HARRISON, J. D. and TILLER, W. A. (1961) *Trans. AIME* **221**, 649.
HAY, D. R. and SCALA, E. (1965a) *Proc. 1st Int. Conference Electron and Ion Beam Sci. and Technology* (Bakish, R., ed.), Wiley, New York, p. 550.
HAY, D. R. and SCALA, E. (1965b) *Trans. AIME* **223**, 1153.
HAYES, A. and CHIPMAN, J. (1939) *Trans. AIME* **135**, 85.
HEIL, R. H. JR. (1968) *Solid State Technol.* **11**, 21.
HELFAND, E. and KORNEGAY, R. L. (1967) *J. Appl. Phys.* **37**, 2484.
HERINGTON, E. F. G. (1960) *Endeavour* **19**, 191.
HERINGTON, E. F. G. (1963) *Zone Melting of Organic Compounds*, Blackwell, Oxford.
HESSE, G. and SCHILDKNECHT, H. (1956) *Angew. Chem.* **68**, 641.
HEYWANG, W. and ZIEGLER, G. (1954) *Z. Naturforsch.* **9a**, 561.
HEYWANG, W. (1956) *Z. Naturforsch.* **11a**, 238.
HISCOCKS, E. S. R. and ELLIOTT, C. T. (1969) *J. Mater. Sci.* **4**, 784.
HOLMES, P. J. (1963) *J. Phys. Chem. Solids* **24**, 1239.
HULME, K. F. (1955) *Proc. Roy. Soc.* **68**, 393.
HURLE, D. T. J. (1961) *Solid State Electron.* **3**, 37.
HURLE, D. T. J. (1969) *J. Cryst. Growth* **5**, 3.
HURLE, D. T. J. and JAKEMAN, E. (1969) *J. Cryst. Growth* **5**, 227.
HURLE, D. T. J., JAKEMAN, E., and PIKE, E. R. (1968) *J. Cryst. Growth* (Suppl. Proc. ICCG, Birmingham) **B-4**, 633.
HURLE, D. T. J., MULLIN, J. B., and PIKE, E. R. (1964a) *Phil. Mag.* **9**, 423.
HURLE, D. T. J., MULLIN, J. B. and PIKE, E. R. (1964b) *Sol. State Commn.* **2**, 197.
HURLE, D. T. J., MULLIN, J. B., and PIKE, E. R. (1964c) *Sol. State Commn.* **2**, 201.
HURLE, D. T. J., MULLIN, J. B., and PIKE, E. R. (1967) *J. Materials Sci.* **2**, 46.

IOFFE, A. F. (1956) *Zhur. Tekh. Fiz.* **26**, 478 (in Russian). Translation *Sov. Phys. Tech. Phys.* **I**, 462.
JINDAL, B. K. (1972) *J. Cryst. Growth*, **16**, 280.
JINDAL, B. K. and TILLER, W. A. (1968) *J. Chem. Phys.* **49**, 4632.
JOHNSTON, W. C. and TILLER, W. A. (1962) *Trans. AIME* **224**, 214.
KAMM, G. K. (1977) *Rev. Sci. Instrum.* **48**, 1222.
KANAYA, K., YAMAZAKI, H., KAWAKATSU, H., OKAZAKI, I., and SHIMIZU, T. (1968) *J. Electron Microsc.* **17**, 229.
KAPITZA, P. (1928) *Proc. Roy. Soc. (A)* **119**, 358.
KARL, N. and PROBST, K. H. (1970) *Mol. Cryst. Liquid Cryst.* **11**, 155.
KECK, P. H. and GOLAY, M. J. E. (1953) *Phys. Rev.* **89**, 1297.
KECK, P. H., GREEN, M., and POLK, M. L. (1953) *J. Appl. Phys.* **24**, 1479.
KIRGINTSEV, A. N. (1960) *Mathematical Theory of Zone Melting Processes*, IZd, CoAN SSSR Novosibirsk (in Russian).
KIRGINTSEV, A. N. (1963) *Sov. Phys. Sol. State* **5**, 1079.
KIRGINTSEV, A. N., KUDRIN, V. D., and KUDRINA, K. N. (1963a) *Sov. Phys. Sol. State* **5**, 686.
KIRGINTSEV, A. N., KUDRIN, V. D., and KUDRINA, K. N. (1963b) *Sov. Phys. Sol. State* **5**, 681.
KOBAYASHI, N. (1978) *J. Cryst. Growth* **43**, 417.
KOOY, P. and COUWENBERG, H. J. M. (1962) *Philips Tech. Rev.* **23**, 161.
KORCZAK, P. and STAFF, C. B. (1974) *J. Cryst. Growth* **24/25**, 386.
KRAMER, H. P., BOGERT, B. P., and HAGELBARGER, D. W. (1961) *J. Appl. Phys.* **32**, 764.
KRUMNACKER, M. and LANGE, W. (1969) *Kristall Tech.* **4**, 207.
LAWLEY, A. (1962) *Introduction to Electron Beam Technology* (Bakish, R., ed.), Wiley, New York, p. 184.
LAWLEY, A. (1968) *Techniques of Metals Research*, vol. 1 (Bunshah, R. F., ed.), Interscience, p. 845.
LEITZ, E. (1950) British Patent 691,335.
LEVICH, V. G. (1962) *Physico Chemical Hydrodynamics*, Prentice-Hall Inc.
LEWIS, G. N. (1907a) *Proc. Am. Acad.* **43**, 259.
LEWIS, G. N. (1907b) *Z. Physik Chem.* **61**, 129.
LORD, N. W. (1953) *Trans. AIME* **197**, 1531.
LOZOVSKI, V. N. (1971) *Izv. Vuz. Fiz.* **12**, 24 (in Russian).
MAIR, B. C., EBERLY, P. E., KROUSKOP, N. C., and ROSSINI, F. D. (1958) *Analytic Chem.* **30**, 393.
MASERJIAN, J. (1963) *Solid State Electron.* **6**, 477.
MASON, D. R. and COOK, J. S. (1961) *J. Appl. Phys.* **32**, 475.
MASON, H. L. (1961) *J. Res. Natn. Bur. Stds.* (USA) C-Engineering and Instrumentation 65 C, 97.
MATLOCK, J. H. (1977) Advances in single crystal growth of silicon in 'Semiconductor silicon', 1977, E. Chem. Soc. Proc. **77**, 2 (Huff, H. R. and Sirtl, E., eds.).
MEMELINK, O. W. (1956) *Philips Res. Report* **11**, 183.
MLAVSKY, A. I. (1961) *Jl. E. Chem. Soc.* **108**, 263 C.
MOLDOVANOVA, M., POPOV, A., STANEV, N., and ZHELEVA, N. (1971) *Ann. Univ. Sofia Fac. Phys.* (Bulgaria) **63**, 145 (in Russian).
MULLIN, J. B. (1975) in *Crystal Growth and Characterisation* (Ueda, R., and Mullin, J. B., eds.), North-Holland, Amsterdam, p. 75.
MULLIN, J. B. and HULME, K. F. (1960) *J. Phys. Chem. Solids* **17**, 1.
MULLIN, J. B., STRAUGHAN, B. W., and BRICKELL, W. S. (1965) *J. Phys. Chem. Solids* **26**, 782.
MULLINS, W. W. and SEKERKA, R. F. (1964) *J. Appl. Phys.* **35**, 444.
NERNST, W. (1904) *Z. Physik Chem.* **47**, 52.
NEUMANN, L. and HUGGINS, R. A. (1962) *Rev. Sci. Instrum.* **33**, 433.
NICOLAU, I. F. (1970) *J. Materials Sci.* **5**, 623.
NICOLAU, I. F. (1971) *J. Materials Sci.* **6**, 1049.
O'CONNOR, J. R. (1960) *J. Appl. Phys.* **31**, 1690.
OKADA, T., MATSUMI, K., and MAKINO, H. (1971) *Solid State Phys.* (*Japan*) **6**, 170.
OLIVER, B. F. (1963) *Trans. AIME* **227**, 960.
OTAKE, S. and MATSUNO, N. (1971) *Japn. J. Appl. Phys.* **10**, 1135.
PAAR, N. L. (1960) *Zone Refining and Allied Techniques*, George Newnes, London.
PAUREAU, J. (1972) *Rev. Phys. Appl.* **72**, 367.
PEYZULAYEV, SH. I., KONOVALOV, E. YE., and KONDRATYEVA, L. I. (1965) *Phys. Metals Metallog.* **19**, 59.
PFANN, W. G. (1952) *Trans. AIME* **194**, 747.
PFANN, W. G. (1955) *Trans. AIME* **203**, 961.
PFANN, W. G. (1957) *Metallurgical Rev.* **2**, 29.
PFANN, W. G. (1962) *Science* **135**, 1101.
PFANN, W. G. (1964) *J. Appl. Phys.* **35**, 258.
PFANN, W. G. (1966) *Zone Melting*, 2nd ed., Wiley, New York.
PFANN, W. G. (1967) *Sci. Am.* **217**, 63.
PFANN, W. G. (1969) US Patent 3,423,189.

PFANN, W. G., BENSON, K. E., and HAGELBARGER (1959) *J. Appl. Phys.* **30**, 454.
PFANN, W. G., BENSON, K. E., and WERNICK, J. H. (1957) *J. Electronics* **2**, 597.
PFANN, W. G. and OLSEN, K. M. (1953) *Phys. Rev.* **89**, 322.
PFANN, W. G. and WAGNER, R. S. (1962) *Trans. AIME* **224**, 1139.
POHL, R. G. (1954a) *J. Appl. Phys.* **25**, 668.
POHL, R. G. (1954b) *J. Appl. Phys.* **25**, 1170.
REED, R. E. (1972) *J. Vac. Sci. Technol.* **9**, 1413.
REED, R. E., UELHOFF, W., and ADAIR, H. L. (1976) ASTP Experiment MA-041, ONRL Report 76-113.
REED, T. B. (1961) *J. Appl. Phys.* **32**, 821 and 2534.
REED, T. B. (1966) *Proc. ICCG Boston*, p. 39.
REISS, H. (1954) *Trans. AIME* **197**, 1054.
REISS, H. and HELFAND, E. (1961) *J. Appl. Phys.* **32**, 228.
ROMANENKO, V. N. (1960) *Sov. Phys. Solid State* **2**, 793.
RONALD, A. P. (1959) *Analytic Chem.* **31**, 964.
RUTHERFORD, J. L., SMITH, R. L., HERMAN, M., and SPANGLER, G. E. (1965) *New Physical and Chemical Properties of Metals of Very High Purity*, Gordon & Breach, New York, p. 345.
SAVITSKI, E. M. and BURKHANOV, G. S. (1978) *J. Cryst. Growth* **43**, 457.
SCHADLER, H. W. (1963) *The Art and Science of Growing Crystals* (Gilman, J. J., ed.), Wiley, New York, p. 343.
SCHILDKNECHT, H. (1966) *Zone Melting*, Academic Press.
SCHILDKNECHT, H. and MANNL, A. (1957) *Angew. Chem.* **69**, 635.
SCHILDKNECHT, H. and MASS, K. (1967) Cited by Pfann, W. G. (1967).
SCHILDKNECHT, H. and VETTER, H. (1959) *Angew. Chem.* **71**, 723.
SCHWABE, D., SCHARMAN, A., PREISSON, F., and OEDER, R. (1978) *J. Cryst. Growth* **43**, 305.
SEIDENSTICKER, R. G. (1966a) *Jl E. Chem. Soc.* **113**, 152.
SEIDENSTICKER, R. G. (1966b) *Proc. ICCG Boston*, p. 733.
SHAH, J. S. (1967) Ph.D. thesis, Bath University, England.
SHAH, J. S. (1974) in *Crystal Growth*, Vol. 1, 1st edn. (B. R. Pamplin, ed.), Pergamon Press, Oxford.
SHAH, J. S. (1981) Environment for crystal growth, in this book, Chapter 4.
SHAH, J. S. (unpublished results) Work done at the H. H. Wills Physics Laboratory, Bristol, UK, in the JEOL (Japan Electron Optics Laboratories) electron beam zone refiner.
SHAH, J. S. and BROOKBANKS, D. M. (1972) *Platinum Metals Rev.* **16**, 94.
SHAH, J. S. and HUNTLEY, D. A. (1970) *J. Cryst. Growth* **6**, 216.
SHAH, J. S. and PAMPLIN, B. R. (1969) *Jl E. Chem. Soc.* **116**, 1565.
SHERMAN, J. and EWELL, R. B. (1942) *J. Phys. Chem.* **46**, 641.
STERLING, H. F. and WARREN, R. W. (1963) *Metallurgica* **67**, 301.
STÖBER, F. (1925) *Z. Krist.* **61**, 299.
STOREY, R. N. and LAUDISE, R. A. (1970) *J. Cryst. Growth* **6**, 261.
SWIGGARD, E. M. (1967) *Jl E. Chem. Soc.* **114**, 976.
TAKAHASHI, M., MANAMATSU, S., and KIMURA, M. (1971) *Proc. ICCG Marseille*, p. 681.
TAKAI, H. (1978) *J. Cryst. Growth* **43**, 463.
TEEDE, N. F. (1967a) *Proc. IRE Australia* **28**, 115.
TEEDE, N. F. (1967b) *Solid State Electron.* **10**, 1069.
TILLER, W. A. (1963a) *J. Appl. Phys.* **34**, 2757.
TILLER, W. A. (1963b) *J. Appl. Phys.* **34**, 2763.
TILLER, W. A. (1965) *J. Appl. Phys.* **36**, 261.
TILLER, W. A., JACKSON, K. A., RUTTER, J. W., and CHALMERS, B. (1953) *Acta Met.* **1**, 428.
THEUERER, H. C. (1956) *Trans. AIME* **206**, 1316.
THEUERER, H. C. (1962) US Patent 3,060,123.
THURMOND, C. D. and STRUTHERS, J. D. (1953) *J. Phys. Chem.* **57**, 831.
TRAINOR, A. and BARTLETT (1961) *Solid State Electron* **2**, 106.
TROUSIL, Z. (1962) *Czech. J. Phys.* **B12**, 227.
VAN DEN BOOMGAARD, J. (1955) *Philips Res. Report* **10**, 319.
VAN DEN BOOMGAARD, J. (1956) *Philips Res. Report* **11**, 27 and 91.
VAN DEN BOOMGAARD, J. KRÖGER, F. A., and VINK, H. J. (1955) *J. Electronics and Control* **1**, 212.
VELICKY, B. (1964) *Phys. Stat. Sol.* **5**, 207.
VERHOEVEN, J. D. (1965) *Trans. AIME* **223**, 1156.
VERHOEVEN, J. D. (1966) *J. Metals* **18**, 26.
VETTER, H. (1960) Dissertation, Erlangen Univ. (Germany).
VOLCHOK, B. A. (1962) *Sov. Phys. Solid State* **4**, 789.
WAGNER, R. S. and ELLIS, W. C. (1964) *Appl. Phys. Letters* **4**, 89.
WAGNER, R. S., MILLER, C. E., and BROWN, H. (1966) *Trans. AIME* **236**, 554.
WARREN, R. W. (1962) *Rev. Sci. Instr.* **33**, 1378.

WEINREICH, O. A. and DERMIT, G. (1963) *J. Appl. Phys.* **34**, 225.

WERNICK, J. H. (1962) *Ultrahigh Purity Metals*, Am. Soc. of Metals, Cleveland, USA, p. 55.

WHITE, E. A. D. (1965) *Br. J. Appl. Phys.* **16**, 1415.

WIEDER, H. H. (1965) *Solid State Comm.* **3**, 159.

WIEDER, H. H. (1978) *J. Vac. Sci. Technol.* **14**, 1292.

WIEDER, H. H. and DAVIS, N. M. (1965) *Solid State Electron.* **8**, 605.

WILCOX, W. R. (1962) *Ultrapurification of Semiconducting Materials* (Brooks, M. S. and Kennedy, J. K., eds.), Macmillan, New York, p. 481.

WILLIAMSON, W. J. (1969) *J. Vac. Sci. Technology* **6**, 765.

CHAPTER 9

Methods of Growing Crystals Under Pressure†

A. G. FISCHER

*Professor, Department of Electrical Engineering, University of Dortmund,
46 Dortmund, Federal Republic of Germany‡*

9.1. Introduction: *Explanation of the Decomposition Tendency of
Compounds Having Mixed Bonding, or Why Do We Need
High Pressure?* (*A. Fischer*, 1958)

Many binary compounds which are known as semiconductors and phosphors decompose into their elements before reaching their melting points. This aggravates crystal growth from the melt so that for a long time only crystal growth from the vapor phase, which is unsatisfactory in certain respects, could be practiced.

A review of the binary compounds of the type AB using the ionic or the covalent bonding proportion as the ordering principle reveals that, as a rule, only crystals with a high proportion of ionic binding (more than 60%) and crystals with a high proportion of covalent bonding (more than 80 or 90%) are melting undecomposed, whereas crystals with predominantly mixed bonding decompose before reaching their melting point. Since in the vapor single molecules of these compounds (e.g. ZnS) are very unstable, this means that true decomposition takes place, even though the impression of a sublimation may be gained owing to the recombination of the decomposition products at colder spots of the vessel. In particular the mixed ionic-covalent crystals composed of light elements, which have the diamond or related structure and which are important as semiconducting phosphors owing to their large forbidden bandgaps, show this disturbing tendency to decompose to a high degree. This creates difficulties in their preparation. For growth from the melt, a pressurized atmosphere is usually needed, as described in this chapter.

The question must be raised: What causes this behavior? With predominantly ionic bonding, the existence of an extended liquid phase range, i.e. the absence of decomposition, is easy to understand. In these crystals, which can be imagined as being constituted from oppositely charged spheres held together by nondirectional Coulomb forces, the surface ions are attracted firmly by their neighbors behind them.

† The research reported in this chapter was sponsored in part by the Air Force Cambridge Research Laboratories, Office of Aerospace Research, under Contract Numbers AF19 (604) 8018 and AF19 (628) 3866, and by US Army Electronics Command, Contract DAAB07–69–C–0290. Published in abbreviated form in *J. Electrochem. Soc.* **117**, 41C (1970). This work was performed at RCA Laboratories, Princeton, NJ.

‡ Paper submitted March 15, 1971, upgraded February 1979.

Contrarily, with covalent crystals of the diamond, zinc blende, or wurtzite type, it is well known that each atom must be connected with its four neighbors by four directional electron pair bonds. Yet there is no atom in existence whose shell can *a priori* fulfill this request of diamond lattice construction (i.e. having four tetrahedrically arranged orbital lobes). For this, a deformation of the electron hull is necessary, whereby deeper terms are raised, e.g. s–p degeneration. Only these "deformed" atoms can then build up the diamond lattice. From this follows: in a crystal surface the atoms which are not clamped from all four sides by neighbors but only from three and which, therefore, are deformed into the quadruvalent orbital configuration only insufficiently, do not possess the correct valency angles for interlocking with their near neighbors. Therefore, they are bound only loosely and tend to fly off when hit by thermal vibrations or when exposed to other stimuli.

With atoms of high polarizability where the upper terms which are to be mixed are closely together, i.e. with heavy atoms, this surface instability effect due to loose bonding is only slight. The tendency to decompose is the stronger the greater the energy becomes which is required to deform the atoms into the tetrahedral orbital shape, i.e. the larger the energetic distance of the atomic terms which have to be mixed, the lower the polarizability, the lower the atomic weight. This tendency of the crystal to decompose superficially is increasing, in particular, with increasing dissimilarity of atoms A and B, i.e. with increasing ionic bond starting from a pure covalent bond. For it is evident that greater hull deformations are needed until, say zinc and sulfur can simultaneously accept and yield four electrons, than gallium and phosphorus, or silicon and silicon.

If this hull deformation energy, which has to be exerted to mold the atoms into the tetrahedral shape, is larger than the energy that is gained by subsequent forming the lattice, then these crystals can only be formed if atomic deformation is facilitated by huge external pressure forces. Crystals which must be formed in this way, such as the cubic modification of boron nitride (borazone) or of carbon (diamond), are surface unstable at high temperature just like the other compounds considered here, but, in addition, they convert into the stable lattice modification (graphite) starting from the surface.

To retard this decomposition, crystals of these decomposable compounds can be grown from their melts only under pressure of the more volatile component. The requisite techniques are the subject of this chapter.

The necessary apparatus, from the simple to the sophisticated, has been investigated. Methods using resistance or r.f. heating, open or sealed ampoules, crucibles of various materials, without or with high pressure steel autoclave, under inert or active gas pressure, and the required control systems, are described. A broad description of the liquid encapsulation method is given, and construction details for the autoclave are contained in the figures and photographs. Special attention is given to the growth of gallium phosphide and gallium arsenide phosphide crystals, and to the preparation of the starting materials from the elements.

Wide bandgap, zincblende type compounds like ZnS (m.p. 1830°C, Addamiano and Dell, 1957), ZnSe (m.p. 1515°C, Fischer, 1959), CdS (m.p. 1475°C, Addamiano and Dell, 1957), ZnTe (m.p. 1300°C, Carides and Fischer, 1964), and GaP (m.p. 1470°C, Richman, 1963) decompose into their volatile components far below their melting points, so that crystals can be grown from the melt only under pressure. Since large crystals can be grown from the melt faster than by other methods, the need for initially costly pressure equipment is no longer a major deterrent.

As usual in many fields, the long period of dormancy in which only a few devoted enthusiasts do the exploratory work, without much response, is over for crystal growth from the melt under pressure. It is now an industrial method. This was caused by the coincidence of an industrial need for large GaP single crystalline substrates for electroluminescent diodes, triggered by the achievement of high efficiency in the visible, and the achievement of large GaP crystals made by pulling through a boric oxide blanket under pressure.

The chapter is arranged by describing the techniques in sequence of increasing complexity and refinement, which coincides frequently with the chronological order of their conception. Since materials synthesis is closely intermeshed with crystal growth proper, the synthesis methods are not described separately but within the context. We shall treat Bridgman and Czochralski methods using high pressure inert gases and active vapors, the necessary equipment, temperature control systems as they become necessary, and materials synthesis from the elements. Since it is of prime importance now, we shall describe in detail crystal growth using liquid encapsulation, first published by Metz *et al.* (1962), applied to GaP first by Bass and Oliver (1968) at SERL, and done now by automated methods as predicted by Fischer and Pruss (1969). At the end we shall describe new methods.

This chapter contains, interwoven with a review, numerous new results, useful practical recognitions, and technological innovations, especially in the figures. Many of them have already been adopted by other crystal growers.

The paper is confined to techniques and methods; properties of the prepared crystals will only be mentioned in passing as far as they are related to their growth.

9.2. Resistance Heater Methods

9.2.1. BRIDGMAN GROWTH UNDER HIGH INERT GAS PRESSURE

Tiede and Schleede (1921) found that ZnS and CdS could be melted by retarding the decomposition with high nitrogen pressure. For instance, ZnSe can be melted in graphite crucibles under 150 bar of argon pressure. GaN and InN decompose before melting even under 200 bar of nitrogen pressure. Nitrogen behaves in this case like an inactive gas, due to the stability of the nitrogen molecule.

To repulse the emanating vapors effectively back to the melt meniscus, the pressure of the inert atmosphere should be as high as possible; more than 100 bar are usually applied to ZnS and CdS. The atmosphere at the melt surface should be still, not turbulent. This precludes crystal pulling because of the strong thermal draughts at the open meniscus of the melt. The crucible should be covered by a lid. GaP crystals were grown under 100 bar of argon pressure in semisealed graphite crucibles (Fahrig *et al.*, 1972).

After examining other constructions (Addamiano and Dell, 1957; Medcalf and Fahrig, 1958) we found that the most advantageous graphite-tube furnace, suitable for 2200°C maximum and 200 bar, is the one shown in Figs. 9.1 and 9.2 (Fischer, 1961, 1963). In this furnace, a straight, high density carbon tube, which can be tapered in its wall thickness along its length at will to influence the thermal gradient, is clamped and contacted at the ends by packings of compressed graphite granules to allow for thermal expansion and contraction. The autoclave lid is one electrode, the bottom vessel the other. Since only 12 V maximum are needed, the pressure gasket seal can also act as an electrical insulator. The bolts are also insulated by sleeves and washers. For reasons of purity, the use of carbon or molybdenum

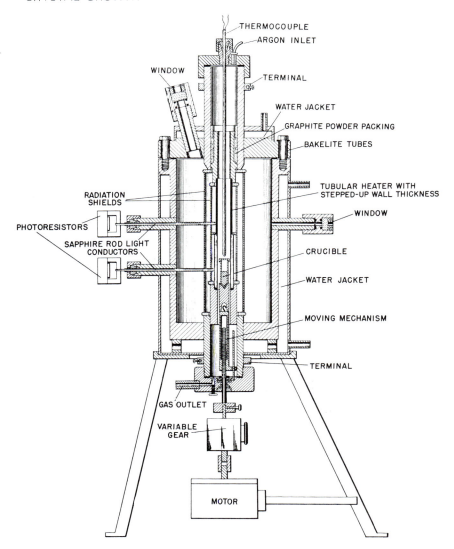

FIG. 9.1. Carbon tube furnace (12 V, 2000 A) for up to 2200°C, 250 bar. The carbon tube with stepped-up wall thickness and radiation shields is held by granular carbon packings at the ends.

cylinders as radiation baffles is preferable to granular ceramic insulation packings used in the beginning (Fischer, 1958). Other graphite furnaces using delicate double helices or double wall graphite tubes with both contacts at the same end are not as rugged and trouble free in use.

Corrosive vapors from the melt attack thermocouples, smoke and thermal storms in the compressed, dense gas make the use of optical pyrometers inaccurate. Therefore, the temperature is sensed best with a sapphire rod light conductor as shown in Fig. 9.1 (Fischer, 1963).

FIG. 9.2. Photograph of pressure furnace of Fig. 9.1.

Stabilization of the input power against line voltage fluctuations† is a must in all crystal growth equipment to be described. In addition to this, automatic temperature control was via saturable reactor and step-down transformer, and a proportional controller. To achieve accurate control with the big, sluggish saturable reactor, load inertia had to be increased by using thick-walled cylindrical graphite radiation baffles, not shown in Fig. 9.1. To prevent oscillations, a control system must be able to respond faster than the process to be controlled. Later we found it easier to coarsely regulate 90 % of the current by handwheel of a variable transformer, and to pass only 10 % through a much smaller saturable reactor for accurate, fast control.

A simple electromechanical melting point detector (a heavy, inert slug submerges as soon as the ingot melts and then activates a signal) was found to be very useful to prevent overshooting the melting point too far, see Fischer (1958).

Crystals of all wide gap II–VI compounds are now grown successfully by this method, at many places all over the world.

† Automatic voltage regulator, General Radio Company, Concord, Mass.

9.2.2. THE CAPILLARY-TIPPED AMPOULE METHOD FOR ZnTe

ZnTe has a relatively low melting temperature and decomposition pressure, but the starting material usually contains a slight, hard-to-remove excess of tellurium. The ZnTe, when it is sealed into a quartz ampoule, therefore develops a higher pressure than its corresponding true stoichiometric decomposition pressure, which is 1.9 bar.† This leads to thermoplastic expansion of the ampoule unless it is externally supported (see below). It was found that ZnTe crystals can be grown from the melt in unsupported quartz ampoules if the ampoules have a capillary extension which protrudes from the furnace into the cold (Figs. 9.3 and 9.4). This capillary traps the excess volatile component and then seals itself. This simple method is applicable to other, similar, compounds (Fischer, 1966).

9.2.3. THE "SOFT AMPOULE" METHOD

In order to really control the stoichiometric composition of the growing crystal one has to apply excess cation or anion pressure. This requires a gas-tight container which is chemically inert at temperatures of about 1500°C where compounds such as GaP, ZnSe, or CdS melt.

A quartz or borosilicate (Vycor) ampoule, even when thermoplastic at the melting point of the compound, can serve well as a gastight liner. By letting the internal decomposition

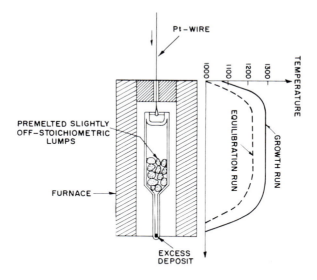

FIG. 9.3. Melting of ZnTe in unsupported quartz ampoules with capillary cold trap to remove excess components.

---

† This value is higher than the published value of 0.6 bar, which was obtained theoretically. We determined the true value by lowering the pressure over a liquid-encapsulated melt (see later in this chapter) until the melt started fuming violently. The corresponding values, determined in the same way for ZnSe (2.2 bar) and for CdS (3.2 bar), also differ from the published values. ZnS could not be measured by this method, due to rapid volatilization of the B_2O_3 encapsulant at 1830°C. The decomposition pressure of GaP (35 bar) could be verified by this method.

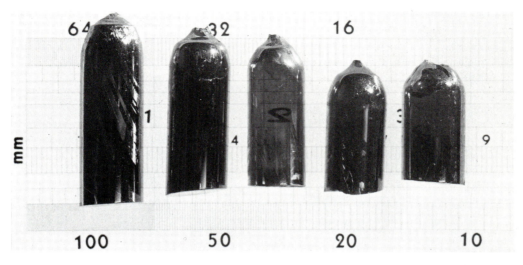

FIG. 9.4. ZnTe boules, grown by the method of Fig. 9.3.

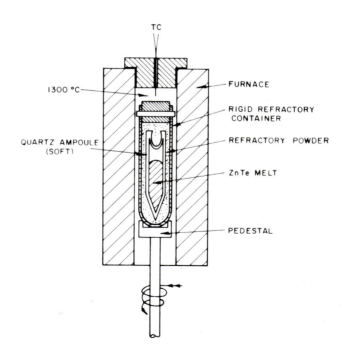

FIG. 9.5. Gradient freezing of ZnTe in sealed Vycor ampoule which is embedded in SiO_2 sand and packed into ceramic tube.

pressure drive it against a tight-fitting graphite enclosure which counteracts the mechanical forces (Figs. 9.5 and 9.6), Fischer (1959) showed that many bars of pressure can be sustained. This method works for ZnSe and CdS, and has been used in modified form for GaP by Blum and Chicotca (1968). It is an inexpensive method, and can also be used for horizontal gradient freezing (Fig. 9.7).

It was found earlier by Fischer (US Pat. 3,033,659, 1959) that during the initial heating of the closed ampoule to the melting point of the material, the internal pressure could be high while the ampoule is still brittle rather than thermoplastic, leading to ampoule failure. This can be prevented by approaching the melting point very fast so that the glass reaches its softening temperature while the charge inside is still relatively cool.

A limitation of this method is the gradual devitrification of the quartz glass, which restricts the growth runs to a few hours. However, this can be improved by carefully avoiding any alkali contaminations of the ampoule; see Holton *et al.* (1969).

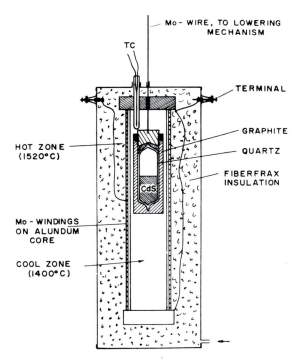

FIG. 9.6. Growth of CdS by "soft ampoule" method, also applicable for ZnSe and GaP.

9.2.4. PREPARING II–VI COMPOUNDS FROM THE PURE ELEMENTS

The II–VI compounds are normally purchased as chemically prepared fluffy powders.† Oxides can be removed and particle size can be enlarged by firing in hydrogen. But if one melts such powders, the volume of the final ingot is only about one-quarter of the original powder filling, leaving too much unused crucible space.

† For instance, from General Electric Co., 21800 Tungsten Road, Cleveland, Ohio 44117.

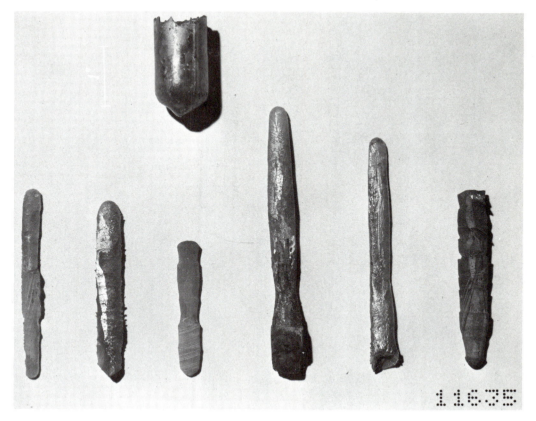

FIG. 9.7. Doped ZnSe ingots. The elongated ones have been subject to horizontal zoning in sealed quartz ampoules.

As compared to the customary laborious synthesis of polycrystalline chunks from the pure element by vapor reaction in open or sealed quartz tubes, the following method is much simpler and yields compact slugs which can be made to completely fill the crucible.

The spectroscopically pure elements are reacted in a small "bomb" (Fig. 9.8) which is pressurized up to full cylinder pressure (250 bar) with purified argon (Fischer, 1961). When the molybdenum wound heater is slowly warmed up to the melting point of the compound, the pressure rises to about 400 bar by thermal expansion of the gas. The mixed lumps of the elements, even if they should react with each other strongly exothermically, cannot create a shock wave, due to the inert gas atmosphere slowing down the rate of reaction. The upper end of the quartz tube reaches out of the furnace into the colder part of the bomb, acting as a trap and "reflux" cooler. Since some anion vapor is usually lost, one has to add more than actually required, at the beginning.

All II–VI compounds can be synthesized quite easily in this way. For ZnS, the quartz tube has to be replaced by carbon. For the III–V compounds, especially the phosphides, better methods are described later.

Incidentally, the easy-to-build "bomb" of Fig. 9.7 proved to be very useful in numerous other applications, as illustrated in Figs. 9.9 and 9.10.

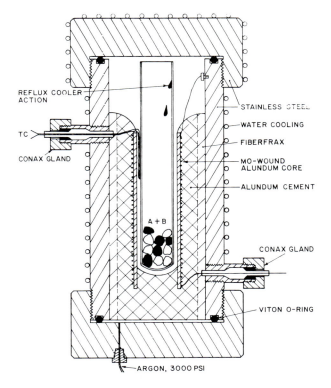

REFLUX COOLER ACTION

TC

CONAX GLAND

STAINLESS STEEL

WATER COOLING

FIBERFRAX

MO-WOUND ALUNDUM CORE

ALUNDUM CEMENT

A + B

CONAX GLAND

VITON O-RING

ARGON, 3000 PSI

FIG. 9.8. Bomb for synthesis of polycrystalline II–VI ingots from the elements.

9.3. Induction Heater Methods

9.3.1. VERTICAL AND HORIZONTAL BRIDGMAN METHOD IN UNSUPPORTED QUARTZ AMPOULES

Once induction heating equipment is available, it is preferable to resistance heating because it allows a carbon crucible to be heated to a very high temperature inside a quartz ampoule, whereas the ampoule itself can be kept below its weakening temperature. Standard-wall quartz tubes of 1 in. outside diameter can stand 30 bar of pressure up to 1000°C. However, safety shields should be used for protection of the operator.

In the arrangement shown in Fig. 9.11, the lower, cooler part of the ampoule extends into a wire-wound heater which is bifilarly wound (two parallel helixes connected at one end, with both terminals at the other end) as should be all wire-wound resistance heaters located near r.f. coils to prevent erratic heating due to r.f. coupling. This lower part of the ampoule contains the excess cation or anion material which provides the required vapor pressure. The upper, hotter part receives an experimentally adjusted amount of heat from the graphite susceptor inside (Fischer, 1966). The temperature of this part can be roughly indicated with thermocrayons.† The construction of the ampoule is shown more detailed in Fig. 9.12, ZnSe crystal boules grown by this method in Fig. 9.13.

---

† Tempilstik Corporation, New York, N.Y. or A. W. Faber-Castell, 8404 Stein bei Nürnberg (West Germany) (Thermocolor).

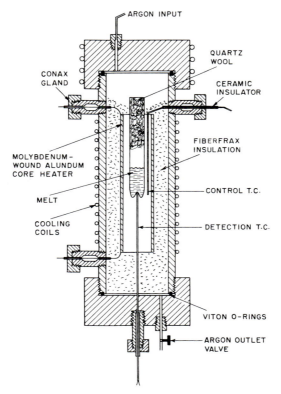

FIG. 9.9. Set up for simplified thermal differential analysis under pressure.

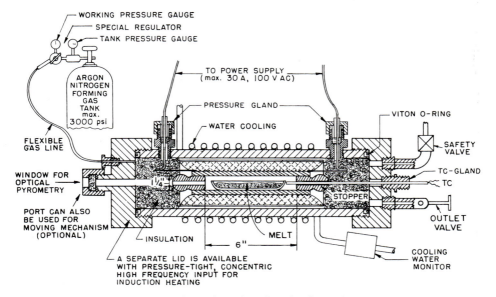

FIG. 9.10. Set up for horizontal gradient freezing under pressure.

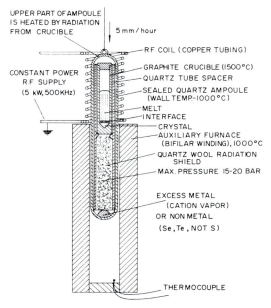

UPPER PART OF AMPOULE
IS HEATED BY RADIATION
FROM CRUCIBLE

5 mm / hour

RF COIL (COPPER TUBING)

GRAPHITE CRUCIBLE (1500°C)

QUARTZ TUBE SPACER

SEALED QUARTZ AMPOULE
(WALL TEMP-1000°C)

CONSTANT POWER
R F SUPPLY
(5 kW, 500 KHz)

MELT
INTERFACE

CRYSTAL

AUXILIARY FURNACE
(BIFILAR WINDING), 1000°C

QUARTZ WOOL RADIATION
SHIELD

MAX. PRESSURE 15-20 BAR

EXCESS METAL
(CATION VAPOR)
OR NON METAL
(Se, Te, NOT S)

THERMOCOUPLE

FIG. 9.11. R. F. Bridgman growth under controlled atmosphere in unsupported quartz ampoule.

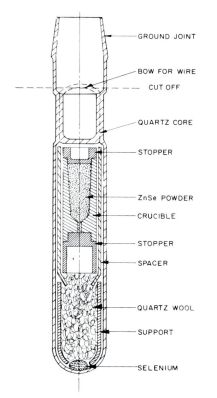

GROUND JOINT

BOW FOR WIRE

CUT OFF

QUARTZ CORE

STOPPER

ZnSe POWDER

CRUCIBLE

STOPPER

SPACER

QUARTZ WOOL

SUPPORT

SELENIUM

FIG. 9.12. Details of ampoule construction.

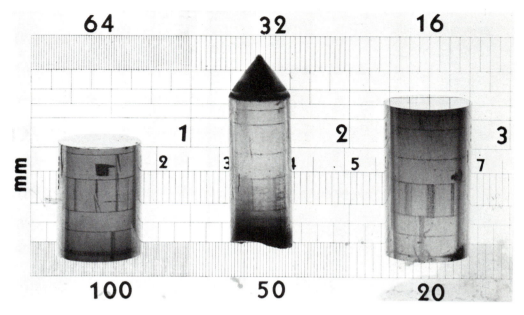

FIG. 9.13. ZnSe crystals grown by the method shown in Fig. 9.11.

To melt-grow ZnS and CdS a sulfur atmosphere cannot be used with porous carbon susceptors because of CS_2 formation which leads to explosions, as discussed by Fischer (1961, 1962). Vitreous carbon is more inert, but growth runs still must be short. Much better results are obtained by melting II–VI compounds in their metal vapor atmospheres. In this way, aluminum-doped ZnS, CdS, and ZnSe have been prepared highly *n*-type and with high mobility by Fischer (1963), in a one-step process, in contrast to other methods which require afterfiring of the crystal submerged in a cation metal melt.

This method has also been used for horizontal zone melting by Fischer (1961), Fig. 9.14. For example, silver impurities have been effectively swept to one end of a ZnSe ingot (Fig. 9.15) rendering it weakly *p*-type there; see Fischer (1962).

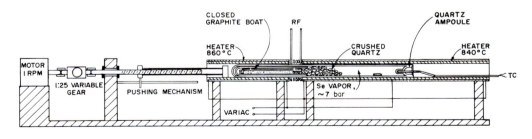

RF – ZONE MELTING OF ZnSe

FIG. 9.14. R.F. zone melting of ZnSe in supported quartz ampoule.

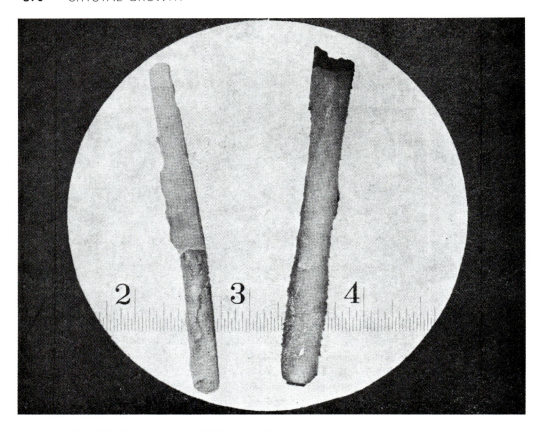

FIG. 9.15. Horizontally zoned ZnSe ingots. The one on the right contains silver impurities.

9.3.2. VERTICAL BRIDGMAN GROWTH IN PRESSURE-RELIEVED AMPOULES

To remove the explosion hazard of quartz glass ampoules, the decomposition pressure inside the ampoule, which reaches a dangerous magnitude in the preparation of GaP (35 bar), can be balanced by external pressure in a steel autoclave. For this purpose, the autoclave of Figs. 9.1 and 9.2 was upgraded into the configuration of Figs. 9.16 and 9.17.

An r.f. work coil placed inside a thick-walled steel vessel does not lead to significant power losses if 2–3 in. of clearance is maintained. The water-cooled r.f. leads should be brought through the steel walls coaxially if magnetic steel walls and/or frequencies above one megacycle are used. For nonmagnetic (stainless steel) walls, which are preferable for cleanliness anyhow, and for frequencies of 500 kHz or lower, simple straight parallel feedthroughs with Teflon bushings,† as shown in later figures, are sufficient.

This vertical Bridgman method has been successfully used for ZnSe crystal growth (Fischer, 1961, 1962, 1963) and can also be used for GaP (Fahrig *et al.*, 1972).

† "Conax Glands", Conax Corp., Buffalo, New York.

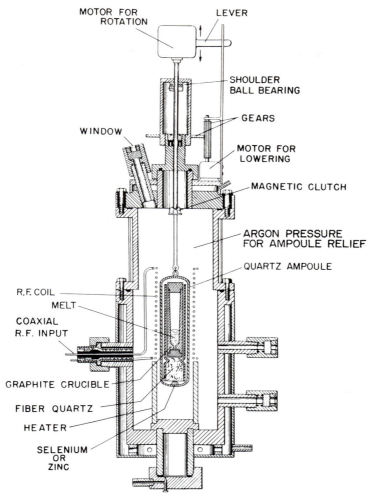

Fig. 9.16. R.F. Bridgman growth of ZnSe in pressure-compensated sealed quartz ampoule in steel autoclave. Note coaxial r.f. input.

In order to simplify this method and do away with ampoule pumping and sealing, a demountable, reusable ampoule with a ground joint can be used; see Fischer (1962) (Fig. 9.18). The internal atmosphere communicates with the external pressurized gas (which, for this reason, must be of ultrahigh purity) through a capillary which can be heated to prevent clogging. The crystal grows, in the case of ZnSe, for example, in an argon–selenium atmosphere. This method has the additional advantage that the added argon pressure prevents bubbling of the solidifying melt caused by exudation of the excess anion concentration of the melt at the interface, which is frequently observed in pure anion atmospheres. The disadvantage is, of course, that some anion vapor escapes through the capillary and contaminates the autoclave. This can be minimized by trapping it on a nearby cold surface.

FIG. 9.17. Photograph of apparatus shown in Fig. 9.16.

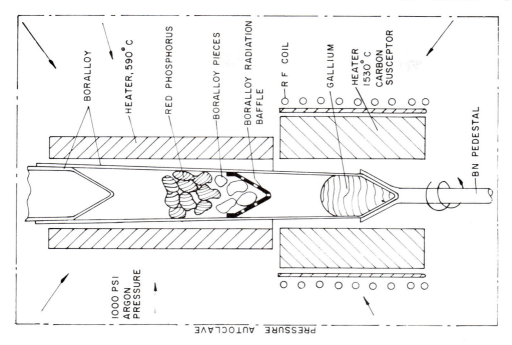

Fig. 9.19. GaP synthesis and crystal growth in a demountable pyrolytic BN ampoule.

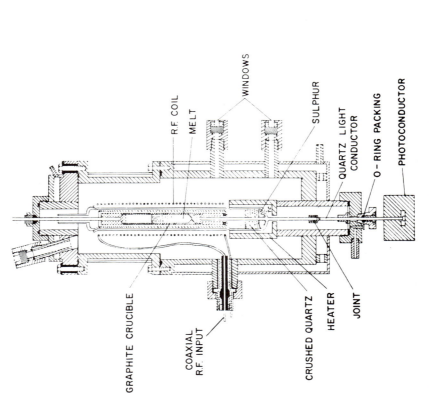

Fig. 9.18. Demountable quartz ampoule with gas communication between inside and outside.

For melt-growth of GaP we found early that boron nitride containers are desirable, since BN does not react with GaP according to Wang *et al.* (1964). "Boralloy" pyrolytic BN crucibles† are better than hot-pressed BN since they are quite pure and can be reused almost indefinitely because GaP does not adhere to the walls. Also, the walls are thin and elastic and thus do not exert strain onto the crystal during cooling to room temperature. For II–VI crystals, vitreous carbon‡ makes an excellent container material.

This growth method, in conjunction with Boralloy containers, has yielded a very economical method of GaP synthesis and crystal growth (Fig. 9.19). Pure gallium and red phosphorus§ are placed in a long, conical Boralloy crucible, separated by a heat shield, as shown. The crucible is closed by a loosely fitting lid, made intentionally nonfitting by an interposed crumb of red phosphorus which volatilizes subsequently. The whole system is pumped, then pressurized with argon to 75 bar. The gallium is heated to 1500°C, the phosphorus to 590°C corresponding to 40 bar of phosphorus pressure. Vapour pressure equilibrium is attained quite slowly since the red phosphorus under argon pressure decomposes into white phosphorus only very gradually.

After the GaP has formed, the crucible is slowly programmed to a lower temperature, for Bridgman type crystallization. At room temperature, the system is depressurized and opened. After the white phosphorus has burned off, the crucible is inverted and the GaP ingot falls right out. This is one-step operation, and it excludes contamination with silicon (no contact with quartz) or carbon, and all parts can be reused. The usefulness of this method has been confirmed by others (Blum and Chicotka, 1973; Woodbury, 1976).

In all of these methods, the necessary r.f. power setting for reaching the melting temperature must be determined in a preliminary "dry" run where a thermocouple is inserted into the crucible which contains a dummy load. Tungsten–rhenium thermocouples‖ are most suitable, even though they drift somewhat, due to phosphorus contamination. This effect is much stronger in platinum–rhodium thermocouples which are extremely sensitive to phosphorus poisoning.

During the actual run, one must rely on the constancy of the r.f. generator which therefore must be highly stabilized. The output is sensed via an r.f. pickup coil and "thermocross" tube (Fig. 9.20), or grid-dip meter. If r.f. susceptors other than the charge itself are used, thermocouples shielded from chemical attack by alundum tubes can be brought near the melt. An additive combination of r.f. pickup signal and thermocouple signal was found to give even better constancy (better than $\pm 1°C$) (see also Fig. 9.34). Very small loads (thimble-sized crucibles) require very fast-responding control systems. We used a controller with a tiny magnetic preamplifier of the signal¶ which proved to work much faster than the customary chopper preamplifier. Whereas pound-sized loads can be controlled with sluggish saturable reactor systems in the line power supply, small loads require thyristor or thyratron input power control of the r.f. generator. Since at the time, SCRs were surge sensitive and failed frequently, we resorted to six small, fast-responding presaturated saturable reactors in the pulsed d.c. of the high voltage to the r.f. tube,†† which were definitely preferable to big, sluggish saturable reactors in the primary circuit, or jerky electromechanical regulators. Today, SCR control is reliable, hence preferable.

---

† Union Carbide Corp., New York, NY.
‡ Beckwith Carbon Company, van Nuys, California.
§ Gallium: Alusuisse Metals Inc., Fort Lee, NJ. Phosphorus: Mitsubishi International Corp., New York, NY.
‖ Engelhard Industries, Inc., Newark, NJ.
¶ Electronic Control Systems Inc., Fairmount, W.Va.
†† McDowell Electronics Co., Metuchen, NJ.

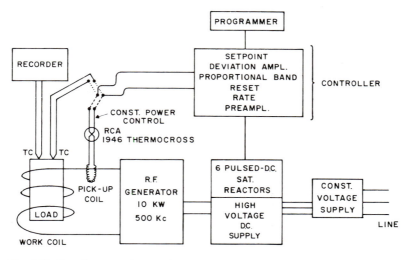

FIG. 9.20. Complete control system for r.f. generator. It can be switched from thermocouple signal input to power signal input.

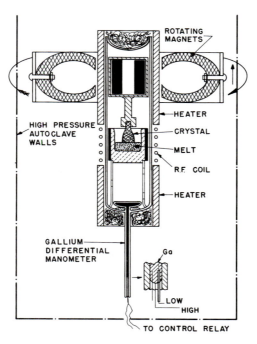

FIG. 9.21. Differential gallium manometer to sense pressure differences between ampoule inside and outside.

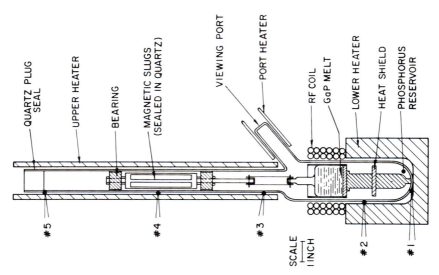

QUARTZ PLUG SEAL

UPPER HEATER

BEARING

MAGNETIC SLUGS (SEALED IN QUARTZ)

VIEWING PORT

PORT HEATER

RF COIL

GaP MELT

LOWER HEATER

HEAT SHIELD

PHOSPHORUS RESERVOIR

#5

#4

#3

SCALE
I INCH

#2

#1

Fig. 9.23. Quartz ampoule with inserts for set up of Fig. 9.22.

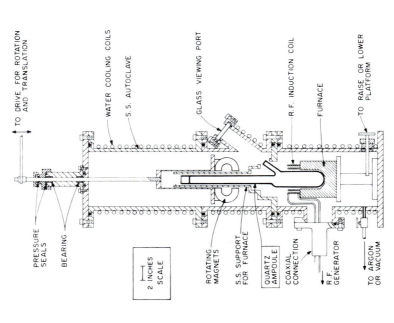

TO DRIVE FOR ROTATION AND TRANSLATION

WATER COOLING COILS

S. S. AUTOCLAVE

GLASS VIEWING PORT

R.F. INDUCTION COIL

FURNACE

TO RAISE OR LOWER PLATFORM

PRESSURE SEALS

BEARING

2 INCHES
SCALE

ROTATING MAGNETS

S.S. SUPPORT FOR FURNACE

QUARTZ AMPOULE

COAXIAL CONNECTION

R.F. GENERATOR

TO ARGON OR VACUUM

Fig. 9.22. Apparatus for Gremmelmeier growth of GaP (Weisberg, 1966).

9.3.3. CZOCHRALSKI-TYPE PULLING UNDER PRESSURE

The method, developed by Gremmelmeier (1956), under pressure in sealed quartz ampoules, with force transfer by rotating magnets, requires rather accurate bucking of the internal decomposition pressure by external autoclave pressure to prevent explosions or implosions of the ampoule; see Fischer (1963). The aforementioned slow attainment of the equilibrium phosphorus pressure over red phosphorus lumps necessitates a differential manometer (Fig. 9.21), but this adds further complications. Except for the auxiliary heaters to maintain the vapor pressure, one needs additional heaters to keep the viewing window unfogged, since visibility is mandatory for crystal pulling. Such systems have been built (Figs. 9.22 and 9.23), and macropolycrystals have been pulled by Weisberg *et al.*, but the liquid encapsulation method (see later) was more successful.

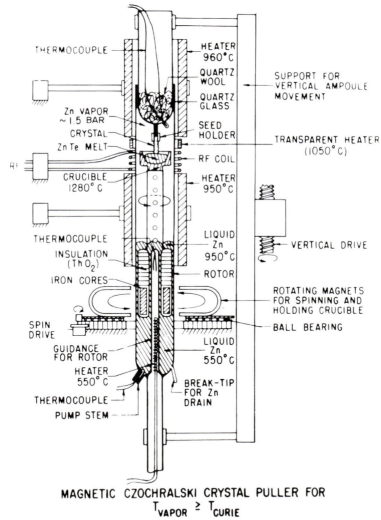

MAGNETIC CZOCHRALSKI CRYSTAL PULLER FOR
$$T_{VAPOR} \geqq T_{CURIE}$$

FIG. 9.24. Gremmelmeier machine for ZnTe. The enclosed magnets rotate in liquid zinc at 500°C; only the meniscus of the zinc bath is heated to 950°C to provide the pressure.

For pulling II–VI compounds, atmospheres consisting of group VI vapors cannot be used, because sulfur, selenium, and tellurium vapors are opaque. Therefore one has to use metal vapor pressure to prevent melt decomposition. The temperatures required to reach the proper vapor pressures are above 700°C, where magnets lose their power. For ZnTe, a modified Gremmelmeier machine was built by Fischer (1963) where the internal magnets rotate in liquid zinc at 500°C (Fig. 9.24). Instead of the Gremmelmeier method one can use demountable ampoules with precision-sealed rotating shafts (Figs. 9.25 and 9.26), or liquid-sealed rotating shafts. This method was abandoned due to lack of support, but it is the only

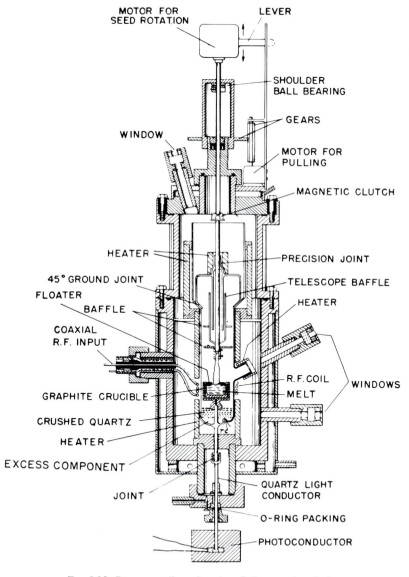

FIG. 9.25. Pressure puller using close-fitting rotating shaft.

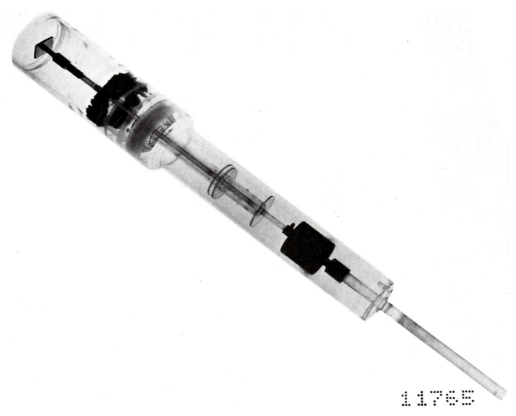

FIG. 9.26. Quartz container for pulling with rotating shaft, in set up of Fig. 9.25.

one which would permit pulling of II–VI crystals, since there is no suitable encapsulant for these materials as yet (see later).

9.4. Crystal Growth Using Liquid Encapsulation (LE)

9.4.1. CZOCHRALSKI PULLING

Crystal pulling using an oriented seed is the most esthetic crystal growth method. By placing a liquid blanket of molten boric oxide on top of the semiconductor melt and applying inert gas pressure exceeding the decomposition pressure, melt decomposition and escape of vapors can be prevented in an elegant and simple way, making auxiliary heaters and ampoule enclosures unnecessary (Metz *et al.*, 1962; Bass and Oliver, 1968; Mullin *et al.*, 1965). Now crystal pulling under pressure through the vitreous blanket is possible. At 1500°C, the B_2O_3 evaporates only very slowly, so that crystal growth runs over many hours are possible. The loss of phosphorus is only slight. The boric oxide glass blanket is also beneficial in reducing the temperature loss of the melt surface. Moreover, it dampens oscillations of the melt meniscus caused by external vibrations (building vibrations). Also it acts as a getter for certain impurities. Temperature control at the beginning of the run is very critical. If the melting point temperature is exceeded too far (e.g. by 30°C), the melt starts fuming and becomes metal-rich. This promotes polycrystallinity. Besides, the B_2O_3 blanket becomes cloudy, reducing the visibility for the operator. To reduce this risk, the inert gas pressure should be as high as possible. We used 75 bar for GaP.

Boric oxide[†] is the only practical encapsulant. It must be vacuum-dried for 24 hr at 1200°C; see Bass and Oliver (1968). To be suitable, an encapsulant must "wet" the melt and the crucible walls, otherwise vapors will escape through the annular slot at the pulled crystal or at the crucible walls (Fischer and Pruss, 1969; Fischer, 1963). Yet they must not mix with the melt, a difficult request. Proposed molten salts such as BaF_2 (Mullin *et al.*, 1965) do not wet carbon or BN. They wet quartz, but heavily attack it. However, these molten salts are "pervious" blankets for II–VI compounds because they are solvents for the melt. Since they are water-liquid instead of viscous, they drip off the seed near the interface and hence do not protect the pulled crystal.

III–V melts do not readily mix with molten boric oxide, but II–VI melts readily emulsify with B_2O_3. ZnSe boules can be pulled with difficulties; see Fischer and Pruss (1969) (Fig. 9.27), but the ingot is heavily contaminated with boron. For satisfactory II–VI pulling, better encapsulants have to be found yet.

Our home-built equipment for LE pulling is shown in Figs. 9.28A, 9.28B, and 9.29. The system was run safely in a normal laboratory, with direct control by viewing through a pentaprism.[‡] It features a closed-loop control system, which permits pulling of long needle-shaped crystals (Fig. 9.30), as a proof of its fine constancy. GaP single crystals pulled with

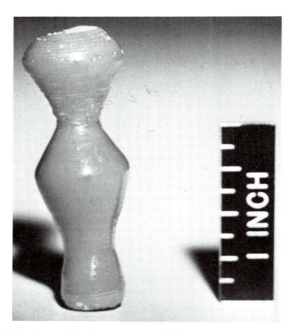

FIG. 9.27. Polycrystalline ZnSe boule pulled from a B_2O_3-encapsulated melt.

[†] Ultrapure B_2O_3, United Minerals & Chemicals Co., New York.
[‡] Built at a total cost of $12,000, including controls shown in Fig. 9.19 and r.f. generator (only 8 kW needed with proper impedance matching).

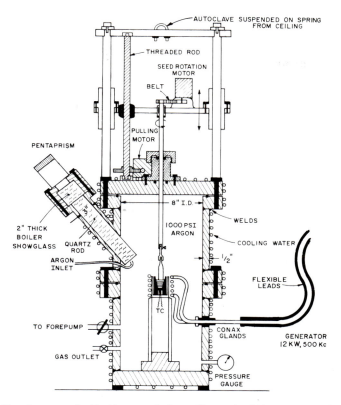

FIG. 9.28A. Apparatus for liquid encapsulation pulling under 75 bar pressure, with direct viewing (1970). The quartz rod window eliminates most of the optical turbulence. Defogging is achieved by magnetically removable glass disks in front of rod (not shown). Vibrations have been minimized by rubber-mounting the motors and by hoisting the whole autoclave to the ceiling on a spring.

the system are shown in Fig. 9.31. Commercially available equipment is shown in Fig. 9.28B.

To minimize loss of visibility due to fogging of the window, the quartz viewing rod (Fig. 9.28A) carried a glass disk at its lower end which could be dropped by a magnetic mechanism if it had become foggy. However, cloudiness in the B_2O_3 blanket was the most disturbing hindrance.

We found that it is important to start with a GaP charge which fills the crucible solidly. If instead one starts with a crucible filled with granular GaP, too much boric oxide is needed to fill the crevices and to cover the tops of the highest chunks to prevent their decomposition. When the GaP finally melts, the resultant boric oxide layer is too thick for convenient pulling, and the visibility is mediocre from the start, due to molten GaP dispersed in the B_2O_3. Once the encapsulant becomes opaque by dispersed GaP, one can see the pulled part of the crystal only minutes after its formation, when it emerges from the encapsulant.

We therefore found it advantageous to synthesize a solid ingot, in the same type of crucible from which the crystal is pulled later on. Demountable, reusable equipment

FIG. 9.28B. Photograph of a commercial high pressure crystal pulling furnace.

Fig. 9.29. Photograph of LE puller, partially disassembled.

FIG. 9.30. GaP needle, pulled to verify constancy of the growth conditions.

FIG. 9.31. GaP single crystals. They were pulled using polycrystalline seeds, and without visibility.

capable of producing sound, stoichiometric GaP ingots of 80 g weight is shown in Fig. 9.32, the ingots prepared with it in Fig. 9.33. These ingots contain single crystal grains of about 1 in. in size. This method is now used everywhere (Gault, 1976).

The crucible arrangement and the accessories for liquid encapsulation pulling of GaP are shown in Fig. 9.34. Note the retractable melting point detector, and the ohmmeter method to sense contact of the seed with melt (not needed after some experience).

The limited visibility of the growth interface through the cloudy B_2O_3 blanket, and the subsequent necessity to grow the crystal with insufficient control, prevented the attainment of the desired cylindrical shape of the crystals. Using an X-ray image of the growing crystal on a screen the operator can instantly perceive any diameter fluctuations and can take corrective action. Even automated crystal growth has become possible in this way (Fig. 9.35) (Roksnoer *et al.*, 1977). Using a seed with the {111} Ga face toward the melt, imperfection-free GaP crystals could be drawn, provided their diameter did not exceed 15 mm.

Another automated method for pulling cylindrical crystals is based on the weight change (Bardsley *et al.*, 1974). Figure 9.36 delineates the method of weighing the crucible during Czochralski pulling under pressure. Several corrections have to be applied to the rate of weight loss, for example the levitation of the crucible by the forces of the r.f. field. The requisite control loop is shown in Fig. 9.37. These are the most sophisticated crystal pulling machines ever built.

9.4.2. LIQUID ENCAPSULATION BRIDGMAN GROWTH

The above-described LE pulling methods, though a great advance over previous methods, require costly equipment. LE Bridgman growth (Fig. 9.38), even though it does

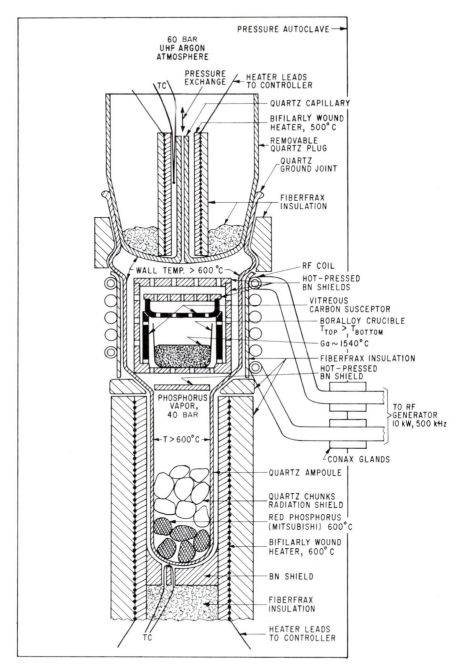

FIG. 9.32. Demountable set up for synthesis of GaP ingots.

FIG. 9.33. Solid GaP ingots, and Boralloy (pyrolytic BN) crucible. If a pointed crucible had been used, single crystals could have been grown.

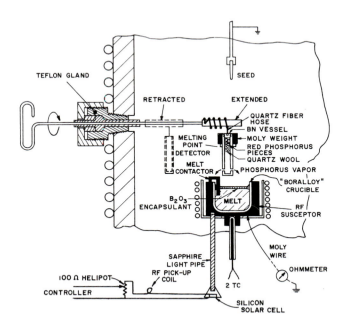

FIG. 9.34. Crucible arrangement and accessories for liquid encapsulation pulling. Note retractable melting point detector and electro-optic temperature sensor, combined with r.f. power signal pickup for control. Initial contact between seed and melt is monitored by an ohmmeter. (Not needed with the X-ray method, or with the weighing method.)

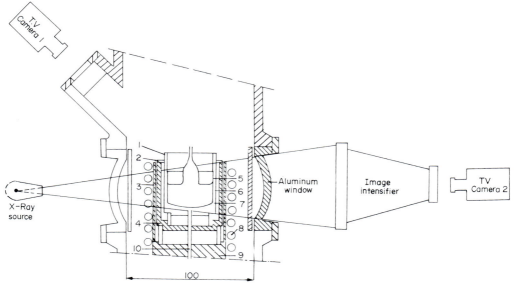

FIG. 9.35. Schematic diagram of the high pressure crystal puller and crucible-melt assembly: (1) afterheater, (2) cover, (3) radiation shield, (4) convection shield, (5) quartz crucible, (6) boron-oxide layer, (7) graphite container, (8) r.f. coil (aluminum), (9) pedestal, (10) thermocouple. (Roksnoer *et al.*, 1977.)

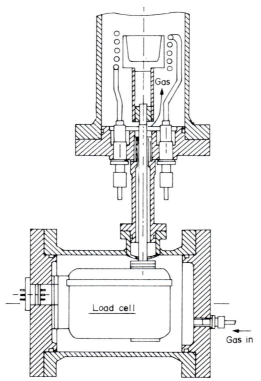

FIG. 9.36. Diagram of crucible weighing assembly for pressure puller. (Bardsley *et al.*, 1974.)

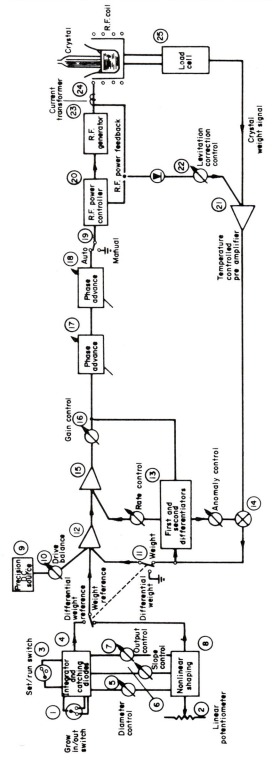

FIG. 9.37. Block schematic diagram of the servo-controller. (Bardsley *et al.*, 1974.)

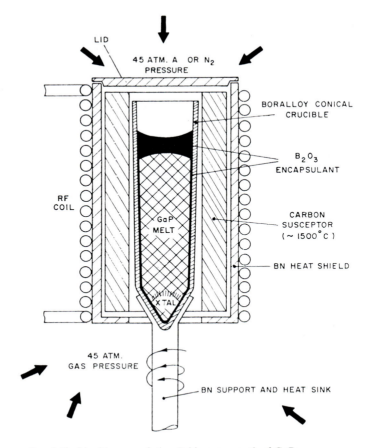

FIG. 9.38. Liquid encapsulation Bridgman growth of GaP.

not easily supply oriented crystals, is more economical. However, compared to the methods described before (see under Vertical Bridgman Growth in Pressure-relieved Ampoules, Figs. 9.16–9.19 and 9.32), it does not permit synthesis and growth in one run, and the GaP crystal cannot be so easily removed from the Boralloy crucible since B_2O_3 pulls off a thin BN skin each time.

9.4.3. LIQUID ENCAPSULATION ZONE LEVELING

The growth of bulk GaP_xAs_{1-x} crystals, and ZnS_xSe_{1-x} crystals, from the melt by pulling is not possible, due to the liquidus–solidus difference of the pseudobinary melt system. Also, horizontal zone leveling was so far impossible, because one could not provide the correct pressurized anion vapor atmosphere mixture. It is suggested that this difficulty can be overcome by covering a long horizontal boat containing the proper alloy with liquid B_2O_3, and by zoning in a pressure autoclave (Fig. 9.39).

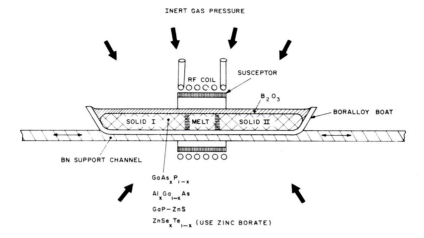

FIG. 9.39. Liquid encapsulation horizontal zone leveling.

9.5. Outlook: New Methods

9.5.1. ISSUING PHOSPHORUS VAPOR INTO THE MELT

During the LE crystal pulling of GaP there is a continuous and noticeable loss of phosphorus through the boric oxide blanket. This leads to gallium-rich conditions, thus limiting the time of the run and the size of the crystal. Preliminary experiments showed that this condition can be cured by injecting phosphorus vapor into the gallium-rich GaP melt. This experience led to the conception of a new unified synthesis and growth method (Fig. 9.40) where the reaction between gallium and phosphorus no longer depends on slow diffusion but on direct injection, resulting in considerable time savings. Preliminary experiments by Fischer (1971) proved that this method is feasible.

9.5.2. CONTINUOUS LIQUID-PHASE EPITAXY GROWTH

During these experiments a new method was discovered which removes the most common flaw of pulled crystals—imperfections caused by strain.† If the gallium melt in the setup of Fig. 9.41 is kept at a temperature below the GaP melting point, the injected phosphorus vapor reacts with the gallium to form clouds of small GaP flakes. They rapidly re-dissolve in the gallium as long as their concentration is low. At colder parts of the flux the solubility is reduced and the GaP reprecipitates. With proper arrangements (Fig. 9.41), well-formed, faceted crystals are obtained which are completely free from imperfections. The growth rate can approach that of melt-growth. Mixed arsenide–phosphide crystals can probably also be grown by this method. A similar method has been found to yield GaP LEDs with high efficiency (Kaneko *et al.*, 1973).

---

† Roksnoer found a way to pull strain-free GaP crystals with maximum 15 mm diameter.

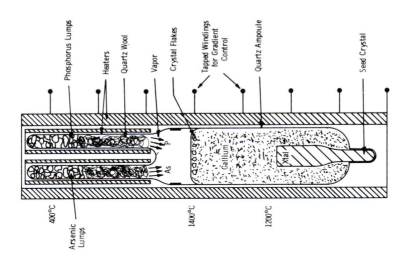

FIG. 9.41. Continuous liquid-phase epitaxial growth of crystals.

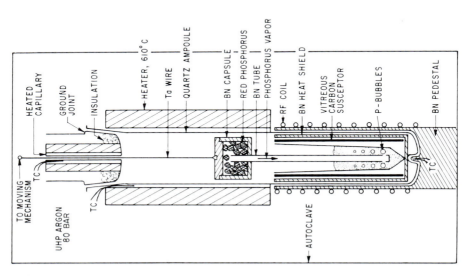

FIG. 9.40. Novel, accelerated GaP synthesis and crystallization method.

9.6. Summary and Conclusions

Melt-growth of crystals under pressure from decomposable materials has come of age, especially for the important wide-gap III–V and II–VI materials. A synopsis of available methods reveals that at present liquid encapsulation pulling is the best method for GaP. Ultimately, unified synthesis and Bridgman growth methods may prove more economical, and continuous liquid-phase epitaxy growth may be preferable since it can yield imperfection-free crystals and has potential for alloy crystals of the III–V compounds.

Liquid-encapsulation pulling of II–VI materials is difficult and must wait for a more suitable encapsulant than B_2O_3. For the II–VI materials, the various Bridgman methods can be employed. Pulling under metal vapor pressure is possible.

For all these methods, the requisite technologies and most suitable methods have been investigated, evaluated, and described.

References

ADDAMIANO, A. and DELL, P. A. (1957) *J. Phys. Chem.* **61**, 1020, 1253.
BARDSLEY, W., COCKAYNE, B., GREEN, G. W., HURLE, D. T. J., JOYCE, G. C., ROSLINGTON, J. M., TUFTON, P. J., and WEBBER, H. C. (1974) *J. Cryst. Growth* **24-25**, 369.
BASS, S. J. and OLIVER, P. E. (1968) *J. Cryst. Growth* **3**, 286.
BLUM, S. E. and CHICOTCA, R. J. (1968) *J. Electrochem. Soc.* **115**, 298.
BLUM, S. E. and CHICOTKA, R. J. (1973) *J. Electrochem. Soc.* **120**, 588.
CARIDES, J. N. and FISCHER, A. G. (1964) *Solid State Comm.* **2**, 217.
FAHRIG, R. H., WEBB, G. N., and HART, R. L. (1972) *Second National Conference on Crystal Growth, Princeton, NJ.*
FISCHER, A. G. (1958) *Z. Naturforsch.* **13**a, 105.
FISCHER, A. G. (1959) *J. Electrochem. Soc.* **106**, 878.
FISCHER, A. G. (1961) *Bull. Am. Phys. Soc.* **116**, 17.
FISCHER, A. G. (1961) Report No. 1, AFCRL 360, Contract AF19 (604) 8018.
FISCHER, A. G. (1961) Reports No. 2 and 3, AFCRL 721 and 979, Contract AF19 (604) 8018.
FISCHER, A. G. (1962) Report No. 2, AFCRL–62–588, Contract AF19 (628) 3866.
FISCHER, A. G. (1963) *Electrochem. Soc. Meeting, Pittsburgh,* Enlarged Abstract No. 53.
FISCHER, A. G., Final Report, AFCRL–63–526, Contract AF19 (604) 8018.
FISCHER, A. G. (1966) Final Report, AFCRL–67–005, Contract No. AF19 (628) 3866.
FISCHER, A. G. (1971) *Internat. Crystal Growth Conf. Marseille.*
FISCHER, A. G., US Pat. 3,033,659 (1959).
FISCHER, A. G., US Pat. 3,117,336.
FISCHER, A. G. and PRUSS, V. T. (1969) *ACCG Conference, Gaithersburg.*
FROSCH, C. J. and DERICK, L. (1961) *J. Electrochem. Soc.* **18**, 251.
GAULT, W. A. (1976) *Electrochem. Soc. Meeting Las Vegas,* Enlarged Abstract No. 344.
GREMMELMEIER, R. (1956) *Z. Naturforsch.* **11**a, 511.
HOLTON, W. C., WATTS, R. K., and STINEDURF, R. D. (1969) *J. Cryst. Growth,* **6**, 97.
KANEKO, K., AYABE, M., DOSEN, M., MORIZANE, K., USOI, S., and WATANABE, N. (1973) *Proc. IEEE* **61**, 884.
MEDCALF, W. E. and FAHRIG, R. H. (1958) *J. Electrochem. Soc.* **105**, 719.
METZ, E. P. A., MILLER, R. C., and MAZELSKY, R. (1962) *J. Appl. Phys.* **33**, 2016.
MULLIN, J. B., STRAUGHAN, B. W., and BRICKELL, W. S. (1965) *J. Phys. Chem. Solids* **26**, 782.
RICHMAN, D. (1963) *J. Phys. Chem. Solids* **24**, 1131.
ROKSNOER, P. J., HUIJBREGTS, J. M. P. L., VAN DE WIJGERT, W. M., and DE KOCK, A. J. R. (1977) *J. Cryst. Growth* **40**, 6.
TIEDE, E. and SCHLEEDE, A. (1921) *Ber. dt. Chem. Ges.* **53B**, 1717.
WANG, C. C., CARDONA, M., and FISCHER, A. G. (1964) *RCA Rev.* **25**, 159.
WEISBERG, L. R. *et al.,* AFCRL–62–540, Contract AF19 (604) 6152.
WOODBURY, H. H. (1976) *J. Cryst. Growth* **35**, 49.

CHAPTER 10

Crystallization from Solution at Low Temperatures

R. M. HOOPER,† B. J. McARDLE, R. S. NARANG, and J. N. SHERWOOD

*Department of Pure and Applied Chemistry, University of Strathclyde, Glasgow
G1 1XL, Scotland*

10.1. Introduction

The primary objective of this chapter is to present an experimentalist faced with the problem of producing a crystal by growth from solution at low temperatures with the basic experimental criteria which the authors believe are essential for the preparation of large, perfect, mono-crystals. We shall deal principally with the apparatus and conditions for the growth of materials of moderate to high solubility in the temperature range ambient to 353 K and at atmospheric pressure. We shall not consider the growth of crystals from low-solubility systems or under higher pressures. This involves specialized techniques details of which are given elsewhere.[1,2]

The principal advantages of crystal growth from solution in this temperature range are the proximity to ambient temperature and, consequently, the degree of control which can be exercised over the growth conditions. Temperatures can be readily stabilized to 0.01–0.001 K. As a result, supersaturations can be accurately and precisely controlled. These factors, coupled with the ease of efficient agitation of the growing crystal and solution, reduce fluctuations of all kinds to a minimum. The proximity to ambient temperature reduces the possibility of major thermal shock to the crystal both during growth and on removal from the apparatus. Rather more important, however, is that the growth of the crystal under closely controlled equilibrium conditions and usually at temperatures far removed from the melting point results in the reduction of both equilibrium and non-equilibrium defects to a minimum and in many cases to virtually zero. The method is particularly suited to those materials which suffer from decomposition in the melt or in the solid at high temperatures and which undergo phase transformations above the present working range. There are numerous organic and inorganic materials which fall within these categories. It also permits the preparation of a variety of different morphologies and polymorphic forms of the same substance by variation of growth conditions or of solvent.

Against these many advantages one must cite the major disadvantages which are the possibility of solvent inclusion and, in many cases, the slow rate of production of the

---

† Present address: Dept. of Engineering Science, University, Exeter EX4 4QD.

crystal. The former can be minimized by growth under controlled conditions. The latter is unavoidable but the high quality of the product may well compensate for this.

10.2. Basic Requirements

As for all other methods of crystal growth, the fundamental requirement is the availability of pure materials. In solution growth it is more easy to succeed when attempting to grow crystals from relatively impure materials than is the case with other growth methods. The product, although visually acceptable is, however, of ill-defined quality. With reasonable, careful, pre-purification such as crystallization, sublimation, and zone refining of the solute and similar treatment, coupled with fractional and spinning-band distillation, of the solvent, materials of high and well-defined purity can be obtained. With the availability of such materials the prime essential is the choice of a suitable solvent.

10.2.1. CHOICE OF SOLVENT

The ideal solvent should:

(1) Yield a prismatic habit in the crystal and also have the following characteristics.
(2) High solute solubility.
(3) High, positive, temperature coefficient of solute solubility.
(4) Low volatility.
(5) Density less than that of the bulk solute.
(6) Low viscosity.

The last two characteristics simplify apparatus design since it is desirable that the growing crystal should not float and that it should be well agitated.

The use of solvents of low volatility reduces the possibility of uncontrolled loss of solvent during lengthy growth periods. This is particularly important where supersaturation is achieved by methods other than solvent evaporation.

The first three properties are the most important characteristics. These require some further qualification and definition.

10.2.1.1. *Crystal Habit*

The most useful crystals are those which grow at approximately equivalent rates in all dimensions. Growth of this type usually results in a large bulk of material from which can be cut samples of any desired orientation. Furthermore, as will be seen later, where dislocations and other defects propagate, they do so from the nucleus or seed along specific directions into the bulk of the crystal. If the crystal grows with a bulky habit these imperfections usually become isolated into defective regions surrounded by large volumes of high perfection. In needle-like or plate-like crystals the growth dislocations follow the principal growth directions and the crystal remains dominantly imperfect. The situation is obviously worse for the needle-like specimens where the defects continuously propagate into the whole of the crystal bulk.

Ionic crystals are more often than not prepared from aqueous solution because of their lower solubility in organic solvents. Consequently, variations in habit due to solvent variation are uncommon. Molecular materials show a more diverse behaviour. Table 10.1 gives some instances of the variations which can be found. Unfortunately, it is not possible to formulate rules for solvent selection. It is a matter of experimentation to find the most suitable liquid. Where success cannot be achieved with single solvents there is no reason why mixed solvents should not be used, thus combining desirable features of each.

10.2.1.2. *Solubility*

Having selected solvents which produce a suitable habit, the next problem is to determine whether or not they will yield a satisfactory rate of growth. The principal experimental factors involved in this decision are the solubility of the material in the solvent and its temperature dependence. The former governs the amount of material which is available for growth and hence defines the total size limit. Both factors define the supersaturation which, as is well established, is the driving force which governs the rate of crystal growth.

The supersaturation δ may be defined as

$$\delta = \Delta C/C_0,$$

TABLE 10.1. *Influence of Solvent on Crystal Habit*

Material	Solvent	Habit
Fluoranthene	Xylene	Needles
	Toluene	Needles
	Benzene	Thick plates
	Ethanol	Thin plates
Anthracene	Toluene	Thin platelets
	Benzene	Thin platelets
	Acetone	Small prisms
	Methylethylketone (MEK)	Thin plates
	MEK/benzene	Thick plates
Fluorenone	Benzene	Needles
	Alcohol	Small prisms
Benzil	Xylene	Prisms
	Ethanol	Needles
Pentaerythritol	Water	Prisms
	Acetone	Plates
Methyl salicylate	Water	Needles
	Benzene	Needles
	Acetone	Plates
Urea	Water	Needles
	Methanol	Needles
	Methanol/water	Prisms
Phlorogucinol dihydrate	Ethanol	Plates
	Water	Needles
Oxalic acid	Acetone/H_2O	Prismatic
	Water	Plates

where C_0 is the equilibrium concentration of solute at the temperature of growth and ΔC the increment by which the true concentration exceeds this.

The solution can be supersaturated in a number of ways. The most commonly used method is to lower the temperature of the solution below the equilibrium saturation temperature. If this is done continuously at a controlled rate (*temperature lowering method*), ΔC is determined by the rate of lowering of the temperature. Linear variations will yield constant values of ΔC over small temperature intervals provided that the solubility–temperature curve is not changing slope too rapidly. Even if this is not the case, it is a relatively easy task to match the temperature lowering rate to the shape of the solubility curve and hence to achieve a constant supersaturation over a wider temperature range. This is probably the most versatile and easy method to operate.

If the temperature is suddenly lowered, ΔC is the difference between the two equilibrium values. ΔC, and consequently the rate of growth, will decrease as growth progresses. This is not desirable and can be overcome by constantly replenishing the solution saturated at the upper temperature as growth proceeds (*constant temperature differential method*). The ways in which this can be achieved will be discussed later. Although more complicated than the previous method it can be more appropriate under some circumstances and it does provide for growth under rigidly constant temperature and supersaturation conditions.

When neither of the above methods is appropriate an alternative method is to allow the slow evaporation of the solvent at constant temperature (*solvent evaporation technique*). If this can be controlled so that the evaporation rate is constant, then again a constant supersaturation can be maintained. In practice, this is more difficult to achieve than is a controlled temperature change. The inevitable fluctuations can lead to poor quality growth.

The choice of method of supersaturation is, to a large extent, dependent on the shape of the solubility curve and the magnitude of the solubility. Consequently, a necessary prerequisite is the availability of accurate solubility–temperature data. If this is not readily available in the literature it can be obtained by analysing solutions carefully saturated at a series of temperatures using any convenient analytical technique, e.g. chemical analysis, density determinations, radiochemical analysis, etc. It is preferable when doing this to saturate the solution at a temperature higher than that for which the information is required and then to cool the solution with vigorous stirring to the equilibrium temperature. Under these conditions the solute will crystallize from solution thus confirming that equilibrium is maintained. Dissolution can take a surprisingly long time and attempts to saturate solutions by dissolving solid at a fixed temperature can yield misleading results.

Figure 10.1 depicts a general and probably quite untypical solubility curve in which the material shows a rapid change of solubility with temperature. Each region (A, B, C, and D) may, however, be typical of the behaviour of any particular solvent in the temperature range which we are considering. Let us assume that region B, which we would define as being of moderate to high solubility and moderate solubility–temperature gradient, is a satisfactory region in which to work with the temperature lowering method. To define this point of commencement we must rely on our past experience. Table 10.2 presents the solubility data for a wide range of materials which we have successfully grown in our laboratory in sizes in the range from 1 to 100 cm^3. We find that, provided the solubility is in the range 200–1000 g solute per 1000 g solvent and that the ratio of

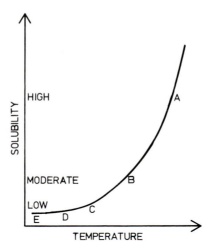

FIG. 10.1. Generalized solubility curve.

solubility–temperature gradient to solubility (solubility ratio) lies in the range 0.03–0.01, then excellent crystals can be grown at temperature lowering rates of 0.5–1 degree per day, i.e. at supersaturations $\sim 2\%$. This implies a certain temperature precision (± 0.005 K) to prevent bursts of rapid and uncontrolled growth which lead to imperfections in the resulting crystal.

When the solubility and solubility gradient exceed this range, e.g. at A, the demands on the precision of the system become too great. Minor fluctuations in temperature yield large fluctuations in solubility, supersaturation, and growth rate. An increase in the precision of the temperature of the apparatus (± 0.001 K) can go some way towards reducing these variations for systems at the lower end of the A range. This involves working on the limit of operability of the system, however, and it is more advantageous to seek a solvent with better characteristics. Where this is not feasible, this precision can be achieved by resorting to the constant temperature differential method where the temperatures of both source and growing crystal are maintained at precisely controlled constant values.

It is obviously not desirable to operate in a region of rapidly changing solubility gradient such as at C (Fig. 10.1). Constant growth conditions are difficult to define for any but short periods of time.

In region D the gradient is shallow and the ΔC attainable by either the temperature lowering or constant temperature differential techniques is significantly reduced. Consequently, growth rates will be low and minor fluctuations would again interfere considerably with growth parameters. Of these two methods the latter could be more suitable where the solubility remains high. Usually, however, it is necessary to resort to solvent evaporation to yield reasonable supersaturations.

The solvent evaporation method is at its best where the overall solubility is low. It is notoriously difficult to control solvent evaporation rates. At low solubilities the inevitable fluctuations yield only small fluctuations in supersaturation and better results are ensured. As the solubility increases, the evaporation rate must be more precisely controlled. Control can also be a problem with the more volatile solvents. Although one

TABLE 10.2. *Solubility (S) and Solubility–Temperature Gradient (dS/dT) for Organic and Inorganic Materials*

Materials	Solvent	Solubility (g per 1000 g solvent (S) at 303 K)	(dS/dT) (S K^{-1})	(dS/dT)/S (K^{-1})	Temperature for Good Growth
Easy to grow					
Sulphur	Carbon disulphide	80	7.6	0.011	35–25
Sodium chlorate	Water	1100	14.7	0.013	45–25
Sodium bromate	Water	425	5.6	0.013	45–25
Lead nitrate	Water	634	9.3	0.015	45–25
Glutamic acid hydrochloride	Water	490	8.0	0.016	45–25
Strontium formate	Water	222.5	3.5	0.016	45–25
Ammonium perchlorate	Water	307 (310 K)	5.3	0.017	45–25
Pentaerythritol tetranitrate	Ethyl acetate	236 (323 K)	4.4	0.019	50–25
Pentaerythritol tetranitrate	Acetate	588 (323 K)	16.9	0.029	50–25
Ammonium dihydrogen phosphate	Water	285.4	5.9	0.021	45–25
Pyrene	Toluene	227	5.25	0.023	45–25
Benzophenone	Acetone or methanol	300	8.0	0.027	35–25
Potash alum	Water	180	5.3	0.029	45–25
Fluoranthene	Benzene	487	15.8	0.032	40–25
Oxalic acid	Water	442	15.0	0.034	40–25
Trinitrotoluene	Benzene	1130	40.0	0.035	35–25
Difficult to grow					
Sodium chloride	Water	360.4	0.22 (<310 K) 0.41 (>310 K)	0.0006 0.001	By evaporation
Hexamethylene-tetramine	Water	1295	−52.6		By evaporation
Anthracene	Methylethyl-ketone	15	0.8	0.053	45–25
Benzil	Ethanol	210	19	0.09	
Calcium carbonate	Water	53.2 × 10^{-3}	−85 × 10^{-3}	−0.016	By diffusion
Silver iodide	Water	48 × 10^{-4}	1.75 × 10^{-4}	0.036	By diffusion
Growth improved by varying conditions					
Urea	Water	1335	12.5	0.01	
	Methanol	70.5	2	0.03	
	Methanol/water	1056	18	0.02	
Pentaerythritol	Water	82 (323 K)	3.9	0.05	
	Water	650 (353 K)	15	0.02	
Adipic acid	Water	100	7.1	0.07	
	Water/acetone	196	8.4	0.04	

might assume that the increased evaporation rates and hence supersaturations which could be attained would yield rapid growth, the superimposed increased fluctuations can be self-defeating. It is easier to control the forced evaporation of a less-volatile solvent.

Using the above general guidelines it is possible to select a solvent and growth method with which to produce a crystal of large dimensions within a reasonable period. It must be stressed, however, that although the methods described are most suited to the solubility conditions noted, they are not exclusive. Even where conditions are less favourable to a particular method, crystals of adequate quality can still be grown using this method. The periods required may be more lengthy than those required for growth by the ideal method, the resulting crystals may be smaller and more stringent control may be necessary, but successful growth can still be achieved.

One region which is exclusive, however, and which provides an excellent example of what can be achieved in extreme situations with sufficient control, is that in which the solute solubility is extremely low. In this region, excessively high supersaturations develop readily and microcrystalline precipitates form. The above methods cannot be employed and novel methods which allow the precisely controlled, slow development of supersaturation must be used. In these, the supersaturation is achieved either by the slow interdiffusion of solutions of two reacting species, which on mixing react to form the solute, or by the interdiffusion of a solution with a solvent in which the solute is insoluble or less soluble. The necessary precise control over the supersaturation is exercised by controlling the flux of the interdiffusing pair, i.e. by varying temperature, concentration gradient, and solvent viscosity. This is the basic principle of *gel growth techniques*[2] for the growth of crystals of virtually insoluble salts and *diffusion growth* for the production of large crystals of charge-transfer complexes such as TTF/TCNQ.[3] With these methods and precise control, small but highly perfect crystals of relatively insoluble materials can be grown.

The preceding discussion has centred around those materials which have positive temperature dependences of solubility. The few materials which have retrograde solubilities can also be grown. Under these circumstances the supersaturation must be achieved by solvent evaporation or constant (negative) temperature differential close to room temperature. Constant temperature increase would be feasible but it is not desirable to complete the growth of a crystal at high temperatures if this can be avoided.

Before leaving this section and as a final guide to particular situations, we should like to return to a brief but more detailed consideration of Table 10.2.

The upper part of the table gives examples of materials which can be relatively easily grown to yield crystals of large size by the temperature lowering method (or constant temperature differential method) and at temperature lowering rates in the range $0.1–1 \, K \, hr^{-1}$ provided that the apparatus has a temperature stability of better than $\pm 0.005 \, K$ over long periods of time.

The remaining materials are much more difficult to grow. Thus the quality of crystals of anthracene, urea (water and methanol), pentaerythritol, and hexamethylenetetramine (ethanol) is much worse than that of the previous materials when they are grown at similar rates close to ambient temperatures.

With increased temperature stability ($\pm 0.001 \, K$) and slower temperature lowering rates, small but adequate crystals of anthracene and hexamethylenetetramine (ethanol) can be produced. Better results can be achieved by solvent evaporation.

Pentaerythritol provides an excellent example of a substance the solubility of which changes rapidly with temperature. At low temperatures (323 K) the solubility is low and the solubility ratio (0.05) is well above the range predicted to yield adequate growth by temperature lowering. The solubility increases with temperature, however, until, in the range 353–368 K we have a solubility ratio within the desired range. Satisfactory growth can now be effected with greater ease by the temperature lowering method.[4] The high temperature of the final crystal presents some problems, but with careful post-growth cooling good specimens can be obtained.

Urea presents problems due to its very high solubility and low solubility ratio in aqueous solutions. Growth by temperature lowering is difficult to control even with high temperature stability. Parallel problems exist with methanol solutions due to the significantly decreased solubility. Addition of a small quantity of methanol to the aqueous solution depresses the solubility without significantly influencing the solubility gradient and the solubility ratio falls to the manageable region. From this mixed solvent, large, prismatic crystals can be produced.[4] Parallel effects have been noted for adipic acid when grown from aqueous and aqueous acetone solutions.[5]

For materials such as dipotassium hydrogen phosphate (L), sodium chloride (L), hexamethylenetetramine (aqueous solution) (N), lithium sulphate monohydrate (N), and lithium iodate (N), where the solubility gradient is low (L) or negative (N), growth by solvent evaporation (L and N) or constant (negative) temperature differential (N) is essential. Similarly, these methods are also necessary where a required phase of a material has a limited temperature range of existence, e.g. dipotassium hydrogen phosphate.[6]

The constant temperature differential technique can only be operated efficiently where the solubility gradient, whether positive or negative, is reasonably high.

The lower limit of the solvent evaporation technique is difficult to place. We would suggest that it lies around solubilities of 20 g per 1000 g solvent. Between this point and that corresponding to the solubilities of those materials which can be grown satisfactorily by the diffusion methods, e.g. calcium carbonate, silver iodide, and TTF/TCNQ, lies a region of solubilities where growth is difficult. The experimentalist requiring crystals of such materials has often to be satisfied with specimens of small size and poor quality. There are many materials with potentially interesting chemical and physical properties which have solubilities in this region, e.g. dye-stuffs, biological materials, and organic charge-transfer complexes. Consequently the development of techniques for growing large, perfect crystals of these materials could be very rewarding.

10.3. Crystallization Apparatus

It will be gathered from the previous section that the most generally useful method of growth for the production of large monocrystals from solution is the temperature lowering method. If an apparatus using this technique could be constructed in such a way that it could readily be adopted for growth by solvent evaporation, then it would have almost universal applicability to all materials of reasonable solubility. We have developed such a system which we and others have used with considerable success over a period of years for the production of crystals of a large range of organic and inorganic materials. We do not claim it to be novel. That would be difficult with the numerous previous designs which are available and on which we have drawn in the design of the present system. It does contain a number of features which together contribute to the easy production of high-quality crystals. We believe that there is more to be gained from a detailed description of this apparatus than a general survey of all other systems. Some alternative methods to accommodate the extreme situations where this type of apparatus cannot be used will also be described.

10.3.1. A GENERAL PURPOSE LABORATORY CRYSTALLIZER

The scientific literature contains numerous reports on apparatus designed for the production of large crystals of organic and inorganic substances from saturated solutions at low temperatures.[7] The variations in design of the apparatus described arise for a number of reasons. Some of these relate to general experimental factors, e.g. the final size of the crystal required, nature of the solvent phase, etc. Others reflect certain specific criteria which are fundamental to the growth of crystals of good quality. These are:

(a) Thermostatic control to better than 0.01 K (preferably 0.005–0.001 K).
(b) Efficient stirring of the solution to prevent layering and spurious nucleation.
(c) Efficient reciprocated stirring of the crystal to prevent local supersaturations (or under-saturations) which lead to sudden increases (or decreases) in growth rate and hence to solvent incorporation and "veiling".
(d) Controlled supersaturation of the solution.

Consideration of these factors has led to the development of the apparatus shown diagrammatically in Fig. 10.2. A complete unit with control devices, etc., is illustrated in Fig. 10.3. Detail of the control box and the associated electrical circuit is given in Fig. 10.4.

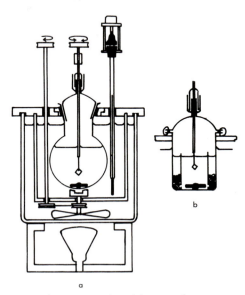

FIG. 10.2. (a) Diagrammatic representation of the general purpose crystallizer in the form used for growth by the temperature lowering technique. (b) The vessel used for growth by solvent evaporation using Forno's technique.[9]

FIG. 10.3. Photograph of the complete crystal growth unit showing support frame and central control box. The latter contains all controls for reciprocated stirring of the thermostat bath and crystal and for the thermostatic control of the bath.

Referring to the points enumerated above, the operation of the system is as follows:

(a) The solution is contained in the sealed central flask. A greaseless vacuum seal (West-glass Corporation, El Monte, California) and attached "ground-in" or mercury seal stirrer (Quickfit and Quartz, ST10/3) prevent the loss of volatile organic solvents and hence uncontrolled evaporation. The efficiency of the seal can be judged from the fact that there is no loss of highly volatile solvents, e.g. carbon disulphide, over periods of several months.

The solution is heated by the external 20 l bath which contains oil or water depending on the temperatures employed. The thermostat bath is heated with the infra-red lamp at

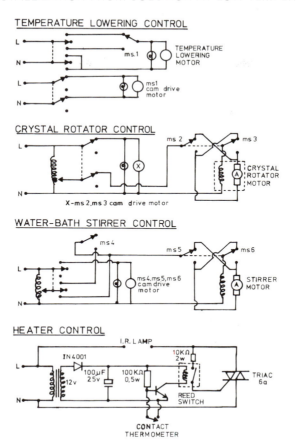

Fig. 10.4. Detailed photograph and circuit diagram of the control unit. Linear temperature lowering rates are varied by adjusting the on/off time of the two microswitches with the adjustable cams. Provision of a series of synchronous motors of speeds in the range 1 rev/min to 1 rev/day attached to the contact thermometers gives a wide range of precise temperature lowering rates.

the base of the unit and control is effected by the long-range contact thermometer. The bath liquid acts as an excellent absorber and thus prevents any radiation damage to the growing crystal by the infra-red lamp. The large bulk of the bath rules out rapid temperature fluctuations and the control can be maintained well within the limits specified above.

(b) and (c). The internal solution is stirred by paddles attached to the seed holder or by an external magnet on a yoke driving an internal sealed stirrer resting on a platform at the bottom of the flask. Both the seed holder and the magnetic stirrer can be rotated at speeds in the range 30–60 rev/min. The motion can be reversed as desired (usually every 30–60 s) to ensure continuous, efficient agitation of the solution.

(d) The supersaturation of the solution is achieved by driving the contact of the long-range contact thermometer to lower temperatures using a synchronous or stepping motor attached to the magnetic setting device. The rate of temperature lowering can be varied by a cam-operated time-sharing device (Fig. 10.4), which switches the contact thermometer drive motor on for varying periods of time depending on the disposition of the cams relative to a microswitch. In its simplest form, and coupled with 1 rev/min and 1 rev/day motors, the device provides a variable, linear, temperature lowering rate in the range 0.1–1 degree/day. This is quite satisfactory for the preparation of crystals of most moderate to high solubility materials since the solubility–temperature profile is sufficiently linear over the small temperature ranges used. Non-linear temperature lowering rates, which allow for curvature of the solubility–temperature relationship over wider temperature intervals and for the increase in size of the growing crystal, can be achieved simply by programming the speed of the motors driving the cam system.

This type of system is versatile, efficient, and easy to operate. It has been used successfully to grow specimens of a wide variety of organic and inorganic materials. The transition to a large-scale apparatus can be done by extrapolation. The basic essentials of control, agitation, etc., are the same, and the prime difference is the larger volume. There are numerous examples in the literature of larger systems for the production of large crystals or larger numbers of crystals.[7]

The crystallizer operates best as a temperature-lowering system. Temperature and supersaturation can be precisely controlled over a wide range. This is to be recommended as the best method, provided that the solubility and solubility gradient are not too low (or the latter negative) and that the phase behaviour of the material is such that growth can be effected over a reasonably long temperature range. Where this is not the case, one must resort to alternative methods of supersaturating the solution. The simplest alternative is to use slow evaporation.

Slow evaporation can be most easily achieved by using a controlled flow gas inlet/outlet attached to the head of the flask and a trap to collect the condensing liquid. The solution is maintained at a temperature higher than ambient, thus exercising one basic control on the evaporation rate. The rate of removal of vapour by the gas stream is the second. This type of system has been used for the growth of hexamethylenetetramine with its retrograde solubility in water and sodium chloride with its virtually zero solubility gradient. The results are adequate but the rate of supersaturation cannot be controlled as well as with temperature lowering. Consequently the perfection of the crystals suffers. Some workers have overcome this problem for systems involving growth from aqueous solutions by resorting to electrolytic decomposition of the slightly acidified solution.[8] This yields very precise control over supersaturation. There are problems, however, with self-heating of the solution. This can interfere with temperature stability and solution saturation, particularly in a low volume crystallizer of the present type. It is less of a problem where larger volumes are concerned and the method has been shown to be particularly successful for KD_2PO_4[6] growth, where one is limited to growth temperatures in the small range ambient to 303 K due to a phase transformation in the solid.

Where the solute solubility is low there is a limit to the size of crystal which can be produced by all of the above methods. This can be overcome in some cases by the constant replenishment of solute during the process. Forno[9] has described an excellent apparatus for replenishment coupled with supersaturation by evaporation which we have successfully adapted (Fig. 10.2b) for use in the crystallizer described above. Due to the temperature differential between the bath and the cooler ambient, evaporation takes place. The vapour condenses on the bell of the system, flows down the walls, and becomes resaturated as it drains through the bed of solid solute in the outer compartment at the base. This resaturated solution then flows through small holes into the growth chamber. Using this method Forno achieved quite rapid growth of 10 cm diameter crystals of hexamethylenetetramine—much larger than those produced previously.

Similarly, replenishment coupled with supersaturation by constant temperature differential can be used. The rearrangement of the basic apparatus used for this purpose is depicted diagrammatically in Fig. 10.5. The solution is saturated at a temperature T_1 by equilibrating it with suspended solid in the right hand of the two flasks (A). The saturated solution is then pumped by the peristaltic pump P, via frits F (to remove suspended solid), and a pre-cooler C, into the growth bath B at temperature T_2 ($< T_1$). The spent liquid is then returned to A for resaturation and recycling. This arrangement has the advantage that the supersaturation can be precisely established and growth again effected at constant temperature. This reduces the possible generation of thermal strains in the crystal during growth. The method has been used as the basis of at least one commercial crystallization system.[10]

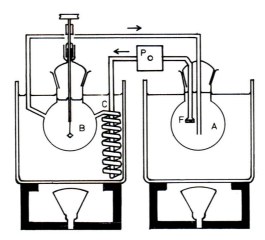

Fig. 10.5. The arrangement of the basic system used for growth by the constant temperature differential method. The stirring and temperature control systems for both baths have been omitted for clarity. These occupy much the same position as shown in Fig. 10.2.

Recycling systems are particularly useful in the extreme where only small quantities of low solubility materials are available. The volume of such systems can be reduced to a minimum and supersaturation can be kept low. Simple miniature apparatus of this type has proved useful for the preparation of small crystals for X-ray crystallography or of charge transfer complexes and dyes for electrical measurements.

10.4. Saturation and Seeding

Having decided on the best method of supersaturation and the arrangement of apparatus to be used, the next problems are associated with the preparation of the saturated solution and the seeding of this to produce a suitable crystal.

10.4.1. SATURATION

The principal objective in saturation is to produce a solution in equilibrium at a set temperature with no spurious nuclei in the solution. There are many ways described in the literature of achieving this state. We find that in most cases the direct method is satisfactory.

The solution is saturated with the solute at a temperature slightly higher than that to be used initially in the experiment. This is usually carried out in a round-bottomed flask equipped with an efficient stirrer. The solution is then decanted from the excess solid through a heated sintered glass funnel into the growth flask. The process may be speeded up if necessary by application of pressure. The flask is then sealed, the solution stirred vigorously, and the temperature increased until the solid, which has precipitated from the cooled solution, has redissolved and the solution is at equilibrium. This process can take a considerable time (days). That the equilibrium has been attained can be tested by suspending a piece of crystal in the solution. If the system is not at equilibrium the crystal will dissolve or solute will crystallize on the seed. As a result, the density of the solution in the vicinity of the seed changes and rising or falling density flow patterns can be observed. Final adjustments of the temperature can then be made.

Once the final equilibrium point has been reached, the seed is withdrawn and the temperature of the bath raised several degrees to dissolve away any remaining spurious nuclei. The temperature is then reduced to the equilibrium temperature, the seed inserted, and growth commenced. This procedure is most directly applicable when subsequent growth is to be carried out by temperature lowering or solvent evaporation. Obvious minor variations are necessary for the recycling methods of supersaturation. The basic requirements are still the same.

10.4.2. SEED SELECTION AND MOUNTING

Vapour and melt-grown crystals suffer from mechanical damage during growth or removal from the growth tube or substrate. This damage can lead to multi-crystalline growth when such crystals are used as seeds for solution growth (see below).

Seed crystals are best prepared by slow cooling or slow evaporation of a saturated solution in a clean, controlled temperature, enclosure kept specifically for this purpose.

Alternatively, an excellent source is the base of the flask in the main crystallizer. Small crystals inevitably form here during the growth of the larger specimens. They are often of excellent perfection.

However prepared, these crystals should be suspended on a smooth support. The use of a rough support provides many centres on which further nucleation can occur. Growth of these additional nuclei can often interfere with the growth of the original crystal. In some cases this effect can be used to advantage, and the suspension of a smoother thread (e.g. nylon line) in the solution can yield a few seed crystals of excellent quality firmly attached to the line. The crystals, still attached to the thread, can be separated and used as seeds for further growth.

Alternative methods of suspension are to tie the crystal with wire, to attach it to the stirring rod by pushing it into a piece of flexible tubing attached to the rod, to drill the crystal and attach it by a glass filament passed through the hole or to glue the crystal directly to the glass support. Of these variations the last two are to be preferred. As will be seen later, in order to prepare crystals of very high quality it is essential to dissolve away the deformed surface layers of the seed from which defects can propagate into the bulk of

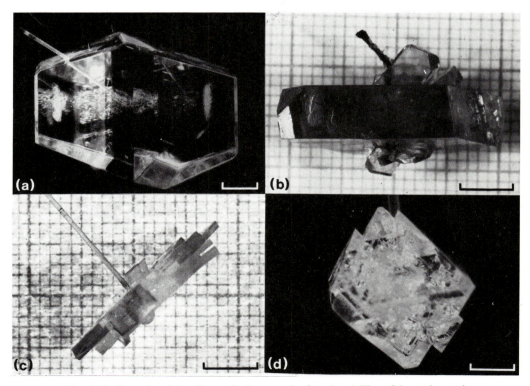

FIG. 10.6. Examples of the effect on further growth of careless drilling of the seed crystal prior to attachment of a suspension: (a) sulphur, being brittle, is unaffected; (b) pyrene, the damage is localized around the hole and could be removed by slight dissolution; (c) stearic acid and (d) adamantane both suffer considerably due to their high plasticity and multi-crystalline growth results. (Scale = 1 cm.)

the subsequently grown large crystal. Consequently, it is advantageous to have the seed crystal fixed firmly to the support. The method chosen will depend upon the individual system to be studied. Obviously, the glue must be insoluble in the solvent. This places some restrictions on the last-mentioned method and makes the drilling method the more generally applicable. There are problems, however, in that the drilling, which is usually done with a hand drill and solvent, can cause some deformation of the seed. For the more brittle crystals this damage is usually restricted to the region of the hole. Some of the more plastic materials undergo an extensive recrystallization which can lead to multi-crystalline growth.

The difference can be seen from Fig. 10.6a–d. Figure 10.6a shows a sulphur crystal grown directly on to a drilled seed. This solid is highly brittle and has few active dislocation slip systems.[11] Thus the damage following drilling is minimal and perfect growth occurs to yield a crystal of the same shape as the seed. Pyrene is an intermediate example. The drilling causes more extensive damage which, unless removed by dissolution, results in the growth of twins from the region around the hole (Fig. 10.6b). The problem becomes more marked with the monoclinic long-chain hydrocarbons, where slip in the basal plane is facile. This leads to polygonization and recrystallization in these planes and the crystal grows in layers (Fig. 10.6c). The damage is even more severe with the highly plastic crystals such as adamantane, where the slightest mishandling causes the total recrystallization of the solid[12] and an apparently perfect seed gives a multi-crystalline product (Fig. 10.6d). However, with careful handling, which for these solids involved gluing the crystal on to the support, excellent single crystals can be obtained.

Although the above examples are all molecular solids, inorganic materials, which are predominantly brittle, show similar, if not quite as wide, variations.

10.5. Factors that Influence the Perfection of the Final Crystal

Using the techniques described above, large, well-faceted, optically clear crystals can be produced. The final problem is to assess and to improve the absolute perfection of these crystals. The ideal method of assessment is X-ray topography[13] although etching techniques can be very rewarding and can be carried out using far less sophisticated apparatus.[14] Care must be taken, however, when using the latter techniques to distinguish between bulk and surface imperfections.[15]

There are four basic factors which determine the perfection of the final crystal:

(1) the purity of the starting materials;
(2) the perfection of the seed;
(3) the growth rate imposed on the crystal;
(4) the efficiency of agitation of the seed and the solution.

The purity of the crystal is governed primarily by the purity of the starting materials. Residual impurities, which do not go into solid solution, can generate imperfections in the solid. When efficient pre-purification has been carried out this does not present a problem. An important secondary factor, however, is the inclusion of solvent during growth. This is more often than not a consequence of the three remaining factors.

The remaining factors are linked in a complex inter-relationship formed round the role of imperfections as growth centres and sources of further imperfection propagating into the crystal, their influence on growth rate and their generation in the growing crystal by excessively fast growth rates and unstable growth conditions.

The growth rate itself depends both on the induced supersaturation and the number of propagating centres. When these are equivalent around the surface of the growing crystal, regular and even growth occurs. The formation of concentration gradients in the growth bath in the vicinity of the surface causes local fluctuations in supersaturation and hence growth rate. This uneven growth leads to localized stresses at the surface and to the generation of further imperfections. The degree to which concentration gradients form depends ultimately on the efficiency of agitation of the crystal and solution and on the control of temperature in the system. It is surprising what influence apparently minor changes such as small fluctuations of bath temperature, stirrers stopping for a short period, etc., can have on the overall perfection of the final crystal. Similar effects result when an excessive growth rate is imposed on the crystal. The measure of the influence of these various factors and their inter-relationship is perhaps best expressed by considering a few typical examples.

During seed preparation the surface of the seed inevitably becomes damaged. One consequence of the damage is to produce structural imperfections in the surface regions of the seed. As noted above, these affect the perfection of the final crystal in two ways. Firstly, they can propagate into the growing crystal. Secondly, they act as growth centres which increase the rate of growth of the crystal.

The initial effect is well demonstrated in Fig. 10.7, which compares a photograph and an X-ray topograph of a pyrene crystal. The seed can be seen as the fuzzy volume in the centre of the photograph and the dark, heavily dislocated area at the bottom centre of the

(a)   (b)

FIG. 10.7. Comparison of (a) a photograph and (b) an X-ray transmission topograph (220 reflection, MoKα) of a pyrene crystal. The topograph is of the upper half of the crystal. (Scale = 1 cm.)

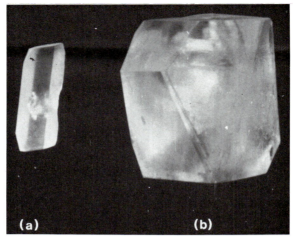

FIG.10.8. Oxalic acid dihydrate crystals grown under identical conditions from seeds of (a) good and (b) poor structural perfection. The large crystal weighs 28 g and the small crystal 0.5 g. The difference in morphology is also noteworthy.

topograph. The resulting dislocations radiate from the seed in the [110] and [1$\bar{1}$0] directions.[16] As can be seen from the topograph, continued growth would eventually yield volumes that are virtually free of dislocations.

The inter-relationship between seed imperfection and the rate of growth is exemplified by the two oxalic acid dihydrate crystals shown in Fig. 10.8. These two crystals were grown simultaneously from seeds of equivalent size in the same growth bath under identical growth conditions. They differ in that the smaller crystal was grown from a carefully selected and handled seed, the surface of which was dissolved prior to growth. That used for the larger crystal was of much lower quality. Due to the much more rapid rate of growth of the latter, resulting from the increased number of growth centres at the surface, some veiling resulted, and the larger crystal contained numerous dislocations and trapped solvent. In this case the solvent inclusion was unavoidable; it could also result from the forced growth of a more perfect seed.

Included solvent of this kind not only increases the total impurity level but also produces stress in the crystal which can result in the generation of dislocations. The effect can be seen in Fig. 10.9, which shows the consequences of solvent inclusion in a rapid growth crystal of benzophenone. When grown with care from a good seed, crystals of this material contain few, if any, dislocations. Rapid growth (the example shown was grown to 1 cm length in a few hours) has caused the inclusion of solvent (reproduced on the topograph as the large black areas). The total dislocation structure of the crystal is predominantly that generated by these inclusions.

Obviously, there is much to be gained by both careful seed selection and removal of the surface damage in order to encourage more perfect growth. Seed selection without detailed examination is rather arbitrary. If X-ray topographic facilities are not available, the most reliable method is to select a good-looking seed and then to dissolve its surface in the crystal growth bath immediately prior to growth. Hence the need to attach the crystal firmly to its support. The resulting influence on the growth rate is shown in Fig. 10.10a[17] which depicts the increase in weight of a sulphur seed as a function of the rate of lowering

FIG. 10.9. X-ray topograph of a benzophenone crystal grown from ethanol solution (002 reflection, MoKα). The large dark areas are due to solvent inclusions. (Scale = 25 cm.)

of the bath temperature (effectively the supersaturation). Curve 1 is for an untreated seed. Partial dissolution of the seed results in a lowering of the growth rate (curve 2) and dissolution to a greater extent to even lower rates (curve 3). In parallel with this decreasing growth rate was an increase in perfection to yield crystals which were virtually free of dislocations (Fig. 10.10b) and included solvent (Fig. 10.10c) other than in the region of the seed. The final state of the crystal is adequately demonstrated in Fig. 10.11, which shows topographs of sections of a typical sulphur crystal (grown under the conditions of Fig. 10.10a, curve 3) cut across the seed and at a distance from the seed.[18] The outer portion of the crystal contains few dislocations and these have been induced mechanically by handling after growth rather than as a consequence of propagation from the seed. The extremely high perfection of the outer portion of the crystal is well demonstrated by the occurrence of *Pendellösung* fringes.[19] Continued growth under these conditions would yield a massive, perfect, sample. Crystals of this type of many materials have been grown in sizes up to 200 g in weight.

In summary, we would recommend that the best course for the preparation of high-quality monocrystals is to use well-selected seeds, to dissolve the surface of these as much as is possible, then to grow them at a slow rate in a well and efficiently stirred solution. Obviously, we can give no hard and fast rules as to the ideal growth rate to choose for all specimens since this is a property of the particular system to be examined. Sufficient information is given above to allow the first steps to be taken. More detailed conditions for a particular system can only be defined after some preliminary experimentation.

A final point is that the crystal when grown should not be removed from the bath and transferred directly to ambient temperature. Even small temperature differentials (1–2 degrees) can lead to strain and cracking on completion. It is wise to lift the crystal from the solution into the air space which is above the solution but usually below the level of the surrounding thermostat bath. If the bath is switched off, then, due to its large bulk, it will cool slowly to room temperature and the crystal will not be subjected to sudden thermal shock.

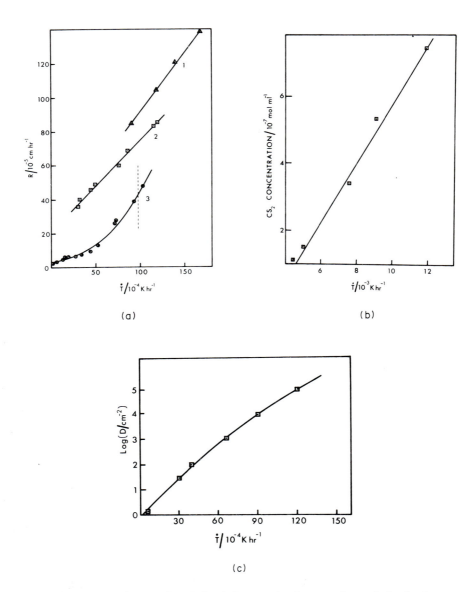

Fig. 10.10. The influence of seed dissolution on: (a) the rate of growth R; (b) the dislocation content D; and (c) solvent inclusion in orthorhombic sulphur crystals grown from solution in carbon disulphide.[17] $\dot{T}$ is the temperature lowering rate.

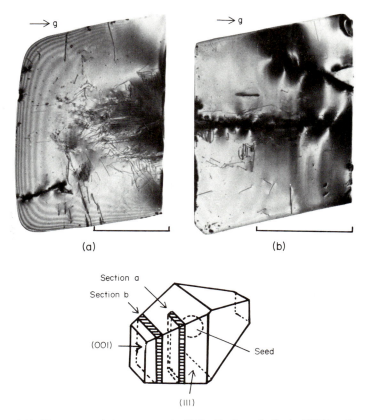

Fig. 10.11. X-ray transmission topographs (222 reflections, $AgK\alpha_1$) of (001) sections of a sulphur crystal cut from (a) near the seed, and (b) at a distance from the seed. The positions of the sections in the crystal are shown in the diagram. The large dark areas are cracks induced during sectioning.[18]

10.6. Control of Crystal Morphology

In some circumstances it is desirable that a crystal be grown in a very specific orientation. This may be simply for convenience. Alternatively, it may be that the particular physical property of the crystal which it is proposed to utilize is unidirectional. There is then a need to cause the crystal to propagate dominantly in such a habit that a reasonable number of satisfactorily sized samples of this particular orientation can be cut for use.

A basic factor in deciding the morphology is the solvent and this has been discussed in the section on solvent selection (section 10.2.1). Other than with this, the morphology of a crystal can vary depending on:

(a) the conditions of growth rate, temperature and supersaturation;
(b) the distribution and nature of dislocations in the crystal;
(c) the nature of impurities and pH.

(a) and (b) are to some extent related since variations in growth conditions can cause the formation of different dislocation systems which in turn cause some faces to propagate more rapidly than others. Thus there are cases where needle-like growth can be associated with the dominant dislocation system with a line direction parallel to the needle axis (e.g. urea which grows in needles with the major axis parallel to [001]).[20] When these dislocations can be encouraged to grow out or to change line direction (by adjustment of growth conditions and stirring) a prismatic habit develops. The critical influence of stirring rate on the redistribution of dislocations and hence changes in morphology of crystals grown from highly dislocated seeds has been beautifully demonstrated in a recent paper by Lefaucheux and his collaborators.[21] With more perfect seeds of the type discussed here, the influence of variation in stirring rate is less critical but none the less very important.

A good, simple, example of the influence of dislocations on habit is found in sodium chlorate.[22] Seeds grown by the slow evaporation of aqueous solution take up a {100} prismatic habit and contain only dislocations with ⟨100⟩ line directions. The crystals are cuboids. The ratio of the lengths of each major axis can be related to the number and the degree of screw character of the dislocations propagating parallel to that axis. Figure 10.12 show a series of topographs of crystals of three different habits. By using dislocation contrast[23] it can be shown that the single dislocations are either of mixed, edge, or screw character and that groups of dislocations (fans) predominantly comprise mixed dislocations.

Figure 10.12c has an equivalent number of fans of mixed dislocations in the two directions in the plane of the topograph and takes up a dominantly cubic habit. Figure 10a and b shows 200 and 020 reflections from a cuboid crystal. From these we identify edge, screw, and mixed dislocations. The growth is obviously dominated by the single pure screw dislocation OO' propagating rapidly in the [100] direction. The rate here is some two to three times faster than that in the [010] direction along which lie only a number of mixed and edge dislocations and ten times faster than in the [001] direction where we can identify only two dislocations of mixed character emanating from points A. Rapid or unstable growth results in the generation of dislocations of [110] and [120] line directions (Fig. 10.12d) and the consequential formation of {110} and {120} faces. As growth is re-stabilized these faces can be caused to grow out yielding once again a cuboid habit. This is a simple but easily characterizable example of this effect. Many other examples could be quoted.

Thus by generating favourable slip systems it should be possible to produce crystals of well-defined morphology.

Such a prospect is, however, limited since it may not be possible to provoke the dislocation slip system which will produce the most desirable morphology. Also, considerable fundamental work is required on the characterization of line defects in the crystal before attempting this. There is, however, scope for development in this area.

An empirical approach which can be equally satisfactory is to select seeds of a particular morphology and which presumably contain the desired dislocation configurations. A slow but uncontrolled recrystallization of a bulk of solution will yield a host of small crystals. Examination of these will soon indicate whether or not one or several habits dominate. The more careful growth of the selected seeds should yield a larger crystal of the same morphology. Figure 10.13 compares crystals of triglycine sulphate of normal and elongated habits. The latter type on further growth yielded much more satisfactorily

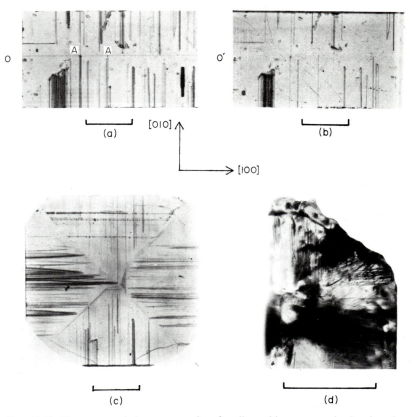

FIG. 10.12. X-ray transmission topographs of sodium chlorate crystals showing the distribution of edge, screw and mixed dislocations: (a) 200 reflection, $AgK\alpha_1$; (b) 020 reflection, $AgK\alpha_1$ of a cuboid crystal; (c) 200 reflection, $AgK\alpha_1$ of a cubic crystal; and (d) 200 reflection, $MoK\alpha_1$ of a crystal showing $\{110\}$ and $\{120\}$ faces. Due to variation in contrast with the orientation of the diffraction vector, screw dislocations of Burgers vector [100] and edge dislocations of line direction in [010] are visible in (a) and not in (b). Similarly, a single edge dislocation of line direction [100] is visible in (b) but not in (a). The remaining dislocations are of mixed character.

oriented crystals from which cuts could be made for the examination of the pyro-electric properties of this material.

With such an approach one is limited to the basic habits which result during seed preparation. There is also the problem that the habit variations may arise from inhomogeneity of supersaturation in the initial solution during seed preparation. Consequently, as noted for sodium chlorate crystals of $\{110\}$ habit, the habit may change on more careful growth. In spite of this, however, the method does provide a possible means of producing pure crystals of specific habit.

A more reproducible manner of obtaining morphological changes is to cause habit modification by the addition of impurities or by changes in pH. This is an area of great industrial importance where such morphological changes are induced during crystallization to yield solids with better filtration and packing characteristics. In view of

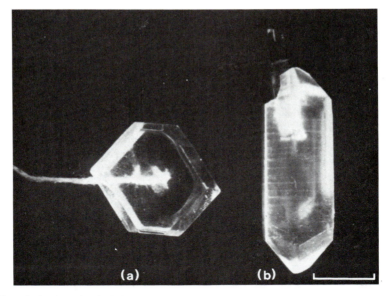

FIG. 10.13. Crystals of tri-glycine sulphate of: (a) normal and (b) elongated habit grown from selected seeds. (Scale = 1 cm.)

this it is rather surprising that little detailed fundamental information is available on the mode and mechanism of habit modification by impurities. It is, however, a well-established technique, and where purity of the sample is not of great importance it can be used. Table 10.3 gives some examples of the variations which can be achieved. The available data refers to specific systems, and it is difficult to make general recommendations of likely habit modifiers for unproved systems. Rather than attempt this we feel that it is better to refer the reader to standard texts on the subject in which a wealth of detail is given for the specific systems which have been examined.[24]

TABLE 10.3. *Effect of Impurities on the Habit of Crystals Grown from Solution*

Sodium chlorate	Cubic	Borax	Octahedral
Ammonium dihydrogen phosphate	Prismatic	Cr^{+3}, Fe^{+3}	Needle-like
Sodium chloride	Cubic	$Fe(CN)_6^{-4}$	Dendritic
Ammonium chloride	Needle-like	Urea	Cubic
Urea	Needle-like	Ammonium chloride	Cuboid
Ammonium sulphate	(010) tablet	Fe^{3+}	Psuedohexagonal prisms
Rochelle salt	Prismatic	Cu^{++}	(001) tablet
Ammonium nitrate	Needle-like	Mn^{++}, Co^{++}	Equiaxed prisms
Anthracene	Platelets	Anthraquinone	Prismatic
Sucrose	Platelets	Raffinose	Needle-like
Triglycine sulphate	Platelets	Fe^{3+}	Needle-like
Lithium sulphate Ammonium perchlorate	(001) platelet	Various dyes	(102) prisms
Sodium borate	Prismatic	Dye orange	Platelet

Acknowledgements

In conclusion we hope that the information given above will be useful and helpful to those wishing to embark on the growth of crystals from solution. It represents the accumulated experience of a number of years' work in this area. The work has had the financial support of a number of bodies. Of those we wish particularly to acknowledge the support of the Science Research Council and the Corporate Laboratory, Imperial Chemical Industries Ltd.

References

1. ELWELL, D. and WANKLYN, B., This book, Chapter 12.
2. HENISCH, H. K. (1970) *Crystal Growth from Gels*, The Pennsylvania State University Press.
3. BEGG, I. D., NARANG, R. S., ROBERTS, K. J., and SHERWOOD, J. N. (1980) Growth and perfection of crystals of TTF/TCNQ, *J. Cryst. Growth* (in press).
4. DAMIEN, J. C., DEVOS, L., and MORE, M. (1972) *Fabrication de monocristaux moléculaires en solution, Cristallogénèse Expérimentale* (Meinnel, J. and Descamps, E. A., eds.), University of Rennes, France, p. 31.
5. NARANG, R. S. and SHERWOOD, J. N. (1978) Crystallization and impurity incorporation in adipic acid, I, Chem. E. Symposium Series No. 54, *Alternatives to Distillation*, p. 267.
6. FRASER, B. C. and PEPINSKY, R. (1953) *Acta Cryst.* **6**, 273.
7. See bibliography given below.
8. ROUSE, L. M. and WHITE, E. A. D. (1976) Crystal growth by electrolytic concentration, *J. Cryst. Growth* **34**, 173; DELFINO, M. (1976) Solution growth of ionic salts by electrolytic solvent decomposition, *J. Cryst. Growth* **32**, 378.
9. FORNO, C. (1974) The growth of large crystals of hexamine from solution, *J. Cryst. Growth* **21**, 64.
10. WALKER, A. C. and KOHMAN, G. T. (1948) Growing crystals of ethylene diamine tartrate, *AIEE Trans.* **67**, 565.
11. DI-PERSIO, J., ESCAIG, B., HAMPTON, E. M., and SHERWOOD, J. N. (1974) Dislocations in α-sulphur, I, *Phil. Mag.* **29**, 732.
12. SHERWOOD, J. N. (1979) Lattice defects and the plasticity of plastic crystals, *The Plastically Crystalline State* (Sherwood, J. N., ed.), Wiley, London, p. 39.
13. TANNER, B. K. (1976) *X-ray Diffraction Topography*, Pergamon Press, Oxford.
14. AMELINCKX, S. (1969) Surface methods, *Solid State Physics*, **6** (Supplement) (Seitz, F., ed.), Academic Press, New York.
15. NARANG, R. S., SHAH, B. S., and SHERWOOD, J. N. (1974) Growth and perfection of phenanthrene single crystals, II, *J. Cryst. Growth* **22**, 201.
16. HOOPER, R. M. and SHERWOOD, J. N. (1976) Dislocations in pyrene crystals, *J. Chem. Soc. Faraday* I, **72**, 2872.
17. HAMPTON, E. M., SHAH, B. S., and SHERWOOD, J. N. (1974) The growth and perfection of orthorhombic sulphur crystals, *J. Cryst. Growth* **22**, 28.
18. HAMPTON, E. M., SHERWOOD, J. N., DIPERSIO, J., and ESCAIG, B. (1974) Dislocations in α-sulphur, II, *Phil. Mag.* **29**, 742.
19. LANG, A. R. (1970) Recent application of X-ray topography, *Modern Diffraction and Imaging Techniques in Materials Science* (Amelinckx, S., Gevers, R., Remant, G., and van Landeryt, I., eds.), North-Holland, Amsterdam.
20. NARANG, R. S. and SHERWOOD, J. N., Unpublished work.
21. GITS-LEON, S., LEFAUCHEUX, F., ROBERT, M. C. (1978) Effect of stirring on crystalline quality of solution grown crystals—case of potash alum, *J. Cryst. Growth* **44**, 345.
22. HOOPER, R. M., NARANG, R. S., and SHERWOOD, J. N. (1980) To be published.
23. LANG, A. R. (1973) The properties and observation of dislocations, *Crystal Growth: An Introduction* (Hartman, P., ed.), North-Holland, London.
24. Articles in *Crystal Growth* (1949), *Discussions of the Faraday Society*. Also J. W. Mullin, this book, Chapter 14.

Bibliography

Since this chapter deals principally with a specific apparatus and its use rather than a survey of all available systems, we cite some general references in this area to provide further information for those wishing to design their own systems. Also included is a bibliography of general texts dealing with growth from solution.

GENERAL REFERENCES TO SOLUTION
GROWTH APPARATUS

BUCKLEY, H. E. (1951) *Crystal Growth*, Chapman & Hall, London.

TORGESEN, J. L., HORTON, A. T., and SAYLOR, C. P., Equipment for single crystal growth from aqueous solution, *J. Res. of the NBS-C, Engineering and Instrumentation* **67C**, 25.

TORGESEN, J. L. and STRAUSSBURGER, J. (1964) Equipment for growing crystals from solutions in volatile solvents, *Science* **146**, 53.

SCHEIBER, M. (1967) The effects of high magnetic fields on the isothermal dissolution and growth rates of $Fe(NH_4)_2(SO_4)_2CH_2O$ and $KAl(SO_4)_2 . 12H_2O$ seed crystals, *J. Cryst. Growth* **1**, 131.

SCHLICHTA, P. J. and KNOX, R. E. (1968) Growth of crystals by centrifugation, *J. Cryst. Growth* **3, 4**, 808.

TICHY, K. and HONZL, J. (1968) Flow crystallisation of organic substances unstable in solution, *J. Cryst. Growth* **2**, 369.

NOVOTNY, J. and MORAVEC, F. (1971) Growth of TGS from slightly supersaturated solutions, *J. Cryst. Growth* **11**, 329.

ACKER, E., HAUSSUHL, S., and RECKER, K. (1972) Zuchtung und physikalische Eigenschaften von Monoklinem Zinndifluorid, *J. Cryst. Growth* **B/14**, 467.

NESSAU, K. (1972) The growth of crystals from boiling solutions, *J. Cryst. Growth* **15**, 171.

BRUTON, T. M. (1973) The growth of single crystals by thermal diffusion, *J. Cryst. Growth* **18**, 269.

GRUGIC, D., ZIZIC, B., NAPIJALO, M., and JANIC, I. (1972) Contribution à l'étude de la croissance de monocristaux de NaCl a partir de solution aqueuse pur, *J. Cryst. Growth* **19**, 122.

DELFINO, M., DOUGHERTY, J. P., TWICKER, W. K., and CHOY, M. M. (1976) Solution growth and characterisation of L-glutamic acid hydrochloride single crystals, *J. Cryst. Growth* **36**, 267.

GENERAL REFERENCES TO CRYSTAL GROWTH
MECHANISMS OF PARTICULAR IMPORTANCE TO
SOLUTION GROWTH

HARTMAN, P. (ed.) (1973) *Crystal Growth: An Introduction*, North-Holland, Amsterdam.

SHEFTAL, N. N. (1956) *Growth of Crystals*, Vol. 1, Consultants Bureau Inc., New York.

SHEFTAL, N. N. (1962) *Growth of Crystals*, Vol. 3, Consultants Bureau Inc., New York.

ELWELL, D. and SCHEEL, H. J. (1975) *Crystal Growth from High-Temperature Solutions*, Academic Press, London.

BRICE, J. C. (1973) *The Growth of Crystals from Liquids*, North-Holland, Amsterdam.

WALTON, A. G. (1967) *The Formation and Growth of Precipitates*, Interscience, New York.

ZETTLEMOYER, A. C. (ed.) (1959) *Nucleation*, Marcel Dekker Inc., New York.

EGLI, P. H. and JOHNSON, L. R. (1963) In *The Art and Science of Growing Crystals* (Gilman, J. J., ed.), Wiley, London, p. 164.

BENNEMA, P. (1966) Techniques for measuring the rate of growth of crystals from solutions, *Phys. stat. solidi* **17**, 555.

BRICE, J. C. (1976) The kinetics of growth from solution, *J. Cryst. Growth* **1**, 218.

BENNEMA, P. (1976) Analysis of crystal growth models for slightly supersaturated solutions, *J. Cryst. Growth* **1**, 278.

BENNEMA, P. (1969) The importance of surface diffusion for crystal growth from solution, *J. Cryst. Growth* **5**, 1969.

CHAPTER 11

Liquid Phase Epitaxy

R. L. MOON†

Varian Solid State Laboratory, Palo Alto,
California, USA

11.1. Introduction

Liquid phase epitaxy (LPE) has recently become an important method of crystal growth, primarily because of its importance to the electronics industry. Besides being of commercial importance, this method also provides a means to study physical–chemical processes occuring during solution growth. Unlike normal solution growth, a seed which may not be of the same composition as that of the depositing epitaxial layer is always present. Basically, LPE differs only in degree from solution growth. The most important difference is that the solution is usually more dilute, leading to slower growth rates, to fewer spontaneous crystallites, and to more stoichiometric layers. The epitaxial layers are often very pure because the dilute solution and favorable segregation coefficients keep unwanted impurities in solution, and their thicknesses are very thin, being microns rather than millimeters in extent. These factors make LPE important to the electronics industry for the growth of non-silicon based materials, which are used to fabricate a variety of devices for microwave, optoelectronic, and magnetic bubble memory applications. Frequently, other methods of epitaxy are also available, but LPE will, in general, produce material that forms devices superior in performance to those grown by the other methods. Some notable devices almost exclusively grown by LPE are the lasers and light-emitting diodes based on either AlGaAs–GaAs or the longer wavelength InGaAsP–InP alloys, the light emitting diodes of GaP and the magnetic bubble memories formed from magnet garnets. These devices attest to the process control now possible and to the state of development to which LPE has evolved.

Although deposition by LPE has grown rapidly in the last 15 or so years, its origins extend back to the nineteenth century. Apparently, sodium nitrate was grown in an oriented fashion on freshly cleaved calcite as early as 1836 (Frankenheim, 1836). By the turn of the century the growth of salts from aqueous solution onto cleaved mineral faces had been thoroughly investigated (Barker, 1906). LPE studies continued to use aqueous solutions as the primary growth media (Pashley, 1956), until 1963 when Nelson fabricated germanium tunnel diodes and GaAs lasers from molten metal solutions. Since that time, LPE usage has grown rapidly as demands for new electronic materials increased. To date, LPE technology has concentrated mainly, but far from exclusively, on the growth of III–V alloys and magnetic garnet materials. Other materials that have been grown by LPE using various solutions are silicon from Al–Ga or tin (Baliga, 1978; Girault *et al.*, 1977), SiC

† Now at Hewlett-Packard Optoelectronics Div., Palo Alto, CA.

from $TiSi_2$ (Pellegrini and Feldman, 1974), $Pb_{1-x}Sn_xTe$ from lead (Longo *et al.*, 1972), CdS from cadmium (Blinnikov and Ralyuzhnaya, 1972), and $LiNbO_3$ from a $Li_2O-V_2O_5$ flux (Baudrant *et al.*, 1978), to name a few.

The process itself is largely limited by diffusion of the solid forming species in solution and at times by surface reactions that are coupled with the diffusion flux. Diffusion occurs over the entire solution of a quiescent isothermal liquid, but during free or forced convection it proceeds most importantly through a thin boundary layer. Solutions are generally confined in a crucible or boat, which is the heart of the system and where most of the design ingenuity has occurred, particularly in the growth of III–V materials. Aside from a means of holding the substrate and a method of moving the substrate into and out of the solution, all that is needed is a well-controlled furnace, a reactor tube, and possibly a gas-handling system. This system is usually minimal, consisting of metering valves, and, if necessary, a gas purifier. Few changes are required as different elements within a major material type are mixed. Thus, unlike vapor or molecular beam techniques, the apparatus does not "grow" with increasing alloy complexity.

Most aspects of LPE have been studied in some detail since 1963 and this has led to an immense number of articles. Fortunately, several reviews have appeared during this time, with the articles by Dawson (1972), Deitch (1975), and Kressel and Nelson (1973) emphasizing growth of III–V materials and the review by Giess and Ghez (1975) concentrating on garnets and some of the mathematical aspects of growth. This chapter endeavors to outline some of the theoretical and experimental results common to most LPE growth and is intended, not as a review, but as a guide which shows the interplay between its various aspects. Specifically, this chapter describes the apparatus, comments on the phase diagrams commonly used, discusses the growth kinetics by the common modes of growth, relates the resulting surface morphology of the epitaxial layer to process variables, and shows the perturbing influence of lattice mismatch.

11.2. Apparatus

A wide variety of techniques have been devised to implement the LPE process, including variations in both the process method and equipment design. Reactors are classified as either vertical or horizontal, depending upon the directions in which the substrate enters the solution. In III–V technology, both types are used, whereas in garnet growth the vertical dipping method completely dominates. Figure 11.1 shows two common reactors—both consist of a furnace, solution, and substrate holders, and, in the case of III–V materials, a quartz tube which contains a high purity H_2 atmosphere. Garnet growth reactors are somewhat simpler since the ambient gas is air and only single layers are required for bubble memory applications.

III–V compounds for optoelectronic or microwave device applications often require that more than one layer be grown, and this has led to a large number of boat designs to accomplish this task. In the original work of Nelson (1963) on the LPE growth of GaAs, a tipping method was used where during equilibration a GaAs substrate was placed at one end of a graphite boat while the solution was placed in the other end. Once equilibration was achieved, the furnace was tipped to enable the solution to contact the substrate, then a controlled cooling cycle followed. At present, because of the obvious limitation of the tipping method, the most extensively used technique is the horizontal slider approach,

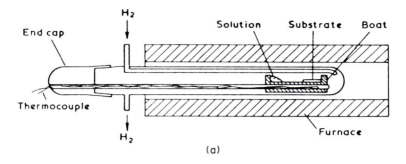

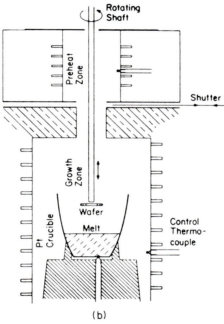

FIG. 11.1. Typical LPE growth systems: (a) horizontal system used in III–V growth (Deitch 1975); (b) vertical system used in magnetic garnet growth (Ghez and Giess, 1974).

which is well suited for multilayer fabrication (Fig. 11.2a). With this technique a number of solutions come sequentially into contact with the substrate for the multilayer growth. There are numerous variants to this approach, including whether the solutions move to the substrate or vice versa. Perhaps the most significant variation is the supercooled sliding boat where a section of an equilibrated solution is "sectioned out" from the remaining solution (Fig. 11.2b). The isolated solution can then be controllably supercooled without interference from an equilibration wafer or dendrites. Another innovation is the Peltier-induced or current-induced LPE in which local Peltier cooling at the solid–liquid interface and electromigration are caused by an electrical current driven through leads embedded in the boat (Fig. 11.2c).

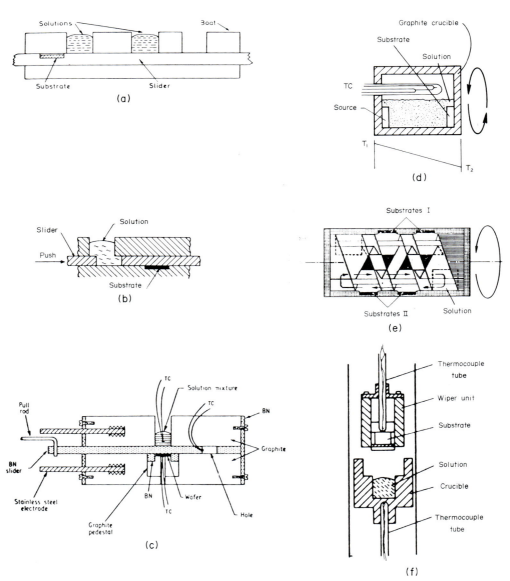

FIG. 11.2. Various III–V LPE boats: (a) multi-bin slider (Nelson, 1971); (b) supercooled (Mihara *et al.*, 1975); (c) Peltier-induced LPE (Daniele and Michel, 1974); (d) steady state (Stringfellow and Greene, 1971); (e) multilayer rotary (Scheel, 1977); (f) vertical dipping (Deitch, 1970).

Rotary type boats that move the substrates or solutions by a rotation rather than translation are also used (Velms and Garrett, 1972; Holonyak, 1977; Donahue and Minden, 1970). In the case of the steady state growth mode, promoted by an imposed temperature gradient, both the substrate and source wafer are rotated in and out of the solution (Fig. 11.2d). One ingenious multilayer rotary boat uses a double screw arrangement resembling an Archemedean screw (Fig. 11.2e). By a rotation about the horizontal axis, a counter-current motion is set up by two screws with opposite senses. As the outer screw turns, the solutions move horizontally across the substrates until they reach a channel to the central screw. The solutions then move back to their starting position because the central screw moves backwards on rotation owing to its reversed sense. The procedure can be repeated as many times as is necessary.

Vertical dipping, another popular technique, used mainly for single layers, immerses the substrate into the solution for growth. The holder designs vary in their method of breaking through any crust on the solution surface and of providing a means of scraping off any residual solution after withdrawing the substrate (Fig. 11.2f).

11.3 Phase Diagrams

Phase diagrams in LPE growth serve primarily as a guide to the solution compositions which will yield the desired solid layer, although they are also necessary for the interpretation of growth kinetic data. In general, most phase diagrams convey only approximate information, so fine tuning of the solution composition is usually necessary. Garnet growth represents one of the more complex systems, having as many as seven components and several possible primary phase fields. However, once the desired compositions are determined, it is relatively easy to produce very high quality garnet layers with defect densities as low as $10 \, cm^{-2}$ or less. The ternary III–V systems, on the other hand, have much simpler phase equilibria, usually containing only a continuous solid solution between the binary compounds. Because solid solutions readily form in these systems, a great deal of effort has been directed towards controlling the resultant solid composition so that a range of bandgaps can be produced. Unlike the garnet systems, calculations of III–V phase diagrams are extensive—but unfortunately not on an *a priori* basis, since some of parameter values used in the calculation must first be obtained experimentally. A sketch of the phase information for both systems is given in the remainder of this section.

The garnet-containing systems crystallize in one of the four primary phases: hematite, magnetoplumbite, garnet, or orthoferrite (Nielsen and Dearborn, 1958). Generally, the fluxed melts contain at least four components: rare earth oxides, Fe_2O_3, and the solvents PbO and B_2O_3. The epitaxial layer is altered by adding mixtures of rare earth oxides to provide a range of magnetic properties. Figure 11.3 shows a pseudoternary diagram representative of the garnet systems; note that the garnet (YIG) does not melt congruently so that a pseudobinary between PbO and garnet does not form. In order to force the garnet to crystallize as the primary phase, Fe_2O_3 in excess of the stoichiometric amount is added and becomes part of the solvent. Too much Fe_2O_3 results in either magnetoplumbite or hematite being the primary phase, whereas too little causes the orthoferrite phase to precipitate. It is interesting to note that the ability to supersaturate the solution drastically changes depending upon the primary phase (Blank and Nielsen, 1972). The maximum supercooling is 5°C when orthoferrite is the primary phase, regardless of how near the garnet–orthoferrite boundary the solution may be; but, the addition of just enough Fe_2O_3 to cross this boundary permits supercooling by 30–60°C without spontaneous nucleation. Exact solution compositions in the garnet systems are best calculated by forming molar ratios of the constituents (Blank and Nielsen, 1972). These ratios thus determine the saturation temperature, primary phases, and layer

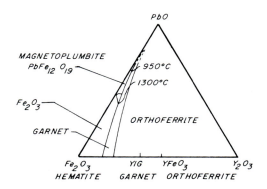

FIG. 11.3. General features of the pseudoternary $PbO-Y_2O_3-Fe_2O_3$ phase diagram (Nielsen and Dearborn, 1958).

composition. Some allowance must be made for the kinetic effects on composition, because both the growth rate and ionic radii of the substituting ions influence the segregation coefficients.

In III–V systems the binary compounds such as GaAs and GaP are very stoichiometric, possessing very narrow existence regions; in addition, they form a low temperature eutectic with the column III element, normally used as the solvent for growth. When two binary compounds combine, a continuous solid solution usually develops. As an example, Fig. 11.4a shows a perspective diagram depicting the liquidus surface, the solid solution plane, and the binary boundaries of the Al–Ga–As ternary system. The actual phase diagram of this and other systems of interest is supported by experimental and theoretical studies. (See reviews by Panish and Illegems, 1972; Kressel and Nelson, 1973; Stringfellow, 1974.)

The theoretical studies are based on a regular solution model (with appropriate choice of adjustable parameters) and generally lead to an accurate description of the phase relationships over the regions of interest. Since the equations necessary are discussed at length in the above reviews, only a brief description regarding the development of the equations now follows. Description of the III–V phase equilibria develops from the equality of chemical potentials of each species in each phase and by assuming that a regular type solution describes the liquid and solid phases. Without a solution model, the behavior of the activity coefficients is impossible to predict unless a complete study of the system is undertaken. In any nonideal system the lack of ideality is accounted for by an excess free energy/mole of solution which is added to the ideal free energy. For a binary regular solution, the excess free energy takes the form

$$G_m^x = \Omega_{ij} \frac{N_i N_j}{n} \tag{11.1}$$

and leads to the activity coefficients from the equation

$$RT \ln \gamma_i = \frac{\partial G_m^x}{\partial N_i}, \tag{11.2}$$

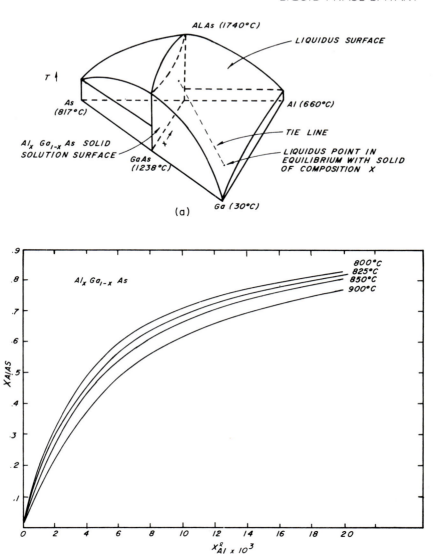

FIG. 11.4. Phase equilibria in the III–V ternary $Al_xGa_{1-x}As$ alloy: (a) perspective of the Al–Ga–As system; (b) results of a phase diagram calculation showing X^s_{AlAs} vs. X^l_{Al}.

where G^x_m is the excess free energy/mole, N_i and N_j are the moles of i and j in a solution totaling n moles, γ_i is the activity coefficient of i, and Ω_{ij} is the interaction parameter. For binary III–V systems, it is usually assumed that

$$\Omega_{ij} = a - bT, \tag{11.3}$$

where a and b are constants and T is the temperature. Such a representation is not strictly a regular solution; if it were, Ω_{ij} would be independent of T. Instead the model is termed a "simple solution" (Guggenheim, 1967; Panish, 1974). Values for Ω_{ij} are obtained from the liquidus data of the binary systems, by the use of the equation

$$\Omega_{ij} = \frac{-RT}{2(0.5 - x_i)^2} \left[\ln 4x_i(1 - x_i) + \frac{\Delta S_F}{R}\left(\frac{T_F}{T} - 1\right) \right], \tag{11.4}$$

where x_i is the mole fraction of the lesser component, R is the gas constant, and ΔS_F and T_F are the heat and temperature of fusion, respectively (Vieland, 1963). The interaction parameter for the solid solution is calculated by fitting the liquidus and solidus lines of the pseudobinary to a regular solution model. Now, in addition to the set of equations generated from purely thermodynamic arguments involving chemical potentials between phases, it is possible to estimate the activity coefficients from the binary systems, then to use this information to calculate a ternary or quaternary phase diagram. Even with this model some adjustment in the parameters is required before the experimental and theoretical data agree.

When all the equations are combined and the appropriate constants are used, an adequate description of the phase equilibria results. Figure 11.4b shows the results of a calculation for the binary $Al_xGa_{1-x}As$ using the data of Panish and Illegems (1972). In this system Al has a high segregation coefficient, $K_{Al} > 10$, which is typical of the column III elements that are the higher melting compound in every III–III′–V binary. As growth proceeds, this leads to a change in the Al composition owing to depletion at the solid–liquid interface. Fortunately, estimates of the change are possible because the phase diagram can be described completely and therefore diffusion limited factors can be added by performing a numerical simulation (Crossley and Small, 1971, Ijuin and Gouda, 1976; Joullie, 1977; and Bryskiewicz, 1978).

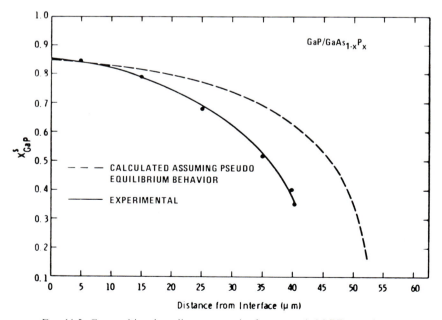

FIG. 11.5. Compositional grading as a result of ramp-cooled LPE growth.

A rough estimate of the solid composition variation during growth is possible by using the calculated phase diagram and by assuming pseudoequilibrium growth, which excludes diffusion effects (Illegems and Pearson, 1969; Ijuin and Gouda, 1976). Growth is thought to proceed over small temperature intervals and the diffusion in the solid is considered to be negligible. After an incremental deposit, caused by a small change in temperature, the solution changes composition in correspondence with the solid just formed to yield a new uniform solution from which the next deposit grows. This incremental stepping of temperature continues until the total temperature interval spans the interval used during growth. As long as the concentration gradients are not too great, a condition ensured by spanning a small total temperature interval or by cooling very slowly, the experimental results generally agree with the predictions. Figure 11.5 shows the calculated and experimental composition versus thickness curve for growth of $GaAs_{1-x}P_x$ on GaP (Moon *et al.*, 1978). The discrepancy between estimate and experiment is caused by a high cooling rate and points out the need for incorporating kinetic terms in the model.

11.4. Growth Kinetics

11.4.1. MODES OF LPE GROWTH

The final layer thickness (or growth rate) for a given system depends upon the mode of creating and relieving supersaturation, which in turn depends on the temperature program used during growth and the time at which the substrate contacts the growth solution during this program. Of the various modes available, supercooling, ramp cooling, steady state, and transient mode LPE are the most often employed.

Supercooled growth perhaps represents the simplest mode of cooling where the temperature of the solution is lowered from an initial value to a new value. This change in temperature yields a supersaturation condition, provided the solution is saturated initially. Placing a substrate into a supercooled solution causes the growth of a layer which thickens with time at $t^{1/2}$, as the data in Fig. 11.6a show. It is worth noting, in this case, that extrapolation of the plotted line to zero time yields a negligible thickness within the experimental error of the data; this is not the case when surface kinetics effects are present. The layer growth is rapid initially and gradually slows. In a sense, the process is self-limiting, an advantage for thin layers, since the ultimate layer thickness is set by the supercooling and is only asymptotically approached. A further advantage of this mode is the enhanced nucleation driving force which the supersaturation produces.

Ramp cooling, sometimes called tipping, uses a temperature program that continuously decreases the temperature throughout the growth period. Usually, the program cools linearly, hence the name. The cooling causes the solution saturation to decrease below the value that existed at the starting temperature. Introduction of the substrate can occur at any time during the program, but is usually near the beginning before any homogeneous nucleation in the solution occurs. If, when the substrate contacts the solution, it is in equilibrium with the solution, then pure ramp cooling growth occurs. However, if the solution has cooled somewhat below the equilibrium temperature, a mixed mode of ramp and supercooled growth takes place. Thickness of a layer grown by ramp cooling increases as $t^{3/2}$ for moderate growth times; Fig. 11.6b shows this behavior, which can also be modified by surface kinetics. Layer thicknesses continue to increase with time in contrast

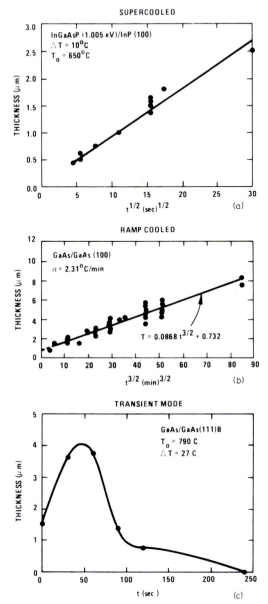

FIG. 11.6. Thickness vs. time for different modes of LPE growth: (a) supercooled (Houng, 1979); (b) ramp-cooled (Moon and Kinoshita, 1974); (c) transient mode.

to supercooled growth. Consequently, this mode is useful when thick layers (10–100 μm) are required.

Transient mode LPE (TMLE), unlike the two previous methods, is a nonisothermal process in which a substrate contacts a warmer saturated solution (Deitch, 1970). Probably this method is used more inadvertently than intentionally. The cool substrate creates a local supersaturation at the solid–liquid interface, initiating layer growth. Growth is very rapid at first, but soon slows, passes through zero, then actually becomes negative as the solution regains its initial temperature. Hence, the layer thickness as a function of time passes through a maximum, as Fig. 11.6c shows. The principal advantage of this method is the enhanced nucleation rate that results from the high local supercooling, which is possible without supercooling the entire solution and running the risk of homogeneous nucleation.

Steady state growth relies on a temperature gradient between a source and substrate wafer to provide the driving force for growth. Layer thickness increases as t after some initial transients. This mode is the best for producing thick layers and is particularly useful when uniform composition is required.

Peltier-induced or current-controlled growth combines diffusion-limited growth with the Peltier cooling at the solid–liquid interface and the differential ion migration resulting from an applied electrical current (Daniele and Michel, 1974; Bryskiewicz, 1978b). Growth rates generally increase linearly, after an initial transit, with the applied current density. At current densities of $\sim 10\,\text{A/cm}^2$, these rates can be 2–10 times faster than the normal diffusion-limited rates at the same temperature. Surprisingly, a uniform composition can be maintained throughout a thick layer; for instance, a variation of x in $Al_x Ga_{1-x}As$ (with $0.1 < x < 0.3$) less than ± 0.005 has been demonstrated across a layer $40\,\mu m$ thick (Daniele, 1975). Doping densities can, however, be modified by changing the current density (Lawrence and Eastman, 1975), and in the case of an amphoteric dopant like silicon in GaAs different changed species in solution are believed responsible (Jastrezelski and Gatos, 1977). The advantages of this method are the same as those for steady state growth with the added feature of higher growth rates. Maintaining good electrical contact with the substrate and the added complexity of the boat are its main drawbacks.

Obviously, these modes can be combined to give a time dependency on thickness that does not clearly correspond to any given mode, but may show different dependencies over different time intervals (Hsieh, 1974). Furthermore, as will be shown later, surface kinetics and solution thickness can also modify these general rules.

11.4.2. DIFFUSION EQUATIONS USED TO DESCRIBE LPE GROWTH

In most cases considered so far, LPE growth is controlled largely by diffusive transport in the solution, although surface kinetics do at times play a role. During growth, a flux of material towards the substrate–solution interface results because of a disturbance in the equilibrium of the solution by the temperature program. The one-dimensional representation in Fig. 11.7 summarizes the conditions during LPE growth by showing the concentration in solution vs. distance from the solid–liquid interface (White and Wood, 1972). Initially, this solution has a concentration C_0 throughout; then, as a result of some temperature change, a new equilibrium value C_e occurs. If the interfacial concentration C_i corresponds to C_e, the growth rate is controlled totally by diffusion in the solution. However, should the surface atoms require some time to find their niche, then $C_e \neq C_i$ and the growth becomes a two-step process where the rate of incorporation of growth units is balanced by their diffusion in solution.

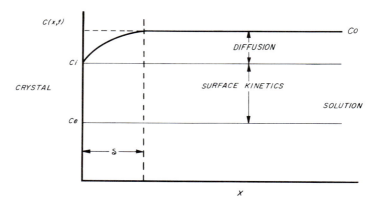

FIG. 11.7. Solution concentration gradients showing the driving forces of diffusion and interfacial kinetics (White and Wood, 1972).

A complete description of LPE growth requires mathematically describing the situation where combined heat and mass transport are present, but this can be simplified. In melt growth the heat of crystallization is an important factor, and heat flow largely determines the growth rate. But in LPE growth this heat is small relative to the diffusion forces (White and Wood, 1972). The thermal diffusivity exceeds the mass diffusivity by 10^3, suggesting that heat dissipation is rapid and that the idea of an isothermal solution is a reasonable approximation. Diffusion in the solid can normally be neglected. As a result only the diffusion in the solution is considered.

The species that are actually diffusing in the solution are often difficult to identify. Multicomponent systems exhibit the behaviors shown in Fig. 11.6 as well as simple binary materials. Often the transport of one constituent, which may form a complex in solution, limits the growth rate; so to circumvent this ambiguity the term growth unit is used (Ghez and Giess, 1974).

The equations necessary for a one-dimensional analysis are relatively simple and apply in most cases. In the solutions, the diffusion equation is

$$\frac{\partial C}{\partial t} = D\frac{\partial^2 C}{\partial x^2} + V\frac{\partial C}{\partial x}, \tag{11.5}$$

where D is the diffusivity, C is the concentration of growth units in solution, x is the distance, and t is the time. Normally this equation simplifies to the first two terms by neglecting the convective term $V(\partial C/\partial x)$. The term V represents the velocity resulting from free and forced convections and the growth rate, and is negligible so long as $D(\partial^2 C/\partial x_2) \gg V(\partial C/\partial x)$. A more detailed analysis dealing with the induced convection arising from the crystal growth itself shows that at low solute concentrations this convection is unimportant relative to diffusion (Westphal and Rosenberger, 1978; Wilcox, 1972). In cases where temperature gradients exist in the solution, free convection is always possible. Whether it develops can be analyzed by the Rayleigh number criterion (Appendix A.1).

To fully analyze the LPE process requires expressions for the growth rate and layer thickness as a function of time. Equation (11.5) describes the situation where the coordinate system is moving with the solid–liquid interface and is not anchored to a fixed reference system such as the end of the boat (Pohl, 1954). Additional assumptions consider that the solid–liquid interface remains planar and the deposition area remains constant. The expression for the growth rate develops from the mass conservation at the solid–liquid interface and is

$$R(t) = \frac{D}{C_s} \left(\frac{\partial C}{\partial x} \right)_{S/L \text{ interface}} \tag{11.6}$$

where $R(t)$ is the growth rate, C_s is the concentration of growth units in the crystal, and the other terms are described above (Pohl, 1954; Small and Barnes, 1969). The calculation of the layer thickness after a time t has elapsed is obtained by integrating the growth rate expression in eqn. (11.6) between $t = 0$ and t and is

$$H(t) = \int_0^t R(t)\,dt, \tag{11.7}$$

where $H(t)$ is the epitaxial layer thickness. With these equations and the expressions for the boundary conditions, a complete description of the diffusion-limited process is possible for finite and semi-finite solutions.

Most situations can be analyzed by using Laplace transform methods or by finding the appropriate solution in either Crank (1957) or Carslaw and Jaeger (1959). However, including explicit expressions for the various parameters, such as D or the shape of the liquidus, forces the use either of Fourier transformers (Ghez, 1973) or of numerical methods (Crossley and Small, 1971). Unfortunately, these mathematical details, albeit interesting, cannot be given, and only the expressions for $H(t)$ for the different modes are presented in the next sections.

11.4.3. SEMI-INFINITE SOLUTIONS WITHOUT INTERFACIAL KINETICS

The behavior of most LPE processes can be described very well by assuming that the solution is semi-infinite in extent and the interfacial kinetics are very rapid, so $C_e = C_i$. Any description of LPE further simplifies by assuming that the diffusivity and the slope of the liquidus line m are constant over the temperature range of growth. Since most growths seldom use temperature excursions $> 50°C$, such assumptions lead to only small errors unless D and m are strong functions of temperature. With these simplifications, the results are identical with more exact formulations as long as the growth times are not too long (Moon and Long, 1976; Ghez, 1973). Whenever the temperature dependencies of D and m are included, the variation with temperature follows an Arrhenius expression of the form $e^{-\Delta H/RT}$, where ΔH is either the activation energy for diffusion or the heat of solution (Ghez, 1973; Minden, 1970; Tiller and Kang, 1968), respectively. However, by using the above simplifications rather than the complete expressions for D and m, the impact of the various controllable parameters emerges more clearly.

The easiest mode of LPE to describe is supercooled growth where the driving force for growth arises from the supersaturation that a step lowering of the temperature produces. The amount of supersaturation in solution that a temperature change ΔT causes is

$$\Delta C = C_0 - C_e \simeq \Delta T/m \tag{11.8}$$

unless it exceeds the supersaturation necessary to promote homogeneous nucleation; in which case the maximum value can be no greater than the critical supersaturation value, regardless how large ΔT may be. A description of this growth has the following boundary and initial conditions:

$$
\left.\begin{array}{llll}
t = 0 & T = T_0 & C = C_0 & \text{for } x > 0, \\[4pt]
t > 0 & T = T_e & C = C_e & \text{for } x = 0, \\[4pt]
& & C = C_0 & \text{for } x = \infty.
\end{array}\right\} \tag{11.9}
$$

Under these conditions, $R \propto t^{-1/2}$ and the layer thickness is given by

$$
H(t) = \frac{2\Delta T}{C_s m} \sqrt{\frac{D}{\pi}}\, t^{1/2}. \tag{11.10}
$$

As a consequence $H(t)$ is directly proportional to the temperature change producing supersaturation and scales as $t^{1/2}$, a behavior showing a smaller and smaller incremental increase in thickness as t increases (Small and Barnes, 1969; Hsieh, 1974).

Ramp cooling, on the other hand, continually changes the value of C_e at the solid–liquid interface by program cooling the temperature of the solution. Normally, the solutions are cooled linearly at cooling rates α between 0.1 and 5.0 °C/min. The upper limit of the cooling rate is set by the heat capacity of the furnace and is typically $\sim 3°$C/min except for radiation furnaces where rates of 12°C/min are possible. For ramp cooling, the boundary conditions differ from the supercooled case only by the concentration variation at the solid–liquid interface, so that they are:

$$
\left.\begin{array}{llll}
t = 0 & T = T_0 & C = C_0 & \text{for } x > 0, \\[8pt]
t > 0 & T = \alpha t & C = C_0 - \dfrac{\alpha t}{m} & \text{for } x = 0, \\[8pt]
\dfrac{\partial C}{\partial x} = C_0 & \text{or} \quad C = 0 & \text{for } x = \infty.
\end{array}\right\} \tag{11.11}
$$

Growth starts at the equilibrium concentration C_0 and continues as long as the temperature program forces a concentration change. The growth rate increases as $t^{1/2}$ so that the layer thickness grows as $t^{3/2}$ (Small and Barnes, 1969; Hsieh, 1974). The complete expression for $H(t)$ is

$$
H(t) = \frac{4}{3} \frac{\alpha}{C_s m} \sqrt{\frac{D}{\pi}}\, t^{3/2}. \tag{11.12}
$$

It is worth noting that both $R(t)$ and $H(t)$ are proportional to the cooling rate and that, unlike supercooled growth, $H(t)$ can increase indefinitely as long as the solution temperature decreases. As the total temperature change increases, the potential for a rough surface from constitutional supercooling also increases, so the total ΔT is usually $\ll 100°$C.

To initiate ramp-cooled growth some supercooling is frequently used. The solution is supercooled by driving the furnace temperature down to a temperature below that at which the substrate and solution are in equilibrium. After ΔT_{sc}, growth commences when the substrate and solution contact each other while the temperature program continues to cool. The total layer thickness is calculated by adding eqns. (11.10) and (11.12) (Hsieh, 1974). In the beginning, supercooled growth completely dominates, with the ratio of ramp-cooled to supercooled layer thickness equal to $\dfrac{2}{3}\dfrac{\alpha}{\Delta T_{sc}}t$.

The next case is one of nonisothermal growth, namely transient mode LPE (TMLE). A complete description of the TMLE requires an analysis of both the temperature and concentration variations during growth, i.e., coupled heat and mass transport. However, a

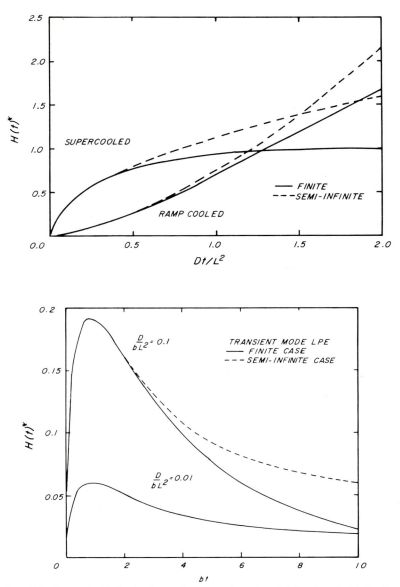

FIG. 11.8. Theoretical behavior for various growth modes with fast interfacial kinetics.

satisfactory description of the process results by assuming that the concentration at the solid–liquid interface immediately achieves a new concentration set by the initial temperature difference between the substrate and the growth solution (Moon and Vander Plas, 1978). Once the contact is made, the system strives to re-establish thermal equilibrium, thus $C(x = 0)$ begins to recover back to the original concentration that existed before the substrate was inserted. The initial and boundary conditions for TMLE are, then:

$$
\left.
\begin{array}{llll}
t = 0 & T = T_0 & C = C_0 & \text{for} \quad x > 0, \\[2mm]
t > 0 & T = T_0 - \Delta T e^{-bt} & C = C_0 - \dfrac{\Delta T}{m} e^{-bt} & \text{for} \quad x = 0, \\[3mm]
& \dfrac{\partial C}{\partial x} = 0 \quad \text{or} \quad C = C_0 & & \text{for} \quad x = \infty,
\end{array}
\right\} \qquad (11.13)
$$

where b is the relaxation constant for the process. From these conditions the equation for layer thickness becomes

$$
H(t) = \sqrt{\frac{4D}{\pi b} \frac{\Delta T}{C_s m}} \left[e^{-bt} \int_0^{\sqrt{bt}} e^{\lambda^2} d\lambda \right], \qquad (11.14)
$$

where the term inside the brackets is the Dawson function of bt. This function, tabulated by Kaporov (1965), increases to a maximum value near $bt \sim 1$, then decreases. As a consequence, the determination of b requires only the knowledge of the time at which $H(t)$ is a maximum.

The results for these three modes of growth can be arranged in dimensionless form and compared to the results obtained from the same mode operating in a finite solution. Figure 11.8 shows this comparison where the dimensionless quantities $H(t)^*$ for thickness and Dt/L^2 for time are used. (The exact expression for $H(t)^*$ is given in section 11.8 and additional details concerning the finite cases are discussed in Appendix A.3.) In all cases after a certain time, the semi-infinite model begins to overestimate $H(t)^*$. For the TMLE the ratio of the diffusive transport to the thermal relaxation, reflected in D/bL^2, also influences, but so long as D is $\lesssim 0.1 b^2$, both cases agree. This difference, in estimates of $H(t)^*$, arises because in the finite cases the nutrient supply is also finite and no longer an influences, but so long as D is $\lesssim 0.1 b^2$, both cases agree. This difference, in estimates of finite solution does not have to be considered because most growth times are short.

11.4.4. SEMI-INFINITE SOLUTIONS WITH INTERFACIAL KINETICS

Growth behavior is modified drastically when $C_i \neq C_e$ and interfacial kinetics become important. Volume diffusion is now either coupled with a surface reaction limitation or becomes relatively insignificant. The occurrence of a growth rate dependency on substrate orientation is a simple example. This effect is not particularly strong in III–V binary compounds, but becomes more apparent in ternary and quaternary alloys. A good example of surface kinetics is illustrated in the growth of garnets where analysis shows that combined diffusion and surface kinetic processes are operating (Ghez and Giess, 1973, 1974). A more detailed analysis of the garnet system separates these effects and shows that interfacial kinetics control at lower temperatures whereas diffusion processes dominate at higher temperatures (van Eck, 1978). This indicates that as T increases, $K \to \infty$ and $C_i \to C_e$.

Mathematically, the surface reaction, which depends on the order of the reaction, can be combined into a new boundary condition at the solid–liquid interface. Transport to the growing surface must equal the reaction rate there to give the equation

$$D \left(\frac{\partial C}{\partial x} \right)_{x=S/L\,\text{interface}} = K(C_i - C_e)^n, \tag{11.15}$$

where K is the reaction constant and n is the reaction order. When $n = 1$, eqn. (11.15) is the same as the radiation boundary condition used in heat transfer analysis (Carslaw and Jaeger, 1959). For garnets, $n = 1$ appears valid only at high but not at low supersaturation, where higher order processes are evident (van Eck, 1978). Only first-order kinetics are analyzed here.

The thickness of a layer grown by supercooling increases more slowly as the interfacial reaction begins to strengthen. The equation, describing the thickness development as a function of time for growth from a semi-infinite solution with interfacial kinetics, is

$$H(t) = \frac{2\Delta T}{C_s m} \sqrt{\frac{D}{\pi}} t^{1/2} - \frac{\Delta TD}{mC_s K} [1 - e^{h^2 Dt} \text{erfc}(hDt)], \tag{11.16}$$

where $h = K/D$ and proceeds directly from solving the diffusion equation with the boundary conditions of eqns. (11.15) and (11.9) (Ghez and Giess, 1974; Crank, 1957; Bolkhovityanov and Zembatov, 1977). Plotting eqn. (11.16) in dimensionless form results in the graph shown in Fig. 11.9a. Clearly, $H(t)^*$, at any fixed time, decreases as the interfacial reaction increases relative to diffusion transport, a tendency reflected by the ratio KL/D. The behavior of $H(t)^*$ *for $KL/D = \infty$* corresponds to the fast kinetic case described in the previous section and serves as a benchmark.

The epitaxial layer now thickens as $t^{1/2}$ minus a transient term that approaches a constant value. The asymptotic solution shows that

$$H(t) = \frac{2\Delta T}{C_s m} \sqrt{\frac{D}{\pi}} t^{1/2} - \frac{\Delta T}{mC_s} \frac{D}{K} \tag{11.17}$$

and that unless $K \to \infty$, $H(t) \neq 0$ at $t = 0$. This offset at $t = 0$ provides a measure of the kinetic constant which is obtained from the negative intercept, $-(D\Delta T)/(mC_s K)$. Experimentally this type of behavior occurs in the growth of silicon from tin (Baliga, 1978) and garnets from PbO flux (Ghez and Giess, 1973) and is shown in Fig. 11.9b. For very short growth times, eqn. (11.16) expands to give a linear growth law which is

$$H(t) \sim K \frac{\Delta T}{C_s m} t, \tag{11.18}$$

so $H(t)$ is controlled by the supercooling and reaction constant. When the reaction is no longer first-order, numerical analysis is the easiest way of describing the situation (Joullie, 1977). The behavior also alters when finite boundary conditions are imposed, and this is discussed in Appendix A.3.

Ramp cooling with surface kinetics produces a somewhat unexpected result because $H(t)$ approaches a limiting value even for semi-infinite solutions (Ghez and Lew, 1973).

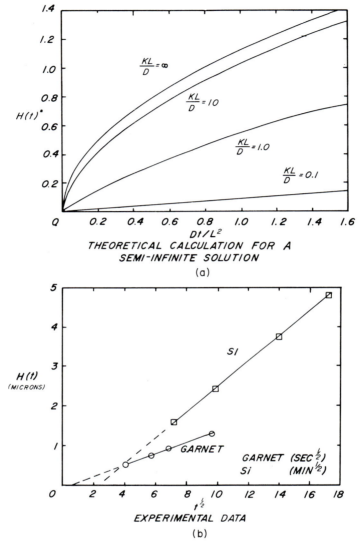

FIG. 11.9. Supercooled growth with interfacial kinetics: (a) theoretical behavior of $H(t)^*$ vs. Dt/L^2; (b) experimental behavior showing the trend expected from the asymptotic solution: silicon (Baliga, 1978), garnet (Ghez and Giess, 1973).

The limiting thickness $H(\infty)$ occurs because as the temperature is driven lower by the ramp cooling process, K decreases at a rate determined by its activation energy. Even though K may be large initially, it can rapidly decrease with temperature causing the growth rate to pass through a maximum and then to decrease. The limiting thickness $H(\infty)$ will decrease as the activation energy for the kinetic process increases relative to the heat of solution, while the relative importance of kinetics to diffusion remains constant; e.g., the activation energy for the surface process is high and the solubility change with temperature is small. On the other hand, increasing the diffusion rate, other things being constant, will increase $H(\infty)$. Finally, when interfacial kinetics totally control the growth process,

$$H(t) = \frac{K_0 \alpha t^2}{2C_s m},$$

(11.19)

where K_0 is the initial value of K. The thickness now scales as t^2 rather than $t^{3/2}$. Furthermore, because $C_i \neq C_e$, the likelihood of interfacial instabilities increases as a result of a localized supersaturation from the concentration buildup.

11.5. Surface Morphology and Lattice Mismatch

Most applications using LPE layers require very smooth surfaces for subsequent processing into devices. For this reason considerable effort has been directed towards examining the factors influencing surface morphology. Unfortunately, the surface appearance is controlled by many things and it is often difficult to decide the exact cause of the various undulations which are observed. Many of the causes are the same for homoepitaxial and heteroepitaxial layers, but the occurrence of lattice mismatch in the heteroepitaxial case adds further complexity. Since it is difficult to separate surface morphology from the effects of lattice mismatch and because of much common ground, they are combined here. This section examines the photomicrographs of some different types of surfaces that develop, the improvements in homoepitaxial surfaces that super-cooling brings, and the effects that mismatch stress has on e growth of epitaxial layers.

11.5.1. SURFACE MORPHOLOGIES

The types of surface defects, commonly seen in LPE, fall into several groups: those arising from latent defects in or on the substrate, such as crystallographic imperfections, misorientation, or poor substrate preparation; those defects forming during growth because of interfacial instabilities, such as constitutional supercooling, or from mismatch dislocations; and those occurring during solution removal. Undoubtedly, there is an interplay between some of these with the most obvious being between the latent defects and the growth of instabilities during growth.

Wavy lines, like those of a surf approaching a beach, are commonly seen on both homo- and heteroepitaxial layers and develop because the substrate is misoriented. Figure 11.10a shows the surface of an AlGaAs layer grown on a GaP substrate which is misoriented between 0.5° and 1° from the (111)B orientation. The spacing between waves can often change over the surface of the wafer because a "pillow-shaped" cross-section sometimes develops as a result of nonplanar polishing. Thus, the surface may be very smooth at one spot where the orientation is perfect and gradually become wavy as the surface becomes more and more misoriented.

Liquid droplets on the surface or embedded in the layer result in part from substrate defects, either from inclusions or from thermal decomposition. Liquid inclusions also occur becuase nucleation is hindered by dirty substrates or severe mismatch between the substrate and epitaxial layer. Spotty nucleation produces a reduced number of growth sites that grow at the expense of further nucleation. As growth proceeds, these islands

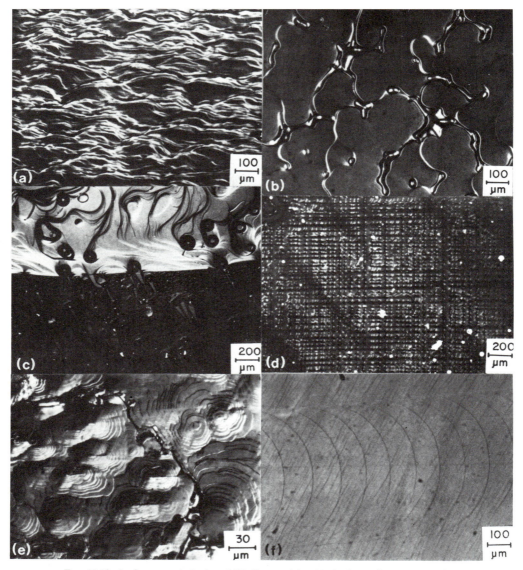

Fig. 11.10. Surface morphologies of III–V materials: (a) ripples and waves caused by substrate misorientation; (b) liquid inclusion resulting from poor nucleation; (c) thermal decomposition of a substrate; (d) growth induced by mismatch dislocations; (e) growth stimulated by dislocations associated with stacking faults of microtwin lamellae; (f) meniscus lines from solution removal.

finally impinge on one another and trap portions of the solution (Astles and Rowland, 1974). The resulting surface in all of these cases is the same and is shown in Fig. 11.10b where GaAs was grown on a GaAs(111)B substrate which had a slight native oxide film. Clean substrates or modes of growth that enhance nucleation reduce the number of these inclusions (Moon and Vander Plas, 1978). When one constituent of the substrate is more volatile than the rest, it can vaporize, leaving a localized undersaturated liquid. Usually the site of the decomposition is an imperfection like a dislocation, and is more common in LPE grown GaP (Michel, 1975) and InP (Pak *et al.*, 1975) than in GaAs LPE. Nevertheless, GaAs can lose As and Fig. 11.10c shows this thermal decomposition on a GaAs(111)B substrate; the dark area is the surface which was exposed to H_2 at 900°C and the light area is a region where the substrate was etched back in solution after the exposure. A gas phase rich in the volatile species generally helps, but may not entirely prevent, this decomposition.

Orderly arrays of ridges and mounds with plateaus having a small hole or slash result from faster localized growth stimulated by the presence of dislocations (Moon and Antypas, 1973; Kimura *et al.*, 1977). Orderly arrays of ridges form when mismatch dislocations occur in a regular pattern. The example in Fig. 11.10d is of a surface of GaAs that was deposited on a freshly grown AlGaAs layer. Because nucleation of GaAs on AlGaAs is more difficult than on GaAs, growth initiated at sites associated with the mismatch dislocations, introduced after the coherency thickness was exceeded. Such examples are only seen when the mismatch is slight, being 0.17% in this case, and when nucleation by other means is hindered. At greater lattice mismatch the dislocations are more dense and disordered and result in the pattern seen in Fig. 11.10e; the slashes are stacking faults or microtwin lamellae (Booker, 1965), which often develop, having partial dislocations of opposite sense about which spiral growth proceeds. These are seen more often at large mismatches.

Very faint lines, which are roughly perpendicular to the direction of solution removal and which cross the surface of the epitaxial layer, are shown in Fig. 11.10f. These are "meniscus lines" which result from a "slip–stick" motion of the edge of the liquid as it moves across the surface during solution removal at the end of growth (Small *et al.*, 1975).

11.5.2. SUPERCOOLING AND SUBSTRATE MISORIENTATION EFFECTS

Ripples, terraces, or more random undulations form mainly from substrate misorientation. A misorientation near a low index surface causes terraces which may lead to the step bunching that is observed (Peters, 1972; Saul and Roccasecca, 1973). When the orientation of GaAs is < 5 min of arc off orientation, smooth flat surfaces grow during LPE (Bauser *et al.*, 1974). Moving away from the exact orientation causes the surfaces to show a wave-like pattern that becomes more apparent as misorientation increases and finally decomposes into a cellular pattern, characterized by furrows several microns deep (Mottram and Peaker, 1974). On the more well-behaved surfaces, misorientation < 0.2°, the terrace heights are ~ 0.1 μm, while widths or wavelengths are several tens of microns. A more detailed study shows that nuclei formation and thermal fluctuation at the solid–liquid interface govern the surface terraces of GaAs (Mattes and Route, 1974). The shape of the nuclei change with substrate orientation and also the imposed temperature gradient.

Fortunately, the sensitivity of LPE layer smoothness to substrate orientation relaxes when supercooling is used (Mihara *et al.*, 1975; Crossley and Small, 1972b). This further shows that the initial stages of growth or nuclei formation are responsible in part for the terraces because supercooling enhances nucleation. Figure 11.11 shows the relationship between supercooling and misorientation, surface morphology, and layer thickness

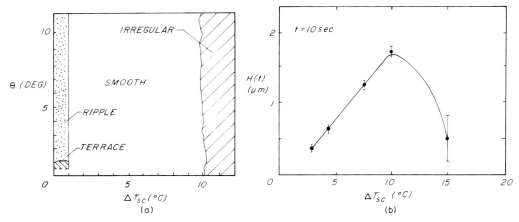

FIG. 11.11. The effect of supercooling on surface morphology and thickness (Toyoda *et al.*, 1975): (a) substrate misorientation and supercooling vs. surface appearance; (b) thickness vs. supercooling.

(Toyoda *et al.*, 1975, 1976). With supercooling, a smooth layer can be grown even when the substrate is misoriented several degrees (Fig. 11.11a). At only a few tenths of a degree misorientation, the width and height of the terraces decreases with increasing supercooling (Toyoda *et al.*, 1976). Increasing ΔT_{sc} and keeping t constant produces an increase in $H(t)$, but only to a point. Above this point, $\Delta T_{sc} \cong 10°C$ in this example (Fig. 11.11b), $H(t)$ actually decreases at precisely the same point that the surface morphology deteriorates. This behavior develops at the onset of homogeneous nucleation (or two-phase growth) in the solution (Hsieh, 1974) and sets an upper limit to ΔT_{sc}, depending on the materials system.

In order to understand the relationship between substrate misorientation and surface morphology, it is useful to imagine that the initial terraces are morphological perturbations, then to ask whether the perturbations will grow with time or not. A misoriented substrate provides the necessary perturbation since steps on the surface have a single sign (+ or −) and can bunch together to form a terrace. In contrast, dislocations or two-dimensional nuclei on low index planes possess equal numbers of + and − steps; these opposite polarity steps annihilate each other when they meet, leading to a macroscopically smooth surface, not a perturbation.

Morphological stability theory proves to be a useful means of describing the situation (Mullins and Sekerka, 1964) and has been developed for supercooled LPE (Nishinaga *et al.*, 1978). Two forces are competing: volume diffusion tending to destabilize and capillary forces acting to stabilize the growing surface. Theory shows that there exists a critical wavelength or minimum stable distance between terraces (see Appendix A.4). For the supercooled case, the critical wavelength is

$$\lambda_0 = \frac{2\pi C_e \Gamma}{(C_0 - C_e)} \sqrt{\pi D t}, \tag{11.20}$$

where Γ is the capillarity constant which is proportional to γ, the interfacial free energy. As a consequence $\lambda_0 \propto \Gamma (\Delta C)^{-1}$, so the critical terrace spacing decreases with increased supercooling and with decreased interfacial energy. A temperature dependency arises potentially through the value of C_e and Γ_0. Wavelengths $> \lambda_0$ are stable and grow, but those ripples with $\lambda < \lambda_0$ must coalesce to form stable ripples with longer wavelengths. The most stable wavelength is not predicted by this theory, but experimental results show a decrease in the ripple spacing with supercooling (Nishinaga *et al.*, 1978).

11.5.3. STRESSES AND SURFACE CHANGES AS A RESULT OF MISMATCH

Lattice mismatch, occurring whenever there is a lattice constant disparity between layer and substrate, can have a profound effect on LPE growth. Mismatch produces elastic strain, and inevitably leads to dislocations or other defects as the layer thickens. The effects of induced strain vary considerably and can be so severe that epitaxy is prevented entirely, even when only a slight mismatch is present. Before proceeding, however, it is necessary to understand what takes place as a mismatched layer grows on a substrate.

Mismatch dislocations occur only after the layer reaches a critical thickness, which depends on the initial mismatch. Thus, the layer first grows pseudomorphically with the substrate before dislocations are introduced. (In cubic systems the unit cell is usually deformed to a tetragonal structure because the planar stresses produce, through the Poisson-induced strain, a movement perpendicular to the plane (Kishino *et al.*, 1974). At a critical thickness h_c the energy in the elastically strained layer becomes greater than the same layer with mismatch dislocations (van der Merwe, 1970; Jesser and Kuhlmann-Wilsdorf, 1967; Matthews, 1975). Mismatch dislocations then form either by gliding in from the surface or by bending over threading dislocations from the substrate. At thicknesses $> h_c$ dislocations, either totally or combined with elastic strain, accommodate the mismatch.

The value of h_c decreases as the mismatch between substrate and layer increases. To calculate h_c, in an idealized case, requires calculation of ε_c, the strain at which the sum of the elastic and dislocation energies is minimal. This strain is a function of thickness and the value of the misfit parameter f, defined as $f = (a_s - a_e)/a_e$, where a_s and a_e are the stress free lattice constants of the substrate and epitaxial layer, respectively. The largest value possible for ε_c is the misfit f. When ε_c is predicted $> f$, the film strains to match its substrate exactly, so ε_c becomes f. But, when $\varepsilon_c < f$, dislocations only partially relieve the strain to the amount equal to $f - \varepsilon_c$. The critical thickness is the thickness where $\varepsilon_c = f$ and for materials with the same elastic properties this is given by

$$h_c = \frac{b_{mf}}{8\pi f(1 + v_p)} \ln\left(\frac{h_c}{b_{mf}} + 1\right), \tag{11.21}$$

where b_{mf} is the edge component of the Burgers vector of the mismatch dislocations and v_p is Poisson's ratio (Matthews, 1975). A useful approximation (Olsen and Ettenburg, 1978) that agrees within a factor of 2 with experimental data is

$$h_c \sim \frac{b_{mf}}{2f}. \tag{11.22}$$

However, a better approximation to eqn. (11.21) uses 4 instead of 2 in the denominator. From this one might expect a pseudomorphic layer $\sim 1 \mu$m thick when $f \sim 10^{-4}$, but at

$f \sim 10^{-2}$, h_c would be only 10^2 Å. Figure 11.12 shows a plot of h_c vs. f for a $b_{mf} \sim 2$ Å (a reasonable value for GaAs) and a photomicrograph showing mismatch dislocations on the surface of an epitaxial layer that has just lost coherency with the substrate (Moon et al., 1978).

Layers can exist in a state of tension or compression depending on the sign of f as the following table shows:

Mismatch Stress State in Epitaxial Layers

f	Lattice Constants	Stress
>0	$a_s > a_e$	Tension
0	$a_s = a_e$	Neutral
<0	$a_s < a_e$	Compression

These stresses can be useful at times as in the garnets where the stresses induce, by magnetostriction, the uniaxial magnetic anistropy needed for bubble memories. In III–V alloys the stresses are usually unwanted and merely complicate heteroepitaxy. In order to compute the actual stress σ in an epitaxial layer the difference in thermal expression coefficients must also be included, so for a uniformly strained isotropic film,

$$\sigma = (1 - \eta)f + \eta(\alpha_s - \alpha_e)\Delta T_{gr}\frac{\varepsilon}{1 - v_\rho}, \qquad (11.23)$$

where η is the fractional stress relief (Miller and Caruso, 1974), α_s and α_e are the substrate and epitaxial film thermal expansion coefficients, respectively, E is Young's modulus and ΔT_{gr} is the difference between growth and room temperature (Besser et al., 1971; Blank and Nielsen, 1972).

Stress influences the final appearance of the film, particularly when it is in tension. Layers in tension are susceptible to cracking, rather than introducing dislocations, as a means of relieving mismatch stress during growth. Garnets are particularly susceptible because of the high dislocation energy owing to a large Burgers vector in these systems ($b \sim 10$ A and the dislocation energy is proportional to Gb^2, where G is the shear modular). Cracks normally will not propagate in a film strongly bound to its substrate unless the film thickness is greater than the Griffith crack length and the misfit stress exceeds the fracture stress in the layer. A relationship for this effect develops by using the expression for the Griffith crack length and by estimating the surface free energy of the crack from the Young's modulus and lattice constant (Matthews and Klokholm, 1972). Providing that crack sources, such as fissures, steps, nicks, inclusions, or other faults, exist inside or at the edge of the layer, the expression relating f to a thickness above which crack propagation might be expected is

$$h_g = \frac{a_0(1 - v_\rho)^2}{5\pi f^2}. \qquad (11.24)$$

This relationship applies to pseudomorphic films, but by replacing f with the actual strain parallel to the film plane it will also work for nonpseudomorphic layers. Good agreement between eqn. (11.24) and experimental data in garnet systems is seen (Matthews and Klokholm, 1972). Comparison of eqn. (11.24) with eqn. (11.21) shows that misfit

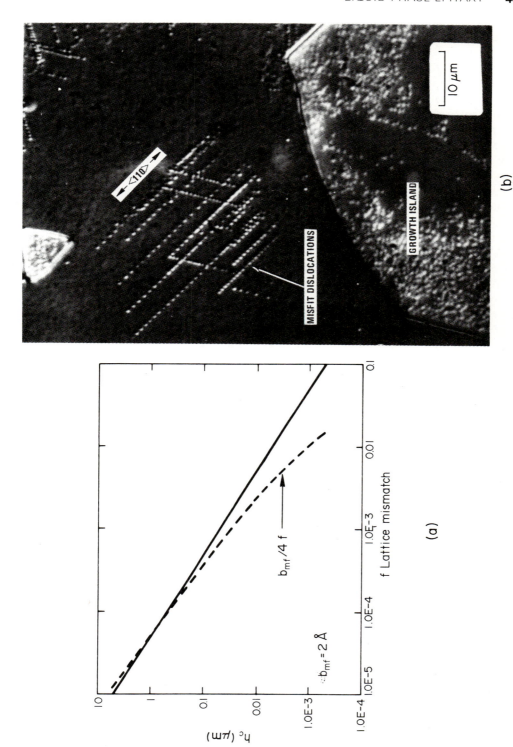

FIG. 11.12. Introduction of dislocations as a result of lattice mismatch: (a) critical thickness above which dislocations are expected vs. mismatch; (b) photomicrograph showing the introduction of dislocations as the layer loses coherency (Moon et al., 1978).

dislocations should relieve the elastic strain before cracking occurs, provided that dislocation can be introduced.

Under compression the behavior is different and leads to film faceting in garnets (Blank and Nielsen, 1972). These facets appear as arrays of hillocks that roughen the surface. Possibly these evolve because under sufficient compressive stress even garnets can nucleate dislocations which serve as rapid growth centers.

One way to control or at least minimize stress in the layer is to maintain a composition that produces a lattice-matched condition at the growth temperature. This is nicely illustrated in the AlGaAs–GaAs heterostructure where introducing the proper amount of P into the AlGaAs layer reduces the stress in that layer to a minimum (Rozgonyi et al., (1974). The amount necessary is small ($x_p^l \sim 10^{-5}$) but the residual stress changes from compression to tension as x_p^l varies about its minimum value as is seen in Fig. 11.13.

One heteroepitaxial system which exemplifies much of the above discussion is the $In_{1-x}Ga_xAs$ ternary alloy grown on InP substrates. Again, epitaxy can occur either in a state of tension, compression, or unstressed by adjusting the composition of the solution. Exact matching with the InP occurs at the composition $In_{0.53}Ga_{0.47}As$, so excess Ga produces tension and excess In produces compression since the lattice parameters are 5.65 Å and 6.06 Å for GaAs and InAs, respectively. Figure 11.14 outlines the conditions in the layer and shows the various surface morphologies that occur as the stress in the layer changes (Nagai and Noguchi, 1976; Sankaran et al., 1976; Hyder et al., 1977). Smooth surfaces are only possible within a narrow range of compositions. Moving away from this range produces an ever-increasing roughening of the surface until, finally, the layer fails to grow.

Near the lattice-matched condition, a phenomenon known as lattice or compositional latching occurs, where the same solid grows from solutions with different compositions (Stringfellow, 1972). The layer that deposits on the substrate is not of the

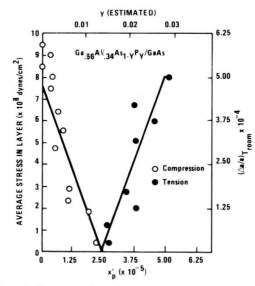

Fig. 11.13. Control of layer stress by adjustment of the alloy composition (Rozgonyi et al., 1974).

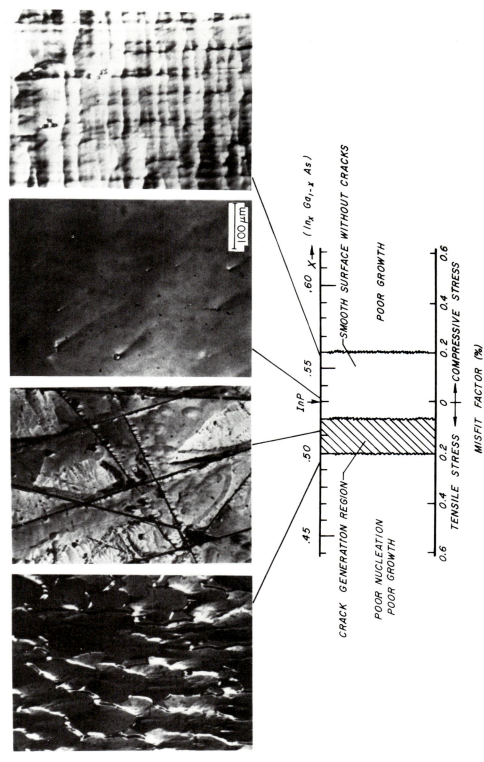

FIG. 11.14. The effect of alloy composition of an epitaxial layer on the surface morphology and state of stress on that layer: $In_xGa_{1-x}As$ grown on InP (adapted from Nagai and Noguchi, 1976; photos courtesy of Hyder and Saxena).

447

composition expected from solution equilibria, instead it is of a lattice-matched composition. The total free energy of the solid apparently increases by a composition-dependent mismatch energy term, consisting of strain energy plus mismatch dislocation energy. This new term perturbs the solid composition in the direction needed to minimize the mismatch energy even though the chemical free energy increases. A kinetic model, which refines this qualitative thermodynamic argument, suggests that in the early stages of heteroepitaxy atoms are selected to reduce local strain buildup, producing as a result a coherent growth, but eventually under sufficient chemical driving force, dislocation loops nucleate (Hirth and Stringfellow, 1977). This latching occurs more readily from compositions that would potentially yield a compressive stress if they grew as an epitaxial layer in normal equilibrium with the solution. This trend is followed in the growth of $In_xGa_{1-x}As$ on InP(111)B (Taheda and Sasaki, 1978) and $InAs_{1-x}Sb_x$ on GaSb(100) (Ludowise and Gertner, personal communication), even though $In_{1-x}GaP$ on GaAs(111)B, where the phenomenon was first observed, does not. The discrepancy must, in part, be due to differences in dislocation energies and tensile limits of the layers, since layers prone to cracking under tension would not be expected to show latching.

Even if the correct solution composition for a given orientation is known, there is no guarantee that the same composition will work on a different orientation. This appears most strongly in situations where only slight variances in liquid composition will totally eliminate the appearance of smooth epitaxial layers. Again a good example is the $In_{1-x}Ga_xAs/InP$ system, the segregation coefficient for gallium has different values on the (111)B and (100) orientations and, in addition, shows a different temperature dependence for each orientation. Figure 11.15 shows the distribution coefficient K_{Ga} vs. $1/T$ for the two substrate orientations (Antypas et al., 1978); the dashed lines represent the calculated values from a thermodynamic model using two sets of data. This behavior extends into the quaternary region of InGaAsP (Oe and Sugiyama, 1978) and illustrates some of the

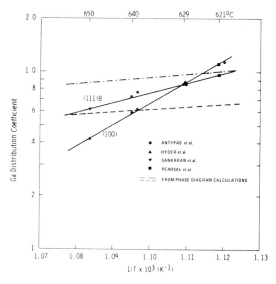

FIG. 11.15. The effect of substrate orientation and temperature on a distribution coefficient during LPE. $In_xGa_{1-x}As$ on InP at the lattice-matched condition (Antypas et al., 1978).

subtle changes that lead to very time-consuming searches for the correct solution composition.

11.5.4. CHANGES IN GROWTH RATE AND SEGREGATION COEFFICIENTS RESULTING FROM MISMATCH

Mismatch, even though very slight, can at times nearly suppress epitaxial growth. The exact value of f at which growth will cease varies with each system, but can qualitatively be understood by assuming the nucleation rate follows an exponential rate expression. The rate decreases by a factor of e^{-Ef^2}, where the argument of the exponential represents the additional volume free energy due to elastic strain of the nuclei (Turnbull and Vonnegut, 1952; Walton, 1969). It is worth noting that regardless of the sign of f, mismatch reduces the nucleation rate. For the InGaAs–InP system in Fig. 11.14, layers will not grow if $|f| \gtrsim 0.2\%$. This sensitivity is even more severe than in some garnet system in which layers fail to grow only when $|f| \gtrsim 1.4\%$ (Blank and Nielsen, 1972).

Intimately connected with nucleation are the growth rate and the segregation coefficients, both of which change by a strain energy term. As a consequence, a dynamic equilibrium exists that is self-adjusting, where the growth rate changes because of misfit and the segregation coefficients change because the growth rate changes. These effects have been partially separated in a study of garnet growth that used a dipping process at high rotation rates to eliminate transient effects. $R(t)$ is, then, expected to decrease from a maximum value at the lattice matched condition by a term proportional to f^2, $(\Delta a)^2$; Fig. 11.16a shows such a decrease for the epitaxial growth of La_2O_3 and Ga_2O_3 doped $Y_3Fe_5O_{12}$ (YIG) films. However, the change in $R(t)$ with composition is not always as clear cut since adding a small amount of gallium to the YIG film increases $R(t)$, but adding

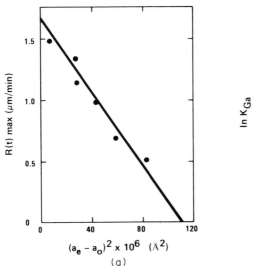

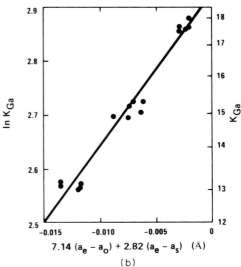

FIG. 11.16. The effect of mismatch on the growth rate and distribution coefficient of Ga in YIG (Brice *et al.*, 1975).

a large amount decreases it (Robertson *et al.*, 1974; Brice *et al.*, 1975). The distribution coefficient varies with both Δa and $R(t)$, and follows an equation of the type

$$\ln K_i = b_i + c_i(a_f - a_0) + d_i(a_f - a_s) + e_i R(t), \tag{11.25}$$

where K_i is the distribution coefficient of solute i, b_i to e_i are constants, and a_0 is the lattice parameter of pure stoichiometric YIG (12.3758 Å) (Brice *et al.*, 1975). The first three terms on the rhs arise from an expression for the free energies in the solid and liquid. At stable equilibrium these energies must be equal and minimal. This criterion leads, through the method of Lagrangian undetermined multipliers, to an expression relating $\ln K_i$ to f. The theoretical expression shows that the constants c_i and d_i are related to the changes in the lattice constant caused by foreign atoms straining the lattice, $\partial a_e / \partial x_i$; Fig. 11.16b shows the behavior of gallium-doped YIG which follows this general trend. The dependency of K_i on $R(t)$ develops from the same expression used for melt growth across a diffusion boundary δ. The effective coefficient K_{eff} changes from the equilibrium value K_0, but always tends towards one as $R(t)$ increases (Burton *et al.*, 1953). Note again that $\ln K_i$ is proportional to f, not f^2, although $R(t)$ is proportional to f^2.

The same trend in III–V alloys is expected, but has not been established. Chances are that the trend might not be observed because of the ease with which dislocations are introduced in these alloys. The growth rate as a function of mismatch is expected to decrease from the matched condition, then to increase as dislocations are introduced and enhance growth. As a result, what might be observed is a growth rate minimum at the lattice matched condition. The surface morphologies shown in Fig. 11.10d and f suggest this and thickness measurements on epitaxial layers of InGaAsP grown on GaAs(111)B show this trend (Moon, unpublished results).

11.6. Outlook for LPE

Diffusion in the solution has, in most cases, been rate limiting, but the effects of interfacial kinetics are now appreciated and further work should lead to an understanding of the surface processes that control the interfacial kinetics. Orientation dependence of the growth rate and segregation coefficients are well known, and partially explained, so a reasonable complete theoretical treatment is expected shortly. The different species occurring in the solution are not so firmly established, but the work on Peltier-induced LPE of III–V has shed some light on the different partially charged species which are present. In turn, this may lead to a more comprehensive explanation of the doping dependence on orientation, although the Schottky barrier model (Casey *et al.*, 1971; Zschauer and Vogel, 1971) now agree well with experimental evidence. Surface morphologies have been studied extensively and many of the important factors are known. Further studies will likely just refine many of the present conclusions. The increasing desire to make heterojunction devices should place additional emphasis on growing lattice matched layers, since all such devices degrade in performance as a result of mismatch.

Some form of supercooled growth has emerged as the preferred growth mode in most materials systems. Supercooled growth in the garnet systems has eliminated the non-uniformities in both the parallel and perpendicular directions relative to the growth axis. The isothermal growth environment and the use of rotation to produce a thin diffusion boundary layer are also partially responsible. In III–V systems some of the beneficial effects have been observed for supercooling. The principal difference between the two materials systems is the large amount of supercooling and stability that is possible in the magnet garnet solutions.

III–V materials are now grown by both LPE and VPE methods. Whether this will continue is open to some question if larger more uniform areas are required for device fabrication. Magnetic garnets, on the other hand, are exclusively grown by LPE and this will likely continue. In the case for LPE, the simplicity of the equipment, the higher deposition rates, the wider choice of dopants, the longer minority carrier diffusion lengths and fewer deep levels seen in III–V LPE material, all favor its usage. But, balanced against these arguments are those favoring VPE, namely, the larger and more uniform areas possible, the potentially more uniform composition, and the ability to deposit layers greatly mismatched from the substrates. Also, the growth of aluminum-bearing compounds, once the domain of LPE, is now possible by VPE since the advent of AlGaAs organometallic VPE (Blakeslee and Bischoff, 1971; Stringfellow and Hall, 1978). The ability to grow LPE layers on greatly mismatched substrates has, however, been enhanced, but still not to the extent of VPE, by the use of transient mode LPE. Uniformity over large areas and large scale production is where LPE has been lacking. Even this has been partially overcome in the production of GaP light-emitting diodes. Often 50 or more substrates are dipped into a solution at one time and then ramp-cooled to produce a single epitaxial layer. These production methods are, nevertheless, oriented towards single layer deposition, so large scale multilayer growth by LPE remains a challenge.

In summary, liquid phase epitaxy now appears to be a firmly established and mature growth method that will remain an important facet of solid state technology.

Acknowledgements

I should like to thank Drs. B. Cooper III, M. Ludowise, and H. Vander Plas for reading and commenting on the manuscript. In addition, the photographs and data provided by Drs. G. A. Antypas, S. B. Hyder, R. Saxena, and Y. M. Houng are greatly appreciated. And finally, the typists, N. Anderson and S. Carter, are thanked for their patience and endurance.

11.7. Appendixes

APPENDIX A.1. FREE CONVECTION DURING LPE

Free convection can develop in LPE boats as a result of temperature or density gradients which may be present across the solution. Normally, the situation is analyzed in terms of the dimensionless temperature gradient, known as the Rayleigh number; when it exceeds a critical value, convection cells are expected, otherwise the solution is still. The Rayleigh number is defined as

$$Ra \text{ (thermal)} = g\frac{\alpha_T \Delta T L^3}{k v}, \tag{A.1}$$

where g is the acceleration caused by gravity, α_T is the volumetric thermal expansion coefficient, k is the thermal diffusivity, v is the kinematic viscosity, and ΔT is the temperature difference between horizontal surfaces L distance apart. Its critical value reflects the condition where the rate of free energy liberated by the buoyant up-rising of the fluid exceeds the rate of energy dissipated by thermal and viscous motion.

The critical value is 1700 for the classical case where two semi-infinite conducting horizontal plates confine a solution heated from the bottom. Many other configurations are possible, so the critical number depends on the geometric shape as well as the thermal conditions (see Carruthers, 1977, for a recent review). Depending on the conditions, laminar often cellular convection, which may be steady or transitory, is expected until Ra reaches a second critical value above which the flow is turbulent. Usually the second number is an order of magnitude or so greater than the first. The apparatus can be stabilized against convection by adjusting the temperature gradient or changing the geometry. Just reducing the thickness of the solution leads to a stabilization since $Ra \propto L^3$.

Density or solute gradient from growth can also exist and drive the convection cells (see Small and Crossley, 1974; Jakeman and Hurle, 1973). To account for this, an equivalent expression for concentration differences exists and is

$$Ra \text{ (solute)} = g\frac{\Delta CBL^3}{Dv}, \tag{A.2}$$

where B is the volumetric change/unit concentration change, ΔC is the concentration difference, and D is the mass diffusivity in the solution. Since the mass diffusivity is normally much less than the thermal diffusivity, the critical Ra may be exceeded at very small density differences. Concentration gradients can be stabilizing or destabilizing depending on their direction and can destabilize a stable thermal situation. In a horizontal LPE system, a substrate is often below a source wafer, and growth results in a depletion of the solution adjacent to the solid. The resultant gradient will be stabilizing if the solvent is denser than the solute (as in GaAs growth) but destabilizing if the opposite. Fortunately, in III–V LPE, the solutal gradient is stabilizing when the substrate is under the solution (Small and Crossley, 1974), and as a result free convection is not expected from thin solutions used in III–V growth.

APPENDIX A.2. OTHER FACTORS—THE CONSEQUENCES OF AN ARRHENIUS EXPRESSION DESCRIBING THE LIQUIDUS AND SPONTANEOUS NUCLEATION

An important assumption in sections 11.4.4 and 11.4.5 is that m is constant throughout the growth. By relaxing this restriction an estimate of the $H(t)$ variation with temperature is possible. The equilibrium concentration for most solutions varies as

$$C_e = B\,e^{-\Delta H/RT}, \tag{A.3}$$

where ΔH is the heat of the solution (Elwell and Neate, 1971). As a consequence of the exponential form $H(t)$, scales for supercooling as $e^{-\Delta H/RT_0}$ and for ramp growth as $T^{-2}\,e^{-\Delta H/RT_0}$ (Giess and Ghez, 1975). The experimental results, which show this trend, for GaAs grown on GaAs by ramp cooling are shown in Fig. 11.17. The same data for $H(t)$ are plotted vs. $1/mC_s$ and vs. $1/T$. In both cases, the expected form is followed and shows that for these short growth times the results are identical.

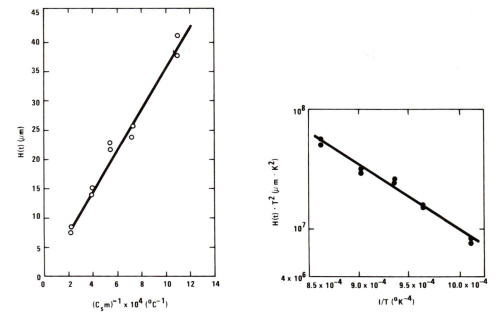

FIG. 11.17. The scaling of layer thickness with the reciprocals of the slope of the liquidus line and temperature for the growth of GaAs by ramp cooling (Moon and Long, 1976).

A complete description of the ramp-cooling process results when eqn. (A.3) is used in conjunction with a linearly decreasing temperature program (Minden, 1970; Tiller and Kang, 1968; Ghez, 1973). The change in concentration with time is now

$$C_e = C_0 e^{-t/\tau}, \tag{A.4}$$

where τ is the relaxation time which can be estimated from the properties of the system. This relaxation time accounts for the nonlinearity of the liquidus and comes from the equation

$$\tau = \frac{mC_0}{\alpha} = \frac{RT_0^2}{\alpha\Delta H}. \tag{A.5}$$

For the case of ramp cooling the greatest error in calculated values of $H(t)$ is expected; however, $H(t)$ that is obtained using eqn. (A.4) is identical to the expression that assumes m is constant so long as $\alpha t/T < 0.1$ (Ghez, 1973). Such a condition exists during most LPE growth, e.g. if $T_0 = 800°C$ and $\alpha = 1°C/min$, $t \sim 80$ min. Deviation from an exact solution to the ramp-cooling problem increases as ΔH becomes smaller compared to the initial thermal energy RT_0 (Ghez).

Spontaneous nucleation inside the solution can drastically reduce the layer thickness expected. Any time during growth the concentration in the solution can become oversaturated, so when $C_e - C(x, t)$ exceeds a critical value, small crystallates form, effectively altering the boundary conditions (Crossley and Small, 1971). Description of this behavior is easily handled by numerical analysis (Crossley and Small, 1972), or by modifying growth efficiency curves to account for initial and volume nucleation which might depend on the previous history of the solution (Malinin and Nevsky, 1978; Malinin *et al.*, 1978).

APPENDIX A.3. FINITE SOLUTIONS

Imposing a finite boundary condition in effect eliminates the continuous source of nutrients located at infinity. Now the concentration profile, which in the semi-infinite case is unlimited, becomes fixed when it reaches the boundary surface. At this point, the dependency of $H(t)$ and $R(t)$ on time alters and no longer follows the trend of the semi-infinite solution.

A mathematical description of this situation usually assumes, depending on the boundary conditions, that the growing interface is located at both $\pm L/2$ of a solution of thickness L or that the flux at one boundary surface is zero, by symmetry, and the growth surface is at $L/2$. The manner of saturating the solution, consequently, governs its effective thickness (Moon, 1974). For instance, placing a saturation wafer or forming a solid crust at the free interface halves the thickness of the solution available for growth, since growth now occurs at both surfaces.

The mathematical expressions for $H(t)$ for growth from finite solutions are in the form of a power series and these are listed in Table 11.1 along with the expressions for the semi-infinite solutions. The comparison between growth in a semi-infinite and finite solution has been shown previously in Fig. 11.8. Clearly evident is the change in behavior as the influence of the finite boundary is felt and the overestimation in $H(t)$ when the expressions for semi-infinite solutions are not valid.

For example, in ramp cooling $H(t)$ is proportional to $t^{3/2}$ but increases only as t after $Dt/L^2 \sim 1$. At this time $C(x, t)$ becomes fixed and therefore advances at the rate that the

TABLE 11.1. *Thickness $H(t)$ as a Function of Time for Different LPE Growth Modes*

Growth Mode	Semi-infinite	Finite
Supercooled $K = \infty$	$\dfrac{2\Delta T}{C_s m}\sqrt{\dfrac{Dt}{\pi}}^{1/2}$	$\dfrac{\Delta T}{C_s m}\left[1 - 2\displaystyle\sum_{n=0}^{\infty}\dfrac{e^{-\lambda_n^2\Phi}}{\lambda_n^2}\right]$ where $\begin{aligned}\Phi &= \dfrac{Dt}{L^2}\\ \lambda_n &= (n+\tfrac{1}{2})\pi\end{aligned}$
Ramp $K = \infty$	$\dfrac{4}{3}\dfrac{\alpha}{C_s m}\sqrt{\dfrac{D}{\pi}}\,t^{3/2}$	$\dfrac{\alpha}{C_s m}\left[Lt + \dfrac{2L^3}{D}\left(\displaystyle\sum_{n=0}^{\infty}\dfrac{e^{-\lambda_n^2\Phi}}{\lambda_n^4} - \dfrac{1}{6}\right)\right]$
TMLE $K = \infty$	$\dfrac{\Delta T}{C_s m}\sqrt{\dfrac{4D}{\pi b}}\left[e^{-bt}\displaystyle\int_0^{\sqrt{bt}} e^{\lambda^2}\,d\lambda\right]$	$\dfrac{2\Delta T D}{C_s m L}\displaystyle\sum_{n=0}^{\infty}\dfrac{e^{-\lambda_n^2\Phi} - e^{-bt}}{b - \dfrac{D}{L^2}\lambda_n^2}$
Supercooled K finite	$\dfrac{2\Delta T}{C_s m}\sqrt{\dfrac{D}{\pi}}\,t^{1/2} - \dfrac{\Delta T}{mC_s h}\left[1 - e^{h^2 DT}\mathrm{erfc}(h\sqrt{Dt})\right]$ where $h = \dfrac{K}{D}$	$\dfrac{L\Delta T}{C_s m}\left[\dfrac{\Phi}{1+r} + 2\displaystyle\sum_{n=1}^{\infty}\dfrac{1 - e^{\alpha n^2\Phi}}{\alpha_n^2(1 + r + r^2\alpha_n^2)}\right]$ where $r = \dfrac{D}{KL}$ and $\tan\alpha_n + r\alpha_n = 0$
Steady state $K = \infty$		$\dfrac{\Delta T}{C_s m}\Phi + \dfrac{2L}{C_s m}\Delta T\displaystyle\sum_{n=1}^{\infty}\dfrac{\cos n\pi(1 - e^{-n^2\Phi})}{n^2}$

boundary condition changes, i.e. at αt. The expression for $H(t)$ after long times is $H(t) \propto \alpha L t - $ constant (Moon, 1974). Even this expression requires correction owing to the changes at the nongrowing interface, but the general conclusions are still valid (Malinin and Nevsky, 1978; Malinin *et al.*, 1978).

The growth rate will increase by simply decreasing the diffusion region thickness. This has been effective in garnet growth where a boundary layer is set up by horizontally rotating a substrate in a manner commonly used in the investigation of electrode processes (Giess and Ghez, 1975). The thickness of the boundary layer reaches a steady state value determined by the angular rotation. From analysis (Levich, 1942; Burton *et al.*, 1953), the boundary layer δ is given by

$$\delta = 1.6 D^{1/3} v^{1/6} \omega^{-1/2}, \tag{A.6}$$

where v is the kinematic viscosity and ω is the rotation rate; this expression is valid so long as the growth rates remain low such that $R^{-3} \omega^{3/2} D^2 v^{-1/3} \ll 1$ (Wilson, 1978). Steady state occurs quickly in times $t < \delta^2/D \propto \omega^{-1}$. Figure 11.18 shows $H(t)$ vs. t and ω for garnet growth where the enhanced growth is clearly evident (Ghez and Giess, 1973). Note that at $t = 0$ the intercept is not zero but has a finite value resulting from interfacial kinetics.

A different behavior arises when interfacial kinetics are present during growth through an established boundary layer so steady state growth exists. Within δ, all the diffusion formed concentration gradients occur, while outside $C(x > \delta, t)$ remains constant. In this case, the appropriate equation in Table 11.1 shows that

$$H(t) = \frac{\Delta T}{mC_s} \frac{D/\delta}{1 + r} t + \text{const},$$

the constant arising from the transient term formed by the sum expression. The term r reflects the relative importance of transport to kinetics and reduces by $(1 + r)^{-1}$ the values of the diffusion coefficient determined in this steady state condition (Ghez and Giess). Growth reverts to transport limitations as $K \to \infty$ or $r \to 0$.

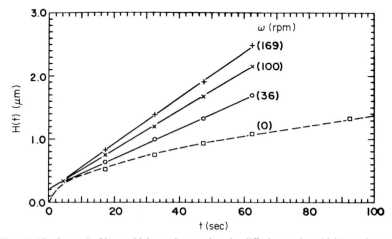

FIG. 11.18. Control of layer thickness by varying the diffusion region thickness through substrate rotation. Growth of $Eu_{1.1}Y_{1.9}Fe_5O_{12}$ on $Gd_3Ga_5O_{12}(111)$ substrates (Ghez and Giess, 1973).

Steady state growth, a mode only possible in a finite system, requires the establishment of a temperature gradient between source and substrate wafers. Initially, the two wafers will be at the same temperature, then the temperature of the source is raised, producing a transient period while the temperature profile is changing to the final uniform temperature gradient (Stringfellow and Greene, 1971). The expression for $H(t)$ develops from expressions in Crank and is also stated in Table 11.1. Immediately after the transient period, $H(t)$ grows as t and is proportional to the temperature gradient. This method is particularly useful if thick layers or growth at a constant temperature are needed and has the added advantage of being more stable than ramp growth to constitutional supercooling.

APPENDIX A.4. STABILITY THEORY AND CONSTITUTIONAL SUPERCOOLING

Morphological stability theory includes the stabilizing effects of the capillary forces and, therefore, proves to be a useful means of describing surface perturbations (Mullins and Sekerka, 1964). In this theory, the growing surface is no longer planar, but has superimposed on it a sinusoidal perturbation described by

$$\delta(t)\sin \omega y, \tag{A.7}$$

where y is perpendicular to the growth direction x, $\delta(t)$ is the amplitude of the perturbation, ω is the angular frequency defined by $\omega = 2\pi/\lambda$, and λ is the wavelength between perturbations. The progress of the ripple amplitude follows the expression

$$\delta(t) = \delta_0 \exp \int_0^t f(\omega)\,dt, \tag{A.8}$$

where δ_0 is the initial perturbation and $f(\omega)$ is $(d\delta/dt)/\delta(t)$. Since any perturbation can be described by a Fourier series, only the most rapidly changing component needs to be examined.

Whether or not $\delta(t)$ grows with time depends on the behavior of $f(\omega)$. A curve of $f(\omega)$ is positive in a region $0 < \omega < \omega_0$ and negative for $\omega > \omega_0$. As a result any perturbation with $\omega > \omega_0$ will die away, but those ripples with $\omega < \omega_0$ will remain and grow.

For the supercooled case (Nishinaga et al., 1978) the critical angular frequency is

$$\omega_0 = \frac{(C_0 - C_e)}{C_e \Gamma} \sqrt{\frac{1}{\pi D t}} = \frac{2\pi}{\lambda_0}, \tag{A.9}$$

where Γ is capillary constant which is proportional to γ, the interfacial free energy. As a consequence, $\omega_0 \propto \Delta C$, so ω_0 increases with increased supercooling and with decreased interfacial energy.

A second consequence of the supercooled analysis is that etch back should lead to smooth surfaces, since an approximate expression for $f(\omega)$ is

$$f(\omega) = \frac{D\omega}{C_s - C_e}\left[-C_e\Gamma\omega^2 + \frac{C_0 - C_e}{\sqrt{\pi D t}}\right]. \tag{A.10}$$

$$\underbrace{}_{\substack{\text{capillary}\\\text{term}}} \quad \underbrace{\phantom{\frac{C_0 - C_e}{\sqrt{\pi D t}}}}_{\substack{\text{vol. diffusion}\\\text{term}}}$$

But for etch back, $f(\omega) < 0$, irrespective of ω because $(C_0 - C_e) < 0$. Taking this one step further, by neglecting the capillary term and by assuming that $\Delta C \propto (-\Delta T)$, $f(\omega) = -A\omega\Delta t^{-1/2}$. This leads, by integrating eqn. (A.8), to an expression for the smoothing kinetics given by

$$\delta = \delta_0 \exp[-2A\omega\Delta Tt^{1/2}] \qquad (A.11)$$

and shows that smoothing depends on $\Delta Tt^{1/2}$, a dependency that agrees with experiments on InP (Nishinaga et al., 1977).

A similar analysis of ramp-cooled growth (Small and Potemski, 1977), with a saturation wafer on top and the substrate on the bottom of the solution, leads to an expression for the expected wavelength given by

$$\lambda = 2\pi\left(\frac{3DC_0\Gamma}{R_0C_s}\right)^{1/2}, \qquad (A.12)$$

where $R_0 = -\alpha L/2mC_s$. Consequently, λ decreases as $R^{-1/2}$ or $\alpha^{-1/2}$. So by increasing the cooling rate the terrace spacing is expected to decrease, and experimentally this is seen (Small and Potemski, 1977).

The normal constitutional supercooling criterion is a special case which only examines the volume diffusion term and has been extensively analyzed (Minden, 1970). By this description, a perturbation will potentially grow, when the temperature gradient at the S/L interface causes a supersaturation condition to exist just ahead of the interface. Mathematically, the stability condition is stated by

$$\left(\frac{dT}{dx}\right) \geq m\left(\frac{dC}{dx}\right)_{x=S/L\,\text{interface}} \qquad (A.13)$$

The concentration gradient during LPE changes with time, thus maintaining the inequality of eqn. (A.13) requires dT/dx to increase with time. For ramp cooling, dT/dx must increase indefinitely in a semi-infinite solution, but by changing to a finite solution the dT/dx requirement approaches a fixed value after times at which dC/dx has a fixed value owing to the diffusion profile becoming constant (Minden, 1970; Moon, 1974). The thinner the solution, the sooner the concentration profile is steady and the lower the value of dT/dx must be. Supercooled growth and TMLE are, in contrast, more stable modes of growth since the dT/dx requirement actually decreases with time.

11.8. Symbols

a_e = lattice parameter of the epitaxial layer.
a_s = lattice parameter of the substrate.
B = the volumetric change/unit concentration change.
b = relaxation constant for the TMLE growth mode.
b_{mf} = Burgers vector of the misfit dislocation used to relieve misfit.
$C(x,t)$ = concentration of growth units.
C_0 = initial concentration of growth units.
C_i = interfacial concentration.
$C_e = C_e(T)$ = equilibrium concentration.
$\Delta C = C_0 - C_e$.
C_j^s or C_s = concentration of j in the solid.

C_j^l = concentration of j in the liquid.

D = solution diffusivity.

E = Young's modulus.

f = lattice mismatch = $\dfrac{a_s - a_e}{a_e}$ or sometimes $2\dfrac{a_s - a_e}{a_e + a_s}$.

$$F(\omega) = \frac{\dfrac{d\,\delta(t)}{dt}}{\delta(t)}.$$

G = shear modulus.

G_m^x = excess free energy/mole of solution.

g = acceleration due to gravity.

$H(t)$ = layer thickness.

$H(t)^*$ = dimensionless thickness:

$$\text{for supercooled growth} = \frac{H(t)C_s m}{L\Delta T_{sc}};$$

$$\text{for ramp growth} = \frac{H(t)C_s m}{\alpha L};$$

$$\text{for TMLE} = \frac{H(t)C_s m}{L\Delta T}.$$

ΔH_s = heat of solution.

h_c = critical thickness at which dislocations are expected to be introduced because of lattice mismatch.

$J = \dfrac{D}{bL^2}$ = ratio of the diffusivity to the relaxation parameter in TMLE.

K = kinetic coefficient.

k = thermal diffusivity.

k_i = segregation or distribution coefficient of species $i = C_i^s/C_i^l$.

L = thickness of solution.

m = slope of the liquidus line dT/dx_j.

N_i = the number of moles of i in solution.

n = total number of moles in solution.

R = gas constant.

$R(t)$ = growth rate.

$r = \dfrac{D}{KL}$ = relative importance of transport to kinetics.

T = temperature.

ΔT or ΔT_{sc} = supercooled temperature interval.

t = time.

V = velocity resulting from free, forced, and growth rate.

x = direction of growth which is perpendicular to the growth plane.

x_i = mole fraction of i.

y = direction perpendicular to the growth direction along which perturbation occurs.

α = cooling rate.

α_T = volumetric thermal expansion coefficient.

Γ = capillary constant = $\gamma\Omega/RT$.

γ = surface free energy.

γ_i = activity coefficient of i in solution.

δ = thickness of diffusion boundary layer.

$\delta(t)$ = amplitude of the surface perturbation.

η = fractional stress relief in an epitaxial layer.

λ = wavelength of surface perturbation.

$\lambda_n = (n + 1/2)\pi$.

v = kinematic viscosity.

v_ρ = Poisson's ratio.

σ = stress in an epitaxial layer.

τ = relaxation time in ramp cooled LPE = $\dfrac{MC_0}{\alpha} = \dfrac{RT_0^2}{\alpha\Delta H}$.

Ω_{ij} = interaction parameter for regular and simple solutions.

Ω = precipitate volume/mole of added solute.

ω = angular frequency of surface perturbation or rotation rate during dipping LPE.

References

ANTYPAS, G. A., HOUNG, Y. M., MOON, R. L., HYDER, S. B., ESCHER, J. S. and GREGORY, P. E. (1978) *International Symposium on GaAs and Related Compounds, Clayton, MO, September 1978.*

ASTLES, M. G. and ROWLAND, M. C. (1974) *J. Cryst. Growth* **27**, 142.

BALIGA, B. J. (1978) *J. Electrochem. Soc.* **125**, 598.

BARKER, T. V. (1906) *J. Chem. Soc. Trans.* **89**, 1120.

BAUDRANT, A., VIAL, H., and DAVAL, J. (1978) *J. Cryst. Growth* **43**, 197.

BAUSER, E., FRIKS, M., LOECHNER, K. S., SCHMIDT, L., and ULRICH, R. (1974) *J. Cryst. Growth* **27**, 148–153.

BESSER, P. J., MEE, J. E., ELKINS, P. E., and HEINZ, D. M. (1971) *Mat. Res. Bull.* **6**, 1111.

BLAKESLEE, A. E. and BISCHOFF, B. K. (1971) *Electrochem. Soc. Ext.* Abstr. No. 181, Spring Meeting, Cleveland, Ohio.

BLANK, S. L. and NIELSEN, J. W. (1972) *J. Cryst. Growth* **17**, 302.

BLINNIKOV, G. A. and RALYUZHNAYA, G. A. (1972) *Inorg. Matls.* **8**, 560.

BOLKHOVITYANOV, YU. B. and ZEMBATOV, H. B. (1977) *J. Cryst. Growth* **37**, 101–106.

BOOKER, G. R. (1965) *Phil. Mag.* **11**, 100.

BRICE, J. C., ROBERTSON, J. M., STACY, W. T., and VERPLANKE, J. C. (1975) *J. Cryst. Growth* **30**, 66.

BRYSKIEWICZ, T. (1978a) *J. Cryst. Growth* **43**, 101.

BRYSKIEWICZ, T. (1978b) *J. Cryst. Growth* **43**, 567.

BURTON, J. A., PRIM, R. C., and SLICHTER, W. P. (1953) *J. Chem. Phys.* **21**, 1987.

CARRUTHERS, J. R. (1977) Thermal convection instabilities, in *Preparation and Properties of Solid State Materials*, Vol. 3 (W. R. Wilcox and R. A. Lefever, eds.), Marcel Dekker, New York and Barel.

CARSLAW, H. S. and JAEGER, J. C. (1959) *Conduction of Heat in Solids*, Oxford University Press, London, Chap. 12.

CASEY, JR., H. C., PANISH, M. B., and WOLFSTIRN, R. B. (1971) *J. Phys. Chem. Solids* **32**, 571.

CRANK, J. (1957) *Mathematics of Diffusion*, Oxford University Press, London, Chap. 4.

CROSSLEY, I. and SMALL, M. B. (1971) *J. Cryst. Growth* **11**, 157.

CROSSLEY, I. and SMALL, M. B. (1972a) *J. Cryst. Browth* **15**, 268.

CROSSLEY, I. and SMALL, M. B. (1972b) *J. Cryst. Growth* **15**, 275.

DANIELE, J. J. (1975) *Appl. Phys. Lett.* **27**, 373.

DANIELE, J. J. and MICHEL, C. (1974) *Proceedings of the 4th International Symposium on GaAs and Related Compounds, Deauville, France*, Inst. of Phys., London, p. 155.

DAVIES, J. E. and WHITE, E. A. D. (1974) *J. Cryst. Growth* **27**, 261–265.

DAWSON, L. R. (1972) Liquid phase epitaxy, in *Solid State Chem.* (H. Reiss, ed.), Pergamon Press, Oxford, vol. 7.

DEITCH, R. (1970) *J. Cryst. Growth* **7**, 69.

DEITCH, R. H. (1975) Molten metal solution growth, in *Crystal Growth* (B. Pamplin, ed.), Pergamon Press, Chap. 11 (1st edition).

CHAPTER 12

High-temperature Solution Growth

D. ELWELL
Center for Materials Research, Stanford University, USA

12.1. Introduction

High-temperature solution growth is probably the most versatile method of growing crystals since some solvent may be found for any material required in crystal form. According to the definition of "high temperature", only those materials are excluded which decompose on heating, e.g. biological materials.

In general, crystal growth from the melt is preferable if the material to be crystallized is stable at its melting point. The major advantage of solution methods is that they permit crystal growth at a temperature well below the melting point, and the materials grown from solution are normally in the following categories:

(1) Those which melt incongruently, i.e. which decompose before melting.
(2) Those which undergo a solid state phase transition which results in severe strain or fracture, so that growth should occur at a temperature below this transition.
(3) Materials of very high vapour pressure at the melting point.
(4) Materials which become non-stoichiometric because of the loss of a volatile constituent.
(5) Refractory materials which are technically difficult to grow from the melt because of crucible or furnace problems.

Occasionally crystals or layers of materials which can be grown from the melt are produced from solution because higher quality material can be produced. This is particularly the case for some technologically important materials like gallium arsenide, where bulk crystals are grown from the melt but the active device is a layer grown from metallic solution (see Chapter 11). A layer of crystal grown at lower temperatures has a lower concentration of point defects and may also be crystallographically superior on grounds of dislocation density, stoichiometry, and uniformity of dopant concentration.

The major disadvantage of solution methods is that the crystal grows in the presence of a major impurity—the solvent. This has the consequence that a slow rate of growth must be used in comparison with growth from the melt if the capture of solvent inclusions by the growing crystal is to be avoided. The solvent may also introduce into the crystal impurities which are major constituents of the solvent or which were initially present as impurities in the solvent.

463

The chapter will be mainly concerned with crystal growth from molten salt solvents, the branch normally called "flux growth". Its major emphasis will be on the growth of bulk crystals, although the growth of thin films by the flux method will be mentioned briefly. Electrocrystallization from molten salt solutions will also be discussed in outline, as will the use of water as a high-temperature solvent. The latter solvent clearly requires high pressures simultaneously with high temperatures and is often treated as a separate category—hydrothermal growth.

The literature dealing with high-temperature solution growth is now quite extensive, mainly in the form of review articles. The only book dealing with the subject is that of Elwell and Scheel (1975), but fairly recent reviews dealing with flux growth are included in the collections of Ueda and Mullin (1975) and Kaldis and Scheel (1976). So far there has not been a book devoted only to hydrothermal growth although a monograph on the subject was produced by Lobachev (1971). For further details on several aspects discussed in this article, the reader is referred to Chapters 6, 7, 11, and 14 of the first edition (Pamplin, 1975).

In view of the limitations imposed by the requirement of compressing a large subject into a few pages, this chapter will deal very generally with the topic although reference will be made to some interesting developments of the last few years.

12.2. Choice of a Solvent

Most reviews of high-temperature solution growth list the properties of an "ideal" solvent for the material to be crystallized. Primarily these are that the solubility for this material should be high and that the phase required should be the only solid phase formed when the solution becomes supersaturated. The solvent should have low viscosity, low melting point, low volatility, low reactivity with some container material, low toxicity, and should be available in high purity at low cost. In addition, residual solvent should be readily separable from the crystals, e.g. by dissolution in water or dilute acid. In practice some compromises are always necessary in choosing a solvent for a particular application.

The major problems in making a choice of a solvent for some new material are the lack of fundamental data (e.g. viscosity, phase diagrams) and also of crystal–chemical principles of bonding in the solution. In general, solubility requires that there be similar chemical bonding in the solvent and crystal. For example, oxide crystals are normally grown from mainly ionic solvents such as $PbO–PbF_2$, while metallic conductors are crystallized from metal solvents, e.g. tungsten carbide from cobalt. Metallic solvents also appear most suitable for covalently bonded materials such as the semiconductors GaAs, Si, or InP, while diamond is normally crystallized from molten Ni–Fe alloys.

Although the optimum solvent will be one with similar bonding to the solute, crystal–chemical differences must exist so that solid solubility between solvent and solute is avoided. Differences in valence or in ionic radii are desirable if the incorporation of solvent ions is to be avoided. As an example, a $PbO–B_2O_3$ solvent is satisfactory for the growth of $Y_3Fe_5O_{12}$ films since the Pb^{2+} ion differs in valence from the Y^{3+} and Fe^{3+}, while B^{3+} is too small to replace Y^{3+} or Fe^{3+} ions in appreciable concentrations. (It should be noted, however, that both lead and boron are present in these films in detectable concentrations and the lead may have an influence on the properties of the material grown.)

If possible, solvents with at least one constituent in common with the solute are desirable. As examples, gallium is used as the solvent to grow GaAs, $Pb_2V_2O_7$ to grow rare-earth vanadates, polysulphide fluxes of general formula Na_2S_x to grow crystals of sulphides such as $NaInS_2$, and excess TiO_2 produces the best crystals of $BaTiO_3$.

It is still not possible to describe the action of a good solvent in terms of processes which occur at the atomic level. The liquid state is extremely complex and only recently has it been possible to compute reasonable values for transport and other properties of simple ionic liquids using molecular dynamics calculations based on ionic pair potentials. The problems involved in seeking a theoretical understanding of what is a good solvent were discussed by Elwell (1975), but the conclusion must be that atomic models of solvent processes are a long way off.

In the meantime, it would be useful to have empirical rules which could be used to predict good solvent compositions. The most detailed approach towards the prediction of melt compositions has been made by Wanklyn (1977), who considered the example of complex oxides such as $ThGeO_4$. Such oxides are often grown from a ternary melt containing a refractory oxide (such as ThO_2), an acidic oxide (GeO_2 in this case), and a basic oxide (e.g. K_2O or PbO). According to experience with a wide range of such systems, an excess is normally required of the acidic oxide, depending upon the difference in melting points of the constituent compounds. If the two components of the complex oxide are of similar melting point, then good crystals can be grown from solutions containing stoichiometric quantities of these constituents. If their melting points differ, then it is desirable or necessary to add the lower melting component in excess. If the melting-point difference is 500–1000°C, the excess is fairly small, but an excess of 40 mole % or higher was found to be required when the difference in melting point exceeded 1000°C. This trend is illustrated in Fig. 12.1a. No correlation was found between the requirement of an excess acidic oxide and incongruent melting behaviour.

This study of the influence of melt composition on the composition, size, and form of the crystals grown led to the development of a generalized pseudoternary diagram to facilitate the choice of melt compositions for crystal growth of materials having large differences between the melting points of their constituents (Fig. 12.1b). The solubility of the

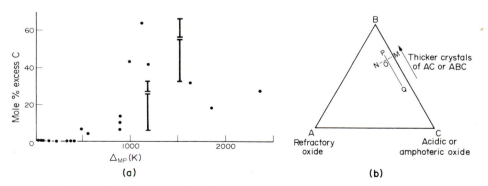

FIG. 12.1. (a) Mole % excess acidic oxide used to grow complex oxides versus difference in melting point of the component oxides. (b) Generalized pseudoternary diagram used for growth of materials with larger melting-point difference between the components A and C (Wanklyn, 1977).

refractory oxide is represented by the distance from PQ to BC and needs to be determined for a chosen ratio of basic oxide B to acidic oxide C and starting temperature. The form of the crystal was found in many cases to vary systematically with the B:C ratio (with other experimental conditions maintained constant). The choice of composition could therefore be simplified to the selection of the point O on the line PQ where crystals of the desired phase had an optimum axial ratio, followed by trial experiments along NOM to find the optimum concentration of refractory oxide A. Alternatively the solubility could be determined by differential thermal analysis or some other technique. Larger crystals were often obtained by choosing a composition close to the point P which is the boundary of the region where the desired phase crystallizes (i.e. by choice of a composition where the excess of the acidic component is minimized).

It was subsequently found (Wanklyn, 1978) that fewer, larger crystals of more equidimensional form could be obtained by replacement of part or all of the basic oxide by the corresponding fluoride. This observation had been appreciated for many years in the case of $PbO-PbF_2$ but evidence was also presented that KF replacing K_2O also has a beneficial effect. Particularly in the latter case, the improvement could have resulted from a lowering in the viscosity of the solution.

Another factor which can favour the growth of large crystals is the complexity of the melt. Small additions of B_2O_3 are considered by many crystal growers to reduce the number of crystals nucleated, although carefully controlled comparisons are not normally available.† Evidence has also been presented that MoO_3 has a beneficial effect in restricting nucleation (Wanklyn *et al.*, 1975), and other examples of additives are mentioned by Elwell and Scheel (1975). The effect of such additives is not clearly understood but is believed to be due to complexing of the solute ions which increases the width of the metastable region and so reduces the probability of further nucleation once crystals have been formed in the solution. Elwell and Coe (1978) recently made measurements of a wide range of properties of the $PbO-PbF_2-B_2O_3$ solvent system and of dilute solutions of yttrium aluminium garnet $Y_3Al_5O_{12}$. They found evidence for a high degree of solute–solvent interaction, and also strong complexing between PbO and B_2O_3 in the solution, with the F^- ions bonded less strongly. Since the $PbO-PbF_2-B_2O_3$ system is the most widely used molten salt solvent for crystal growth of oxides, these measurements support the postulate that fairly strong solute–solvent bonding is a desirable property of good solvent.

The use of additives to affect nucleation is less common in metallic solutions but examples are known. For instance, Logan and Thurmond (1972) added bismuth to the gallium used to grow GaN and reported a reduction in the number of crystals nucleated.

Hydrothermal growth also makes extensive use of additives, called mineralizers. In this case the main effect of the mineralizer is not so much to affect the nucleation characteristics but to increase the solubility, and sometimes to produce a new solid phase rather than that which crystallizes from pure water. The effect of the mineralizer is clearly associated with its interaction with the solute (or by its modification of the water–solute interaction). The best known mineralizer is sodium hydroxide (NaOH) which is used for the growth of quartz, the only material to be grown hydrothermally on a large, industrial scale. The solubility of SiO_2 in a 1 M NaOH solution is higher by an order of magnitude than that in pure water.

---

† This is an example of a general problem in crystal growth, that because of the production requirements, the grower rarely has time to perform experiments primarily aimed at a better understanding of his system.

12.3. Experimental Techniques

12.3.1. GENERAL REQUIREMENTS

The essential requirement for crystal growth is the creation of supersaturation, which means that the solution will contain a solute concentration higher than the equilibrium value. The three most important methods for achieving supersaturation are: (1) slow cooling of the solution, (2) solvent evaporation, and (3) temperature gradient transport. These are illustrated in Fig. 12.2, which shows a simple binary phase diagram for an idealized system. Although most solutions used for crystal growth are of complex composition, they often behave as pseudo-binary at least over the range where crystal growth occurs. It is certainly convenient to consider a single solid phase—the solute—which crystallizes from a liquid phase containing solute plus solvent.

In addition to these most popular techniques, several alternatives exist which may be preferable for some applications. For example, the supersaturated state may be created by a chemical reaction which results in the formation of a relatively insoluble phase. This is, of course, the situation which occurs during precipitation from solution, but normally the reaction takes place too quickly for macroscopic crystals to be produced. Only rarely is this method used for crystal growth, the best example being the IG Farben method for the growth of emeralds; in this case BeO and Al_2O_3 are dissolved in a lithium molybdate or other solution, while silica SiO_2 floats on the surface and dissolves very slowly and so regulates the rate at which the emerald ($Be_3Al_2Si_6O_{18}$) can form. An alternative and normally preferable way of providing a slow, controllable reaction is to supply one component via the vapour phase. This vapour–liquid–reaction–solid method involves one more step than the more familiar vapour–liquid–solid technique in which the solute is supplied via the vapour phase to create a supersaturated state, usually in a small volume of solution to grow crystalline materials in the form of "whiskers".

An alternative means of crystal growth involves the transport of ions to a pair of electrodes and electron transfer, leading to the deposition of a conducting species at the cathode. This process of electrodeposition will be considered in section 12.3.6.

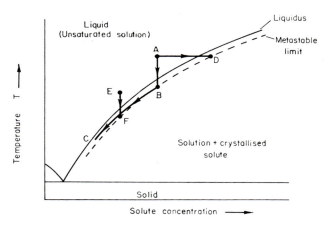

FIG. 12.2. Methods of achieving supersaturation in high-temperature solutions: *ABC*, slow cooling; *AD*, evaporation; *EF*, thermal gradient transport.

12.3.2. GROWTH STABILITY

Whatever method is used to achieve a supersaturation, the conditions of growth should be such that the crystal can attain the required size under stable conditions. Crystals grown from high-temperature solutions are normally faceted, so that stable growth implies the development of large, plane facets. Conversely unstable growth is often indicated by more rapid growth at the edges and corners of the crystal where the supersaturation is higher than at the face centre since the edges are effectively supplied with solute from a greater volume of solution. A high degree of instability leads to dendritic growth, with branch-like protuberances projecting from the edges or corners of the crystal. Very good examples of this can be seen, for example, in the electrolytic growth of LaB_6 at high-current densities (Elwell *et al.*, 1975). A lower degree of instability is indicated by terraced growth, with raised edges on the crystal facets and depressions in the centre. Apart from these macroscopic indications of unstable growth, a practical criterion that the growth of a crystal has been unstable is the presence within the crystal of solvent inclusions, which may vary in size from several millimetres (normally at the centre of the crystal or a face centre) to very tiny inclusions which cannot be seen using an optical microscope.

Theoretical treatments of growth instability usually concern two major aspects. The first is that a plane interface, the crystal facet, should advance without change in shape. This condition can be met if the supersaturation decreases with distance from the growing crystal interface, so that any protuberance which develops by random fluctuations will tend to decay rather than to grow with respect to the plane interface. A simple treatment of this "supersaturation gradient" problem leads to an expression for the maximum stable growth rate of the form

$$v_{max} \simeq \frac{D\phi n_e}{\rho RT^2}\frac{dT}{dz},\qquad(12.1)$$

where D is the solute diffusion coefficient, ϕ the heat of solution, n_e the equilibrium concentration of solute at temperature T, ρ the crystal density, R the universal gas constant, and dT/dz the temperature gradient normal to the crystal–liquid interface. The latter variable can be controlled to some extent by the experimenter and there is some evidence that a high-temperature gradient does favour more stable growth (much stronger evidence is available in the case of crystal growth from a slightly impure melt, where the growth rate is controlled mainly by the thermal transport rather than by solute transport).

Equation (12.1) is, however, of limited value since the predicted values of v_{max} are much lower than stable growth rates than can be achieved in practice. Furthermore, stable growth from solution often occurs in a zero or even a negative temperature gradient. More sophisticated treatments of the stability of a plane interface are available as discussed in Chapter 3, and an outstanding success of perturbation theory is that it predicts that an unstable interface will develop protuberances of a particular wavelength, in the region of some tens of microns. The presence of periodic inclusions has been observed in several flux-grown crystals, in confirmation of this prediction. However, as in the case of the simplified treatment which leads to eqn. (12.1), predicted values of the maximum stable growth rate are well below those found by experiment. The reason for this discrepancy lies in the difficulty of taking into account the stabilizing effect due to faceting. The so-called habit planes which bound a solution-grown crystal are of substantially lower surface energy than alternative planes of the same crystal, so that the development of protuberances which involve other crystal faces is energetically unfavourable. An alternative and complementary way of viewing this stabilization is that it results from the mode of growth of habit faces, which is by the lateral propagation of layers from "active" sites, normally taken to be those where dislocations with a screw component intersect the surface.

The second problem which is treated theoretically concerns the conditions for avoiding the breakdown in polyhedral shape due to the supersaturation inhomogeneity—the difference in supersaturation between the centre and edges of a facet. This problem has been given particular attention by Chernov (1972) (see also Kuroda *et al.*, 1977; Wilcox, 1977) who pointed out that a crystal may develop a slight curvature to the habit plane such that the decrease in supersaturation at the centre is compensated by a greater departure from the habit plane. In this way the kinetic coefficient (the growth rate per unit supersaturation) is higher at the centre and so the growth rate is constant over the whole surface. The tendency of the edges to grow faster than the face centres becomes more pronounced as the crystal becomes larger, so that the maximum rate at which the crystal can grow decreases as growth proceeds. Calculations of the limiting growth rate of polyhedral crystals are normally inaccurate because they treat the relatively simple case where transport of solute across the crystal occurs only by diffusion. In practice there is always some transport by convection which tends to reduce the supersaturation inhomogeneity. A theory by Carlson (1958) based on the breakdown in growth stability downstream from the corner of a crystal in a flowing solution gives the maximum stable growth rate for a crystal of side l as

$$v_{max} = [0.24 Du\sigma^2/S^{1/3}\rho^2]^{1/2} n_2 l^{-1/2}. \tag{12.2}$$

Here u is the solution flow rate, σ the relative supersaturation, S the Schmidt number of the solution, and ρ the crystal density. Evidence has recently been presented (Janssen-van Rosmalen and Bennema, 1977) to the effect that Carlson's theory does not accurately describe the process by which the crystal develops inclusions, but eqn. (12.2) does appear to give at least approximate values for the maximum stable growth rate and its dependence on the crystal size.

Although improvements in both major aspects of stability theory are desirable, both must be taken into account if crystals are to be grown under conditions close to optimum.

12.3.3. SLOW COOLING

The rate of growth v of an established crystal of area A is determined by the supersaturation. This in turn is governed by the process used to produce the supersaturated state which is necessary for crystal growth. In the most widely used technique, the solution is cooled at a rate dT/dt and the growth rate v is given by

$$v = \frac{V}{\rho A} \left(\frac{dn_e}{dT}\right) \left(\frac{dT}{dt}\right). \tag{12.3}$$

The crystal density ρ and the slope dn_e/dT of the solubility curve are fixed by the system. The solution volume V is chosen according to the total mass of crystals required and the area A of the crystal faces will change as the crystals grow. The experimental parameter which is generally most important is the cooling rate dT/dt, which directly governs the rate of mass deposition. This may be varied during the experiment and will directly affect the instantaneous growth rate. In general a constant cooling rate of about $1°C\,hr^{-1}$ has been adopted, but this is largely a question of experimental convenience.

Scheel and Elwell (1972) have proposed programs for slow cooling based on the requirement that the stable growth rate should decrease as the crystal gets larger (see Pamplin, 1975). These programs do not differ strongly from a constant cooling rate except during the initial period where the crystal area is small.

Experimental determinations of the maximum stable growth rate in flux growth are still scarce. Hergt and Görnert (1974) determined the concentration of solvent inclusions in yttrium iron garnet $(Y_3Fe_5O_{12})$ crystals cooled at different rates. The inclusion concentration was observed to fall rapidly from $1.4\,deg\,hr^{-1}$ to $0.4\,deg\,hr^{-1}$ although an even slower cooling rate led to an increase in inclusion concentration which was attributed to instability in the temperature controller. The maximum rate of stable growth in unstirred solutions was determined as $150\,\text{Å}\,s^{-1}$ or $1.3\,mm\,day^{-1}$. A size dependence of the stable growth rate was not detected in these experiments. An important variable appearing in eqn. (12.3) is the volume of solution V. The growth of fairly small crystals for physical measurements can be achieved using quite small crucibles, typically $50–100\,ml$ in volume. At one extreme, very expensive materials can be prepared in single crystal form using even smaller crucibles. For example, Garton et al. (1972) grew crystals of MgO and $LaAlO_3$ doped with ^{17}O from PbF_2 solution using crucibles only $10\,ml$ in volume. The largest crystals were up to $5\,mm$ in size.

At the other extreme, the economics of commercial production requires a much larger scale and experiments using crucibles several litres in volume have been grown by the Airtron Division of Litton Industries (Nielsen, 1964) and at Bell Laboratories (Grodkiewicz et al., 1967). The disadvantage of this approach is that nucleation is uncontrolled and many crystals are normally nucleated, usually with a large core having a very high concentration of inclusions. An alternative approach is to use a smaller crucible, typically $500\,ml$ in volume, and to restrict nucleation to one or a few crystals with the aid of stirring of the solution. This approach will be discussed in more detail in section 12.3.8.

12.3.4. EVAPORATION

Solvent evaporation is much less popular than slow cooling as a means of producing crystallization and is in general used only when compound formation occurs at low temperatures. For example, according to the reaction

$$PbO + TiO_2 \xrightarrow{T < 1200°C} PbTiO_3,$$

TiO_2 may be crystallized only by evaporation of PbO at temperatures above $1200°C$.

There is an advantage that growth may be isothermal so that crystals (especially solid solutions or doped crystals) are more homogeneous. The growth rate is given by

$$v = \frac{n_e}{\rho A}\frac{dV}{dt} \tag{12.4}$$

so that the growth rate depends on n_e (and hence on T) and especially on the evaporation rate dV/dt.

Control of the growth rate is more difficult in practice than for slow cooling. The most popular technique is to vary the size of the hole in an otherwise sealed crucible, but the evaporation rate and hence the growth rate then varies during growth as the hole becomes smaller due to deposit of solute transported in the vapour. It is preferable to use a closed system with the net evaporation rate controlled by the difference between the surface temperature of the solution and the lowest temperature within the system. A number of alternative designs are described by Elwell and Scheel (1975) and it is important to control carefully the temperature profile in order to avoid crystallization at the surface of the solution where the local supersaturation tends to be high because of solvent evaporation.

12.3.5. GRADIENT TRANSPORT

Gradient transport techniques rely upon transport of solute from a hotter to a cooler region, the rate of growth being balanced by dissolution of nutrient material. This steady

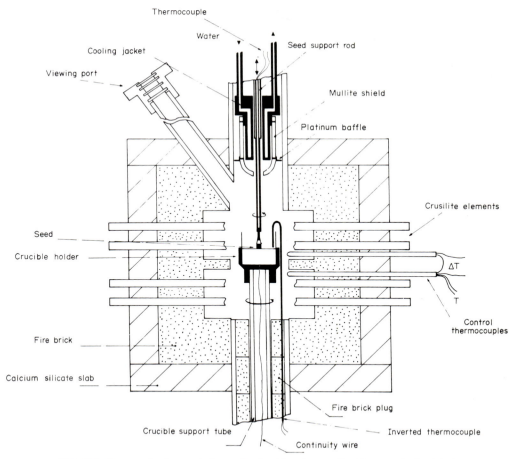

FIG. 12.3. Furnace for top-seeded growth of oxide using gradient transport of nutrient material located at the crucible base.

state method is particularly suitable for the growth of solid solutions and the crystals are in general much more homogeneous than those grown by other methods.

If the form of the interface kinetic law may be written as $V = F(n_i - n_e)^q$, with n_i the interface concentration of solute and B and q constants at constant temperature, then the growth rate is given (Dawson *et al.*, 1974) by

$$v\left(\frac{\rho\delta_c}{D} + \frac{\rho\delta_N A}{DA_N}\right) + \left(\frac{v}{B}\right)^{1/q} = n_e\frac{\Delta H\,\Delta T}{RT^2}, \tag{12.5}$$

where δ_c is the solute boundary layer thickness at the crystal and δ_N that at the nutrient material, A_N the area of nutrient, and ΔT the temperature difference between nutrient and crystal. The growth rate is primarily regulated by adjustment of ΔT and of the crystal rotation rate which determines δ_c. Gradient transport is often used in top-seeded solution growth with the nutrient material located at the base of the crucible and the seed immersed with one face just below the melt surface, although horizontal gradient transport has also been used successfully.

A furnace designed for the growth of oxide crystals by gradient transport in fluxed melts is shown in Fig. 12.3. A practical problem in the application of this method is that the low volatility solvents such as BaO/B_2O_3, which can be used in open crucibles over long periods, have rather high viscosities, typically of the order of 10 poise. In such liquids the mixing achieved both by seed rotation and thermal convection is very poor. The solution splits into non-mixing cells and the crystal growth rate falls since the solute in its vicinity is not replenished by flow from the nutrient (Lawrence and Elwell, 1976). This problem can be overcome only by the choice of a less-viscous but low-volatility solvent.

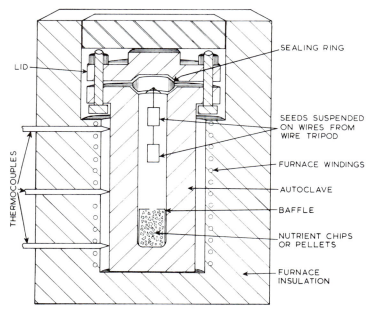

FIG. 12.4. Typical arrangement of furnace and autoclave for hydrothermal growth of crystals.

An alternative approach to top seeding which may be applied to low-viscosity but volatile solvents involves the use of a stirred, sealed crucible (Tolksdorf and Welz, 1973). In this case transport occurs downwards in a stirred solution between polycrystalline material held just below the melt surface and a seed crystal secured at the bottom of the solution. An air-cooled "cold finger" located adjacent to the seed is used to provide the required temperature gradient.

The problem of solvent evaporation is also avoided in hydrothermal growth where a sealed container is inevitable. A typical arrangement used for hydrothermal growth is shown in Fig. 12.4. The nutrient material is maintained in a compartment at the bottom of the solution, and a number of seed crystals are suspended in the upper section of the liquid. The temperature gradient must be carefully controlled since this determines the crystal growth rate, which is typically in the region of a few millimetres per day as in molten salt solution growth. An important practical requirement in hydrothermal growth is an effective seal which is capable of extended operation at kilobar pressures and several hundred degrees. Further details of hydrothermal growth conditions are given in Chapter 14 of Pamplin (1975).

12.3.6. THIN SOLVENT ZONE METHODS

A second alternative, which permits growth under constant conditions, utilizes a thin solvent zone. Low-volatility solvents such as BaO/B_2O_3 may then be used without problems associated with transport across a bulk solution. The width of a thin zone is typically of the order of 1 mm or less, so that the rate of diffusive flow of solute may be several mm per day. The growth rate is given, at least approximately, by

$$v = \frac{D\rho}{m(1 - n_e)\rho_s}\frac{dT}{dz} \tag{12.6}$$

with ρ_s the density of the solution and m the slope of the solubility curve.

Crystal growth is promoted either by moving a heater (travelling heater method), by moving the crystal and polycrystalline source material with the heater fixed (travelling solvent method), or by allowing the solvent zone to move in a temperature gradient towards the hotter surface (temperature gradient zone melting). Other terminology has also been used. The parameter which determines the growth rate is the temperature gradient, unless the solvent zone is made to migrate at a rate which should be equal to or less than that of eqn. (12.6).

If the zone is extremely thin, the supersaturation gradient becomes unimportant and fast stable growth is possible, with growth rates approaching those used for melt growth. This condition appears difficult to achieve in practice since the solvent zone tends to break up. More recent work has been concentrated on zones of thickness comparable with that of the diffusion boundary layer, and good quality crystals of CdTe (Wald and Bell, 1975) and $CaCO_3$ have been grown. The most interesting development in recent years has been the use of the Gasson (1965) strip heater to locate the position of the molten zone (Brissot and Belin, 1971; Belin, 1976). For the growth of calcite from a Li_2CO_3 solvent, the strip heater was used as the only source of heat to grow crystals 25 mm in diameter at 750°C. A major problem with this technique is the temperature inhomogeneity, which results from heat conduction through the strip, and a system of holes was used to minimize this effect.

High optical quality calcite was prepared by a two-stage process, since gas bubbles were introduced into the crystal when growth occurred directly from porous polycrystalline feed material.

The Brissot–Belin method was used by Tolksdorf (1974) in preliminary experiments to grow yttrium iron garnet from a BaO/B_2O_3 solvent. These early experiments were relatively unsuccessful, since the maximum length grown was 3 mm, and inclusions were found in the crystal even with a growth rate as low as $0.05\,mm\,hr^{-1}$.

More encouraging results were reported by Henson and Pointon (1971) who used a furnace to provide background heating to raise the temperature range in which the method can be applied. They grew crystals of $Ba_{0.65}Sr_{0.35}TiO_3$, 8 mm in diameter and 6 mm in length, at a temperature of 1540°C, and this appears to be a very promising method for the more refractory materials.

12.3.7. ELECTROCRYSTALLIZATION

The driving force for crystallization in all the methods described above is thermal in origin. In electrocrystallization the driving force is Faradaic and involves ionic migration in solution and electron transfer at two electrodes. In the simplest case of electrolysis of a solution of a metal oxide MO in a solvent which is stable against deposition, the application of a sufficiently high voltage leads to deposition of the metal M at the cathode:

$$M^{2+} + 2e \rightarrow M$$

and simultaneously to the liberation of oxygen at an inert anode:

$$2O^{2-} - 4e \rightarrow O_2\uparrow.$$

The reaction will proceed only if a sufficiently large potential is applied between the electrodes. The rate v at which material is deposited on the cathode depends upon the current density i through Faraday's law, which leads to the relation

$$v = \frac{\Sigma \Omega i}{Fz}, \tag{12.7}$$

where Ω is the molar volume, F is Faraday's constant, z the number of electrons required to deposit one mole, and Σ is a factor between 0 and 1 which is necessary to take into account departures from 100% efficient deposition.

More precise control of the current density and so of the growth rate is possible using a third or reference electrode which is maintained at its equilibrium potential with respect to the anode. The supersaturation is then determined by the potential difference between the reference electrode and the cathode, the so-called over-potential. Three-electrode measurements are useful in attempts to determine the rate-controlling process, which will be solute diffusion to the cathode, charge transfer, or the interface kinetic processes by which atoms on the surface become integrated into the material deposited. The procedures used in the determination of the rate-controlling process and examples of their application are discussed by Elwell et al. (1976).

The simplicity of electrical measurements is one of the great advantages of electrocrystallization, since the mechanism of crystallization can be studied using simple and rapid procedures. In addition, the growth rate may be changed very quickly and conveniently, and the use of pulse or a.c. techniques may enhance greatly the maximum rate at which stable growth is possible.

Electrocrystallization is used to grow crystals of conducting compounds, and the method is particularly useful for materials such as the rare earth borides which cannot be easily grown by alternative methods. A large number of materials have been synthesized by electrolysis of molten salt solutions, although relatively few have been prepared as single crystals. Recent reviews of electrocrystallization have been given by Elwell (1977) and by Feigelson (1978).

The most interesting recent development in electrocrystallization is the deposition of semiconductor films. Epitaxial films of silicon, gallium phosphide, and zinc selenide have been grown by molten salt electrolysis, and indium phosphide and gallium arsenide have also been synthesized by this method. At the present time sufficient data are not available for a prediction of whether it will be possible to produce material of sufficiently high purity for electronic applications, but there is interest in electrocrystallization as a means of depositing large area films for solar energy conversion.

12.3.8. STIRRING

It is widely accepted that stirring the solution normally leads to the growth of more perfect crystals. The evidence for the beneficial effect of stirring is particularly strong in the case of crystal growth from aqueous solutions, where the growing crystals are mounted on a "spider", the direction of rotation being reversed at regular intervals.

In the case of top-seeded growth from fluxed melts (or in electrocrystallization—see De Mattei et al., 1976) the rotation of a seed located just below the melt surface stirs the solution provided that the viscosity is not too high, as mentioned in section 12.3.2. The top-seeded geometry has the advantage that crystal rotation produces a uniformly accessible boundary layer of thickness given by

$$\delta = 1.6 D^{1/3} v^{1/6} \omega^{-1/2}. \tag{12.8}$$

Crucible rotation may be detrimental to good mixing in the solution, particularly by the formation of stagnation surfaces. Counter-rotation at more than 20% of the crystal rotation rate or co-rotation at a rate faster than that of the crystal is particularly detrimental to the flow patterns. However co-rotation at about 12% of the crystal rotation rate appears to be optimum (Capper and Elwell, 1975). A general expression for the optimum crystal rotation rate has not yet been given, but stirring by crystal rotation tends to improve the growth stability by increasing the degree of interface kinetic control of the growth rate.

In the alternative case, where crystals are grown in the bulk solution, the major benefit of stirring is likely to be the decrease in the supersaturation inhomogeneity between crystal edges and face centres. The use of an immersed stirrer is generally impracticable since platinum, often the only material inert against attack by molten salt solvents, is too soft for prolonged operation. The only successful method to date, especially for sealed crucibles, is the accelerated crucible rotation technique (Scheel, 1972). Stirring is achieved by continuously changing the rotation rate which promotes mixing in the horizontal plane and which causes the crystal surface to be swept by the Ekman layer flow. Scheel succeeded in growing a 200 g crystal of $GdAlO_3$ in a stirred solution under conditions

There has been a very extensive volume of material published on the growth of magnetic garnet films from a flux consisting mainly of PbO, with a few per cent of B_2O_3. In one short chapter it is not possible to do justice to this major topic, but several reviews are now available, e.g. those of Robertson (1973), Ghez and Giess (1974), Elwell and Scheel (1975), and Brice (1976).

The theory of deposition differs from that of bulk crystal growth primarily in that in bulk growth the initial transients are normally neglected whereas epitaxial growth often occurs under time-dependent conditions. In addition, an important factor in liquid phase epitaxy is the lattice mismatch between the substrate and the deposited layer, since very small misfits can exert a significant influence on the growth and especially on the incorporation of impurities into the layer grown (Brice *et al.*, 1975).

The hydrodynamics of the transients associated with the process by which a wafer is introduced into a supersaturated liquid have been treated by Brice (1976), who described four processes with time constants ranging from less than a second to 1–2 hr. Thermal transients are also possible, and experimental evidence exists for at least six district regions in garnet films grown by the normal "dipping" method.

At the present time there is a fairly good understanding of the mechanisms by which a single wafer is grown. The major problems facing industrial users of magnetic "bubble" garnet films are primarily those of molecular engineering—the selection of an optimum film composition for the particular device to be fabricated. So far as the choice of experimental variables is concerned, the major choice is of whether the substrate wafer should be mounted horizontally or vertically. Horizontal mounting has the advantage of a uniformly accessible boundary layer assuming that the wafer is rotated, so that the layer deposited is normally more uniform than for vertical mounting. Vertical mounting has the advantage that excess solution drains off more easily following removal from the melt.

Cost requirements in the commercial growth of garnet films favour the simultaneous growth of several films, and holders for ten or more horizontally mounted wafers have now been in use for several years. The close specifications on film uniformity are more difficult to meet than for single wafers, but multiple film growth is a practical reality which promises to become routine for application to an ever-growing range of materials.

12.4. Studies of the Growth Mechanism

Crystals can normally be grown under optimum conditions only if there is understanding of the mechanisms involved in their growth, and especially if the rate-determining step can be ascertained and controlled. In addition, the factors which critically affect the quality of the crystals should be known.

Studies of the mechanisms by which crystals grow from fluxed melts are of two major types—measurements of the growth kinetics under various conditions, and studies of the grown crystals. The most informative studies of growth kinetics are those where the growth rate is measured for various known supersaturations. This ideal is virtually impossible in growth by slow cooling since the supersaturation at which growth occurs is unknown. The top-seeding geometry with thermal gradient transport is, however, particularly convenient for kinetic studies since the supersaturation is controlled by the temperature difference between the seed crystal and the nutrient. A thermobalance for measurement of the rate of weight increase of a rotating crystal has been described by

Elwell *et al.* (1975). Thermal gradient transport has also been used by Watanabe *et al.* (1977) to study the growth mechanism of alumina crystals from Na_3AlF_6 (cryolite). The growth rate was found to be proportional to the supersaturation, as expected for diffusion control. A typical result of similar studies is that solute diffusion is the rate-determining step in unstirred solutions, and that interface kinetics become rate-determining in stirred solutions. In the latter case, the growth rate is often proportional to the square of the supersaturation, as expected for growth by a screw dislocation process at low supersaturations.

The problem of making kinetic measurements on crystals grown by slow cooling can be overcome by using the induced striation method (ISM), in which the growth rate is perturbed at regular intervals by the application of a temperature pulse of short duration. The pulse may be positive, leading to partial dissolution followed by more rapid than average growth, or negative, resulting in temporary rapid growth followed by relaxation, assuming that a constant average cooling rate is maintained. Particularly when doped crystals or solid solutions are grown, the growth rate perturbations result in banding or striations due to compositional inhomogeneities in the crystals, and these bands can be observed by X-ray topography or optical microscopy.

The first reported use of induced striations or "time markers" in flux growth was by Damen and Robertson (1972), and subsequent studies have been reported by Lillicrap and White (1976), Damen (1976), and Lawrence and Elwell (1976). The latter used an electrical potential to induce chromium-doping striations in yttrium aluminium garnet in the top-seeding arrangement, but experienced problems because of the low electrical conductivity of this material. The most intensive investigations using the ISM have, however, been made by Görnert and co-workers on substituted yttrium iron garnet crystals. By applying a positive thermal pulse, Görnert *et al.* (1976) were able to determine that the supercooling under which crystals grew was larger than 5.7 K. Although simplifying assumptions are often necessary before parameters associated with crystal growth can be determined, these experiments can generate values of the diffusion coefficient and the thickness of the boundary layer (Görnert and Wende, 1976; Wende and Görnert, 1977). Ideally, these experiments should be combined with more direct measurements of solution parameters and, in the absence of such confirmation, reservations must be expressed about some of the conclusions, but the ISM is an extremely promising tool in an otherwise difficult experimental area. Perhaps the most important contribution of this method to date is in understanding the non-induced striations which are commonly observed in flux-grown crystals: Görnert and Hergt (1973) identified striations of 2.5–5 μm period which they attributed to the growth process itself, in addition to those due to temperature fluctuation in the solution. The origin of these striations is currently unknown.

The most powerful techniques for studies of as-grown crystals are optical microscopy (including dark-field microscopy) and X-ray topography. X-ray topographic studies have made important contributions to the understanding of factors affecting the perfection of bubble garnet films. A review of this field has been given by Pistorius *et al.* (1975).

In the case of bulk crystals, X-ray transmission topography shows that the major defects are solvent inclusions, growth striations, growth sector boundaries, termination facet strain, and dislocations. Dislocation densities are, however, normally much lower than those obtained in alternative methods except for those few examples where dislocation-free crystals have been grown from the melt. The use of X-ray topography in the study of the growth history of crystals becomes more informative when multiple topographs are taken of the same crystal. For example, Tolksdorf (1977) showed reflection topographs from the facets of a large yttrium iron garnet crystal, showing the strain due to defects intersecting these facets (Fig. 12.7). Roberts (1978) sliced yttrium aluminium garnet (YAG) crystals parallel to a large facet and so was able to trace the path of line defects which mainly originated at the nucleation centre (Fig. 12.8). It is normally found that the

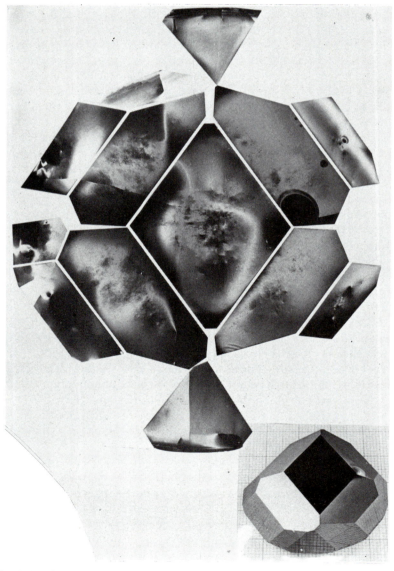

Fɪɢ. 12.7. Photomontage of X-ray surface topographs of the yttrium iron garnet crystal shown at bottom right. (Courtesy Dr. W. Tolksdorf.)

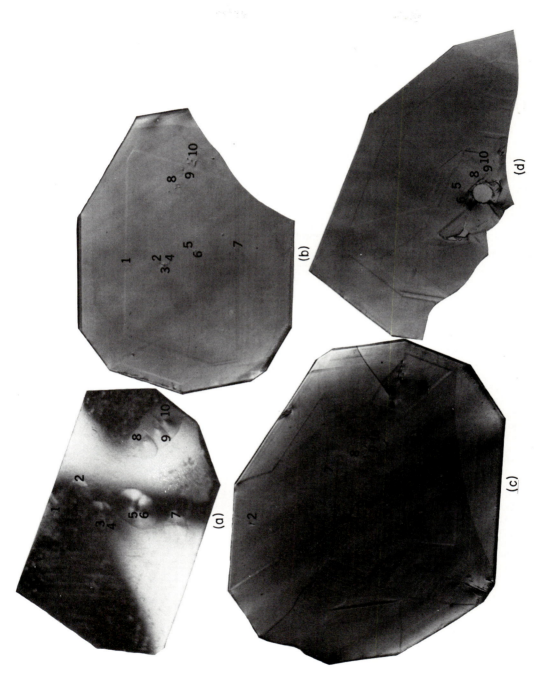

FIG. 12.8. Series of topographs of parallel sections through yttrium aluminium garnet crystal showing relation between surface features (optical micrograph in (a)) and dislocations labelled 1–10 (Roberts, 1978).

centres of vicinal features (growth hillocks) on the facets are intersected by line defects (see also Komatsu *et al.*, 1974), but topographs taken using different reflections showed that these line defects in YAG are mainly edge dislocations. It is still not clear whether the most "active" defects (in the sense of generating the largest hillocks and hence of dominating growth of a large area of a facet) are those having the greatest screw character or whether they are multiple-edge dislocations.

12.5. Summary

In this broad review of crystal growth from high-temperature solutions an attempt has been made to give an impression of the wide applicability of this method and of its commercial importance. An indication has also been given of the background knowledge which has now been accumulated concerning crystal growth by this method. The subject is in a gradual transition from an art to a science, and the present state of technology would not have been possible without considerable advances in understanding the fundamental mechanisms involved.

In addition, emphasis must be placed on the importance of good temperature control in the production of crystals of good size and quality. The major difference between nineteenth-century experiments on flux growth, which could yield 100,000 tiny crystals, and an experiment of today, which gives one crystal several hundred grams in weight, is that close regulation is now possible of furnace temperatures and of cooling rates.

References

BELIN, C. (1976) *J. Cryst. Growth* **34**, 341.

BORNMANN, S., GLAUCHE, E., GÖRNERT, P., HERGT, R., and BECKER, C. (1974) *Krist. Technik* **9**, 895.

BRICE, J. C. (1976) in *Crystal Growth and Materials* (E. Kaldis and H. J. Scheel, eds.), North-Holland, Amsterdam, p. 571.

BRICE, J. C., ROBERTSON, J. M., STACEY, W. T., and VERPLANKE, J. C. (1975) *J. Cryst. Growth* **30**, 66.

BRISSOT, J. J. and BELIN, C. (1971) *J. Cryst. Growth* **8**, 213.

CAPPER, P. and ELWELL, D. (1975) *J. Cryst. Growth* **20**, 352.

CARLSON, A. E. (1958) in *Growth and Perfection of Crystals* (R. H. Doremus, B. W. Roberts, and D. Turnbull, eds.), Wiley, New York, p. 421.

CHERNOV, A. A. (1972) *Soviet Phys. Crystallog.* **16**, 734.

DAMEN, J. P. M. (1976) *J. Cryst. Growth* **33**, 266.

DAMEN, J. P. M. and ROBERTSON, J. M. (1972) *J. Cryst. Growth* **16**, 50.

DE MATTEI, R. C., HUGGINS, R. A., and FEIGELSON, R. S. (1976) *J. Cryst. Growth* **34**, 1.

ELWELL, D. (1975) in *Crystal Growth and Characterization* (R. Ueda and J. B. Mullin, eds.), North-Holland, Amsterdam, p. 155.

ELWELL, D. (1977) in *Crystal Growth and Materials* (E. Kaldis and H. J. Scheel, eds.), North-Holland, Amsterdam, p. 606.

ELWELL, D., CAPPER, P., and D'AGOSTINO, M. (1975) *J. Cryst. Growth* **29**, 321.

ELWELL, D. and COE, I. M. (1978) *J. Cryst. Growth* **44**, 553.

ELWELL, D., DE MATTEI, R. C., ZUBECK, I. V., FEIGELSON, R. S., and HUGGINS, R. A. (1976) *J. Cryst. Growth* **33**, 232.

ELWELL, D. and SCHEEL, H. J. (1975) *Crystal Growth from High-temperature Solutions*, Academic Press, London.

ELWELL, D., ZUBECK, I. V., FEIGELSON, R. S., and HUGGINS, R. A. (1975) *J. Cryst. Growth* **29**, 65.

FEIGELSON, R. S. (1978) *Proceedings of the Laramie Conference on Solid State Chemistry* (to be published).

GARTON, G., HANN, B. F., WANKLYN, B. M., and SMITH, S. H. (1972) *J. Cryst. Growth* **12**, 66.

GASSON, D. (1965) *J. Scient. Instrum.* **42**, 114.

GHEZ, R. and GIESS, E. A. (1974) in *Epitaxial Growth* (J. E. Matthews, ed.), Academic Press, New York.

GÖRNERT, P., BORNMANN, S., and HERGT, R. (1976) *Phys. Stat. Sol.* (a) **35**, 583.

GÖRNERT, P. and HERGT, R. (1973) *Phys. Stat. Sol.* (a) **20**, 577.

GÖRNERT, P. and WENDE, G. (1976) *Phys. Stat. Sol.* (a) **37**, 505.
GRODKIEWICZ, W. H., DEARBORN, E. F., and VAN UITERT, L. G. (1967) in *Crystal Growth* (H. S. Peiser, ed.), Pergamon Press, Oxford, p. 441.
HENSON, R. M. and POINTON, A. J. (1971) *J. Cryst. Growth* **26**, 174.
HERGT, P. and GÖRNERT, P. (1974) *Phys. Stat. Sol.* (a) **21**, 77.
JANSSEN-VAN ROSMALEN, R. and BENNEMA, P. (1977) *J. Cryst. Growth* **42**, 224.
KALDIS, E. and SCHEEL, H. J. (1976) *Crystal Growth and Materials*, North-Holland, Amsterdam.
KOMATSU, H., HOMMA, S., KIMURA, S., MIYAZAWA, Y., and SINDO, I. (1974) *J. Cryst. Growth* **24/25**, 633.
KURODA, T., IRISAWA, T., and OOKAWA, A. (1977) *J. Cryst. Growth* **42**, 41.
LAWRENCE, C. M. and ELWELL, D. (1976) *J. Cryst. Growth* **32**, 287.
LILLICRAP, B. J. and WHITE, E. A. D. (1976) *J. Cryst. Growth* **32**, 250.
LOBACHEV, A. N. (1971) *Hydrothermal Synthesis of Crystals*, Consultants Bureau, New York.
LOGAN, R. A. and THURMOND, C. D. (1972) *J. Electrochem. Soc.* **119**, 1727.
MORRIS, A. W. and ELWELL, D. (1979) *J. Mater. Sci.* **14**, 2139.
NIELSEN, J. W. (1964) *Electronics*, Nov. 30.
PAMPLIN, B. R. (1975) *Crystal Growth*, 1st edn., Pergamon Press, Oxford.
PISTORIUS, J. A., ROBERTSON, J. M., and STACEY, W. T. (1975) *Philips Tech. Rev.* **35**, No. 1.
ROBERTS, K. J. (1978) PhD thesis, Portsmouth Polytechnic; K. J. ROBERTS and D. ELWELL (to be published).
ROBERTSON, J. M. (1973) Philips Research Laboratories Report MS8324.
SCHEEL, H. J. (1972) *J. Cryst. Growth* **13/14**, 560.
SCHEEL, H. J. and ELWELL, D. (1972) *J. Cryst. Growth* **12**, 153.
TOLKSDORF, W. (1974) *Acta Electronica* **17**, 57.
TOLKSDORF, W. (1977) *J. Cryst. Growth* **42**, 275.
TOLKSDORF, W. and WELZ, F. (1973) *J. Cryst. Growth* **20**, 47.
UEDA, R. and MULLIN, J. B. (1975) *Crystal Growth and Characterization*, North-Holland, Amsterdam.
WALD, F. V. and BELL, R. O. (1975) *J. Cryst. Growth* **30**, 29.
WANKLYN, B. M. (1977) *J. Cryst. Growth* **37**, 334.
WANKLYN, B. M. (1978) *J. Cryst. Growth* **43**, 336.
WANKLYN, B. M., MIDGLEY, D., and TANNER, B. K. (1975) *J. Cryst. Growth* **29**, 281.
WATANABE, K., SUMIYOSHI, Y., and SUNAGAWA, I. (1977) *J. Cryst. Growth* **42**, 293.
WENDE, G. and GÖRNERT, P. (1977) *Phys. Stat. Sol.* (a) **41**, 263.
WILCOX, W. R. (1977) *J. Cryst. Growth* **38**, 73.

CHAPTER 13

Dendritic Growth

R. D. DOHERTY

*Department of Metallurgy, Delft University of Technology, Delft, The Netherlands,
on leave of absence from University of Sussex*

13.1. Introduction

The term dendrite was apparently first introduced to the world of crystal growth by Tschernoff (1879) (Smith, 1965). He used it to describe the highly branched structure that he found in the centre of metal ingots (Fig. 13.1). The word dendritic means "treelike", a not inaccurate description of this curious structure, particularly if the tree visualized is a symmetrical conifer such as the Norwegian spruce (*Picea abies*) rather than a somewhat more disorganized deciduous tree such as the oak (*Quercus robur*). The highly branched dendritic structure is not too difficult to explain for a tree which is trying to expose its leaves to obtain the maximum amount of sunlight; but this idea is not a plausible one for the formation of a crystalline "tree" growing from a melt. The crystalline dendrite, by virtue of its extended surface, has a considerably increased surface free energy compared to the equilibrium shape of the crystal and is therefore *thermodynamically* unstable compared with this equilibrium shape. The origin of the dendrite must therefore result from the kinetics of crystal growth.

The kinetic advantage possessed by a dendrite is its ability to lose easily by enhanced diffusion in three dimensions from its sharp leading tip, the rejected solute, and latent heat; the so-called "point effect of diffusion" (Lehmann, 1888). It therefore follows that dendritic growth is to be expected in *diffusion-controlled* crystal growth but not in the *interface-limited* crystal growth processes, that give faceted crystals.

Dendritic growth appears to follow the diffusion-induced, interfacial instability (constitutional undercooling) discussed in Chapter 3 of this book.

Similar dendrites to the ones first described by Tschernoff have been found in almost every type of crystal growth process. The flat hexagonal snowflakes and the dendritic crystals found in the ice-making compartments of domestic refrigerators result from the crystal growth of ice directly from the vapour phase (Bentley and Humphries, 1931; Mason, 1962). Countless metallurgists examining the microstructure of cast metals (Fig. 13.2) have found dendrites produced by crystal growth from the melt. Ice forming from liquid water (Chalmers, 1964) and cyclohexanol and a few other organic crystals (Jackson *et al.*, 1966), forming from their own liquid phase, frequently grow as dendrites particularly when the liquid contains some impurity (Fig. 13.3).

485

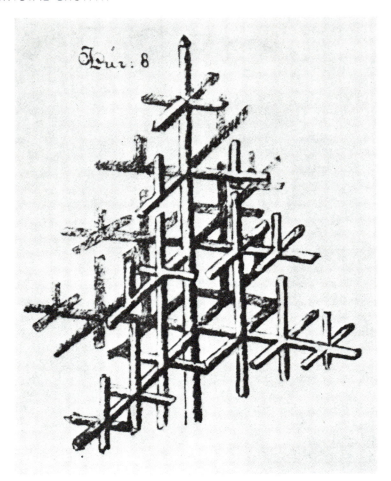

FIG. 13.1. Original sketch of a dendrite in a cast-metal ingot by Tschernoff, published in *A History of Metallography* by C. S. Smith and reproduced by courtesy of the publishers, The University of Chicago Press.

Aqueous solutions of some ionic salts also produce dendrites of the ionic material when cooled so that the solution is supersaturated with the salt as found by Kahlweit (1969). A system which demonstrates dendritic growth particularly clearly is ammonium chloride in water. This salt has a very high solubility in water, but one which falls rapidly as the temperature is reduced. A few minutes spent observing the dendritic growth of ammonium chloride, either in a test-tube or better still with a low-powered transmission microscope, will illustrate much of the following chapter.

Crystal growth from an already crystalline matrix, usually called "solid-state precipitation", is also known to produce precipitates with a dendritic morphology, e.g. Bainbridge and Doherty (1969) and Malcolm and Purdy (1967). Such a solid-state dendrite is shown in Fig. 13.4. However, this is unusual in that most solid-state precipitations in metallic systems produce precipitates that have a needle-like, or

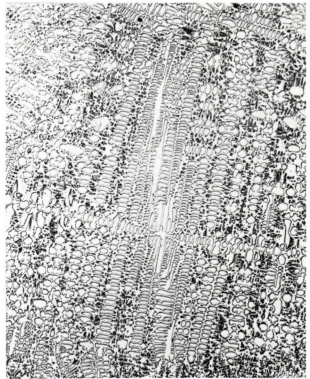

Fig. 13.3. Solidification dendrites of cyclohexanol with fluorescin as added impurity which is being rejected by the solute. The light colour around the dendrite is due to the locally higher concentration of the rejected impurity. Notice that the concentration gradients are steepest near the dendrite tips and also that dendrites are coarsening —the competitive growth of the larger arms at the expense of the smaller is occurring (Jackson et al., 1966). This micrograph is reproduced by courtesy of the Metallurgical Society of AIME. Magnification ×180.

Fig. 13.2. Solidification dendrite that formed in a small ingot of Cu–10%Sn when the melt was supercooled by 6°C during dendrite growth. During the period of dendrite growth the sample was quenched into water. The solid dendrite at the time of the quench has etched white while the fine dendrites formed during the quench are black. The primary dendrite appears to be slightly bent. (Dauncey, 1973.) Magnification ×40.

Fig. 13.4. Solid-state dendrite of γ brass produced in an alloy of Cu–49.9wt.%Zn solution treated in the β single phase region and quenched into the $\beta + \gamma$ phase field at 505°C for 244 s and quenched to room temperature. The structure has fully developed side branches. Magnification × 400.

alternatively a plate-like, morphology. A typical needle-like precipitate is shown in Fig. 13.5. The growth kinetics of the tip of such a needle crystal are, however, now thought to have much in common with the growth of the dendritic tip and to differ only in not producing side branches (Bainbridge and Doherty, 1969; Trivedi, 1970; Purdy, 1971).

Crystal growth from all three states of matter, vapour, liquid, and solid, is therefore known to produce this type of morphology. In addition dendritic growth has been found during cathodic deposition in electrocrystallization by Faust (1968), and has also been reported for so-called "negative dendrites" produced by growth of a liquid phase in superheated crystals (Chalmers, 1964). Dendritic growth is clearly a very widespread phenomenon and, in addition, it has a wide range of important consequences for the crystals that have grown with this morphology. This applies both to crystals that still have the external shape of dendrites at the end of their growth process (e.g. snowflakes, some

Fig. 13.5. Typical Widmanstätten needles of α brass produced in an alloy of Cu–43.7wt.%Zn solution treated in the β single phase region, quenched into the $\alpha + \beta$ phase field at 552°C for 300 s and then quenched to room temperature. α has nucleated at β grain boundaries and grown into one of the β grains. Notice that there is no sign of side branches in this case. Magnification ×400.

electrodeposits, etc.) and also to those crystals which were initially dendritic but which subsequently lose their shape by crystal growth into the spaces between the dendrite arms (e.g. solidification of liquid alloys).

13.2. Consequences of Dendritic Growth

13.2.1. MORPHOLOGY

For the first type of crystal, where the product will be a very fragile low-density mass of interlocking dendrites (e.g. precipitated snow), this structure can be very unstable. This arises for all the obvious reasons and, in addition, as described later, the dendrite arms can "melt off" during a rise in temperature. Such a melt-off converts the dendrite into a set of isolated arms that will have almost no mechanical strength. This is likely to be a major cause of some snow avalanches, as discussed, for example, by Fraser (1966).

13.2.2. CRYSTAL DEFECTS

For the second type of dendritic structure, with a full density at the end of the crystal growth process, dislocations can be easily introduced. This introduction of dislocations can happen since the initial dendritic skeleton is very fragile and minor displacements of the dendrite arms can occur while the interdendritic material is still fluid. As sketched in Fig. 13.6, bent dendrite arms will be misoriented from the rest of the dendritic crystal—the misorientation will be that of a low-angle grain boundary composed of dislocations (see, for example, Shewmon, 1966). The completion of solidification will then give a dislocation boundary surrounding the bent dendrite arm. However, as far as is known to the author no study has yet been published of the dislocation content or the sub-grain misorientations, within freshly solidified dendritic metal crystals.

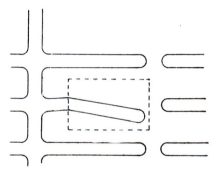

FIG. 13.6. Sketch of means by which a bent dendrite arm could produce a low-angle boundary. The arm is shown bent at an early stage in its growth; further solidification will cause the material to become entirely solid and produce a low-angle boundary (shown as a dashed line) at the position at which the bent dendrite will meet the adjacent dendrites.

13.2.3. SOLUTE SEGREGATION

13.2.3.1. *Micro or Dendritic Segregation*

Following dendritic growth from a liquid containing one or more alloy additions partially soluble in the crystal, the resulting composition will vary over the distance of a dendrite arm. If the solute lowers the melting point of the crystal the alloying element will be concentrated in the liquid so that the first part of the crystal to form—the dendritic skeleton—will be much purer than the last material to freeze in the interdendritic spaces (see, for example, Doherty and Melford, 1966 or, more recently, Criach *et al.*, 1978). The result of this segregation will be a continuous and periodic variation of the solute content in the crystal, and this will outline the original dendritic pattern. This is the universally observed "cored" structure of cast metals examined metallographically (Fig. 13.7)—the etchant has preferentially attacked the higher solute content of the interdendritic regions rather than the purer dendrite arms. An additional result caused by such segregation arises

Fig. 13.7. Microsegregation in a cast ingot of Cu–40wt. %Ni. The dendrite arms are high in nickel and are therefore less etched than the interdendritic "pools" rich in copper. The different orientations of different grains can be clearly seen even though the etch has not revealed the grain boundaries. Magnification × 40.

if the lattice parameter of the alloy varies with the composition, thereby producing a built-in strain from dendrite core to exterior, probably in the form of arrays of dislocations.

13.2.3.2. *Macrosegregation*

Solute segregation over distances greater than the dendrite arm spacing have been reported in cast ingots. Two different origins have been found for this behaviour:

(1) The difference in density between crystal and melt will cause the usually heavier solid dendrites to sink to the bottom of an ingot, resulting in the last region to solidify (i.e. the top of the ingot) being richer in solute, as shown by Kohn (1968).

(2) Where the solid forms in a temperature gradient the contraction caused by growth of higher density crystal from the lower density liquid can bring about flow of the interdendritic liquid into the colder regions of the ingot.

For a review of this phenomenon, see Mehrebian and Flemings (1971) and Flemings (1974).

A particularly serious form of macrosegregation, called the formation of A segregates in steel ingots or "freckle" formation in other ingots such as those of the nickel alloys used as gas-turbine materials, was identified as arising from rapid interdendritic flow by McDonald and Hunt (1969, 1970) and also by Mehrebian *et al.* (1970). What occurs is flow of interdendritic liquid from *colder to hotter* regions, due to differences in density arising from differences in solute content, at a rate *faster than the movement of the solidification isotherms.* This flow initially dispersed in many paths between the dendrite arms is "corrosive" since as the flowing liquid becomes *warmer* it starts to dissolve the already solidified material.

As with the percolation of water in limestone, such a dispersed corrosive flow is *unstable* and can rapidly become concentrated in a few channels. The reason for this instability is that any channel that is slightly more open than the surrounding channels is more corroded by the flow and so becomes larger and more open, thereby concentrating the flow still further.

In limestone the channel formation gives us the entry and exit caves explored by cavers, but the corrosive flow in ingots gives us the channel A seggregates or freckles that are full of solute and are very undesirable.

McDonald and Hunt (1969, 1970) studied channel formation in ingots of ammonium chloride "alloyed" with water and were able, in this transparent analogue of a metal ingot, to map the fluid flow giving the channel segregates.

13.2.4. VOID FORMATION

Voids are frequently found between the arms of the dendrites in, for example, completely solidified metal ingots, e.g. by Tzavaras and Flemings (1965). This occurs because of the increase in density on solidification. The dendritic skeleton fills the liquid volume, but as the arms thicken they occupy less space than the liquid from which they are forming. Unless extra liquid can be "fed" into the solidifying mass, voids almost invariably form as discussed by Campbell (1968). This is a major problem in the foundry industry (Ruddle, 1957), where the ingot moulds are provided with so-called "hot tops" in which liquid is kept molten in an attempt to provide extra liquid to be fed into the ingot. A similar strategy is adopted with the casting of intricate shapes where numerous "feeders" are introduced at different points into the casting.† Where the feeding process has dramatically failed the full dendritic structure can be revealed—this is how Tschernoff first observed his dendrites. A similar structure is shown in Figs. 13.8 and 13.9 for a copper–nickel ingot. The contraction has drained the liquid away from the dendrites which are seen emerging from a hole in the sample. Above this hole (Fig. 13.8) other

---

† Eutectic alloys, which freeze at a fixed temperature and do not show dendritic growth, are therefore much favoured by the foundry industry to ease this problem since feeding of liquid through a mass of "pasty" dendritic crystals is then avoided.

FIG. 13.8. Small ingot of Cu–40wt.%Ni slowly cooled in a crucible. The region shown is that around a central cavity formed at the top of the ingot due to the increase of density on solidification. A set of parallel dendrite arms with the orthogonal secondary arms is seen. Notice the severely bent dendrites in the top left of picture, also other dendrite tips just emerging from the surface on either side of the void. Magnification ×15.

dendrite tips are seen just standing proud of the surface. Note also that one dendrite is very considerably bent. Figure 13.9, which shows a part of this structure at a higher magnification (some oxide is visible), clearly illustrates the primary branches on which the orthogonal secondary arms are easily seen together, in some cases, with the tertiary arms, one set of which is parallel with the primary arms. This illustrates the fact that the dendrite arms in cubic metal crystals lie along orthogonal $\langle 100 \rangle$ directions.

The parallel alignment of all the dendrites in the field of view results from their origin from a single crystal nucleus, that is they are all part of the same metal crystal.

As discussed, e.g. by Flemings (1974), fine-scale microporosity will be located in the spaces between dendrites; however, this can be due either to solidification shrinkage alone, but it is often accentuated by small amounts of dissolved gas, e.g. hydrogen in aluminium alloys, carbon monoxide in steels. See, for example, Thomas and Gruzleski (1978) and Hirschfield and Weinberg (1978). This gas will have been subject to interdendritic segregation, being less soluble in the crystal than in the liquid, and so such gas bubbles are the equivalent of the common solid inclusions that result from microsegregation of the deliberately added solute additions and of the unremoved impurities—the ubiquitous oxide and sulphide inclusions in steels. Since gas has a very low density, gas inclusions or voids have a much larger volume fraction than the equivalent weight of interdendritic solid particles. Removal of gaseous impurities is therefore vital in casting technology.

Fig. 13.9. Same sample as Fig. 13.8 viewed at higher magnification with a scanning electron microscope. The secondary arms and some ternary arms are visible (notice that the tips of the dendrites have become hemispherical); this is probably due to the effect of surface tension after the original growth and not an accurate picture of the true shape during growth; compare with that seen in Figs. 13.3 and 13.4. Some oxide debris is present. Magnification × 70.

13.2.5. CRYSTAL MULTIPLICATION

The highly branched dendrite is a rather fragile structure, especially in the early stages of growth. It has been reported by Jackson *et al.* (1966) that the dendrite structure can be broken up, particularly in the case of dendrites growing from the liquid where there is fluid flow and local changes of temperature (e.g. in convection). This situation arises during the casting of metal ingots, where hot liquid metal is poured into cold moulds. The fragmentation of the dendrites allows each part of the dendrite to act as a nucleus for the growth of differently oriented crystals, and this is likely to be the origin of the so-called "equiaxed" crystals in metal castings (Jackson *et al.*, 1966; Southin, 1967). An exactly similar process has been reported by Melia and Moffitt (1964) for the case of salt dendrites growing in aqueous solutions, though here the process was described as secondary nucleation.

The same type of process is also called "collision breeding" and can occur both in growth from a melt and from solution (Garabedian and Strickland-Constable, 1974). They reported that for collision breeding of ice crystals actual collisions between ice crystals and solid objects were needed; fluid flow itself was not sufficient to cause fragmentation of the primary ice crystal.

An important new application of this crystal multiplication phenomenon is in the process called variously "stir casting" or "rheo casting"—the rapid deformation of partially solidified metal alloys (Spencer *et al.*, 1972; Joly and Mehrebian, 1976; Young *et al.*, 1979; Flemings, 1974b; Vogel *et al.*, 1979). In the stir-casting process, alloys that would normally solidify in a dendritic structure, after being subject to high shear rates ($100\,\mathrm{s}^{-1}$), give numerous, small, nearly spherical grains. These small crystals appear to come from the arms of the original dendrites.

Vogel *et al.* (1979) have proposed that the fragmentation of metallic dendrites could occur by a process of dendrite arm bending, giving a grain boundary within the dendrite (Figs. 13.6 and 13.8). As Vogel *et al.* (1979) point out for *high angles* of misorientation at a grain boundary, the energy of the boundary (γ_{gb}) may be larger than twice the solid liquid interfacial energy γ_{sl}. For $\gamma_{\mathrm{gb}} > 2\gamma_{\mathrm{sl}}$ the grain boundary would be completely "wetted" by the liquid phase leading to "melt-off" of the bent dendrite arm.

The rapidly stirred alloys which transform from dendrites to a slurry of solid particles in a liquid matrix have very interesting properties—notably that of thixotropy. The apparent viscosity of such slurries can vary greatly depending on the stirring rate, the fraction solid, and the previous thermo-mechanical history (Spencer *et al.*, 1972; Joly and Mehrebian, 1976). Flemings (1974b) and Young *et al.* (1979) have discussed possible applications of these thixotropic metal slurries. The applications include improved pressure die casting of complicated shapes in high-melting-point alloys, the fabrication of novel composite materials, and possible purification techniques (Flemings and Mehrebian, 1973). The advantages of using the thixotropic slurries for die castings includes the fact that the alloy is pushed into mould at a low temperature (reduced thermal shock to the mould), partially solidified (so an increased production rate is possible), and in a viscous form (so giving a reduced entrapment of air) (Backman *et al.*, 1977; Ramati *et al.*, 1978). Possible composite materials include dispersions of hard particles in a soft metal such as silicon carbide in aluminium for use as wear-resistant materials (Mehrebian *et al.*, 1974), and also the recent suggestion of making dispersions of soft particles in a strong matrix (e.g. low solute content aluminium in a matrix of high solute content aluminium) as a possible means of increasing the fracture toughness of precipitation-hardening alloys (Vogel *et al.*, 1979).

13.2.6. IMPORTANCE AND ORIGIN OF DENDRITIC GROWTH

These examples of some of the consequences of dendritic growth indicate its importance for the crystallization of commonly used materials. Some idea of the scale of this can be judged from the fact that *every day* nearly two million tons of steel are frozen: all of this steel has undergone a process of dendritic growth. From the earliest days of scientific interest in technology, dendritic growth has attracted interest (Grignon, 1775; Tschernoff,

1879; Lehmann, 1888). Other important steps in the development of a scientific understanding of this topic include the discovery by Northcott and Thomas (1939) that the orthogonal dendrite arms in cubic metal crystals did in fact come from their $\langle 100 \rangle$ symmetry, the development of a quantitative theory of constitutional supercooling by Tiller *et al.* (1953), and the development of a theory of diffusional growth of parabolic dendrites by Ivantsov (1958).

In the following sections a brief discussion will be given of the shape instabilities that give rise to dendritic growth (section 13.3), the various diffusion equations developed to account for the linear growth of dendrites (section 13.4), an account of the experimental investigations of the velocity of dendrite growth (section 13.5) and of other investigations of dendritic growth phenomena (section 13.6).

13.3. Instabilities that Cause Dendritic Growth

As shown by Tiller *et al.* (1953) for unidirectional crystal growth from an alloyed melt, the rejection of low-melting alloys elements by the crystal leads to a local constitutional undercooling unless the temperature gradient in the liquid is greater than

$$\frac{mC_0}{D} \frac{(1 - K_0)}{k_0} v,$$

where m is the slope of the liquidus, C_0 is the alloy content, K_0 the alloy partition coefficient, v the rate of crystal growth, and D the diffusion coefficient of the solute in the liquid.

Constitutional supercooling, if only slight, gives rise to cellular perturbations of the crystal surface, but larger supercoolings lead to fully developed dendrites as discussed by Chalmers (1964).

This will not occur in pure elements ($C_0 = 0$), so that a pure element would be stable against perturbation even for a zero temperature gradient. However, Chalmers (1964) shows that if the temperature gradient in the liquid is negative (temperature falling as distance from the crystal increases)—and this will occur if the latent heat of crystal growth is to be removed into the liquid—then the material is thermally supercooled and again dendrites are found.

The simple instability theory of constitutional supercooling was extended by Mullins and Sekerka (1963, 1964) to include the capillarity of the surface energy. This has been discussed by Sekerka (1968) and also in an earlier chapter in this book, and so need not be repeated here. However, the major results can be briefly given:

(1) For an undercooled crystal with an initially flat solid–liquid interface the surface tension (capillarity effect) will stabilize it against sinusoidal perturbations that have a wavelength less than a critical value λ_0.

(2) An equivalent effect of capillarity for a spherical crystal growing in a supersaturated solution is that if ρ_c is the critical radius for nucleation and growth (section 13.4.1), then the crystal must grow to a radius of $7\rho_c$ before perturbations can develop.

(3) There are two influences on the kinetics of the development of a perturbation: (i) the effect of capillarity, which favours large wavelengths, and (ii) the point effect of diffusion, which favours small wavelengths. The balance of these two effects is such that a wavelength of $\sqrt{3} \cdot \lambda_0$ gives the fastest developing perturbation.

The usual conclusion from this last idea is that the spacing of the cells or dendrite arms is determined by the $\sqrt{3} \cdot \lambda_0$ distance, and there are several experimental investigations that appear to fit this hypothesis (Malcolm and Purdy, 1967; Hardy and Coriell, 1968; Townsend and Kirkaldy, 1968). However, there is some reason for caution in accepting this experimental agreement—the Mullins and Sekerka theory only applies to the initial development of the perturbation and the analysis has not been extended to describe the transition from the initial perturbation to the fully developed dendrite whose spacing can be experimentally measured.

One vitally important constraint on dendritic perturbation has been discussed by Shewmon, 1965, i.e. the effect of the atomic interface kinetics. With a low mobility interface, where a large supercooling is required to drive the interface reaction, there is a reduction of the heat or solute gradients, which feed the "point effect of diffusion". A low-mobility interface is therefore considerably stabilized against dendritic perturbations.

This conclusion is strongly supported by the work of Hunt and Jackson (1965) who, using the analysis of Jackson (1958) of interface structure, were able to divide materials into two classes: (i) those with an atomically rough interface and a high mobility, and (ii) those with an atomically smooth interface and a low mobility. Materials falling into the first category, e.g. metals and some organic compounds growing from their own melt, were found to show cellular and dendritic perturbations, and those in the second category, e.g. salol, grew with faceted morphologies, i.e. they resisted dendritic perturbations.

An interesting example of the same crystal giving either dendrite or a faceted crystal shape is provided by sodium chloride. Rock-salt crystals forming from their melt at 800°C grow as dendrites, but the crystals that precipitate from brine solutions at room temperature are faceted cubes.

The inhibition to the development of instabilities provided by capillarity can be enhanced if significant diffusion can occur either within the growing crystal or along the interface between the growing crystal and the matrix (Shewmon, 1965; Nichols and Mullins, 1965). While diffusion within the matrix phase acts both to promote the growth of the perturbations by the "point effect of diffusion" and to damp out the perturbations in the interests of minimizing interfacial energy (capillarity) the alternative diffusion paths act solely in the interest of capillarity. Shewmon gives the following equation for the rate of growth of the amplitude $\varepsilon(t)$ of a sinusoidal perturbation of wavelength $\lambda = (2\pi/\omega)$:

$$\varepsilon = \frac{\left[2D_\beta G_c \omega - D_\beta C_\beta \omega^3 \Gamma \left(1 + \frac{D_\alpha C_\alpha}{D_\beta C_\beta} \right) - \omega^4 \Gamma' D_I \delta_I \right]}{2(C_\beta - C_\alpha)}, \varepsilon(t), \tag{13.1}$$

where D_α is the diffusion coefficient in the α precipitate phase, D_β that in the β matrix phase, D_I that in the interface, which has a thickness δ_I, C_α and C_β are the concentrations in the two phases (Fig. 13.10), G_c is the concentration gradient in the matrix phase, and Γ' is the capillarity constant which following the derivation of the appropriate form of the Gibbs–Thomson equation (13.2) by Purdy (1971) is given by eqn. (13.3).

$$C_\beta(\rho) = C_\beta \left\{ 1 - \frac{2\gamma V_\alpha}{RT\rho\alpha} \frac{(1 - C_\alpha)}{(C_\beta - C_\alpha)} \right\}, \tag{13.2}$$

where γ is the interfacial free energy per unit area, V_α is the mol. volume of the α-phase, R and T have their usual meanings, and α is the Darken non-ideality factor

$$\alpha = \left[1 + \frac{d \ln f_\beta^B}{d \ln C_\beta} \right]$$

with f_β^B the activity coefficient of B in the β-phase.

So
$$\Gamma' \equiv \frac{2\gamma V_\alpha}{R T \alpha} \frac{(1 - C_\alpha)}{(C_\beta - C_\alpha)} . \tag{13.3}$$

It may be noticed that the equation given by Shewmon is in a slightly different form from that of eqn. (13.1). This is due to the fact that we are considering here the growth of a dendrite of α that is depleted in solute compared to the matrix phase β (see Fig. 13.10) and are using eqn. (13.2) to define the capillarity constant since this is a more useful form for subsequent discussion of the growth of a dendrite crystal.

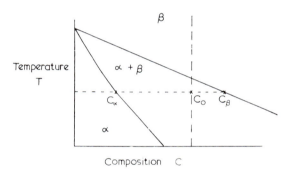

FIG. 13.10. Schematic phase diagram showing β phase undercooled by cooling to temperature T.

Inspection of eqn. (13.1) shows that the first term in the square brackets is that of the destabilizing "point effect of diffusion" in the matrix, the second term is the stabilizing effect of capillarity via diffusion within the matrix phase α *and* diffusion within the precipitate phase β, while the final term is the stabilizing capillarity effect via diffusion along the interface. For crystals growing from a fluid phase the diffusivity in the growing crystal or in the interface are both likely to be negligible compared to that in the fluid matrix so the neglect of these corrections is likely to be justified. However, for growth of a crystal from a crystalline matrix (solid state precipitation) the stabilizing effects of the other diffusion paths is likely to be important. For example, Nichols and Mullins (1965) pointed out that introduction of the correction due to interface diffusion for precipitation of an incoherent precipitate increased the critical wavelength of the perturbation by a factor of 100. (The critical radius of instability of a spherical precipitate in a 10% supersaturated solid solution increased from 0.1 to 10 μm.)

A further possible constraint on the development of dendritic type perturbations during solid state precipitation has been briefly mentioned by Shewmon (1965), Bainbridge and Doherty (1969), and by Cline (1971). This constraint arises where a particular part of the crystal interface has a low-energy or cusp orientation.† These cusps may be very deep for crystal–crystal interfaces, e.g. Hu and Smith (1956) showed interfaces between f.c.c.$_\alpha$ brass and b.c.c.$_\beta$ brass where the interface plane containing both the close-packed $(110)_\alpha$ and $(111)_\beta$ directions has surface tension of one-third of the general surface tension between these crystalline phases. Both Shewmon (1965) and Townsend and Kirkaldy (1968) developed precipitates of b.c.c. iron in a matrix of f.c.c. iron (in a dilute iron–carbon alloy) which they allowed to come to solute equilibrium. These b.c.c. precipitates were then undercooled by rapidly cooling a small amount. Precipitates that had had only a short time to come to equilibrium gave many perturbations, but those precipitates that had had longer equilibrating times (36 h) were very reluctant to perturb.

The most likely explanation for this difference given by Townsend and Kirkaldy (1968) is that after long annealing times the precipitate–matrix interfaces had rotated to achieve low-energy (cusp) orientations. If this explanation is correct it would also account for the frequent absence of side perturbations on precipitates formed from the crystalline matrices particularly in the metallurgical area. These precipitates frequently grow as needles or plates (Purdy, 1971; Shewmon, 1969) with the long axis of the needle or the plane of the plate having interfaces with good crystallographic fit. ‡

The explanation of the absence of dendritic growth in most solid-state precipitation reactions in metal systems in terms of the high anisotopy of the interfacial free energy is supported by the observation of dendrites in the precipitation of the γ-phase (an ordered structure derived from a b.c.c. base) from the β-phase (b.c.c.) in the alloys of copper, silver, and gold with the B subgroup elements such as zinc, aluminium, etc.

The precipitates form with a "cube–cube" relationship that gives a nearly constant interfacial energy, as shown by the nearly spherical equilibrium shape of the precipitates (Stephens and Purdy, 1975). However, this cannot be the sole explanation since other precipitation systems with a similar cube–cube orientation relationship do not give dendrites. For example, in a recent investigation Doherty (1976) looked at the shape changes during precipitation in copper–silver alloys and found no evidence for dendritic perturbations during precipitation of silver-rich precipitates from supersaturated copper solid solutions. The origin of the reluctance to form dendrites in copper–silver alloys compared to that in copper–zinc alloys is likely to be due in part to the ideas developed by Shewmon as expressed in eqn. (13.1).

In the copper–silver alloy there is a large misfit ($>12\%$) between the lattice parameters of the two phases, giving a high density of lattice matching and edge dislocations (Aaronson, 1974). Such a dislocation array is likely to give a high diffusivity path D_I along the interface, hence giving a strong inhibition to dendritic perturbations. For the copper–zinc (β/γ) system, however, the misfit is very small, only 0.3% (Stephens and Purdy, 1975), so there D_I is likely to be negligible. Moreover, the diffusion within the precipitate is also stabilizing, and for a silver-rich precipitate this will be effective while for

---

† The cusps mentioned are those on a γ-plot where the surface tension of a crystal interface is plotted as the radial coordinate of a graph where the other variable is the orientation of the crystal interface. The form of the γ-plot for a crystal–vapour interface is described by, for example, Mykura (1966) and Kelly and Groves (1970). For crystal precipitates within another crystal there is a different three-dimensional γ-plot for each possible relative orientation between the two crystals.

‡ This arises from the nucleation process since precipitates oriented in such a way with respect to the crystal matrix that the interface has this good fit and a low surface tension would have a much lower nucleation barrier to overcome; see Smith (1953) and Shewmon (1969).

Therefore

$$J_B = D \left(\frac{C_\beta - C_0}{\rho} \right) 2\pi\rho^2.$$

But the growth velocity v requires a flux of

$$v\pi\rho^2(C_\beta - C_\alpha).$$

$v\pi\rho^2$ is the volume of precipitate formed per second and $(C_\beta - C_\alpha)$ is the change in composition (atoms/unit volume) that occurs during growth. Therefore

$$v\pi\rho^2(C_\beta - C_\alpha) = \{D(C_\beta - C_\rho)2\pi\rho^2\}/\sigma$$

$$v = (2D/\rho)(C_\beta - C_0)/(C_\beta - C_\alpha) = (2D/\rho)\Omega_0, \tag{13.4}$$

where $(C_\beta - C_0)/(C_\beta - C_\alpha)$ is the solute supersaturation Ω_0.

For a given supersaturation and diffusion coefficient the velocity rises as the tip radius is reduced. But as the tip radius falls the capillary effect of the surface tension γ on the value of C_β must be considered. That is, C_β is a function of ρ, eqn. (13.2).

Eqn. (13.4) now becomes†

$$v = \left\{ \frac{2D}{\rho} \left[C_\beta \left(1 - \frac{2\gamma V_\alpha(1 - C_\alpha)}{RT\rho(C_\beta - C_\alpha)\varepsilon} \right) - C_0 \right] \right\} \bigg/ (C_\beta - C_\alpha)$$

$$= (2D/\rho) \left[\Omega_0 - \frac{2\gamma V_\alpha C_\beta(1 - C_\alpha)}{RT\rho(C_\beta - C_\alpha)^2\varepsilon} \right].$$

Redefining Γ, the capillarity constant, $\Gamma'C_\beta/(C_\beta - C_\alpha)$, gives

$$v = (2D/\rho)|\Omega_0 - (\Gamma/\rho)|.$$

The critical radius ρ_c for growth (and nucleation) occurs when $C_\beta(\rho)$ becomes equal to C_0 (any smaller value of ρ would cause the precipitate to dissolve). This condition implies that

$$\Omega_0 - (\Gamma/\rho_c) = 0. \tag{13.5}$$

Therefore

$$\Gamma = \Omega_0\rho_c$$

and

$$v = (2D/\rho)\Omega_0 |1 - (\rho_c/\rho)|. \tag{13.6a}$$

This is illustrated in Fig. 13.13, where the maximum value of the velocity v_{max} occurs at $dv/d\rho = 0$, which happens when $\rho = 2\rho_c$. The value of v_{max} is given by

$$v_{max} = (2D\Omega_0/2\rho_c)[1 - (\rho_c/2\rho_c)] = (2D\Omega_0)/(2\rho_c) = (D\Omega_0^2)/(2\Gamma), \tag{13.6b}$$

that is, v_{max} is proportional to the square of the undercooling and inversely proportional to the capillarity constant. As pointed out by Chalmers (1964), eqn. (13.6b) gives the right

---

† We are making the simplification that for small values of Ω_0 (0–0.2) the changes in $(C_\beta - C_\alpha)$ due to capillarity can be neglected compared to changes in $(C_\beta - C_0)$. As is shown by Purdy (1971) both C_β and C_α are changed in the same direction by capillarity.

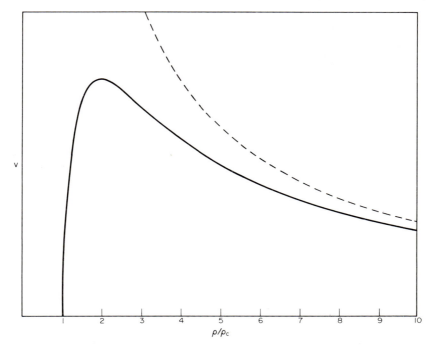

FIG. 13.13. Variation of growth velocity v of the dendrite of Fig. 13.11. Broken line shows what the velocity would have been if there were no capillarity correction and the full line allows for the effect of capillarity.

form of the relationship to account for the experimental results on dendritic growth in undercooled liquid metal systems, but the rates predicted are as much as seventy times larger than the observed growth velocities.

13.4.2. IMPROVEMENTS TO THIS ELEMENTARY TREATMENT

13.4.2.1. *Considerations of Interface Mobility*

If the interface is not perfectly mobile then its movement cannot quite keep up with the diffusion flux. The effect of this is that the value of $C_\beta(\rho, v)$ at the interface falls below the value $C_\beta(\rho)$, which would occur if the interface were perfectly mobile. This is illustrated in Fig. 13.14. Making the simple assumption that the interface velocity is controlled by a linear local undercooling, that is,

$$v = m\{C_\beta(\rho) - C_\beta(\rho, v)\},$$

where m is the interface mobility, or

$$v/m = C_\beta(\rho) - C_\rho(\rho, v),$$

$$C_\beta(\rho, v) = C_\beta(\rho) - (v/m).$$

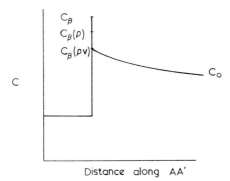

Fɪɢ. 13.14. As for Fig. 13.12 except the concentration at the interface has been modified by both capillarity and velocity effects, thereby reducing the concentration gradient into the β phase, which produces a reduction in the growth velocity.

This further modifies eqn. (13.4) to†

$$v = (2D/\rho)\{[C_\beta(\rho, v) - C_0]/(C_\beta - C_\alpha)\} = (2D/\rho)\{[C_\beta(\rho) - v/m - C_0]/(C_\beta - C_\alpha)\}$$

$$= (2D/\rho)\Omega_0[1 - (\rho_c/\rho)] - (2D/m\rho)v(C_\beta - C_\alpha)$$

$$= (2D/\rho)\Omega_0\{1 - (\rho_c/\rho) - v/[m(C_\beta - C_\alpha)]\}. \tag{13.7}$$

The effect of this last term is twofold:

   (1) The velocity is reduced for all values of the tip radius.
   (2) The radius that supports the maximum velocity increases to larger values.

13.4.2.2. *A More Realistic Shape—The Paraboloid of Revolution*

The various micrographs of the growing dendrites suggest that the previously discussed shape of a cylinder with a hemispherical cap is rather unrealistic and a more "pointed" shape is found (Figs. 13.3 and 13.4) such as would arise if the dendrite tip were a paraboloid of revolution ($z = (x^2 + y^2)/2\rho$). This has been confirmed by Glicksman and Schaefer (1968) from observations on dendrites growing on a liquid tin surface. Glicksman *et al.* (1972) subsequently used a striking holographic technique and were able to show that dendrites, within the bulk of a transparent organic melt, has a shape that was almost exactly a paraboloid of revolution. Bainbridge (1972) has also supported this with measurements on solid-state dendrites in the copper–zinc system.

The diffusion equation around a paraboloid of revolution has been treated by Ivantsov (1958) and also by Horvay and Cahn (1961); these authors showed that eqn. (13.4)

$$(v\rho)/(2D) = \Omega_0$$

becomes for a paraboloid

$$-\left(\frac{v\rho}{2D}\right)\exp\left(\frac{v\rho}{2D}\right)E_i\left(-\frac{v\rho}{2D}\right) = \Omega_0 \tag{13.8}$$

† We are again making the simplification that the change in C_β only affects $(C_\beta - C_0)$ significantly and changes in $(C_\beta - C_\alpha)$ can be neglected for small undercoolings.

where $E_i(-v\rho/2D)$ is the integral exponential function

$$E_i(-x) = \int_x^\infty (e^{-a}/a)\,da.$$

For values of Ω_0 between 0 and 0.2 the effect of this is to decrease the value of $v\rho/2D$ by somewhat more than 50%, i.e. the dendrite grows more slowly than with the simple cylindrical shape.

Defining p, the Peclet number, as $v\rho/2D$ and introducing the capillarity and velocity modification as before,

$$-p\exp\{p[E_i(-p)]\} = \Omega_0\left\{1 - \frac{\rho}{\rho_c} - \frac{v}{m(C_\beta - C_0)}\right\}. \tag{13.9}$$

Let

$$-p\exp\{p[E_i(-p)]\} = f_1(p).$$

Then

$$\Omega_0 = f_1(p)\left\{1 + \frac{\rho_c\Omega_0}{\rho f_1(p)} + \frac{v\Omega_0}{m(C_\beta - C_0)f_1(p)}\right\}.$$

But

$$\Omega_0 = \frac{C_\beta - C_0}{C_\beta - C_\alpha} \quad \text{and} \quad \Gamma = \rho_c\Omega_0,$$

$$\Omega_0 = f_1(p)\left\{1 + \frac{\Gamma}{\rho f_1(p)} + \frac{v}{m(C_\beta - C_\alpha)f_1(p)}\right\}. \tag{13.10}$$

Equation (13.10) is often described as the modified Ivantsov solution.

13.4.2.3. *The Dendrite with the Non-isoconcentrate Interface*

As recognized by Bolling and Tiller (1961), the capillarity and interface velocity modification to the Ivantsov solution raises a fundamental difficulty. The Ivantsov solution is for a dendrite at which C_β is constant around the paraboloid, but at the tip of the dendrite the curvature $2/\rho$ and the normal velocity v are both larger than anywhere else on the dendrite; this means that $C_\beta(v,\rho)$ is smaller at the tip than elsewhere on the dendrite; the dendrite no longer has an "isoconcentrate" surface. A result of this is to cause movement of solute β to the dendrite tip by diffusion in the matrix β, through the dendrite α, and by surface diffusion along the $\alpha\beta$ interface. This arrival of solute will slow down the growth velocity compared to the isoconcentrate situation. There are various models which attempt to take this into account (Bolling and Tiller, 1961; Temkin, 1960; Tarshis and Kotler, 1968; Trivedi, 1970) of which the one due to Trivedi can be expressed in a form similar to eqn. (13.10):

$$\Omega_0 = -p\exp\{p[E_i(-p)]\}\left[1 + \frac{v}{m(C_\beta - C_\alpha)}R_1(p) + \frac{\Gamma}{\rho}R_2(p)\right], \tag{13.11}$$

where diffusion in the dendrite and along the interface has been neglected.† $R_1(p)$ and $R_2(p)$ are functions of p given by Trivedi (13.10).

From a comparison of eqns. (13.10) and (13.11) it is clear that the velocity for any tip radius will be smaller for the non-isoconcentrate Trivedi model than for Ivantsov's isoconcentrate model. This is illustrated in Fig. 13.15, which shows the typical velocity against tip radius curves for the Fisher, the Ivantsov, and the Trivedi relations, all for the same supersaturation, Γ values, etc. This figure illustrates that the velocities fall as the theories become more realistic. It also illustrates the universal phenomenon of this type of kinetic analysis, i.e. a wide range of velocities and tip radii are possible.

In the past it has been conventional to make the obvious, but arbitrary, assumption that the dendrite grows at the maximum possible velocity v_{max} with the tip radius p^{opt}, the

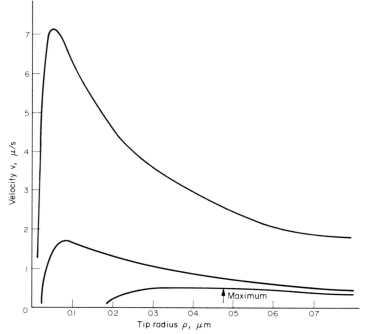

FIG. 13.15. Computed velocity–radius relationship for growth of solid-state γ dendrites. The upper curve is for a simple cylindrical dendrite model, middle curve for a dendrite assumed to be a paraboloid of revolution but with an isoconcentrate interface (Ivantsov model), and lowest curve the Trivedi model, a non-isoconcentrate paraboloid of revolution. Calculated by B. G. Bainbridge from the following data:

$$\Omega_0 = 0.084$$

$$D = 8.8 \times 10^{-12}\,\mathrm{m^2\,s^{-1}}$$

$$\Gamma = 2.18 \times 10^{-9}\,\mathrm{m}$$

$$1/\{m(C_\beta - C_\alpha)\} = 1.36 \times 10^4\,\mathrm{s\,m^{-1}}$$

---

† Trivedi (1970) also shows how to include diffusion in the dendrite.

‡ For $p = 0.1$, $f_1(p) = 0.2$, $1/\{f_1(p)\} = 5$, but $R_1(p) = 7.4$ and $R_2(p) = 9.7$, so the Trivedi analysis gives, as would be expected, a smaller p, that is, $vp/2D$, than does the modified Ivantsov equation. The dendrites, if they obey eqn. (13.11) rather than eqn. (13.10), will grow more slowly at any given tip radius.

optimum value that gives the maximum velocity. On the basis of her recent experimental evidence, Fairs (1978) suggested that for solid-state dendrites a stability criterion might be the correct way of determining the tip radius, and thus the growth velocity. The experimental values of the tip radius appeared to be quite close to the radius for *relative instability* (see Chapter 3) of a spherical precipitate experiencing the same supersaturation as the dendrite. This suggestion gained much support when subsequently applied to the very accurate dendrite growth data of Glicksman *et al.* (1976) for succinonitrile. Doherty *et al.* (1978), section 13.5. A more detailed consideration of the stability criterion for the dendrite's choice of its tip radius and therefore velocity has been recently given by Langer and Müller-Krumbhaar (1978); their analysis gives a very similar result to that suggested empirically by Fairs (1978) and by Doherty *et al.* (1978).

13.4.2.4. *Relaxation of the Assumption of a Parabolic Tip Shape*

Although the assumption of a parabolic tip shape, a paraboloid of revolution, allows a reasonably tractable analysis of the dendrite growth problem in the presence of interface kinetics and capillarity effects, there is no reason to expect that in the presence of these effects a parabolic shape would remain valid (Trivedi, 1970). Nash and Glicksman (1974) considered this problem and developed a complex analysis for the correct shape. In its present form the Nash and Glicksman analysis has been developed only for a maximum velocity dendrite and without interface kinetics.

As discussed by Glicksman *et al.* (1976), the Nash and Glicksman analysis predicts a somewhat more pointed dendrite, and therefore one that grows faster than predicted for the parabolic Trivedi analysis.

Very recently, Trivedi and Tiller (1978) have discussed Nash and Glicksman's analysis and point out that if the side branches that develop behind the tip are considered, then the heat (and solute) released by them will partially offset the changes introduced by Nash and Glicksman so that for a normal branching dendrite the original Trivedi solution might be preferable.

13.5. Experimental Observations of Dendritic Growth Velocities

There is a range of data on the growth of various metals and for white phosphorus, ice, and transparent organic materials growing from their melts, also for the growth of ionic dendrites from aqueous solutions, and for solid-state dendrites and needle-like solid-state precipitates. In only a few of the examples has the tip radius as well as the dendrite velocity been reported, and in only a few experiments has the capillarity constant been properly established. In no case has there been any independent attempt to determine the mobility of the interface, it being usually assumed that the reaction is purely diffusion controlled.

As previously mentioned, Chalmers (1964) applied the Fisher theory to the growth of "thermal" dendrites of pure metals and showed that the predicted velocities were up to seventy times too large. This would clearly be improved by the use of either the Ivantsov or Trivedi analyses. Bolling and Tiller (1961) used the Ivantsov equations and were able to get some sort of fit with the experimental data, but only by using not a linear mobility equation ($v \propto \Lambda\Omega_I$)† but a square-law relationship ($V \propto \Lambda\Omega_I^2$) and, moreover, a relationship that required a large $\Lambda\Omega_I$ to drive the attachment process. There is very strong evidence, summarized by Jackson *et al.* (1967), based on a wide range of theoretical and experimental data on the solidification of metals, that suggests that the attachment process is not likely to behave in this way. Moreover, the Bolling and Tiller assumptions for nickel and tin would require a thermal undercooling of 1.7°C for growth rates of the order of 1 cm s^{-1}. But experiments described by Chalmers (1964) have indicated that at such velocities the undercooling would be less than 0.1°C. This suggests that the isoconcentrate Ivantsov model is unreliable since it is predicting velocities that are larger than the experimental results. (When Bolling and Tiller, 1961, assumed an infinite mobility, their predicted velocities were much larger than the experimental results.) Kotler and Tarshis (1968) applied the Temkin analysis (which is very similar to the Trivedi analysis described here) to the growth of tin, nickel, and ice dendrites in undercooled melts, and were able to get a good agreement using a large assumed interface mobility but using the maximum velocity criterion.

An equivalent agreement was also achieved with the Trivedi analysis for the data on the dendritic growth of phosphorus obtained by Glicksman and Schaefer (1967). However, for both the ice dendrites and the phosphorus dendrites the values of the solid–liquid surface tension required to make the theory fit the results were somewhat larger than direct experimental values obtained by Jones and Chadwick (1970) (by 2 × for ice and by 10 × for phosphorus). However, in these discussions as well the maximum velocity criterion was used.

Doherty *et al.* (1973), and subsequently Dauncey (1973), studied the growth of dendrites in small isothermal melts in copper–nickel and copper–tin alloys. The alloys were undercooled by various amounts and then seeded by addition of small copper particles. The samples after seeding reheated rapidly to an "arrest" temperature that was found to be below the equilibrium liquidus temperature, i.e. the dendrites grew at some residual undercooling. This residual undercooling was confirmed by microanalysis of the dendrites quenched from the arrest temperature; the solute content of these dendrites was close to that expected from the measured arrest temperatures.

From the observation of the lengths of the dendrites in the quenched samples an estimate could be made of the growth velocities of the dendrites as a function of the measured undercooling. In both the copper–nickel and copper–tin systems the relationship was close to that predicted using the Trivedi analysis, but the now suspect maximum velocity criterion was used to obtain this agreement.

In a rather different set of experiments Sharp and Hellawell (1970) grew dendrites in the aluminium–copper system in a fine capillary tube at a controlled velocity and with an imposed positive temperature gradient that partially stabilized the process of dendrite growth. This was, then, the so-called "constrained" dendrite growth rather than the "free" dendrite growth discussed so far. The striking feature of the results of Sharp and Hellawell's study was that as the growth velocity increased, the undercooling, as measured from the copper contents of the dendrite tips, *fell*. This is qualitatively the wrong result for all the free dendrite growth theories previously discussed (section 13.4). The apparent conflict between these results and the expected behaviour as shown in the results on the

---

† $\Lambda\Omega_I$ is that part of the total supersaturation required for the interface process section (see 13.4.2.1).

copper–nickel alloys was beautifully resolved in a set of experiments carried out by Burden and Hunt (1974). They investigated the velocity–undercooling relationships in aluminium–copper alloys over a wide range of growth velocities and temperature gradients (Fig. 13.16). As can be seen from this figure with a *high-temperature gradient*, there is an increase in the growth temperature (a decrease in undercooling) as the velocity increased—the result previously obtained for constrained dendritic growth by Sharp and Hellawell (1970). For the low-temperature gradient the opposite result is reported—an increase of undercooling with the faster dendrite growth—the correct result for free dendritic growth. For growth velocities greater than $100\,\mu\mathrm{m\,s}^{-1}$ the "free" dendrite behaviour dominates even the large gradient of $6°\mathrm{C\,mm}^{-1}$. Burden and Hunt (1974) were able to analyse the whole set of their results and show by a simple analysis that both the constrained results (temperature gradient dominant) and the free dendrite results (velocity dominant) were compatible with theory. Using a simple Zener–Fisher analysis for the free dendritic growth they obtained the following expression for the total undercooling (ΔT) in temperature:

$$\Delta T = \frac{GD}{v} + 2\,[mv(1 - k_0)C_0\Gamma/D]^{1/2} \qquad (13.12)$$

The first term in eqn. (13.7) is the situation for constrained dendritic growth—the undercooling decreases with larger dendrite velocities in the presence of a temperature gradient G (an identical relationship was derived independently by Flemings, 1974a, for constrained dendritic growth, first term only). The second term is the approximate expression for free dendritic growth; the free dendrite term will clearly dominate the undercooling at sufficiently high growth velocities. The analysis is only approximate since the free dendrite analysis was based on the Zener–Fisher model, which assumed a hemispherical tip; this, therefore, *underestimates* the undercooling required for a given

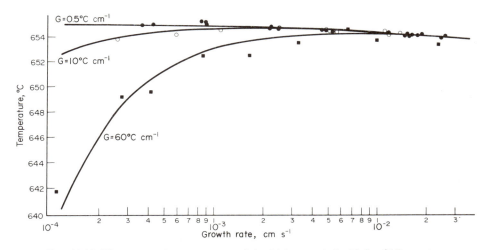

FIG. 13.16. The measured temperatures of dendritic growth in Al–2wt.%Cu against imposed growth velocity for different temperature gradients G, temperature increasing into the liquid. The measured liquidus temperature is a little over 655°C, so the undercooling is the difference between the liquidus temperature and the measured temperatures. (From Burden and Hunt, 1974; courtesy of *J. Crystal Growth.*)

velocity. Burden and Hunt also used the Trivedi analysis (with a maximum velocity criterion) and were then able to get quantitative agreement with their experimental results at high velocities.

Data from solid-state dendritic and needle growths can also be used for tests of the dendrite growth theories—and in this case with slower growth, due to solid-state diffusion of solute being slow, it is possible both to make the precipitation occur at a fixed undercooling and to quench efficiently to arrest the process for examination. The high ratio of thermal diffusivity to solute diffusivity in the metallic solid state allows the complete suppression of thermal (recalescence) effects that make the investigation of solute-dominated dendritic growth from the liquid state difficult; moreover, the good quenching response of a metallic solid-state precipitation reaction in metals allows subsequent investigation by metallographic techniques of the tip radius of the dendrite.

Purdy (1971) investigated the growth of needle crystals (f.c.c. from b.c.c.) in the copper–zinc system and from his measurements of growth velocity and tip radii at two supersaturations was able to show that his observations agreed with the Trivedi predictions. No arbitrary assumptions were necessary apart from taking the incoherent interface at the needle tip to be perfectly mobile. From an analysis of dendrite growth of γ from β in the same system, Bainbridge (1972) and Stephens and Purdy (1974) were also able to obtain a close agreement between experimental results and the Trivedi theory, using reasonable values of the interfacial surface tension, but a finite mobility had to be assumed. (The interface between γ and β brass is partially coherent.) In both these discussions a maximum velocity criterion was used. Fairs (1978) extended the range and accuracy of the measurements of velocities and tip radii of γ dendrites forming from β brass in the copper–zinc system. She was able to show agreement with the Trivedi analysis but only after relaxation of the maximum velocity criterion. In all cases the tip radius appeared to be given better by a stability criterion (equivalent to the spherical radius of *relative instability*) than by the maximum velocity criterion. As in the earlier analysis of Bainbridge (1972) and Stephens and Purdy (1974), Fairs found evidence that there was either *some* interface immobility or alternatively that the diffusion coefficient was being reduced by a flux of vacancies (Bainbridge, 1972). The precipitates require about 4 atomic % vacancies in their structure, but the diffusion of zinc towards the precipitates is faster than of copper away (Resnick and Ballufi, 1955), so that normal diffusion will pump vacancies *away* from the precipitate.

The best and most complete data on dendritic growth velocities were obtained by Glicksman *et al.* (1976) for succinonitrile ($H_2N–CH_2–CH_2–NH_2$), a transparent low entropy of melting compound that solidifies below 59°C to a b.c.c. dendritic crystal. Glicksman *et al.* prepared extremely high-purity material to avoid any conflicting solute effects and observed the growth velocities and tip radii of succinonitrile dendrites under a wide range of thermal undercoolings (see Appendix).

They had previously obtained all the necessary parameters, surface energy, latent heat of solidification, interfacial energy, etc., so that a complete test of the various theories was possible. They reported that the observed velocities were less than predicted by the Trivedi analysis *using a maximum velocity criterion* and even further away from the Nash and Glicksman (1974) analysis. Doherty *et al.* (1978) showed, however, that use of the data, obtained simultaneously by Glicksman *et al.* on the tip radii, allowed the Trivedi theory to be tested after relaxing the arbitrary maximum velocity criterion. Figure 13.17 shows the predicted Trivedi analysis for two of the succinonitrile undercoolings, together with the

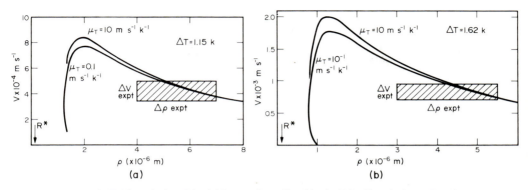

FIG. 13.17. The velocity of dendritic growth predicted by the Trivedi analysis as a function of the dendrite tip radius for thermally undercooled succinonitrile. Two values of the thermal interface mobility μ_t are given, equivalent to the compositional mobility m. The higher value of μ_t is, in each case, equivalent to an infinite mobility. The cross-hatched areas are the range of experimental results reported by Glicksman et al. (1976). Two undercoolings are shown 1.15 and 1.62 K. The radius R^* is the critical radius for nucleation ρ_c. (From Doherty et al., 1978; courtesy of Met. Trans.)

observed spread in the experimental results. It is clear that the Trivedi analysis, as a diffusion equation, is clearly compatible with the experimental results provided the maximum velocity criterion is discarded. Langer et al. (1978) have used the stability theory for dendritic growth, developed very recently by Langer and Müller-Krumbhaar (1978), not only for the velocity data for succinonitrile of Glicksman et al. (1976) but also for their own velocity data for ice dendrites growing from very slightly undercooled water (Fig. 13.18). They point out that if interface kinetics are neglected (an infinite mobility being assumed), then since the stability criterion gives a very high ratio of ρ to ρ_c (Doherty et al., 1978), the non-isoconcentrate theory due to Trivedi is not needed; the simple Ivantsov equation can be used, Ivantsov (1958):

$$\Omega_0 = -p \exp\left[(p)E_i(-p)\right].$$

The value of the Peclet number p is given by the equation suggested by Langer and Müller-Krumbhaar (1978):

$$p = v\rho/2D_\beta = (vd_0/2D_\beta\sigma)^{1/2}$$

with the stability parameter $\sigma = 0.025$ and the capillarity length d_0† as defined by Langer and Müller-Krumbhaar (1978). These two ideas allow a universal plot of the dimensionless velocity $V(vd_0/2D_\beta)$ against the dimensionless undercooling Ω_0. Langer et al. (1978) use their model for thermal dendrites (see Appendix). Their universal plot is shown in Fig. 13.18 together with the experimental data for succinonitrile (Glicksman et al., 1976) and both their own data and some earlier data for ice dendrites. The agreement is clearly very good, though there is some departure at the highest undercoolings. This

---

† d_0 is given for solute supersaturation, discussed here, as $d_0 = \gamma V_\alpha C_\beta / RT(C_\beta - C_\alpha)$; for thermally undercooled dendrites such as pure ice and succinonitrile, d_0 is $\gamma T_m / L_v(T_m - T_\infty)$, where L_v is the latent heat per unit volume and T_∞ is the bath temperature.

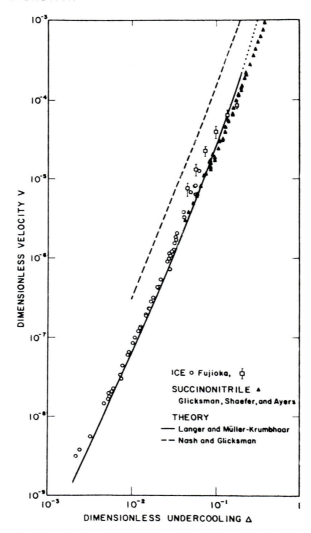

FIG. 13.18. The dimensionless velocity V of succinonitrile and ice dendrites growing in thermally undercooled melts over a range of dimensionless undercoolings. The experimental results are given as points and the full line is the theoretical prediction of the Langer and Müller-Krumbhaar (1978) analysis. The dashed line is the theoretical prediction of the Nash and Glicksman (1974) analysis. (From Langer *et al.*, 1978; courtesy of *J. Crystal Growth.*)

departure may indicate that at larger undercoolings the Trivedi equation might be more appropriate, as was used by Doherty *et al.* (1978). Langer *et al.* (1978) did not report how successful their analysis was in predicting the experimental values of tip radii—this would be interesting to know. They also did not mention if there were any of the convection effects reported by Glicksman *et al.* (1976) for low undercoolings.

It nevertheless appears from the current experimental results that we may now have at last a successful theory for predicting free dendritic growth velocities.

13.6. Other Studies of Dendritic Growth

13.6.1. DENDRITE ARM SPACING

The spacing of the secondary dendrite arms, which branch off the main or primary arms, has been studied for many years, mainly in metallic systems (e.g. Kattamis *et al.*, 1967), but also for ionic dendrites (Papapetrou, 1935). The two variables that have received attention are the alloy content and the cooling rate. There was considerable confusion concerning the effect of alloy additions (Alexander and Rhines, 1950; Horwarth and Mondolfo, 1962), but the role of cooling rate is now well understood.

There are two effects on the arm spacing of a given alloy which arise from the cooling rate—that on the initial formation of the dendrite and that of the subsequent arm "coarsening" before the dendrite is completely solidified.

The initial formation. The instability theory of Mullins and Sekerka (1963, 1964), previously discussed, predicts that at greater undercoolings the arm spacing should be less. This is supported by the quantitative agreement between the theory and the experimental result (Malcolm and Purdy, 1967; Hardy and Coriell, 1968; Townsend and Kirkaldy, 1968) and also by other less-controlled experiments on organic dendrites by Jackson *et al.* (1966) and ionic dendrites by Kahlweit (1969) and by experiments by Feest and Doherty (1973) on metallic melts quenched during the formation of the dendrite skeleton. The last authors found, however, that the small differences in initial arm spacing rapidly disappeared during cooling when extensive arm coarsening occurred.

Arm coarsening. Some years ago Kattamis *et al.* (1966, 1967) demonstrated that during the dendritic solidification of metal alloys the arm spacing continuously increased while the alloy was in the solid and liquid, two-phase state.[†] In other words many of the initial secondary dendrite arms, which had originally grown, disappeared, while the alloy remained in the two-phase region. This process can be observed even during the formation of the dendrite skeleton. For example, in Fig. 13.3 many of the initial perturbations near the dendrite tip do not develop into dendrite arms and some of those that have developed are starting to disappear again. Jackson *et al.* (1966) also demonstrated another but closely related process, that of dendrite arm melt-off. This is the separation of some dendrite arms at the junction (armpit?) with the main dendrite trunk.

This is found to happen only after a local increase in temperature such as occurs after an imposed growth rate fluctuation. Such an arm melt-off will increase the arm spacing as well as giving rise to secondary crystallization nuclei as previously discussed.

Kahlweit (1968) shows for ammonium halide dendrites that there was no melt-off in the absence of local increases in temperature but that the main process in isothermal dendrite arm coarsening was of the shrinkage of the thinner dendrites back from their exposed tip. This is reasonable since the curvature $(1/\rho_1 + 1/\rho_2)$ is largest at the tip of the thinnest arms.[‡] The observed rates of arm shrinkage were compatible with a model by Kahlweit (1968) based on this idea, but the model was not extended to predict the expected rate of arm coarsening since this would require knowledge of the distribution of arm thicknesses. However, there should in principle be no difficulty in doing this since similar calculations

---

[†] Jackson *et al.* (1966) also reported this effect.

[‡] In Fig. 13.9 the tips of the dendrites are seen to be almost the hemispherical caps of the Fisher theory rather than the paraboloids. This is almost certainly a result of capillarity action after the growth of the dendrite skeleton had gone to completion.

exist for the equivalent process of the coarsening of precipitates in solid or liquid matrices (Wagner, 1961).

The mean radius r of such precipitates obeys an equation of the form $r_{(t)}^3 = r_0^3 + At$, where $r_{(t)}$ is the radius after time t, r_0 the radius when $t = 0$, and A a constant involving the precipitate–matrix surface tension and the diffusion coefficient. The experiments on dendrite arm coarsening by Kattamis *et al.* (1966, 1967) and by Matya *et al.* (1968) yield a similar equation for the dendrite arm spacing d as a function of time—both for isothermal holding and for steady cooling in the two-phase region. The dendrite equation is

$$d_t^n = d_0^n + Bt_s, \qquad (13.13)$$

where n is between 2 and 3. This equation has been found to hold over many orders of magnitude of t_s, the local solidification time. There is as yet no theory available which can account for this equation, although variation in the parameter B might be estimated from Kahlweit's model. Attempts to do this, Dauncey (1973), have not been very successful.

One factor that has not been fully discussed is the coalescence of secondary dendrite tips with adjacent dendrites as is sketched in Fig. 13.19. This type of coalescence will occur more frequently as the volume fraction of solid increases; another equivalent type of coalescence process, that of rows of adjacent arms becoming plates, has also been found to occur as the solidification proceeds by Feest and Doherty (1973). Examples of rows of arms forming plates can be seen in Fig. 13.7.

Kattamis *et al.* (1967), who studied the arm coarsening in aluminium–copper alloys, found that, as would be expected from this type of argument, the rate of arm coarsening fell as the volume fraction of the solid dendrite increased.

In addition to the clearly established effect of increased holding or solidification time t_s (eqn. (13.13)) in increasing the dendrite arm spacing, there is now clear evidence that for *the same solidification time* the dendrite arm spacing falls with increased solute content (B is reduced) (Kattamis *et al.*, 1967; Dauncey, 1973; Feest, 1973; and Young and Kirkwood, 1975). Table 13.1, due to Dauncey (1973), shows how for two systems, copper–tin and

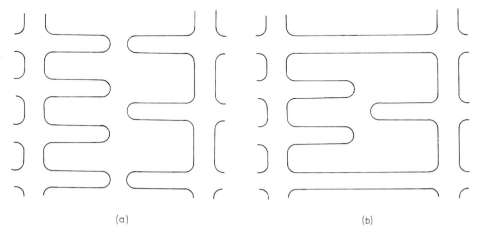

(a)   (b)

FIG. 13.19. Showing how an increase in the volume fraction of solid dendrite will reduce the number of free dendrite tips. (a) Two adjacent dendrites, none of whose secondary arms have coalesced; (b) same structure of the further crystallization at which two tip coalescence events have occurred. Compare with a real structure (e.g. Fig. 13.2).

TABLE 13.1. *Variation of Secondary Dendrite Arm Spacing with Alloy Content at a Holding Time of* 1000 s (*Dauncey,* 1973)

Alloy	d_t ($t_s = 1000$ s) (μm)	Freezing Range (°C)
Cu–5wt%Sn	100	125
Cu–10wt%Sn	63	170
Al–2wt%Cu	170	44
Al–4.5wt%Cu	110	80
Al–15wt%Cu	67	240* (70)

*This is the expected freezing range, by extrapolation of the liquidus and solidus lines, that would exist if no eutectic $CuAl_2$ formed; the actual freezing range to the eutectic temperature is only 70°C.

aluminium–copper, with a solidification time t_s of 1000 s, the secondary dendrite arm spacing d_t falls with increasing solute (and *equivalently with the alloy freezing range*).

The change of d_t with alloy content will be largely offset by the increase in freezing range, since with a given cooling rate an alloy will have a longer complete freezing time (t_s = freezing range/cooling rate). As a result the dendrite arm spacing found after complete solidification will be almost a constant in an alloy, at a given cooling rate, *for any solute content*; this is commonly found (Flemings, 1974a). The only exception to this trend will be partially eutectic alloys such as the Al–15wt%Cu alloy of Table 13.1. This type of alloy will be expected to have a small final dendrite arm spacing since the small value of B in eqn. (13.13) will not then be offset by a large t_s. Commercial casting alloys such as Fe–3wt%C and Al–8wt%Si should benefit from this effect.

Dendrite arm spacing has at least two important consequences for the structure and properties of dendritic crystals.

1. *Microsegregation.* The previously mentioned variation in solute content between the first and last liquid to solidify leads to microsegregation with a "wavelength" of the order of the dendrite arm spacing. Homogenization of this segregation by diffusion within the dendrite will be able to occur given sufficient time t. This can be calculated from the arm spacing d and the diffusion coefficient in the solid D, using the approximate relationship for effective homogenization (Shewmon, 1963),

$$d^2 \approx Dt. \tag{13.14}$$

Taking a typical value of D as 10^{-13} m^2 s^{-1} for metal crystals close to their melting point, the value of t will be 200 s for a 10 μm arm spacing typical of very rapidly solidified alloys, but many days for spacing of 1 mm typically found in the centres of large metal ingots. The removal of microsegregation is therefore impracticable for such ingot material in the as-cast state.

2. *Distribution of second-phase precipitates or inclusions.* Most commercial metals have significant impurity contents and the solidification of these materials causes a concentration of these impurities in the last pools of liquid to freeze between the dendrite arms. The result of this is that inclusions of second-phase material occur on the scale of the

dendrite arm spacing. (Oxide and sulphide inclusions in steels are important examples of this.) By a reduction in the dendrite arm spacing these inclusions can be reduced in size and made into a finely dispersed form with considerable improvements in mechanical properties.

A simple analysis of this can be given. If the volume fraction of the insoluble inclusion is V_f and the dendrite arm spacing is d_s, the volume of an inclusion formed in one interdendritic pool V_i will be given by

$$V_i \approx V_f d_s^3 = \frac{4\pi r^3}{3},$$

where r_i is the radius of the inclusion assumed to be spherical,

$$r_i \approx d_s \left(\frac{3V_t}{4\pi}\right)^{1/3}. \tag{13.15}$$

Equation (13.15) suggests that a reduction in dendrite arm spacing will be much more effective in reducing the size of inclusions than will an equivalent reduction in the amount of impurity by chemical purification.

Grant (1970) has shown that rapid solidification to produce a fine dendrite arm spacing can improve existing alloys and allow the development of new alloys with desirable properties, which by conventional slow ingot casting would be impossible to manufacture. In the last few years there has been considerable technological interest in the use of rapidly cooled alloys, as reviewed by Grant (1978). Grant points out that since this has commercial interest many of the investigations carried out have not been published. The normal route is the production of rapidly cooled alloy powders normally by "gas atomization"; these are then compacted by powder metallurgical routes often by the "hot isostatic pressing" (HIP) process, followed by subsequent heat treatment to give strengthening by precipitation and grain-size effects (see, for example, Lyle and Cebulak, 1975; Cox and van Reuth, 1978; Grant, 1978).

Honeycombe (1978) has pointed out that with really high-speed solidification such as is now achieved by "melt spinning" it is often possible to suppress dendritic growth completely. In the melt-spinning process a stream of liquid metallic alloy is pushed onto a rapidly spinning copper wheel, producing ribbons 20–$100\,\mu m$ thick at speeds of up to $100\,km\,h^{-1}$. The cooling rates achieved are believed to be of the order of $10^{6}\,^{\circ}C\,s^{-1}$, which for some alloys is sufficient to give non-dendritic microstructures. What is assumed to be happening is that the alloy is solidifying at a low temperature —below the so-called T_0 temperature at which it is possible for a liquid alloy to transform to a crystal with the same composition without an increase of free energy. Such a process is, of course, a very long way removed from the normally very small departures from equilibrium conditions that crystal growers try to achieve. In some alloys it is now possible, however, to make an even larger departure from equilibrium by the production of non-crystalline metallic glasses, e.g. Davies (1978). A discussion of that topic is clearly not appropriate for this book.

13.6.2. INFLUENCE OF FLUID FLOW ON DENDRITIC GROWTH PHENOMENON

In addition to the effect of fluid motion on dendrite fragmentation previously discussed in section 13.2.5 there are various other fluid-flow effects that have recently been discussed.

Vogel and Cantor (1977) have considered the influence of turbulent fluid flow on the condition for shape instability. For an isolated spherical crystal growing in a turbulent fluid, the existence of a thin boundary layer around the growing crystal, across which the concentration (or temperature) changes, gives a sharper composition (or temperature) gradient at the interface than would exist in the absence of turbulent flow. As a consequence of these steeper gradients the spherical precipitates become unstable at *smaller radii*. Cantor and Vogel (1977) have also analysed dendritic growth in the presence of fluid flow impinging normally on the dendrite, in terms of similar boundary layer ideas. The effect of this fluid flow is to increase the predicted velocity at any given tip radius (Fig. 13.20). Moreover, if the operating tip radius is controlled by a stability criterion, then the decrease in this tip radius predicted by Vogel and Cantor (1977) will further increase the predicted velocity, since the tip radius will be expected to fall with turbulent fluid flow. Some evidence that supports these ideas put forward by Cantor and Vogel has been provided by the report by Glicksman *et al.* (1976) that at low undercoolings the downward pointing dendrites grew faster than dendrites pointing in other directions. This has been briefly discussed by Doherty *et al.* (1978) in terms of convective flow past a growing dendrite.

It is also possible that the effect of fluid flow will for similar reasons, that of enhanced diffusion, increase the rate of dendrite arm coarsening (whether fragmented or not)— Vogel *et al.* (1979) have provided some limited evidence for this effect.

There are many other important consequences of dendritic growth some of which have been mentioned in the introduction. Each of these is worthy of a detailed discussion by itself, but to do so would go beyond the restrictions of space and of this author's

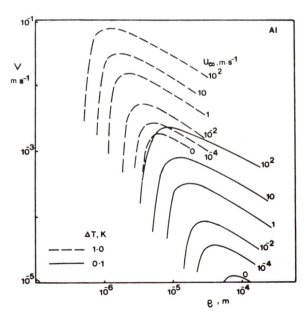

FIG. 13.20. The predicted relationship between dendrite velocity V and tip radius ρ for thermally undercooled aluminium for two different undercoolings, 0.1 and 1.0 K, and for a range of fluid flow velocities U_x impinging normally on the dendrite. (From Cantor and Vogel, 1977; courtesy of *J. Crystal Growth.*)

knowledge. However, it is hoped that the limited discussion presented here will encourage other scientists, who are working (or plan to work) in the area of crystal growth, to share the present author's enthusiasm for studies of dendritic growth.

13.7. Appendix

For a dendrite growing in a thermally undercooled pure melt where heat conduction is rate controlling, the simple treatment leading to eqn. (13.4) is modified as follows: thermal flux from the dendrite tip $= \{K(T_m - T_\infty)2\pi\rho^2\}/\rho$, where K is the thermal conductivity, T_m the melting point of the pure material, and T_∞ the temperature of the undercooled liquid. This thermal flux is balanced by the heat produced per second by the growing dendrite, which is

$$v\pi\rho^2 L_v,$$

where L_v is the latent heat per unit volume. If $\Delta T = T_m - T_\infty$, then

$$2K(\Delta T/\rho) = vL_v$$

$$v = \left(\frac{2K}{\rho}\right)\frac{\Delta T}{L_v},$$

where $\Delta T/L_v$ is the thermal undercooling—to make this a dimensionless undercooling equivalent to Ω_0, it should be multiplied by C, the specific heat of the liquid, per unit volume.

$$v = \left(\frac{2\alpha'}{\rho}\right)\left(\frac{\Delta TC}{L_v}\right)$$

where α' is K/C the thermal diffusivity and $(\Delta TC/L_v)$ is the dimensionless thermal undercooling Ω_0. This equation is the equivalent to eqn. (13.4) in the text. The relationship can be modified for the capillarity and interface mobility effects on the interface temperature and for the non-isothermal nature of a parabolic dendrite tip in exactly the same way as was done for the solute-controlled dendrite (Trivedi, 1970).

Acknowledgements

I gratefully acknowledge the contributions made to my understanding of dendritic growth by many people with whom I have discussed the subject, but a special debt is owed to Dr. B. G. Bainbridge, who has done more than anyone else to enable me to fully appreciate the various dendritic growth theories. I am also grateful for many fruitful discussions with Dr. B. Cantor, Dr. A. Vogel, and Dr. S. Fairs. Any errors in the presentation, however, are all mine.

I would also like to thank Mr. D. Price for the micrographs of Figs. 13.8 and 13.9; Dr. B. G. Bainbridge for permission to use the data calculated by him for Fig. 13.15; The University of Chicago Press for permission to use Fig. 13.1; the Metallurgical Society of AIME for permission to use Figs. 13.3 and 13.17; and the *Journal of Crystal Growth* for permission to use Figs. 13.16, 13.18, and 13.20.

References

AARONSON, H. I. (1962) *Decomposition of Austenite by Diffusional Processes*, Interscience, New York.
AARONSON, H. I. (1974) *J. Microsc.* **102**, 275.
ALEXANDER, B. H. and RHINES, F. N. (1950) *TMS-AIME* **188**, 1267.
BACKMAN, D. R., MEHREBIAN, R. and FLEMINGS, M. C. (1977) **8B**, 471.
BAINBRIDGE, B. (1972) D.Phil. thesis, Sussex University.
BAINBRIDGE, B. G. and DOHERTY, R. D. (1969) *Quantitative Relation Between Properties and Microstructure* (Brandon, D. G. and Rosen, A. eds.), Israel University Press.
BENTLEY, W. W. and HUMPHRIES, W. J. (1931) *Snow Crystals*, McGraw-Hill, New York.
BOLLING, G. F. and TILLER, W. A. (1961) *J. Appl. Phys.* **32**, 2587.
BURDEN, M. H. and HUNT, J. D. (1974) *J. Cryst. Growth* **22**, 99, 109.
CAMPBELL, J. (1968) *The Solidification of Metals*, Iron and Steel Institute, London.
CANTOR, B. and VOGEL, A. (1977) *J. Cryst. Growth* **41**, 109.
CHALMERS, B. (1964) *Principles of Solidification*, Wiley, New York.
CLINE, H. E. (1971) *Acta Met.* **19**, 481.
COX, A. R. and VAN REUTH, E. C. (1978) *Rapidly Quenched Metals*, The Metals Society, London, Vol. II, p. 225.
CRIACH, H. DUKIET-ZAWADZKA and CRIACH, T. D. (1978) *J. Mat. Sci.* **13**, 2676.
DARKEN, L. S. (1948) *TMS-AIME* **175**, 184.
DAUNCEY, P. A. (1973) D.Phil. thesis, University of Sussex.
DAVIES, H. A. (1978) *Rapidly Quenched Metals*, The Metals Society, London, Vol. I, p. 1.
DOHERTY, R. D. (1976) Unpublished research, Departamento do Materiais, Centro Technico Aerospacial, SP, Brazil.
DOHERTY, R. D. and MELFORD, D. A. (1966) *J. Iron Steel Inst.* **204**, 1131.
DOHERTY, R. D., FEEST, E. A., and HOLM, K. (1973) *Met. Trans.* **4**, 115.
DOHERTY, R. D., CANTOR, B., and FAIRS, S. (1978) *Met. Trans.* **9A**, 621.
FAIRS, S. (1978) D.Phil. thesis, University of Sussex.
FAUST, J. W. (1968) *J. Cryst. Growth* **3**, 433.
FEEST, E. A. (1973) *J. Inst. Metals* **101**, 279.
FEEST, E. A. and DOHERTY, R. D. (1973) *Met. Trans.* **4**, 115.
FISCHER, J. C. (1964) Quoted by Chalmers (1964).
FLEMINGS, M. C. (1974a) *Solidification Processing*, McGraw-Hill.
FLEMINGS, M. C. (1974b) *Met. Trans.* **5**, 2121.
FLEMINGS, M. C. and MEHREBIAN, R. (1973) *Trans. Am. Foundrymen's Soc.* **81**, 81.
FLEMINGS, M. C. and NERRO, G. E. (1968) *TMS-AIME* **242**, 50.
FRASER, C. (1966) *The Avalanche Enigma*, Murray, London.
GARABEDIAN, H. and STRICKLAND-CONSTABLE, R. F. (1974) *J. Cryst. Growth* **22**, 188.
GLICKSMAN, M. E. and SCHAEFER, R. J. (1967) *J. Cryst. Growth* **1**, 297.
GLICKSMAN, M. E. and SCHAEFER, R. J. (1968) *The Solidification of Metals*, Iron and Steel Institute, London.
GLICKSMAN, M. E., SCHAEFER, R. J., and BLODGETT, J. A. (1972) *J. Cryst. Growth* **13**, 68.
GLICKSMAN, M. E., SCHAEFER, R. J., and AYERS, J. D. (1976) *Met. Trans.* **7A**, 1747.
GRANT, N. J. (1970) *Fizika* **2**, suppl. 2, 16.
GRANT, N. J. (1978) *Rapidly Quenched Metals*, The Metals Society, London, Vol. II, p. 172.
GRIGNON, P. C. (1775) *Mémoire de Physique, sur l'Art de Fabriquer le Fer*, Paris.
HARDY, S. C. and CORIELL, S. R. (1968) *J. Cryst. Growth* **3**, 599.
HIRSCHFIELD, D. A. and WEINBERG, F. (1978) *Met. Trans.* **9B**, 321.
HONEYCOMBE, R. W. K. (1978) *Rpaidly Quenched Metals*, The Metals Society, London, Vol. I, p. 73.
HORVAY, G. and CAHN, J. W. (1961) *Acta Met.* **9**, 695.
HORWARTH, J. A. and MONDOLFO, L. F. (1962) *Acta Met.* **10**, 1037.
HU, H. and SMITH, C. S. (1956) *Acta Met.* **4**, 638.
HUNT, J. D. and JACKSON, K. A. (1965) *Acta Met.* **13**, 1212.
IVANTSOV, G. P. (1958) *Growth of Crystals*, Consultants Bureau Inc., New York.
JACKSON, K. A. (1958) *Liquid Metals and Solidification*, ASM, Cleveland.
JACKSON, K. A., HUNT, J. D., UHLMANN, D. R., and STEWARD, T. P. (1966) *TMS-AIME* **236**, 149.
JACKSON, K. A., UHLMANN, E. R., and HUNT, J. D. (1967) *J. Cryst. Growth* **1**, 1.
JOLY, P. A. and MEHREBIAN, R. (1976) *J. Mat. Sci.* **11**, 1393.
JONES, D. R. H. and CHADWICK, G. A. (1970) *Phil. Mag.* **22**, 291.
KAHLWEIT, M. (1968) *Scripta Met.* **2**, 251.
KAHLWEIT, M. (1969) *J. Cryst. Growth* **5**, 391.
KATTAMIS, T. Z. and FLEMINGS, M. C. (1966) *TMS-AIME* **136**, 1523.
KATTAMIS, T. Z., COUGHLIN, J. M., and FLEMINGS, M. C. (1967) *TMS-AIME* **239**, 1504.
KATTAMIS, T. Z., HOLMBERG, U. T., and FLEMINGS, M. C. (1967) *J. Inst. Metals* **95**, 343.

KELLY, A. and GROVES, G. W. (1970) *Crystallography and Crystal Defects*, Longman, London.
KOHN, A. (1968) *The Solidification of Metals*, Iron and Steel Institute, London.
KOTLER, G. R. and TARSHIS, L. A. (1968) *J. Cryst. Growth* **4**, 603.
LANGER, J. S. and MÜLLER-KRUMBHAAR, H. (1978) *Acta Met.* **26**, 1681.
LANGER, J. S., SEKERKA, R. F., and FUJICKA, T. (1978) *J. Cryst. Growth* **44**, 197.
LEHMANN, O. (1888) *Molekularphysik*, Leipzig.
MCDONALD, R. J. and HUNT, J. D. (1969) *TMS* **24**, 1993.
MCDONALD, R. J. and HUNT, J. D. (1970) *Met. Trans.* **1**, 1787.
MALCOLM, J. A. and PURDY, G. R. (1967) *TMS-AIME* **239**, 1391.
MASON, B. J. (1962) *Clouds, Rain and Rainmaking*, Cambridge.
MATYA, H., GIESSEN, B. C., and GRANT, N. J. (1968) *J. Inst. Metals* **96**, 30.
MEHREBIAN, R. and FLEMINGS, M. C. (1971) *Solidification*, ASM Metals Park, Ohio.
MEHREBIAN, R., KEANE, M., and FLEMINGS, M. C. (1970) *Met. Trans.* **1**, 1209.
MEHREBIAN, R., RIEK, R. G., and FLEMINGS, M. C. (1974) *Met. Trans.* **5**, 1899.
MELIA, T. P. and MOFFITT, W. P. (1964) *Indust. Engng. Chem.* **3**, 313.
MORRIS, L. R. and WINEGARD, W. C. (1967) *J. Cryst. Growth* **1**, 245.
MULLINS, W. W. and SEKERKA, R. F. (1963) *J. Appl. Phys.* **34**, 323.
MULLINS, W. W. and SEKERKA, R. F. (1964) *J. Appl. Phys.* **35**, 444.
MYKURA, H. (1966) *Solid Surfaces and Interfaces*, Routledge & Kegan Paul, London.
NASH, G. E. and GLICKSMAN, M. E. (1974) *Acta Met.* **22**, 1283.
NICHOLS, F. A. and MULLINS, W. W. (1965) *TMS-AIME* **233**, 1840.
NORTHCOTT, L. and THOMAS, D. E. (1939) *J. Inst. Metals* **65**, 205.
PAPAPETROU, A. (1935) *Z. Krist.* **92**, 89.
PURDY, G. E. (1971) *Met. Sci. J.* **5**, 81.
RAMATI, S. D. E., ABBASCHIAN, G. J., BACKMAN, D. G., and MEHREBIAN, R. (1978) *Met. Trans.* **9B**, 279.
RESNICK, R. and BALLUFI, R. W. (1955) *TMS-AIME* **203**, 1004.
RUDDLE, R. W. (1957) *Solidification of Castings*, Institute of Metals, London.
SAUVEUR, A. and CHOU, C. H. (1930) *TMS-AIME* **90**, 100.
SEKERKA, R. F. (1968) *J. Cryst. Growth* **3**, 71.
SHARP, R. M. and HELLAWELL, A. (1970) *J. Cryst. Growth* **6**, 253.
SHEWMON, P. G. (1963) *Diffusion in Solids*, McGraw-Hill, New York.
SHEWMON, P. G. (1965) *TMS-AIME* **233**, 736.
SHEWMON, P. G. (1966) *Recrystallisation, Grain Growth and Textures*, ASM Metals Park, Ohio.
SHEWMON, P. G. (1969) *Transformation in Metals*, McGraw-Hill, New York.
SMITH, C. S. (1953) *ASM Trans.* **45**, 533.
SMITH, C. S. (1965) *A History of Metallography*, The University of Chicago Press.
SMITHELLS, C. J. (1967) *Metals Reference Book*, Butterworths, London, Vol. III, p. 483, 647, and 665.
SOUTHIN, R. T. (1967) *TMS-AIME* **239**, 220.
SPENCER, D. E., MEHREBIAN, R., and FLEMINGS, M. C. (1972) *Met. Trans.* **3**, 1925.
STEPHENS, D. E. and PURDY, G. R. (1974) *Scripta Met.* **8**, 323.
STEPHENS, D. E. and PURDY, G. R. (1975) *Acta Met.* **23**, 1343.
TARSHIS, C. A. and KOTLER, G. R. (1968) *J. Cryst. Growth* **2**, 222.
TEMKIN, D. E. (1960) *Dokl. Akad. Nauk SSSR* **1326**, 1307.
THOMAS, P. M. and GRUZLESKI, J. E. (1978) *Met. Trans.* **9B**, 139.
TILLER, W. A., JACKSON, K. A., RUTTER, J. W., and CHALMERS, B. (1953) *Acta Met.* **1**, 428.
TOWNSEND, R. D. and KIRKALDY, J. S. (1968) *ASM Trans.* **61**, 605.
TRIVEDI, R. (1970) *Acta Met.* **18**, 287.
TRIVEDI, R. and TILLER, W. A. (1978) *Acta Met.* **26**, 671.
TSCHERNOFF, D. K. (1879) *Investigations on the Structure of Cast Ingots*, Zapinski Imperatorskago Russkago Tekhnicheskago Obshestva (Eng. Trans., W. ANDERSON, *Proc. Inst. Mech. Engrs.* 1880).
TZAVARAS, A. A. and FLEMINGS, M. C. (1965) *TMS-AIME* **233**, 355.
VOGEL, A. and CANTOR, B. (1977) *J. Cryst. Growth* **37**, 309.
VOGEL, A., CANTOR, B., and DOHERTY, R. D. (1979) Solidification and casting of metals, Conference held in Sheffield, 1977, to be published by The Metals Society, paper 73.
WAGNER, C. (1961) *Z. Electrochem.* **65**, 581.
YOUNG, K. P. and KIRKWOOD, D. H. (1975) *Met. Trans.* **6A**, 197.
YOUNG, K. P., RIEK, R. G., and FLEMINGS, M. C. (1979) Solidification and casting of metals, Conference held in Sheffield, 1977, to be published by The Metals Society, paper 74.
ZENER, C. (1946) *TMS-AIME* **167**, 550.

CHAPTER 14

Bulk Crystallization

J. W. MULLIN

Professor of Chemical Engineering, University College London, England

CRYSTALLIZATION is one of the major processing techniques of the chemical industry. Vast quantities of crystalline substances are manufactured commercially, e.g. sodium chloride, potassium chloride, sodium sulphate, ammonium sulphate, sucrose, etc., all of which have worldwide production rates around 100 Mtonnes per year. Many other crystalline substances are also manufactured on a large tonnage scale, ranging from inorganic fertilizer chemicals to products of the pharmaceutical and organic fine chemicals industries.

Even liquids are now being purified on a large scale by low-temperature crystallization instead of by distillation. In certain cases crystallization has a lot to offer, especially where azeotropes and close-boiling mixtures have to be separated. Enthalpies of crystallization are generally very much lower than enthalpies of vaporization, and crystallization operations are usually carried out much nearer the ambient temperature than are distillation processes, so the energy requirements are lower (Table 14.1).

The most topical example in Table 14.1 is water. In the food industry, for example, the freeze concentration of fruit juice has already been established as an important technique. In the field of desalination, the production of drinking water from sea water,

TABLE 14.1. *Energies of Crystallization and Distillation*

Substance	Crystallization		Distillation	
	Melting Point (°C)	Enthalpy of Crystallization (kJ kg^{-1})	Boiling Point (°C)	Enthalpy of Vaporization (kJ kg^{-1})
o-cresol	31	115	191	410
m-cresol	12	117	203	423
p-cresol	35	110	202	435
o-xylene	− 25	128	141	347
m-xylene	− 48	109	139	343
p-xylene	13	161	138	340
o-nitrotoluene	− 4.1	120	222	344
m-nitrotoluene	15.5	109	233	364
p-nitrotoluene	51.9	113	238	366
Benzene	5.4	126	80	394
Water	0	334	100	2260

521

crystallization is potentially one of the more successful methods; the heat energy requirement for crystallization is about one-seventh that of distillation, although it must be appreciated that "cooling energy" is more costly than "heating energy". There are many engineering problems still to be overcome, but the potential of crystallization as a desalination technique is in no doubt, and desalination is one of the great challenges of the present day.

The type of crystallization being referred to here is sometimes known as "mass" or "bulk" crystallization—the growth of millions. or billions, of crystals all at the same time in large industrial crystallizers. This type of crystallization demands very different techniques from those normally associated with the growth of single crystals.

The unit operation of crystallization is governed by some very complex interacting variables. It is a simultaneous heat and mass transfer process with a strong dependence on fluid and particle mechanics. It takes place in a multi-phase, multi-component system. It is concerned with particulate solids whose size and size distribution, both incapable of unique definition, vary with time. The solids are suspended in a solution which can fluctuate between a so-called metastable equilibrium and a labile state, and the solution composition can also vary with time. The nucleation and growth kinetics, the governing processes in this operation, can often be profoundly influenced by mere traces of impurity in the system; a few parts per million may alter the product beyond all recognition.

It is, perhaps, no wonder that crystallization has been called an art rather than a science. Nevertheless, crystallizers have to be designed, constructed and operated successfully, despite the insecure foundations. Broadly speaking, information is needed in four main areas for the design of a crystallizer:

(1) solubility and phase relationships;
(2) hydrodynamics of crystal suspensions;
(3) limits of metastability and nucleation characteristics;
(4) crystal growth rates.

The solubility and phase relationships influence the choice of crystallizer and method of operation. It is important to obtain these data on the actual crystals and liquor to be encountered in the plant; data taken from the literature, normally measured with reagent-grade solutes and solvents, may be totally unsuitable. Traces of impurity can often affect the phase relationships considerably.

It is necessary to know the crystal suspension velocities in order to specify liquor circulation rates in fluidized bed crystallizers and agitation rates in stirred vessels. The crystals, being present in large quantities, are subjected to hindered settling, and further complications can arise if their shapes are irregular. Stokes's law should not be used to predict crystal suspension velocities despite the fact that it is frequently recommended in the literature. Stokes's law strictly applies only to spheres in laminar flow conditions with particle Reynolds numbers less than about 0.3. Drag coefficients on crystalline particles at particle Reynolds numbers around 100, a typical value for a fluidized bed crystallizer, are very much higher. Standard relationships are available for predicting particle suspension characteristics in dense fluidized beds and in agitated vessels, but it is relatively easy to measure these data and it is certainly much more reliable.[1]

The nucleation and growth processes are both exceedingly complex. They are influenced greatly by the solution supersaturation, and impurities in the system may have a profound effect. These processes will now be looked at in some detail.

14.1. Supersaturation

14.1.1. SUPERSATURATION AND METASTABILITY

The state of supersaturation is essential for crystallization to occur, and one of the ways in which the degree of supersaturation can be expressed is by the ratio

$$S = \frac{c}{c^*}, \tag{14.1}$$

where c is the concentration of the solution and c^* is the equilibrium saturation concentration at the same temperature. Thus, for a saturated solution $S = 1$, $S < 1$ denotes undersaturation and $S > 1$ indicates supersaturation. Other common expressions of supersaturation are the concentration driving force Δc and the relative supersaturation σ defined by

$$\Delta c = c - c^* \tag{14.2}$$

and

$$\sigma = \frac{\Delta c}{c^*} = S - 1. \tag{14.3}$$

Of these three expressions for supersaturation, only Δc is dimensional unless the solution composition is expressed in mole fractions. The magnitudes of these quantities depend on the units used to express concentration, as shown by the simple example in Table 14.2. In this example there are very considerable changes in Δc, as expected, but $>10\%$ changes occur in S and $>200\%$ in σ. The situation becomes even more confused if the salt is hydrated, since composition can be expressed in terms of either the hydrate or anhydrous salt.

It is not possible to say, with any firm conviction, which type of expression of solution composition or supersaturation is best; different expressions suit different circumstances. For mass balance calculations, e.g. for the estimation of crystal yield, concentration units such as kg anhydrous salt per kg solvent or kg hydrate per kg free solvent are generally most convenient. The advantage of the former is that it is not affected by any phase changes which might occur over the range of temperature encountered in the crystallization process. The advantage of the latter is that mass balance calculations are simplified.

TABLE 14.2. *Various Expressions of the Supersaturation of an Aqueous Solution of Sucrose (mol. wt. = 342) at 20°C. $c^* = 2040\,g\,kg^{-1}$ of Water (density $= 1.33\,g\,cm^{-3}$) and $c = 2450\,g\,kg^{-1}$ of Water (density $= 1.36\,g\,cm^{-3}$)*[1]

Solution Composition	c	c^*	Δc	S	σ
g kg^{-1} water	2450	2040	410	1.20	0.20
g kg^{-1} solution	710	671	39	1.06	0.06
g litre^{-1} of solution	966	893	73	1.08	0.08
kmol m^{-3} of solution†	2.82	2.61	0.21	1.08	0.08
mol fraction of sucrose	0.114	0.97	0.017	1.18	0.18

† 1 k mol m^{-3} = 1 mol litre^{-1}.

On the other hand, it is quite clear that none of the above expressions coincide exactly with the true thermodynamic supersaturation. The fundamental driving force for crystallization is the difference between the chemical potential of the given substance in the transferring and transferred state, e.g. in solution (state 1) and in the crystal (state 2). This may be written, for the case of an unsolvated solute crystallizing from a binary solution, as

$$\Delta\mu = \mu_1 - \mu_2. \tag{14.4}$$

The chemical potential μ is defined in terms of the standard potential μ_0, and the activity a by

$$\mu = \mu_0 + RT\ln a. \tag{14.5}$$

The fundamental dimensionless driving force for crystallization may therefore be expressed as

$$\frac{\Delta\mu}{RT} = \ln(a/a_{eq}) = \ln S \tag{14.6}$$

where a_{eq} is the activity of a saturated solution and S is the activity supersaturation, i.e.

$$S = \exp(\Delta\mu/RT). \tag{14.7}$$

For electrolyte solutions it is more appropriate to use the mean ionic activity $a_\pm$ defined by

$$a = a_\pm^v, \tag{14.8}$$

where v is the number of moles of ions in 1 mole of solute ($v = v_+ + v_-$), therefore

$$\Delta\mu/RT = v\ln S_a, \tag{14.9}$$

where

$$S_a = (a_\pm/a_{\pm eq})^v. \tag{14.10}$$

Alternatively, the relative supersaturation may be used:

$$\sigma_a = S_a - 1 \tag{14.11}$$

and eqn. (14.9) becomes

$$\Delta\mu/RT = v\ln(1 + \sigma_a). \tag{14.12}$$

For low supersaturations (say $\sigma_a < 0.1$) the approximation

$$\Delta\mu/RT \simeq v\sigma_a \tag{14.13}$$

is valid.

However, for practical purposes, supersaturations are generally expressed directly in terms of solution concentrations, e.g.

$$S_c = \frac{c}{c_{eq}}, \quad S_m = \frac{m}{m_{eq}}, \quad \text{and} \quad S_x = \frac{x}{x_{eq}}, \tag{14.14}$$

where c = molarity (mol per litre of solution), m = molality (mol per kg of solvent), and x = mole fraction.

The relationship between these concentration-based supersaturations and the fundamental (activity-based) supersaturation may be expressed through the relevant concentration-dependent activity coefficient ratio $A = \gamma/\gamma_{eq}$, i.e.

$$S_a = S_c A_c = S_m A_m = S_x A_x, \tag{14.15}$$

where $A_c = \gamma_c/\gamma_{c,eq}$, $A_m = \gamma_m/\gamma_{m,eq}$, and $A_x = \gamma_x/\gamma_{x,eq}$.

If the relevant activity coefficients can be evaluated, it is possible to establish how the different supersaturations differ from one another and, more importantly, from the fundamental supersaturation S_a. The decisive factor is the activity coefficient ratio A. The more it deviates from unity, the greater is the incurred inaccuracy. In general when $A_m > 1$, m-based concentration units are preferred, but when $A_m < 1$, x- or c-based units are better than m-based. The choice between x- and c-based units in this case again depends on the activity coefficient ratio: if $A_x > A_c$, the x-based units are preferred, and vice versa.[2]

In the absence of any information on the activity coefficient ratio, preference should be given to supersaturations based on molal units because of their more practical utility compared with mole fractions and their temperature-independence compared with molar units. In other words, a concentration scale based on mass of solvent is generally preferred to one based on volume of solution.

An extension of the above analysis to more complex cases[2] leads to the conclusion that the dimensionless driving force for crystallization of hydrate should always be expressed in terms of the hydrate and not of the anhydrous salt, i.e. one should use

$$(\Delta\mu/RT)_H = v \ln S_H A_H, \tag{14.16}$$

where the solution concentrations and activity coefficients both relate to the hydrate. The difference between the hydrate H and anhydrous A quantities, $(\Delta\mu/RT)_H$ and $(\Delta\mu/RT)_A$, can be very considerable. When, for lack of information, the quantity A_H cannot be evaluated, the approximation

$$(\Delta\mu/RT)_H \approx v \ln S_H \tag{14.17}$$

may be used, but it has not yet been possible to derive any general rules about the preference for c-, m-, or x-based concentration units for the expression of the supersaturation.

14.1.2. MEASUREMENT OF SUPERSATURATION

If the concentration of a solution can be measured at a given temperature, and the corresponding equilibrium saturation concentration is known, then it is a simple matter to calculate the supersaturation (eqns. (14.1)–(14.2)). Just as there are many methods of measuring concentration, so there are also many ways of measuring supersaturation, but not all of these are readily applicable to industrial crystallization practice.[3]

Solution concentration may be determined directly by analysis, or indirectly by measuring some property of the system that is a sensitive function of concentration. Properties frequently chosen for this purpose include density, viscosity, refractive index, and electrical conductivity, and these can often be measured with high precision, especially if the actual measurement is made under carefully controlled conditions in the laboratory. However, for the operation of a crystallizer under laboratory or pilot plant conditions the demand is usually for an *in situ* method, preferably capable of continuous operation. In these circumstances, problems may arise from the temperature dependence of the property being measured. Nevertheless, several of the above properties can be measured, more or less continuously, with sufficient accuracy for supersaturation determination. In general, density and refractive index are the least temperature-sensitive properties.

For industrial crystallization, where temperature and feedstock conditions cannot be controlled with precision, very crude methods of supersaturation measurement may have to be employed. The most common method consists of a mass balance coupled with feedstock, exit liquor, and crystal production rates taken over a suitable period, e.g. several hours, to smooth-out fluctuations.

14.2. Nucleation

For the design of a crystallizer it is essential to determine the working limits of metastability. If at all possible, uncontrolled nucleation should be avoided in a crystallizer; the solution should not be allowed to become labile. Most modern crystallizers are operated under "controlled" conditions where crystal growth occurs from a solution maintained in the metastable condition.

The classical theories of nucleation do not help to determine the limits of metastability to be expected. According to these theories the rate of nucleation J may be expressed as a function of the supersaturation S by a relationship of the form

$$J = A \exp[-K(\log S)^{-2}], \tag{14.18}$$

which predicts an explosive increase in the nucleation rate above a certain level of supersaturation. However, the constants A and K are usually inapplicable to industrial conditions where the primary nucleation is predominantly heterogeneous. A perfectly clean solution, free from foreign impurities, is virtually impossible to attain, even in the laboratory. Therefore, so far as industrial crystallization is concerned, relationships such as eqn. (14.18) are of little use, and all that can be justified are simple empirical relationships such as

$$J = K_n (\Delta c)^n. \tag{14.19}$$

The nucleation rate constant K_n and the "order" of the nucleation process n depend on the physical properties and hydrodynamics of the system. Values of n are usually greater than 2 and may be as high as 8 or 9. Viscosity can also play an important role in the nucleation process; in highly viscous solutions nucleation is suppressed.

However, it is not *primary* nucleation that has the greatest influence on an industrial crystallizer. The greatest hazard comes from *secondary* nucleation, which can be defined as the generation of nuclei by the crystals already present in suspension (Fig. 14.1). Here, again, only empirical relationships such as eqn. (14.19) can be justified, but the information should be available before a crystallizer can be designed with confidence.

Metastable zone widths and other nucleation data can be measured quite easily in the laboratory with a very simple apparatus consisting of a $50 \, \text{cm}^3$ flask fitted with a thermometer and magnetic stirrer, located in an external cooling jacket. Both primary and secondary nucleation can be studied in such an apparatus. Some typical results[1] are shown in Fig. 14.2, where it can be seen that seeding has a considerable influence on the nucleation process. For a cooling rate of $10°C \, \text{h}^{-1}$, for example, the width of the metastable zone for ammonium sulphate solutions is reduced from about 4 deg C to 2 deg

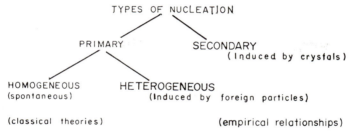

TYPES OF NUCLEATION

PRIMARY SECONDARY
(Induced by crystals)

HOMOGENEOUS HETEROGENEOUS
(spontaneous) (Induced by foreign particles)

(classical theories) (empirical relationships)

FIG. 14.1. Types of nucleation.

C. The difference in slope of these two lines indicates a difference in nucleation mechanism for the two cases (primary and secondary).

Solution turbulence also has an effect on the nucleation process, and this effect should also be measured. In general, agitation reduces the metastable zone width, as shown in Fig. 14.3. These data were measured with solutions of potassium sulphate in a laboratory-scale agitated vessel crystallizer in the absence of crystals, i.e. heterogeneous primary nucleation. Vigorous agitation reduces the metastable zone width from about 12 deg C to about 8 deg C. Incidentally, the presence of crystals induces nucleation (secondary) at supercoolings of about 4°C.

A list of maximum allowable supercoolings for some common aqueous salt solutions, measured in the presence of crystals, is given in Table 14.3. It should be noted, however, that the working value of the supercooling in an actual crystallizer will be lower than 50% of these values. The relation between supercooling $\Delta\Theta$ and supersaturation Δc is

$$\Delta c = \left(\frac{dc^*}{d\Theta}\right)\Delta\Theta,\qquad(14.20)$$

where c^* is the equilibrium saturation concentration.

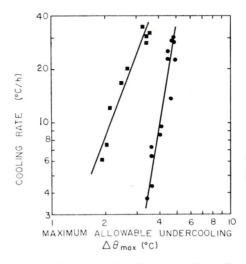

FIG. 14.2. Maximum allowable undercooling $\Delta\theta_{max}$ versus the cooling rate b for seeded and unseeded aqueous solutions of ammonium sulphate[1] (■ seeded, ● unseeded).

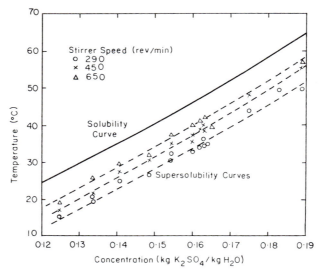

Fig. 14.3. Effect of agitation on the metastable limits of aqueous potassium sulphate solutions.

Secondary nucleation in industrial crystallizers arises mainly from crystal–crystal or crystal–equipment (especially the agitator) contact. Nuclei up to $20 \mu m$ can be formed, and these appear to have variable growth rates (see section 14.3.3), the smaller crystals growing very slowly or not at all. A recent study has been reported[4] on the production of secondary nuclei by low-energy contacts with crystals of potash alum and magnesium sulphate in their respective aqueous solutions: crystal attrition occurred easily and large numbers of particles in the $1–10 \mu m$ size range were produced in both supersaturated and unsaturated solutions. No general theory of secondary nucleation has yet been developed, but several comprehensive reviews of this subject have recently been made.[5, 6]

TABLE 14.3. *Maximum Allowable† Supercooling, $\Delta\theta_{max}$, for some Common Aqueous Salt Solutions at 25°C Measured in the Presence of Crystals under Conditions of Slow Cooling and Moderate Agitation*[1]

Substance	°C	Substance	°C	Substance	°C	Substance	°C
NH_4 alum	3.0	$MgSO_4.7H_2O$	1.0	NaI	1.0	KBr	1.1
NH_4Cl	0.7	$NiSO_4.7H_2O$	4.0	$NaHPO_4.12H_2O$	0.4	KCl	1.1
NH_4NO_3	0.6	$NaBr.2H_2O$	0.9	$NaNO_3$	0.9	KI	0.6
$(NH_4)_2SO_4$	1.8	$Na_2CO_3.10H_2O$	0.6	$NaNO_2$	0.9	KH_2PO_4	9.0
$NH_4H_2PO_4$	2.5	$Na_2CrO_4.10H_2O$	1.6	$Na_2SO_4.10H_2O$	0.3	KNO_3	0.4
$CuSO_4.5H_2O$	1.4	NaCl	4.0	$Na_2S_2O_3.5H_2O$	1.0	KNO_2	0.8
$FeSO_4.7H_2O$	0.5	$Na_2B_4O_7.10H_2O$	3.0	K alum	4.0	K_2SO_4	6.0

† The working value for normal crystallizer operation may be 50 per cent of these values, or lower. The relation between $\Delta\theta_{max}$ and Δc_{max} is given by eqn. (14.20).

14.3. Crystal Growth

As in the case of nucleation, the classical theories of crystal growth have not led to convenient working relationships for industrial crystallization. For crystallizer design and assessment the mass rate of crystallization R is most conveniently expressed in terms of the supersaturation Δc by the empirical relationship

$$R = K_g(\Delta c)^g. \tag{14.21}$$

The "order" g of the growth process is generally between 1 and 2, but the full significance of the overall growth rate constant K_g is not yet understood. There are at least two major contributing factors corresponding to the effects of solute bulk diffusion and the crystal surface "integration" reaction, but in crystallization from aqueous solution the ion-dehydration process is an additional step that should be considered. The complex interactions of these processes have yet to be unravelled.

14.3.1. OVERALL GROWTH RATES

Growth rate data suitable for crystallizer design purposes can be measured in the laboratory in fluidized beds or agitated vessels of many types. A review of the assessment of crystal growth kinetics has been made by Garside.[7]

A typical laboratory fluidized-bed crystallizer is shown in Fig. 14.4. The technique of measuring crystal growth rates by growing large numbers of carefully sized seeds in

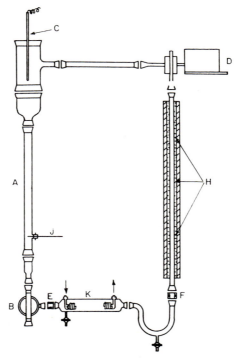

Fig. 14.4. A laboratory fluidized-bed crystallizer[8] (*A*, growth zone; *B*, outlet cock; *C*, resistance thermometer; *D*, pump; *E* and *F*, orifice plates; *H*, heating tapes; *J*, thermometer; *K*, water cooler).

fluidized suspension under carefully controlled conditions is best described with reference to an actual run made with potassium aluminium sulphate (potash alum) crystals.[8] A warm undersaturated solution of potash alum of known concentration is circulated in the crystallizer and supersaturated by cooling to the working temperature. About $5 \text{ g} \pm 1$ mg of closely sized seed crystals (e.g. 22–25 mesh BS sieves; 600–710 μm) are introduced into the crystallizer and the upward solution velocity is adjusted so that the crystals are maintained in a reasonably uniform fluidized state in the growth zone. The crystals are allowed to grow at a constant temperature ($\pm 0.03°$C) until their total mass is about 10–15 g.

At the end of the run the crystals are removed from the crystallizer, washed, dried, and weighed. The final solution concentration is measured and the mean of the initial and final supersaturations is taken as the average of the run. This does not involve any significant error because the solution concentration is not usually allowed to change by more than about 1% during a run. The overall crystal growth rate is then calculated in terms of mass deposited per unit area per unit time at a given supersaturation.

Figure 14.5 shows the growth rates of potash alum crystals at 32°C as a function of supersaturation and crystal size. Several interesting characteristics, typical of many inorganic salts, can be seen. First, the growth rate is not linearly dependent on the supersaturation (concentration driving force). In other words $g \neq 1$ in eqn. (14.21). In this case $g \sim 1.6$. In fact on some six or seven systems studied recently the value of g has ranged from 1.4 to 2.2. First-order growth is apparently more rare than many workers have previously assumed. The second feature seen in Fig. 14.5 is that crystal growth does not appear to commence until a certain minimum level of supersaturation is exceeded. This minimum supersaturation may represent an energy barrier to be surmounted before significant growth can proceed. The third interesting feature is that large crystals of alum grow faster than small crystals. This is really a solution velocity effect (large crystals have higher settling velocities than small crystals), as will be described below in the section dealing with the growth of individual crystal faces.

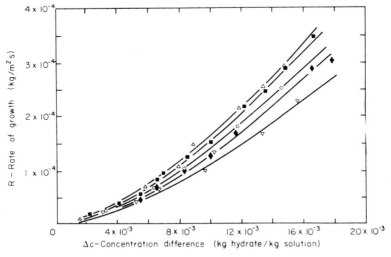

FIG. 14.5. Growth rates of potash alum at 32°C (mean crystal size: △ 1960, ■ 1400, ○ 990, ◆ 750, ▽ 530 μm).[8]

14.3.2. FACE GROWTH RATES

There are two aspects of crystal growth that are of interest to the chemical engineer. First there are the overall mass deposition rates that can be measured by the methods just described. These are needed for the design of the crystallizer itself. Secondly, the rates of growth of individual crystal faces are needed for specification of the operating conditions.

Under different environmental conditions different crystal faces grow at different rates. In general, the high index faces grow faster than the low, and changes in the environment (temperature, supersaturation, pH, impurities, etc.) can have a profound effect on the individual face growth rates. This information is of vital interest to the crystallizer designer because the different face growth rates give rise to habit changes in the crystals.

An apparatus for the precise measurement of individual crystal face growth rate measurements[1, 8, 9] consists of a glass cell in which a crystal mounted on a wire can be observed with a travelling microscope. The solution temperature, supersaturation, and velocity are controlled with precision.

Some typical results for potash alum at 32°C are given in Fig. 14.6. Again it is shown that the growth of this salt is not first-order with respect to supersaturation and that a certain level of supersaturation must be exceeded before significant growth can commence. Furthermore, the solution velocity is an important parameter. In fact this is the cause of the so-called size effect noted in the work with the fluidized bed crystallizer (Fig. 14.5). Large crystals have higher settling velocities than have small crystals. Salts which have been shown to exhibit size-dependent (i.e. velocity-dependent) growth are the alums, nickel ammonium sulphate, and potassium sulphate. Ammonium sulphate and ammonium and potassium dihydrogen phosphate, for example, are unaffected by solution velocity.

Alum crystals grow as almost perfect octahedra exhibiting eight (111) faces. It is a simple matter, therefore, using the solid density ρ to convert the linear face velocity v to an overall mass deposition rate $R = \rho v$. For the case of potash alum, coincidence between the measured rates in the fluidized bed and the predicted rates from the face growth rates is extremely good.[8]

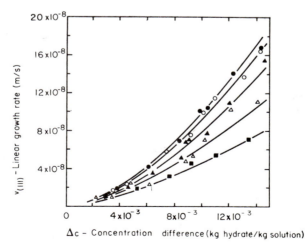

FIG. 14.6. Effect of solution velocity on the (111) face growth rates of single crystals of potash alum at 32°C. (Solution velocity: ● 0.217, ○ 0.120, ▲ 0.0220, ■ 0.066 m s^{-1}.[8]

14.3.3. SIZE-DEPENDENT GROWTH

A considerable amount of experimental evidence is now available which indicates that crystal growth kinetics can often appear to be dependent on the crystal size.[7, 8, 10–14] This condition could arise if the integration kinetics were truly size-dependent[10, 12, 13] or it could be the result of "growth dispersion",[14] i.e. the exhibition of variable growth rates by crystals of a given size due, for example, to differences in surface structure and perfection.[15]

In addition to the above size effect, it would appear that very small crystals (say $< 50\,\mu m$) of many substances grow much more slowly than larger crystals and some do not grow at all.[16–18] The behaviour of very small crystals has a considerable influence on the performance of continuously operated industrial crystallizers because new crystals are constantly being generated, by secondary nucleation, into the $1–10\,\mu m$ size range. These subsequently grow to populate the full crystal size distribution. It is important, therefore, to be able to predict small-sized crystal growth rates in order to assess the performance of crystallizers.

14.3.4. EXPRESSION OF CRYSTAL GROWTH RATES

There is no simple or generally accepted method of expressing the rate of growth of a crystal since it has a complex dependence on temperature, supersaturation, size, habit, system turbulence, and so on. However, for carefully defined conditions, crystal growth rates may be expressed as a mass deposition rate R (e.g. $kg\,m^{-2}\,s^{-1}$), a mean linear velocity $\bar{v}\,(=dr/dt)\,(m\,s^{-1})$, or as an overall linear growth rate $G\,(=dL/dt)\,(m\,s^{-1})$. The relationships between these quantities are

$$R = K_g(\Delta c)^g = \frac{1}{A}\frac{dm}{dt} = \frac{3\alpha}{\beta}\rho G = \frac{6\alpha}{\beta}\rho\frac{dr}{dt} = \frac{6\alpha}{\beta}\rho\,\bar{v}, \tag{14.22}$$

where L is some characteristic size of the crystal, e.g. the equivalent sieve aperture size, and r is the radius corresponding to the equivalent sphere. The volume and surface shape factors, α and β respectively, are defined by

$$m = \alpha\rho L^3 \tag{14.23}$$

and

$$A = \beta L^2, \tag{14.24}$$

where m and A are the particle mass and area. For spheres and cubes, $6\alpha/\beta = 1$. For octahedra, $6\alpha/\beta = 0.816$. Some typical values of the linear growth velocity $\bar{v}\,(=\frac{1}{2}G)$ are given in Table 14.4.

14.3.5. MASS TRANSFER CORRELATIONS

Diffusional mass transfer rates obtained under conditions of forced convection may be correlated by a dimensionless equation of the form

$$Sh = 2 + \phi\,Re_p^a Sc^b, \tag{14.25}$$

TABLE 14.4. *Some Mean Overall Crystal Growth Rates Expressed as a Linear Velocity*[1]

The supersaturation is expressed by $S = c/c^*$ with c and c^* as kg of crystallizing substance per kg of free water. The significance of the mean linear growth velocity, $\bar{v}$, is explained by eqn. (14.22), and the values recorded here refer to crystals in the approximate size range 0.5–1 mm growing in the presence of other crystals. An asterisk (*) denotes that the growth rate is probably size dependent.

Crystallizing Substance	°C	S	$\bar{v}$ (m s^{-1})	Crystallizing Substance	°C	S	$\bar{v}$ (m s^{-1})
$(NH_4)_2SO_4.Al_2(SO_4)_3.$ $24H_2O$	15	1.03	1.1×10^{-8}*	K_2SO_4	20	1.09	2.8×10^{-8}*
	30	1.03	1.3×10^{-8}*		20	1.18	1.4×10^{-7}*
	30	1.09	1.0×10^{-7}*		30	1.07	4.2×10^{-8}*
	40	1.08	1.2×10^{-7}*		50	1.06	7.0×10^{-8}*
					50	1.12	3.2×10^{-7}*
NH_4NO_3	40	1.05	8.5×10^{-7}				
$(NH_4)_2SO_4$	30	1.05	2.5×10^{-7}	KH_2PO_4	30	1.07	3.0×10^{-8}
	60	1.05	4.0×10^{-7}		30	1.21	2.9×10^{-7}
$NH_4H_2PO_4$	20	1.06	6.5×10^{-8}		40	1.06	5.0×10^{-8}
	30	1.02	3.0×10^{-8}		40	1.18	4.8×10^{-7}
	30	1.05	1.1×10^{-7}				
	40	1.02	7.0×10^{-8}	$NaCl$	50	1.002	2.5×10^{-8}
$MgSO_4.7H_2O$	20	1.02	4.5×10^{-8}*		50	1.003	6.5×10^{-8}
	30	1.01	8.0×10^{-8}*		70	1.002	9.0×10^{-8}
	30	1.02	1.5×10^{-7}*		70	1.003	1.5×10^{-7}
$NiSO_4.(NH_4)_2SO_4.6H_2O$	25	1.03	5.2×10^{-9}				
	25	1.09	2.6×10^{-8}	$Na_2S_2O_3.5H_2O$	30	1.02	1.1×10^{-7}
	25	1.20	4.0×10^{-8}		30	1.08	5.0×10^{-7}
$K_2SO_4.Al_2(SO_4)_3.24H_2O$	15	1.04	1.4×10^{-8}*				
	30	1.04	2.8×10^{-8}*	Citric acid monohydrate	25	1.05	3.0×10^{-8}
	30	1.09	1.4×10^{-7}*				
	40	1.03	5.6×10^{-8}*	Sucrose	30	1.13	1.1×10^{-8}*
KCl	40	1.01	6.0×10^{-7}		30	1.27	2.1×10^{-8}*
KNO_3	20	1.05	4.5×10^{-8}		70	1.09	9.5×10^{-8}
	40	1.05	1.5×10^{-7}		70	1.15	1.5×10^{-7}

where the Sherwood number $Sh = KL/D$, particle Reynolds number $Re_p = \rho u L/\eta$, Schmidt number $Sc = \eta/\rho D$ and ϕ is a constant. ρ is the solution density, η is the viscosity, L is the crystal size, D is the diffusivity, and K is a mass-transfer coefficient. For crystal growth $K = K_g$ as defined in eqns. (14.10) and (14.11). The constant 2 in eqn. (14.25) (frequently referred to as the Froessling equation) is the limiting value of Sh as $Re_p \to 0$, representing mass transfer in the absence of natural convection. However, for reasonably high values of Sh (say >100) it is common practice to use the simpler expression

$$Sh = \phi \, Re_p^a Sc^b, \tag{14.26}$$

and this equation is commonly used for the correlation of dissolution and crystallization rate data.[1] The exponents a and b are usually taken to be $\frac{1}{2}$ and $\frac{1}{3}$, respectively, and for freely suspended granular solids the constant ϕ generally falls in the range 0.3–0.9.

14.3.6. "FILMS" AND "BOUNDARY LAYERS"

When a fluid flows past a solid surface there is a thin region near the solid–liquid interface where the velocity becomes reduced due to the influence of the surface. This region, called the "hydrodynamic boundary layer" δ_h, may be partially turbulent or entirely laminar in nature, but in the case of crystals suspended in their liquor the latter is most probable.

For mass transfer processes, another boundary layer may be defined, viz. the "mass-transfer or diffusion boundary layer" δ_m. This is a region close to the interface across which, in the usual case of a laminar hydrodynamic boundary layer around the crystal, mass transfer proceeds by molecular diffusion. Under these conditions, the relative magnitudes of the two boundary layers may be roughly estimated from

$$\frac{\delta_h}{\delta_m} = S_c^{1/3}, \tag{14.27}$$

where Sc is the dimensionless Schmidt number as defined above.

The ratio of the thicknesses of the two layers depends considerably on the solution viscosity and diffusivity. For example, for ammonium alum crystals in near-saturated solution at 25°C, $\eta = 1.2 \times 10^{-3} \ \text{kg m}^{-1} \ \text{s}^{-1}$, $D = 4 \times 10^{-10} \ \text{m}^2 \ \text{s}^{-1}$, $\rho = 1.06 \times 10^3 \ \text{kg}$ m^{-3}. Therefore, $Sc = 2.8 \times 10^3$ and $\delta_h/\delta_m \simeq 14$. However, for sucrose in water at 25°C, $\eta = 10^{-1}$, $D = 9 \times 10^{-11}$, $\rho = 1.5 \times 10^3$, giving $Sc = 7.4 \times 10^5$ and $\delta_h/\delta_m \simeq 90$.

In the description of mass-transfer processes another fluid layer is frequently postulated, viz. the "effective film for mass transfer" δ. This hypothetical film is not the same thing as the more fundamental diffusion boundary layer, but it may probably be considered to be in the same order of magnitude, i.e.

$$\delta_h > \delta_m \simeq \delta. \tag{14.28}$$

The effective film for mass transfer δ is defined by

$$\delta = \frac{\rho D}{K}, \tag{14.29}$$

where ρ is the solution density, D is the diffusivity, and K is a mass-transfer coefficient expressed as mass per unit time per unit area per unit concentration driving force, e.g. $\text{kg s}^{-1} \ \text{m}^{-2} \ (\text{kg/kg})^{-1}$.

As described above, mass-transfer data are frequently correlated by relationships such as eqn. (14.26) with $a = \frac{1}{2}$ and $b = \frac{1}{3}$. Values of the constant ϕ for growing and dissolving crystals may range from about 0.5 to 0.9. So writing a simple, arbitrary form of eqn. (14.26) as

$$Sh = \tfrac{2}{3} Re_p^{1/2} Sc^{1/3} \tag{14.30}$$

and expressing $Sh = L/\delta$ (using eqn. (14.29)), we get

$$\delta = \frac{3L}{2} \left(\frac{\rho u L}{\eta}\right)^{-1/2} \left(\frac{\eta}{\rho D}\right)^{-1/3}, \tag{14.31}$$

and this equation may be used to give a rough estimate of the value of δ. It should be noted, however, that eqn. (14.31) can only be used if the mass-transfer process is first-order with respect to the concentration driving force ($g = 1$ in eqn. (14.23)), otherwise Sh is not dimensionless and eqn. (14.30) is invalid.

Example

Estimate the diffusion and hydrodynamic boundary-layer thicknesses for the dissolution of 2 mm crystals of ammonium alum in near-saturated solution at 25°C using the following data:

$\eta = 1.2 \times 10^{-3} \, \mathrm{kg\,m^{-1}\,s^{-1}}$, $\rho = 1.06 \times 10^3 \, \mathrm{kg\,m^{-3}}$, $D = 4 \times 10^{-10} \, \mathrm{m^2\,s^{-1}}$, $L = 2 \times 10^{-3} \, \mathrm{m}$, $u = 10^{-1} \, \mathrm{m\,s^{-1}}$. Therefore, $Sc = 2.8 \times 10^4$ and $Re_p = 180$, so from eqn. (14.31), $\delta = 16 \, \mathrm{mm}$. Therefore, from eqn. (14.28) $\delta_m \simeq \delta \simeq 16 \, \mu\mathrm{m}$.

From eqn. (14.27), $\delta_n \simeq \delta_m \, Sc^{-1/3} \simeq 14 \, \delta_m \simeq 220 \, \mu\mathrm{m}$.

It is thus possible to calculate values for the thicknesses of these various films, but it is perhaps worthwhile at this point to question meaning and utility of this quantity in crystallization. The concept of a liquid film at an interface is undoubtedly useful in providing a simple pictorial representation of the mass-transfer process, but in the case of crystals growing or dissolving in multi-particle suspensions the actual existence of stable films, of the magnitude normally calculated as shown above, around each small particle, is debatable.

The thickness of the liquid film δ can only be deduced indirectly from the mass-transfer coefficient and diffusivity (eqn. (14.29)), and it is difficult to select the appropriate value of D to use in any given situation. The question arises, therefore, as to whether or not δ is a meaningful quantity to calculate in these circumstances. In any case, the hypothetical nature of the film should be clearly appreciated, and calculated values of its thickness should be used with considerable caution.

14.4. Habit Modification

It is a regrettable fact that crystals do not always grow readily in ideal forms; needles and plates, both undesirable shapes, are very often produced. Consequently, the crystallizer designer is frequently called upon to advise on habit modification, and this he can only do with confidence if he knows the growth characteristics of the individual faces.

Figure 14.7 shows the effect of pH on the growth of ammonium dihydrogen phosphate (ADP) crystals at 25°C.[9] The pH of a supersaturated solution of ADP at 25°C is about 3.8, and under these conditions crystals grow mainly along the Z-axis, on the pyramid faces, i.e. in the (001) direction. Growth on the prism faces, i.e. in the X-direction, is virtually zero. However, when the solution pH is increased to about 5 by the addition of a small amount of ammonia, growth in the Z-direction is increased slightly while that in the X-direction increases considerably. Thus it may be inferred that ADP crystals grown at "normal" pH will be elongated (needle or acicular habit) while those grown at pH 5 will be short prisms (prismatic habit). The behaviour has been confirmed on a laboratory fluidized-bed crystallizer.

Supersaturation can influence crystal habit. The individual crystal face growth kinetics usually depend to a different extent on supersaturation, so by raising or lowering the operating level it is sometimes possible to effect a considerable control over the crystal habit. It may be, of course, that the desired habit can only be grown at a high

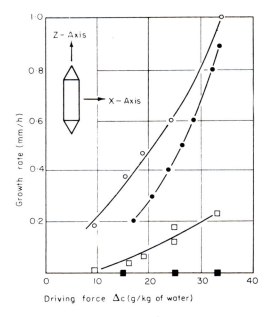

FIG. 14.7. Effect of pH on the X- and Z-axis growth rates of ADP crystals at 25°C.

supersaturation, above the metastable limit, and in such cases a nucleation inhibitor may have to be added to allow growth to proceed as planned.

One of the most common causes of habit modification is the presence of impurities in the crystallizing solution. These impurities may already be present as, for example, in the crystallization of beet-sugar, where the presence of raffinose induces a characteristic flat crystal, or they may be added deliberately, e.g. traces of borax which change the habit of $MgSO_4.7H_2O$ from needles to prisms. In some cases minute traces (<1 ppm) of an impurity can cause startling changes in the crystal habit, but this is not always the case; sometimes habit modification only occurs in the presence of large quantities of impurity in the crystallizing system, e.g. $>5\%$ of biuret is needed to change the habit of urea from needles to brick shapes. Tervalent ions such as Cr^{3+}, Fe^{3+}, and Al^{3+} are particularly active impurities and these are frequently used at levels of ~100 ppm for habit modification purposes. Surface active agents are commonly used to change crystal habits. Anionic surfactants include the alkyl sulphates, alkane sulphonates, and aryl alkyl sulphonates. Quaternary ammonium salts are used as cationic agents. Some polyelectrolytes, e.g. polyacrylamides and sodium carboxymethylcellulose, can exert a habit-modifying effect at very low concentrations.

14.4.1. INDUSTRIAL IMPORTANCE

The majority of reported cases of habit modification have been concerned with laboratory investigations, but the phenomenon is of the utmost importance in industrial crystallization and by no means a mere laboratory curiosity. Certain crystal habits are disliked in commercial crystals because they give the crystalline mass a poor appearance; others make the product prone to caking, induce poor flow characteristics, or give rise to difficulties in the handling or packaging of the material. For most commercial purposes, a granular or prismatic habit is usually desired, but there are specific occasions when plates or needles may be wanted.

In nearly every industrial crystallization some form of habit modification is necessary to control the type of crystal produced. This may be done by controlling the rate of crystallization, e.g. by adjusting the rate of cooling or evaporation, the degree of supersaturation, or the temperature at which crystallization occurs, by choosing the correct type of solvent, adjusting the pH of the solution, or deliberately adding some impurity which acts as a habit modifier to the system. A combination of several of these methods may have to be used. It is also worth remembering that the results of small-scale laboratory trials on habit modification may not always prove of value for large-scale application; they may even be misleading. Pilot-plant trials, however, on batches greater than about 100 l will usually yield reliable information.

The suppression of $CaSO_4$ scaling in boilers by the use of phosphates owes as much to the habit modification of the deposited crystal as it does to nucleation inhibition. The same is true for the suppression of ice nucleation and growth in ice cream by the addition of sodium carboxymethylcellulose. Lecithin is widely used an an oil soluble wetting agent in the manufacture of chocolate confections; it lowers the viscosity and reduces "bloom" on the chocolate surfaces by its dispersive effect on the chocolate fat crystals. Lecithin is also used as a crystallization inhibitor in cotton seed oil. Mullin[1] has reviewed many reported cases of the industrial application of habit modification (Table 14.5) and also discussed the factors that must be considered when selecting a suitable habit modifier.

14.5. Crystallization Methods and Equipment

The many hundreds of different industrial crystallizers in existence may be classified into a few general categories. Terms such as batch or continuous, agitated or non-agitated, controlled or uncontrolled, classifying or non-classifying, circulating liquor or circulating magma, etc., are useful for this purpose. Classification of crystallizers according to the method by which supersaturation is achieved is still probably the most widely used method; thus we have cooling, evaporating, vacuum, reaction, etc., crystallizers.

Many of these classes are self-explanatory, but some require definition. For example, the term *controlled* refers to supersaturation control. The term *classifying* refers to the production of a selected product size by classification in a fluidized bed of crystals. In a *circulating-liquor* crystallizer the crystals remain in the crystallization zone; only the clear mother liquor is circulated. In the *circulating-magma* crystallizer the crystals and the mother liquor are circulated together. Any given crystallizer may well belong to several of the above types.

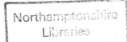

TABLE 14.5. *Some Habit Modifications of Industrial Interest*

Substance	Normal Habit	Habit Modifier	Changed Habit
NH_4 alum	Octahedra	Borax	Cubes
NH_4Cl	Dendrites	Cd^{2+}, Ni^{2+}	Cubes
		$MnCl_2 + HCl$	Granules
$NH_4H_2PO_4$	Needles	Al^{3+}, Fe^{3+}, Cr^{3+}	Tapered prisms
NH_4NO_3	Short crystals	Acid Magenta	Needles
$(NH_4)_2SO_4$	Prisms	$FeCl_3$	Irregular crystals
H_3BO_3	Needles	Gelatin, casein	Flakes
$CaCO_3$		Alkyl aryl sulphonates	Granular precipitate
$CaSO_4.2H_2O$	Needles	Sodium citrate	Prisms
		Alkyl aryl sulphonates	Prisms
$MgSO_4.7H_2O$	Needles	Borax	Prisms
$AgNO_3$	Plates	Sodium oleate	Dendrites
K alum	Octahedra	Borax	Cubes
KBr	Cubes	Phenol	Octahedra
KCN	Cubes	Fe^{3+}	Dendrites
KCl	Cubes	$Fe(CN)_6^{4+}$	Dendrites
K_2SO_4	Rhombic prisms	$FeCl_3$	Irregular crystals
NaBr	Cubes	$Fe(CN)_6^{4+}$	Dendrites
NaCN	Cubes	Fe^{3+}	Dendrites
NaCl	Cubes	$Fe(CN)_6^{4+}$	Dendrites
		Formamide	Octahedra
		Pb^{2+}, Cd^{2+}	Large crystals
		Polyvinylalcohol	Needles
		$Na_6P_4O_{13}$	Octahedra
$Na_2B_4O_7.10H_2O$ needles		Carboxymethyl cellulose	Flakes

14.5.1. COOLING CRYSTALLIZERS

14.5.1.1. *Unstirred Tanks*

These are the simplest types of cooling crystallizer in use. A hot concentrated solution is charged to the open vessel where it is allowed to cool, often over several days, by natural convection. The batch may be given an occasional stir to prevent the formation of hard crystalline lumps on the bottom of the crystallizer. No seeding is required. Sometimes thin rods or strips of metal are hung in the solution on which crystals grow, thus preventing some of the product falling to the bottom of the crystallizer. The magma (crystal slurry) can be removed by hand, or the mother liquor may be drained off.

Because of the slow cooling, large interlocked crystals are usually obtained and these retain mother liquor. The dried crystals, therefore, are generally impure. No control over the product size is possible; the crystals will range from fine dust to large agglomerates, but experience will indicate the size of vessel and cooling time necessary to produce the desired type of crystalline mass.

Handling labour costs are generally high, but for small batches the method is economic because of the small capital outlay and negligible operating and maintenance costs. The main disadvantage is that the equipment is generally bulky and occupies valuable floor space.

14.5.1.2. *Agitated Vessels*

When a stirrer is employed in an open-tank crystallizer, smaller and more uniform crystals are formed and the batch time is reduced. A somewhat purer product results due to the retention of less mother liquor by the crystals, and more efficient washing of the crystals is possible.

The vessel may be equipped with a water jacket or cooling coils. Jackets are generally preferred because coils tend to become encrusted with a hard crystalline deposit and cease to function efficiently. The inner cold surfaces of the crystallizer should be as smooth and flat as possible to minimize encrustation. Polished stainless steel is a good material of construction for this purpose.

The operating cost for an agitated-cooler is higher than that for a simple tank crystallizer, but it is small in comparison with the advantages gained by the quicker through-put. Labour costs for handling the product may still be rather high.

Tank crystallizers, stirred or otherwise, vary in design from shallow pans to large cylindrical tanks, according to the needs of the particular process. A large modern agitated cooling crystallizer is shown in Fig. 14.8a. This vessel has an upper conical section which reduces the upward velocity of liquor and prevents the crystalline product being swept out with the spent-liquor outflow. The magma is circulated in the growth zone of the crystallizer by an agitator located in the lower region of a draft tube. If required, a cooling device may be located inside the crystallizer.

Good mixing within the crystallizer and high rates of heat transfer between the liquor and coolant can be achieved by the use of external circulation. Because of the high liquor velocity in the tubes, low-temperature differences suffice for cooling, and scaling may be considerably reduced. The unit shown in Fig. 14.8b can be used for batch or continuous operation.

14.5.1.3. *Scraped-surface Crystallizers*

High heat transfer coefficients ($50–700 \text{ W m}^{-2} \text{ K}^{-1}$), and hence high production rates, are obtainable with double-pipe, scraped-surface crystallizers. Spring-loaded internal

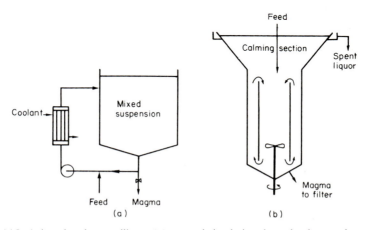

FIG. 14.8. Agitated tank crystallizers: (a) external circulation through a heat exchanger, (b) internal circulation with a draft tube.

agitators scrape the heat-transfer surfaces and create turbulent flow inside the tube. The units vary in size from about 75 to 600 mm in diameter and 0.3 to 3 m long and may be connected in series. Scraped-surface crystallizers are employed mainly for processing fats, waxes, and other organic melts, although several applications to inorganic salts from solution have also been reported.[19]

14.5.2. CONTROLLED CRYSTALLIZATION

The deliberate addition of selected seeds is widely used in crystallization practice. The effect of cooling an unseeded solution rapidly is shown in Fig. 14.9a. The solution cools at constant concentration until the labile zone is penetrated. It then nucleates and a shower of tiny crystals is deposited. The temperature rises slightly, but cooling reduces it and more nucleation occurs. Eventually the temperature and concentration fall as shown. In such a process no control is possible over the nucleation or growth processes.

Figure 14.9b demonstrates the slow cooling of a seeded solution; the temperature is controlled so that the system remains metastable throughout the operation. Growth occurs at a controlled rate only on the added seeds; no spontaneous nucleation occurs because the system is never allowed to become labile. This method of operation is called "controlled crystallization" and many large-scale modern crystallizers operate on this principle.

The mass of seeds M_s of size L_s that can be added to a crystallizer, assuming that crystallization occurs only on the added seeds, depends on the required crystal yield Y (see eqn. (14.18)) and the product crystal size L_p:

$$M_s = Y \left(\frac{L_s^3}{L_p^3 - L_s^3} \right). \tag{14.32}$$

The excessive production of nuclei should not be permitted in a crystallizer. The trouble arises not from the mass of material precipitated, which may be quite negligible, but from the number of seeds produced. A simple example will show this: if 100 g of 100 μm seed are added to a crystallizer, the mass of solute that has to grow on these seeds (about 100 million in number) to produce crystals of a reasonable size is very considerable (Fig. 14.10). For example, if 1 mm crystals are required, 100 kg of solute has to be deposited; 800 kg has to be deposited to produce 2 mm crystals, and for 2.5 mm crystals the necessary deposition increases to 1500 kg. Of course, crystal seeds much smaller than 100 μm can

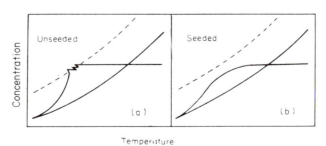

FIG. 14.9. The effect of seeding on a cooling-crystallization process.

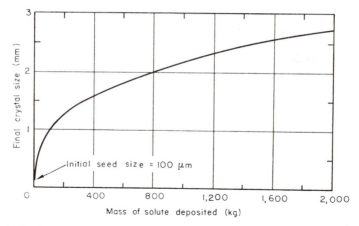

FIG. 14.10. Depositions of crystalline material on 100 g of 100 μm seed crystals.

exist in a crystallizer. A mere 1 g of 10 μm seeds, for example, contains $> 10^9$ particles, and each one is a potential crystal in a supersaturated solution.

It is abundantly clear, therefore, that if a controlled growth is required, nucleation must be avoided or counteracted. Vigorous agitation, thermal shocks, and excessive turbulence should not be permitted, and the supersaturation level should be reduced to the working minimum. High magma densities also help by providing ample crystal surfaces to receive the deposited solute. If these precautions fail to control nucleation, some system of false-grain removal must be installed in the crystallizer.

Control can be exercised over the product crystal size from a batch crystallizer by controlling the cooling rate. For example, natural cooling (Fig. 14.11a) gives a

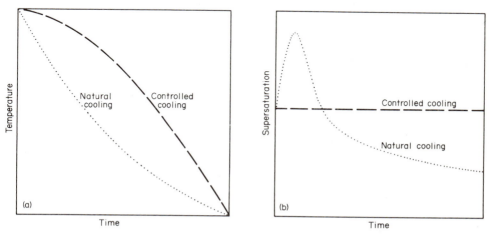

FIG. 14.11. Cooling modes for a batch crystallizer: (a) natural cooling and (b) controlled cooling (constant supersaturation).

supersaturation peak which induces heavy nucleation. However, by following a cooling path which maintains the supersaturation at a constant low level (Fig. 14.11b), nucleation can be controlled within acceptable limits. The calculation of optimum cooling curves for different operating conditions presents a complex problem, but for industrial purposes the following simplified relationship is generally adequate:

$$\theta_t = \theta_0 - (\theta_0 - \theta_f)(t/\tau)^3, \tag{14.33}$$

where θ_0, θ_f, and θ_t are the temperatures at the beginning, end, and at any time t during the process. τ is the overall batch time.

14.5.3. DIRECT CONTACT COOLING

Cooling crystallizers operated with conventional heat exchangers, e.g. coils, jackets, shell and tube exchangers, etc., often suffer from encrustation on the heat-transfer surfaces, which severely reduces the crystallizer performance. One method for avoiding the use of a conventional heat exchanger is to employ "vacuum cooling" (see below), but encrustation on the inner walls of the crystallizer can still be substantial. Another technique is to use direct contact cooling (DCC), where supersaturation is achieved by contacting the process liquor with a cold heat transfer medium.

Some of the advantages of DCC over the more conventional indirect contacting methods include better heat transfer, smaller coolant requirement, and the elimination of heat exchanger encrustation. Problems associated with DCC crystallization arise from the possibility of product contamination and the difficulty of separating and recovering the coolant.

The coolant may be solid, liquid, or gaseous, and heat may be exchanged by the transfer of sensible and/or latent heat. The coolant may or may not boil during the operation, and it may be miscible or immiscible with the process liquor. Thus several types may be envisaged: (i) immiscible, boiling—solid or liquid coolant: transfer of latent heat of sublimation of vaporization is the main source of heat removal, (ii) immiscible, non-boiling—solid, liquid, or gaseous coolant: mainly sensible heat transfer, (iii) miscible, boiling—liquid coolant: mainly latent heat transfer, (iv) miscible, non-boiling—liquid coolant: mainly sensible heat transfer.

Several successful DCC crystallization processes have been operated in recent years, such as the separation of close-boiling hydrocarbons, notably the production of p-xylene[20] and the desalination of sea water.[21] In addition there have been a few applications to the production of inorganic salts from aqueous solution.[1,22]

A continuous crystallizer which utilizes immiscible, non-boiling DCC crystallization for the production of calcium nitrate tetrahydrate, the Cerny process,[23] is shown in Fig. 14.12. Aqueous feedstock enters at the top of the crystallizer at 25°C and cools as it flows countercurrently to an upflow of immiscible coolant (e.g. petroleum at $-15°C$) introduced as droplets into a draft tube. The low-density coolant collects in the upper layers, but the high-density aqueous solution circulates up the draft tube and down the annulus, keeping the small crystals in suspension. Crystals >0.4 mm settle to the lower regions and are discharged in the magma at about $-5°C$. A slow-speed agitator prevents consolidation of the crystals in the magma outlet. The coolant is passed to a cyclone to remove traces of aqueous solution, and recycled through the cooler.

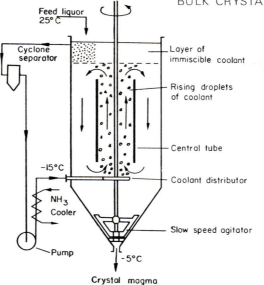

FIG. 14.12. Cerny direct-contact cooling crystallizer.

14.5.4. CLASSIFYING CRYSTALLIZERS

The discovery in Norway in 1919 by Isaachsen and Jeremiassen of a method for maintaining a stable suspension of crystals within the growth zone of a crystallizer led to the development of the first continuous classifying crystallizer, now known as the Oslo-Krystal apparatus. A concentrated solution, which is continuously cycled through the crystallizer, is supersaturated in one part of the apparatus and the supersaturated solution is conveyed to another part where it is released into a mass of growing crystals. A large number of modifications of this unit are available.[1, 24, 25]

The operation of the Oslo-Krystal cooling crystallizer can be described with reference to Fig. 14.13. The hot concentrated feed solution enters the vessel at point *A*, located

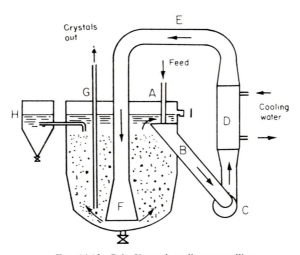

FIG. 14.13. Oslo-Krystal cooling crystallizer.

directly above the inlet to the circulation pipe *B*. Saturated solution from the upper regions of the crystallizer, together with the small amount of feedstock, is circulated by pump *C* through the tubes of heat exchanger *D*, which is cooled rapidly by a forced circulation of water or brine. On cooling, the solution becomes supersaturated, but not enough for spontaneous nucleation to occur; great care is taken to prevent it.

14.5.5. EVAPORATING AND VACUUM CRYSTALLIZERS

When the solubility of a solute in a solvent is not appreciably decreased by a reduction in temperature, supersaturation of the solution can be achieved by removal of some of the solvent. Evaporation techniques for the crystallization of salts have been used for centuries, and the simplest method, the utilization of solar heat, is still a commercial proposition in many parts of the world.

A good example of the modern application of solar evaporation is provided by the activities of the Dead Sea works. The Dead Sea occupies some 900 km^2, and some 100 km^2 of the shallow waters are divided by dykes into compartments where the mixtures of salts crystallize. Brine is pumped from the deep water into these evaporating basins and transferred from one to another as it gets more concentrated. The operating procedures and the sequences of salt recovery are quite complex; in fact, the whole operation is an excellent example of the industrial exploitation of the principles of the Phase Rule.

The Dead Sea is unusual because it contains a highly concentrated brine (27%) with a high percentage of magnesium salts (Table 14.6). One of the principal salts crystallized from Dead Sea water is carnallite ($KCl.MgCl_2.6H_2O$). By comparison, ordinary sea water is quite dilute (3.3% dissolved salts), and its main soluble constituent is sodium chloride (2.6%). Magnesium chloride only accounts for about 0.3%.

Fishery salt, hard saucer-shaped crystals, 5–10 mm diameter, is produced in large quantities by the evaporation of brine in long, shallow, open pans heated by direct fire, hot gases, or steam coils. Grainer salt, used in the manufacture of cheese and butter, consists of rather smaller crystals (1–3 mm), is also made in open-trough evaporators. In these operations, the growing crystals are held at the evaporating surface by the surface tension of the solution until they become large enough to sink.

For most other purposes, common salt is produced from brine in enclosed calandria evaporators, appropriately called salting evaporators. Little control over the crystal size is

TABLE 14.6. *Composition of Dead Sea Water at a Depth of 75 m (Density = 1240 kg m^{-3})*

Salt	Percentage
Sodium chloride	7.2
Potassium chloride	1.3
Magnesium chloride	13.7
Calcium chloride	3.8
Magnesium bromide	0.6
Calcium sulphate	0.1
Total dissolved solids	26.7

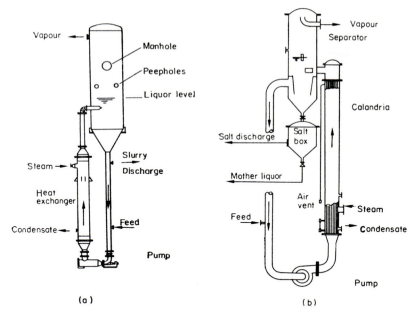

FIG. 14.14. Forced-circulation evaporator crystallizers: (a) Swenson, (b) APV-Kestner.

possible, and fine crystals (0.2–0.5 mm) are formed. Evaporating crystallizers of the calandria type, often in multi-effect series, are also used in sugar refining.

14.5.6. FORCED CIRCULATION EVAPORATORS

Two widely used forced circulation evaporator crystallizers are shown in Fig. 14.14. They are both essentially circulating magma units, with external heat exchangers, operated under reduced pressure. In the Swenson crystallizer (Fig. 14.14a) the crystal magma is circulated from the conical base of the evaporator body through the vertical tubular heat exchanger and re-introduced tangentially into the evaporator below the liquor level to create a swirling action and prevent flashing. Feedstock enters on the pump inlet side of the circulation system. In the APV-Kestner unit (Fig. 14.14b), usually called a long-tube salting evaporator, the feedstock also enters the circuit on the pump inlet side. The liquor level in the separator (the evaporator body) is kept above the top of the heat exchanger to prevent boiling in the tubes. Liquor enters tangentially, giving a swirl, but a baffle in the separator creates a relatively quiescent growth zone. The larger crystals settle to the conical base and are discharged through the salt box or by some other suitable means. The fine crystals are recirculated through the heat exchanger. Both these crystallizers are widely used for a variety of substances such as NaCl, $(NH_4)_2SO_4$, Na_2SO_4, $NiSO_4$, citric acid, etc.

14.5.7. VACUUM OPERATION

The use of reduced pressure in an evaporator to aid the removal of solvent, to minimize the heat consumption, or to decrease the operating temperature of the solution is common practice. However, such units are best described as "reduced-pressure evaporating crystallizers". The true vacuum crystallizer operates on a slightly different principle.

In a vacuum crystallizer supersaturation is achieved by the simultaneous evaporation and adiabatic cooling of the feedstock. A hot saturated solution is fed into a lagged vessel, which is maintained under vacuum. If the feed temperature is higher than the temperature at which the solution would boil under the low pressure existing in the vessel, the liquor cools adiabatically to this temperature. The sensible heat liberated by the solution, together with any heat of crystallization liberated, causes the evaporation of some of the solvent and concentrates the solution.

In a continuously operated vacuum crystallizer, the feedstock should reach the surface of the liquor in the vessel quickly. Otherwise, evaporation and cooling will not take place because the boiling-point elevation due to the hydrostatic head of solution becomes appreciable at the low pressures (10–20 mbar) used in these vessels. Precautions have to be taken, therefore, either to introduce the feed near the surface of the liquor in the vessel or to provide some form of agitation. A large number of different types of vacuum crystallizer are available.[1, 24, 25]

One popular type is the Swenson draft-tube-baffled (DTB) crystallizer (Fig. 14.15). A relatively slow-speed propeller agitator is located in a draft tube which extends to a few inches below the liquor level in the crystallizer. Hot, concentrated feedstock enters at the

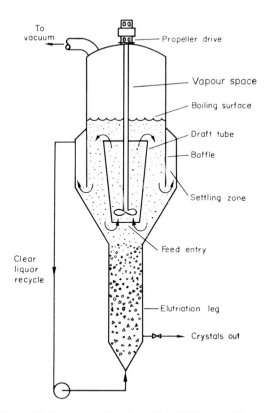

FIG. 14.15. Svenson draft-tube-baffled (DTB) crystallizer.

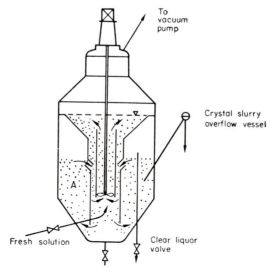

FIG. 14.16. Standard–Messo turbulence crystallizer.

base of the draft tube. The steady movement of magma and feedstock up to the surface of the liquor produces a gentle and uniform boiling action over the whole cross-sectional area of the crystallizer. The degree of supercooling thus produced is very low (about $\frac{1}{2}°C$), and in the absence of violent vapour flashing both excessive nucleation and salt build-up on the inner walls are minimized. The internal baffle in the crystallizer forms an annular space in which agitation effects are absent, thus providing a settling zone which permits regulation of the magma density and control over the removal of excess nuclei.

The Standard–Messo turbulence crystallizer is shown in Fig. 14.16. Two concentric pipes, an outer "ejector tube" with a circumferential slot and an inner "guide tube", create two liquor flow circuits. Circulation is created by a variable-speed agitator in the guide tube. The principle of the Oslo crystallizer is utilized in the growth zone A; partial classification occurs in the lower regions and fine crystals segregate in the upper. There is a fast upward flow of liquor in the guide tube and a downflow in the annulus; this causes liquor to be drawn in through the slot setting up a secondary flow circuit in the lower region of the vessel. Feedstock is introduced into the guide tube and passes into the vaporizer section where flash evaporation takes place. Nucleation, therefore, occurs in this region, and the nuclei are swept into the primary circuit. Mother liquor can be drawn off by means of a control valve, thus affecting a control over the salt slurry density.

The Escher–Wyss DP crystallizer, also known as the Tsukishima crystallizer, contains certain novel features (Fig. 14.17). Although essentially a draft-tube agitated crystallizer, the DP unit contains an annular baffled zone and a double propeller agitator which maintains a steady upward flow inside the draft tube and a downward flow in the annular region. Very stable suspension characteristics are claimed. The internal and external flow circuits may be followed in Fig. 14.20.

Oslo-Krystal vacuum units, operating on similar principles to those described above for the cooling crystallizer (Fig. 14.13), are also widely used. These crystallizers (Fig. 14.18)

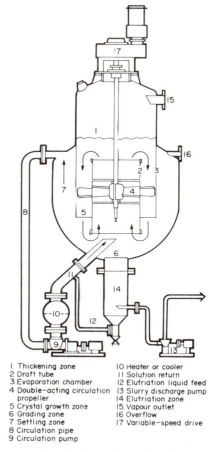

1 Thickening zone
2 Draft tube
3 Evaporation chamber
4 Double-acting circulation propeller
5 Crystal growth zone
6 Grading zone
7 Settling zone
8 Circulation pipe
9 Circulation pump
10 Heater or cooler
11 Solution return
12 Elutriation liquid feed
13 Slurry discharge pump
14 Elutriation zone
15 Vapour outlet
16 Overflow
17 Variable-speed drive

FIG. 14.17. Escher–Wyss DP crystallizer.

may be operated with either (a) a classified suspension (circulating liquor), or (b) a mixed suspension (circulating magma).[26]

Classified operation, while capable of producing large regular crystals, limits productivity because both the liquor velocity and the mass of crystals in suspension have to be restricted to keep the fines level below the pump inlet. Modification to magma circulation can improve the productivity considerably because higher circulation rates and magma densities can be employed. Furthermore, the suspension volume is increased because magma circulates through the vaporizer and downcomer. In this type of operation, however, the bulk classifying action is lost, and it is necessary to provide a secondary elutriation zone in the suspension to permit segregation and removal of excess nuclei. Fines can be redissolved with live steam and the resulting solution fed to the vaporizer.

One of the major factors in the successful operation of any controlled suspension crystallizer is the incorporation of a suitable fines trap. The earlier the excess nuclei are collected and destroyed the more efficient will be the process. In practice, fines are most

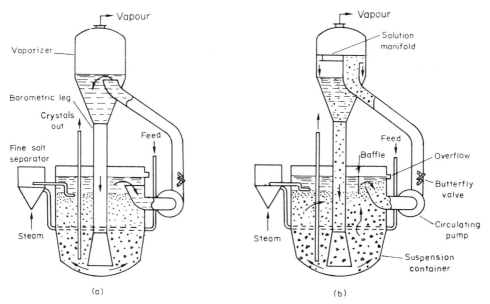

FIG. 14.18. An Oslo-Krystal vacuum crystallizer showing two different methods of operation: (a) classified suspension, (b) mixed suspension.[26]

economically removed when they reach 5–10% of the average product size. Saeman[26, 27] has demonstrated the importance of this aspect of crystallizer design and has shown that the key to effective crystal size control is the segregation time of nuclei in the fines trap. Size classification by hydraulic elutriation cannot be effective unless the segregation time requirements are also satisfied.

14.5.8. SALTING-OUT CRYSTALLIZATION

A solution can be supersaturated by the addition of a substance, preferably a liquid, which reduces the solubility of the solute in the solvent. The added component should be miscible in all proportions with the solvent and the solute should be relatively insoluble in it. This process, known by such names as salting-out, watering-out, precipitation, etc., is widely employed for the crystallization of organic substances from water-miscible organic solvents, e.g. from alcoholic solution by the controlled addition of water. The precipitation of inorganic salts from aqueous solution by an alcohol has also been applied commercially. For example, iron-free alums have been prepared in this way and so have coarse-grained anhydrous precipitates of normally hydrated salts.[1] Various salts can be crystallized selectively from natural brines by alcoholic precipitation.[28]

Salting-out processes offer several advantages. For instance, highly concentrated feedstock solutions can be prepared in a suitable solvent and a high recovery of solute can be effected by the choice of a suitable additive. All this can be done at around room temperature, which is highly desirable if heat labile substances are being processed. Frequently the mixed-solvent mother liquor has a high retention capacity for the impurities, and very pure crystals result. The disadvantage of the process is the need for a solvent recovery unit to separate the mixed solvents.

14.5.9. REACTION CRYSTALLIZATION

The deposition of a solid phase by the chemical interaction of gases or liquids is a common method of preparing many industrial chemicals. Precipitation occurs when the fluid phase becomes supersaturated with respect to the solute. A crude deposition process can be transformed into a crystallization operation if the degree of supersaturation can be controlled below the level at which spontaneous nucleation occurs.

Reaction crystallization is widely employed in industry, especially when valuable waste gases are produced. For example, sodium bicarbonate can be prepared from flue gases containing 10–20% of carbon dioxide by countercurrent contact with brine in packed towers. Ammonia can be recovered from coke-oven gases by interaction with sulphuric acid to give crystalline ammonium sulphate. Many conventional crystallizers can readily be adapted for use as reaction crystallizers. The reactants are usually fed into a zone where intimate mixing takes place quickly. Any heat of reaction may be utilized for the partial evaporation of the solution.[24]

14.5.10. SPRAY CRYSTALLIZATION

Strictly speaking, the term "spray" crystallization is a misnomer. The process is not a true crystallization; it has more in common with spray drying. Solids are simply deposited from a highly concentrated solution or a melt by spraying droplets into a large chamber where they fall countercurrently to an upflowing stream of hot air. In some cases, particularly with molten feedstocks, the product consists of near-spherical particles called *prills*, which have good flow properties, a high crushing strength, and good storage characteristics. Hygroscopic prills can be coated with an inert anti-caking agent. Fertilizer chemicals, particularly ammonium nitrate and urea, are manufactured on a large scale by this method.[29]

Substances with an inverted solubility, e.g. Na_2SO_4, $FeSO_4$, etc., which cause trouble in conventional evaporative crystallization due to scale formation on heat-transfer surfaces, are often manufactured by spray crystallization. In these cases the product usually consists of granules of agglomerated microcrystals.

14.5.11. MELT CRYSTALLIZATION

A considerable number of melt crystallization processes have been developed in recent years.[30] Active interest in the technique is currently being stimulated by the possibility of energy saving in large-scale industrial processing, compared with distillation, e.g. for the separation of close-boiling organic substances (see Table 14.1).

In the Newton Chambers process[31] for the purification of benzene from a coal-tar benzole fraction, an impure feedstock is mixed with refrigerated brine. The slurry is centrifuged to yield benzene crystals (freezing point 5.4°C) and a mixture of brine and mother liquor. After settling, the brine is returned for refrigeration and the mother liquor is reprocessed for motor fuel. The success of the method depends on the efficiency of removal of impure mother liquor adhering to the benzene crystals. There are several possible methods of operation.

In the thaw-melt method (Fig. 14.19a) the benzene crystals are washed in the centrifuge with brine at a temperature above 6°C. This partially melts or "thaws" some of the benzene

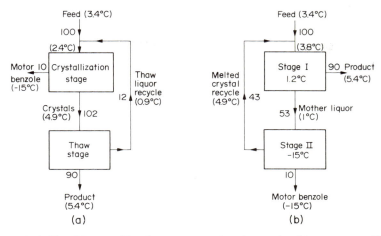

FIG. 14.19. The Newton Chambers process: (a) thaw-melt, (b) two-stage. The temperatures in parentheses indicate the freezing points of the various streams. The numbers indicate typical mass flow rates for a mass feed rate of 100.

crystals, and the adhering mother liquor is washed away. The thaw liquor can be recycled. Multi-stage operation is also possible (Fig. 14.19b). The first crop of crystals is taken as the product and the second, from the liquor, is melted for recycle. The purity of the crystals from the second stage should not be less than that of the original feedstock.

The Proabd Refiner[31] is essentially a batch-cooling process in which a static liquid feedstock is progressively crystallized on to extensive cooling surfaces, e.g. fin-tube heat exchangers, located inside the crystallization tank. As solidification proceeds, the remaining liquid becomes progressively more impure. In some cases, crystallization may be continued until virtually the entire charge has solidified. The crystallized mass is then slowly melted by heating the circulating heat transfer fluid. The impure fraction melts first and drains out and, as melting proceeds, the melt run-off becomes progressively richer in the desired component. Fractions may be taken off during the melting stage if required. A typical flow diagram, based on a scheme for the purification of naphthalene, is shown in Fig. 14.20. The circulating fluid is usually cold water which is heated during the melting stage by steam injection.

The MWB process acts effectively as a multi-stage countercurrent scheme, illustrated for a four-stage operation in Fig. 14.21a. The cycle starts at stage 1, which is fed with melt L_2 and recycle liquor L_1-L, where L is the reject impure liquor stream. A quantity of crystals C_1 is deposited in stage 1. In stage 2 the melted crystals C_1 are contacted with melt L_3 and fresh feedstock F. Crystallization yields crystals C_2 and melt L_3. Stages 3 and 4 follow similar patterns and, in this example, the final high-purity stream C_4, after being remelted, is split into product C and recycle C_4-C. Only one crystallizing vessel is needed (Fig. 14.21b). The crystals are not transported; they remain inside the vessel, deposited on the internal heat exchanger surfaces, until they are melted by the appropriate warm, incoming liquor stream. The control system linking the storage tanks and crystallizer consists of a program timer, actuating valves, pumps, and the cooling loop. The process has found commercial applications in the purification of a wide range of organic substances, including benzoic acid and caprolactam.

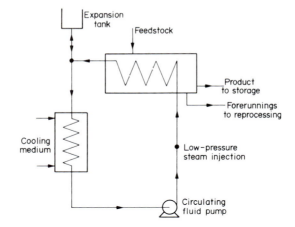

FIG. 14.20. Flow diagram of the Proabd Refiner.

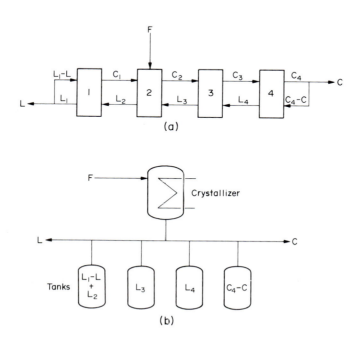

FIG. 14.21. The MWB process: (a) multi-stage scheme, (b) practical lay-out.

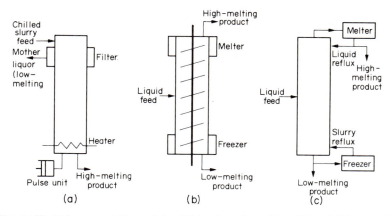

FIG. 14.22. Column crystallizers: (a) end-fed pulse column (Arnold type), (b) centre-fed column with spiral conveyor (Schildknecht type), (c) centre-fed column with reflux at both ends.

A process for fractional crystallization in a column crystallizer by the countercurrent contact between crystals and their melt was first patented by Arnold in 1951. The principles of column crystallization are shown in Fig. 14.22.

In the end-fed pulse unit (Fig. 14.22a) the slurry feedstock enters at the top of the column and the crystals fall countercurrently to a pulsed upflow of melt. There is a heat and mass interchange between the solid and liquid phases, and pure crystals migrate to the lower zone where they are remelted to provide a high-purity liquid for the upflow stream.

The centre-fed crystallizer (Fig. 14.22b), developed by Schildknecht in 1961, utilizes a spiral conveyor to transport the solids through the purification zone countercurrently to the melt. The mode of action is similar to that in a centre-fed distillation column. Laboratory crystallizers of this type are frequently operated batchwise, under total reflux, but reflux may be provided at both ends (Fig. 14.22c) when the unit functions as a complete fractionator. Several comprehensive reviews of the theory and practice of column crystallizers have been made.[32, 33]

An example of the successful commercial application of countercurrent column fractional crystallization is the Phillips process. The essential features of the crystallization zone are shown in Fig. 14.23a. Chilled slurry feed, from a scraped-surface chiller, enters at the top of the column. Crystals are forced downwards by means of a piston, and impure liquor is removed through a wall filter. Wash liquor, produced by melting pure crystals at the bottom of the column, is transported upwards countercurrently to the crystals. The wash liquor may be pulsed upwards. The Phillips process has been used for the large-scale production of *p*-xylene and for the freeze-concentration of beer and fruit juices.

The Brodie Purifier[35] is another successful melt crystallizer. It consists basically of a horizontal centre-fed column with the special feature (Fig. 14.23b) that the column cross-section reduces in the direction of diminishing flow, i.e. towards the cold end, thus maintaining a constant axial flow velocity and preventing back-mixing. The Brodie Purifier has been applied to the large-scale production of high purity *p*-dichlorbenzene.

In the recently developed TNO process[36] separation is effected by countercurrent washing coupled with repeated recrystallization which is facilitated by grinding the

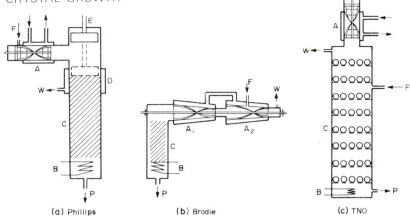

FIG. 14.23. Some column crystallizers used industrially: (a) Phillips, (b) Brodie, (c) TNO. A, scraped surface crystallizer (A_1 refining section, A_2 recovery section); B, heater; C, washing column; D, wall filter; E, reciprocating piston; F, feed inlet; P, product outlet; W, waste (low-melting) outlet.

crystals during their transport through a vertical column. The small crystalline fragments melt more easily. Grinding is achieved by balls rolling over sieve plates in the column (Fig. 14.23c). Good agitation and interphase transport are promoted, but undesirable top to bottom mixing is prevented. Pilot-scale trials with benzene–thiophene separations have been very successful, and scale-up to industrial operation looks promising.

14.6. Design and Operation of Crystallizers

14.6.1. CRYSTALLIZER SELECTION

The temperature–solubility relationship between the solute and solvent is of prime importance in the selection of a crystallizer. For solutions which deposit appreciable quantities of crystals on cooling, the choice of equipment will lie between a simple cooling crystallizer and a vacuum unit. For solutions which do not, an evaporating crystallizer would normally be used, although the method of salting-out could be employed in certain cases.

Other important factors to consider are the required shape, size, and size distribution of the crystalline product. For the production of reasonably large uniform crystals the modern trend is towards the controlled suspension crystallizers, fitted with suitable fines traps, which permit the discharge of a partially classified product. The subsequent washing and drying operations are carried out much more easily, and no sieving or grading of the crystal product is normally required.

The cost and space requirements must also be taken into account. Unfortunately, few comparative performance data or up-to-date cost data are available for the many units commonly used in practice. However, if the cost of a small unit is known, a rough estimate of the cost of a larger one is readily made by the "six-tenths rule"

$$\text{cost} \propto (\text{capacity})^{0.6}$$

for capacities in the range 0.5–50 ton h^{-1}.

Continuous crystallizers are generally more economical in operating and labour costs than the batch units, especially for large production rates. Batch crystallizers are usually cheaper in initial capital cost. One of the main advantages of a continuous unit is that the amount of mother liquor needing reworking is usually small, sometimes less than 5% of the feedstock handled. In batch units, as much as 50% of the mother liquor may require reworking.

Cooling crystallizers are generally less expensive than the vacuum or evaporating units, although the initial cost of agitation equipment can be fairly high, but no costly vacuum-producing or condensing equipment is required. Heavy crystal slurries can be handled in cooling crystallizers not requiring liquor circulation. Unfortunately, cooling surfaces can become coated with a hard crust of crystals with the result that cooling efficiency is reduced considerably. As vacuum crystallizers have no cooling surfaces they do not suffer from this disadvantage, but they cannot be used when the liquor has a high boiling point elevation. Vacuum and evaporating crystallizers generally require a considerable amount of head room.

Once a particular class of crystallizer is decided upon, the choice of a specific unit will depend upon such factors as the initial and operating costs, the space availability, the type and size of crystals required, the physical characteristics of the feed liquor and crystal slurry, the need for corrosion resistance, and so on. The production rate and supply of feed liquor to the crystallizer will generally be the deciding factors in the choice between a batch and continuous unit. Production rates > 1 ton day^{-1} or liquor feed rates $> 20 \, \text{m}^3 \, \text{h}^{-1}$ are best handled on a continuous basis.

14.6.2. INFORMATION FOR DESIGN

It is useful to know what sort of information a crystallizer manufacturer needs before he can begin to specify or design a piece of equipment. While most large fabricators have facilities to determine the necessary physical data, the customer can usually provide this information much more precisely, quickly, and cheaply.

A detailed description of the product should be given, including its full chemical name and formula, specifying if a hydrate is required or not. A realistic purity specification should be laid down The production rate should be given, e.g. as tons yr^{-1} or kg h^{-1} or preferably as both. Of course, other convenient units can be used. Very generous allowances should be made for maintenance and other shutdown periods.

The shape and size of the crystalline product should be specified. An actual sample of the desired crystals, if available, is very helpful. It is important to remember that the more rigid the size specification the more difficult the crystallizer design becomes. Vague statements such as "about 500 μm" should not be made as they have no useful meaning, but test-sieve sizes may be quoted as allowable upper and lower size limits. For example, a size specification such as "90% between 600 and 300 μm test sieves" would be quite acceptable.

A solubility curve of the product in the liquor should be drawn. Care should be taken in doing this because solubility data reported in the literature usually refer to pure solutes and solvents. These may be quite inapplicable. If impurities are known to be present in the working liquors, the relevant solubility data should be measured.

The working temperature range for batch-operated crystallizers, or the operating temperature for continuous units, has to be specified. The optimum temperature of operation is not an easy quantity to determine; it is dependent, for example, on the required crystal yield and the energy expenditure incurred in the operation of heaters, coolers, vacuum pumps, etc. However, since the crystal growth process is temperature dependent, the quality of the crystalline product can depend significantly on the temperature of operation.

The following liquor and solids data are also useful.

Liquor: feedstock analysis and pH; density of the feedstock at the feed temperature; boiling-point elevation of the saturated solution; density and viscosity of the saturated solution over the working range of temperature.

Crystals: true density of the substance; heat of crystallization; settling velocity of crystals of the desired size and shape in saturated solutions over a range of temperatures; crystal growth and nucleation rates as functions of supersaturation and temperature; maximum allowable undercooling (metastable zone width) in the presence of crystals in agitated solution.

14.6.3. CRYSTAL YIELD

An estimate of the crystal yield for a simple cooling or evaporating crystallization may be made from a knowledge of the solubility characteristics of the solution. The general equation may be written[1]

$$Y = \frac{WR[c_1 - c_2(1 - V)]}{1 - c_2(R - 1)}, \tag{14.34}$$

where c_1 is the initial solution concentration (kg anhydrous salt per kg solvent), c_2 is the final solution concentration (kg anhydrous salt per kg solvent), W is the initial weight of solvent (kg), V is the solvent lost by evaporation (kg per kg of original solvent), R is the ratio of molecular weights of solvated (e.g. hydrate) and non-solvated (e.g. anhydrous) solute, and Y is the crystal yield (kg).

In practice the actual yield may differ slightly from that calculated from eqn. (14.34). For example, if the crystals are washed with fresh solvent on the filter, losses may occur due to dissolution. On the other hand, if mother liquor is retained by the crystals an extra quantity of crystalline material will be deposited on drying. It must be remembered that published solubility data usually refer to pure solvents and solutes. As pure systems are rarely encountered industrially, it generally is advisable to check the solubility data with the actual working system.

Before eqn. (14.34) can be applied to the case of vacuum crystallization the quantity V must be estimated. This may be done by using the equation[1]

$$V = \frac{qR(c_1 - c_2) + C(t_1 - t_2)(1 + c_1)[1 - c_2(R - 1)]}{\lambda[1 - c_2(R - 1)] - qRc_2} \tag{14.35}$$

where λ is the latent heat of evaporation of solvent (J kg^{-1}), q is the heat of crystallization of product (J kg^{-1}), t_1 is the initial temperature of solution (°C), t_2 is the final temperature of solution (°C), and C = heat capacity of solution (J kg^{-1} K^{-1}). The symbols R, c_1, and c_2 have the same meaning as in eqn. (14.34).

14.6.4. SCALE-UP AND OPERATING PROBLEMS

The design of a large-scale industrial crystallizer from data obtained on a small-scale unit is not a simple matter. The scaling-up of crystallization equipment is more difficult than that for any of the other unit operations of chemical engineering. If a small crystallizer produces the required type of product, then the proposed large crystallizer should simulate a large number of different conditions obtaining in the small unit. The five most important conditions are:

(1) identical flow characteristics of liquid and solids;
(2) identical degrees of supersaturation in all equivalent regions of the crystallizer;
(3) identical initial seed sizes;
(4) identical magma densities;
(5) identical contact times between growing crystals and supersaturated liquor.

The scaling-up of agitation equipment has long been recognized as a difficult problem, and the two dimensionless numbers most frequently encountered in the analysis of agitated vessels are the Reynolds number Re and Froude number Fr. The former gives the ratio of the inertia and viscous forces, the latter the ratio of centrifugal acceleration and acceleration due to gravity. For a stirrer blade of diameter d rotating at n revs per unit time in a liquid of density ρ and absolute viscosity η, these two numbers may be written

$$Re = \frac{\rho n d^2}{\eta} \quad \text{and} \quad Fr = \frac{n^2 d}{g}. \tag{14.36}$$

For exact scale-up, values of both Re and Fr should be kept constant, but this is exceptionally difficult if not impossible in most cases. For example, if the stirrer diameter is increased by a factor of 4, the stirrer speed must be decreased by a factor of 2 if the Froude number is to be kept constant ($n \propto d^{-1/2}$), but by a factor of 16 if the Reynolds number is to remain the same ($n \propto d^2$). Some compromise must therefore be made, and as the Reynolds number greatly influences fluid friction, heat, and mass transfer, crystallizers are generally scaled-up on the basis of Re rather than on Fr.

The power input P to an agitator is related to Re and Fr by

$$\frac{P}{\rho n^3 d^5} = f(Re, Fr). \tag{14.37}$$

At high values of Re ($> 10,000$) the group $P/\rho n^3 d^5$ (the power number) is roughly constant, so under these conditions if the agitator speed is doubled the power input would have to be increased eight-fold. If the agitator diameter is doubled, its speed remaining constant, the power input would have to be increased by a factor of 32.

14.6.5. MODES OF OPERATION

A large proportion of industrial crystallizers are of the mixed suspension type in which crystals of all sizes are dispersed reasonably uniformly throughout the working zone. Such crystallizers may be divided, according to the type of product discharge, into "mixed suspension, mixed product removal" (MSMPR) and "mixed suspension, classified product removal" (MSCPR) units. Both types can be operated so that either all the

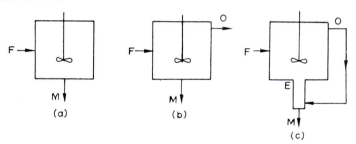

FIG. 14.24. Some basic types of continuously operated mixed suspension crystallizer, (a) mixed product removal (MSMPR), (b) liquor overflow, (c) liquor overflow and recycle through an elutriating leg (classified product removal, MSCPR). [F = feed, E = elutriating leg, M = magma outlet, O = overflow.]

mother liquor leaving the crystallizer leaves together with the product crystals or a part of the mother liquor is allowed to overflow separately.

Three possibilities are shown diagrammatically in Fig. 14.24. A simple MSMPR crystallizer is shown in Fig. 14.24a. In this kind of unit the crystal and liquor residence times are identical. However, by allowing some of the liquor to overflow from the crystallizer (Fig. 14.24b) the crystal residence time can be increased and rendered independent of the liquor residence time. Alternatively, the liquor overflow can be recycled back into the crystallizer through an elutriating leg, thus imparting some classifying action on the product crystals (Fig. 14.24c). The MSCPR crystallizer may also be operated with some liquor overflow (not depicted here). These different modes of action can be seen in the crystallizers illustrated in section 14.5.

As mentioned above, one reason for allowing liquor to overflow from a continuously operated crystallizer is to increase the residence time of crystals in the growth zone. Another is to permit excess fine crystals to be removed from the system. Most crystallizers suffer from excessive nucleation, and to produce reasonably large crystals these excess fines must be removed. If the liquor overflow is to be recirculated it may be passed through a fines trap, e.g. a steam-heated dissolver, for this purpose.

14.6.6. CONCEPT OF THE POPULATION BALANCE

In order to achieve a complete description of the crystal size distribution in a continuously operated crystallizer it is necessary to quantify the nucleation and growth processes and to apply the conservation laws of mass, energy, and crystal population. The importance of the population balance, in which all particles must be accounted for, has been stressed in the work of Randolph and Larson.[38-40]

The crystal population density n (number of crystals per unit size per unit volume of system, e.g. $\mu m^{-1} m^{-3}$) is defined by

$$\frac{\Delta N}{\Delta L} = \frac{dN}{dL} = n,$$
(14.38)

where ΔN is the number of crystals per unit volume in an arbitrary size range L_1 to L_2, i.e.

$$\Delta N = \int_{L_1}^{L_2} n \, dL.$$
(14.39)

Application of the population balance is best demonstrated with reference to the simple case of a continuously operated MSMPR crystallizer (Fig. 14.24a) assuming (a) steady-state operation, (b) no crystals in the feed stream, (c) all crystals of the same shape, characterized by a chosen linear dimension L, (d) no break-down of crystals by attrition, and (e) crystal growth rate independent of crystal size.

A population balance (input = output) in a system of volume V for a time interval Δt and size range $\Delta L = L_2 - L_1$ is

$$n_1 G_1 V \Delta t = n_2 G_2 V \Delta t + Q \bar{n} \Delta L \Delta t, \tag{14.40}$$

where Q is the volumetric feed and discharge rate, G the crystal growth rate (dL/dt), and $\bar{n}$ the average population density. As $\Delta L \to 0$,

$$\frac{d(nG)}{dL} = -\frac{Qn}{V}. \tag{14.41}$$

Defining the liquor and crystal mean residence time $T = V/Q$ and assuming $dG/dL = 0$, then

$$\frac{dn}{dL} = -\frac{n}{GT}, \tag{14.42}$$

which on integration gives

$$n = n^0 \exp(-L/GT), \tag{14.43}$$

the fundamental relationship between crystal size L and population density n characterizing the crystal size distribution. The quantity n^0 in eqn. (14.43) is the population density of nuclei (zero-sized crystals).

Expressions of the rates of nucleation J and growth G ($=dL/dt$) written in terms of supersaturation as

$$J = k_1 \Delta c^m \tag{14.44}$$

and

$$G = k_2 \Delta c^g, \tag{14.45}$$

respectively, may be combined to give

$$J = k_3 G^{m/g}. \tag{14.46}$$

Furthermore, since the nucleation rate

$$J = \left.\frac{dN}{dt}\right|_{L=0} = \left.\frac{dN}{dL}\right|_{L=0} \frac{dL}{dt} \tag{14.47}$$

the relationship between the two processes may be expressed as

$$J = n^0 G \tag{14.48}$$

$$n^0 = k_4 G^{(m/g)-1}. \tag{14.49}$$

Evaluation of the nucleation and growth processes may be made from experimental measurement of the crystal size distribution in a crystallizer operating under steady-state conditions.

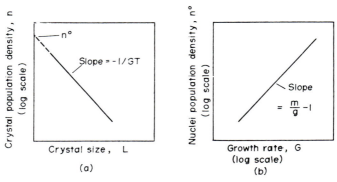

For instance, eqn. (14.43) indicates that a plot of $\log n$ versus L should give a straight line of slope $-1/GT$ and intercept at $L = 0$ equal to n^0 (Fig. 14.25a). If the residence time T is known, then the crystal growth rate G can be calculated.

Similarly, a plot of $\log n^0$ versus $\log G$ should give a straight line of slope $(m/g) - 1$ (eqn. (14.49) and Fig. 14.25b). Thus the kinetic order of nucleation m may be evaluated if the order of the growth process g is known, e.g. from measurements such as those described in section 14.3.

14.6.7. CRYSTAL SIZE DISTRIBUTIONS

Size distributions of particulate substances may be described in various ways. Equation (14.43), for example, represents a distribution on a number basis, and this has obvious utility for population balances. On the other hand, a distribution on a mass basis is often required, e.g. for mass balance purposes.

From eqns. (14.39) and (14.43) the number of crystals N up to size L is given by

$$N = \int_0^L n \, dL$$

$$= n^0 GT [1 - \exp(-L/GT)], \tag{14.50}$$

which for $L \to \infty$ gives the total number of crystals in the system:

$$N_T = n^0 GT. \tag{14.51}$$

The mass of crystals M up to size L is given by

$$M = \alpha\rho \int_0^L nL^3 \, dL, \tag{14.52}$$

where α is a volume shape factor, defined by $\alpha = \text{volume}/L^3$, and ρ is the crystal density. For $L \to \infty$, eqn. (14.52) gives the total mass of crystals in the system:

$$M_T = 6\alpha\rho n^0 (GT)^4. \tag{14.53}$$

The magma density M_T is controlled by the feedstock and operating conditions of the crystallizer. As the population density of nuclei n^0 is related to growth and nucleation rates (eqn. (14.48)) eqn. (14.53) may be rewritten in an alternative form:

$$M_T = 6\alpha\rho JG^3 T^4. \tag{14.54}$$

The mass of crystals dM in a given size range dL is

$$dM = \alpha\rho nL^3 \, dL \tag{14.55}$$

so the mass fraction in a particular size range dL is dM/M_T. Therefore, from eqns. (14.53) and (14.55) the mass distribution is given by

$$\frac{dM(L)}{dL} = \frac{dM}{M_T}\frac{1}{L} = \frac{nL^3}{6n^0(GT)^4} \tag{14.56}$$

which, from the population density relationship (eqn. (14.43)); becomes

$$\frac{dM(L)}{dL} = \frac{\exp(-L/GT)L^3}{6(GT)^4}. \tag{14.57}$$

The peak of this mass distribution (the dominant size L_D of the crystal size distribution) is found by maximizing eqn. (14.57), which gives

$$L_D = 3GT. \tag{14.58}$$

Nucleation in a crystallizer depends on crystal contacts as well as on the level of supersaturation, so the empirical nucleation eqn. (14.46) can be expanded to include the magma density

$$J = k_5 M_T G^i, \tag{14.59}$$

where

$$i = m/g. \tag{14.60}$$

Combining eqns. (14.54), (14.58), and (14.59) gives

$$G = \left[\frac{27}{2\alpha\rho k_5 L_D^4}\right]^{1/(i-1)} \tag{14.61}$$

and combination of eqns. (14.58) and (14.61) yields the interesting relationship

$$L_D \propto \tau^{(i-1)/(i+3)} \tag{14.62}$$

which enables the effect of changes in residence time to be evaluated. For example, if $i = 2$, a typical value for many inorganic salt systems, a doubling of the residence time would only increase the dominant product crystal size by about 15%. However, to double the residence time it would be necessary either to double the crystallizer volume or halve the volumetric feed rate, and hence halve the production rate. So, contrary to popular belief, residence time manipulation is not usually very effective for controlling the product crystal size.

Although discussion has been confined in this section to the MSMPR configuration, the concepts of the population balance can be applied to crystallizers having other flow patterns and corresponding design relationships have been developed.[38-41]

14.6.8. APPLICATIONS OF THE POPULATION BALANCE

Example 1: Evaluation of Nucleation and Growth Rates

The size analysis of potash alum crystals (density, $\rho = 1770 \, \text{kg m}^{-3}$; volume shape factor, $\alpha = 0.471$) leaving a continuous MSMPR crystallizer operated with a residence time $T = 0.25 \, \text{h}$ (for both liquor and crystals) is given in Table 14.7.

TABLE 14.7. *Calculation of Population Densities from a Crystal Size Distribution*

Size Range (μm)	ΔL (μm)	$\bar{L}$ (μm)	$\bar{L}^3$ (μm^3)	$dW^{(a)}$ (g)	$\alpha\rho\bar{L}^3\Delta L$ (g μm)	Population Density $n^{(b)}$ (μm^{-1})
710–500	210	602	2.12×10^8	1.20	3.88×10^{-2}	3.09×10^5
500–355	145	428	7.84×10^7	2.72	9.45×10^{-3}	2.88×10^6
355–250	105	303	2.78×10^7	4.27	2.43×10^{-3}	1.75×10^7
250–180	70	215	9.94×10^6	3.15	5.80×10^{-4}	5.43×10^7
180–125	55	153	3.58×10^6	1.69	1.64×10^{-4}	1.03×10^8
125– 90	35	108	1.26×10^6	0.75	3.68×10^{-5}	2.04×10^8
90– 63	27	77	4.57×10^5	0.28	1.03×10^{-5}	2.73×10^8
63– 45	18	54	1.58×10^5	0.12	2.36×10^{-6}	5.08×10^8

$^{(a)}$ per 100 ml of slurry. $^{(b)}$ per m^3 of slurry. $n = dW/\alpha\rho\bar{L}^3\Delta L$.

When these data are plotted as $\log n$ versus $\bar{L}$ (as in Fig. 14.25a but not given here) the value of n at $\bar{L} = 0$ gives the population density as $n^0 = 8.2 \times 10^8 \, \mu\text{m}^{-1} \, \text{m}^{-3}$.

The slope of the line $(-1/GT)$ is $-0.0129 \, \mu\text{m}^{-1}$, which gives a value of the growth rate $G = 310 \, \mu\text{m h}^{-1}$ ($8.6 \times 10^{-8} \, \text{m s}^{-1}$).

The nucleation rate J (eqn. (14.48)) $= n^0 G = 2.54 \times 10^{11} \, \text{h}^{-1} \, \text{m}^{-3}$ ($7.06 \times 10^7 \, \text{s}^{-1} \, \text{m}^{-3}$).

Other runs with the same crystallizer, made with different residence times, gave the following results:

T (h)	n^0 (μm^{-1} m^{-3})	G (μm h^{-1})	J (h^{-1} m^{-3})
0.167	1.1×10^9	480	5.28×10^{11}
0.25	8.2×10^8	310	2.54×10^{11}
0.5	6.7×10^8	152	1.02×10^{11}

A plot of $\log n^0$ versus $\log G$ (as in Fig. 14.25b but not given here) in accordance with eqn. (14.49) gives a line of slope $(m/g) - 1$. Alternatively, a plot of $\log J$ versus $\log G$ gives a line of slope m/g. In the present case m/g is estimated as 1.2, and as the order of the growth process $g = 1.6$ (section 14.3.1), then the order of the nucleation process $m = 1.9$.

Example 2: Specification of a Crystallizer

An MSMPR crystallizer is required to produce potassium sulphate crystals (density $\rho = 2660 \, \text{kg m}^{-3}$, volume shape factor $\alpha = 0.7$) at the rate $P = 500 \, \text{kg h}^{-1}$. Pilot plant trials indicate that the crystallizer will operate with a steady-state population of nuclei $n^0 = 10^7 \mu m^{-1} m^{-3}$ (i.e. $10^{13} m^{-4}$) and a growth rate $G = 5 \times 10^{-8} m s^{-1}$. The feed liquor rate $Q = 3 m^3 h^{-1}$.

Estimate the operating magma density, crystallizer volume, and dominant size of the product crystal.

Residence time $T = V/Q = 0.333V h$.

Magma density $M_T = P/Q = 500/3 = 167 kg m^{-3}$.

From a mass balance (eqn. (14.53)):

$$M_T = 6\alpha\rho n^0 (GT)^4,$$

$$167 = 6(0.7)(2660)(10^{13})(5 \times 10^{-8} \times 0.333V \times 3600)^4.$$

Therefore the required crystallizer volume $V = 3.3 m^3$.

The dominant size L_D is given by eqn. (14.58):

$$L_D = 3GT$$

$$= 3(5 \times 10^{-8})(0.333 \times 3.3 \times 3600)$$

$$= 5.9 \times 10^{-4} m$$

$$= 590 \mu m.$$

Example 3: Effect of Changes in Operating Procedure

The $3.3 m^3$ MSMPR crystallizer specified in Example 2 is to be operated with a reduced feed liquor rate of $2 m^3 h^{-1}$. If the magma density M_T is kept at $167 kg m^{-3}$, calculate the new nucleation, growth and production rates, and the dominant crystal size. Laboratory tests have indicated that the ratio of the kinetic exponents $m/g = 2.1$.

The proposed reduction in the feed liquor rate will cause an increase in the residence time ($T_2 = V/Q_2 = 1.65 h$ compared with $T_1 = 1.1 h$). This will reduce the operating supersaturation and hence reduce both the nucleation and crystal growth rates. Therefore, eqn. (14.46) should be used in conjunction with eqn. (14.54), which, if M_T is kept constant, gives

$$\left(\frac{T_1}{T_2}\right)^4 = \left(\frac{G_2}{G_1}\right)^{[(m/g)+3]}$$

or

$$G_2 = G_1(T_1/T_2)^{0.784}.$$

So the new crystal growth rate

$$G_2 = 5 \times 10^{-8}(1.1/1.65)^{0.784}$$

$$= 3.64 \times 10^{-8} m s^{-1}.$$

Therefore the new nucleation rate (eqn. (14.48))

$$J_2 = n^0 G_2$$

$$= 10^{13}(3.98 \times 10^{-8})$$

$$= 3.64 \times 10^5 s^{-1} m^{-3}$$

compared with that for the previous case: $J_1 = 10^{13}(5 \times 10^{-8}) = 5 \times 10^{-5}\,\mathrm{s}^{-1}\,\mathrm{m}^{-3}$.
From eqn. (14.57)

$$L_{D2} = L_{D1}(G_2/G_1)(T_2/T_1)$$

$$= 590(3.64/5)(1.65/1.1)$$

$$= 644\,\mu\mathrm{m}.$$

The new production rate

$$P_2 = M_{T2}Q_2 = 167 \times 2$$

$$= 344\,\mathrm{kg\,h}^{-1}.$$

Example 4: Effect of Magma Density

The $3.3\,\mathrm{m}^3$ crystallizer, operating with the reduced feed rate of $2\,\mathrm{m}^3\,\mathrm{h}^{-1}$ (residence time $T = 1.65\,\mathrm{h}$) as specified in Example 3, is to be operated with a magma density of (a) $100\,\mathrm{kg\,m}^{-3}$, and (b) $300\,\mathrm{kg\,m}^{-3}$.
Determine the effect on the product size in each case.
From eqns. (14.46) and (14.54), with T constant and $m/g = 2.1$:

$$M_{T2} = M_{T1}(G_2/G_1)^{5.1}$$

or

$$G_2 = G_1(M_{T2}/M_{T1})^{0.196}$$

(a)

$$M_{T2} = 100\,\mathrm{kg\,m}^{-3},$$

$$G_2 = (100/167)^{0.196}(5 \times 10^{-8})$$

$$= 4.52 \times 10^{-8}\,\mathrm{m\,s}^{-1},$$

and

$$L_{D2} = 3G_2\,T_2$$

$$= 3 \times 4.52 \times 10^{-8} \times 1.65 \times 3600$$

$$= 8.05 \times 10^{-4}\,\mathrm{m}$$

$$= 805\,\mu\mathrm{m}.$$

(b)

$$M_{T2} = 300\,\mathrm{kg\,m}^{-3},$$

$$G_2 = (300/167)^{0.196}(5 \times 10^{-8})$$

$$= 5.61 \times 10^{-8}\,\mathrm{m\,s}^{-1}.$$

Then

$$L_{D2} = 3G_2\,T_2$$

$$= 3 \times 5.61 \times 10^{-8} \times 1.65 \times 3600$$

$$= 1000\,\mu\mathrm{m}.$$

So both cases result in an increase in the dominant size. The corresponding production rates are:

$$P_1 = M_{T1}Q_1 = 100 \times 2 = 200\,\text{kg h}^{-1},$$
$$P_2 = M_{T2}Q_2 = 300 \times 2 = 600\,\text{kg h}^{-1}.$$

References

1. MULLIN, J. W., *Crystallization*, 2nd edn., Butterworth, London, 1972.
2. MULLIN, J. W. and SÖHNEL, O., *Chem. Engng Sci.* **32** (1977) 683 and **33** (1978) 1000.
3. MULLIN, J. W. (ed.) *Industrial Crystallization*, Plenum, New York, 1976.
4. GARSIDE, J. and LARSON, M. A., *J. Cryst. Growth* **43** (1978) 694.
5. BOTSARIS, G. D., p. 3 in ref. 3.
6. ESTRIN, J., in *Chemical Vapour Transport, Secondary Nucleation and Mass Transfer in Crystal Growth* (W. R. Wilcox, ed.), Dekker, New York, 1976, p. 1.
7. GARSIDE, J., *Crystal Growth and Materials* (E. Kaldis and H. J. Scheel, eds.), North-Holland, Amsterdam, 1977, pp. 484–513.
8. MULLIN, J. W. and GARSIDE, J., *Trans. Inst. Chem. Engrs* **45** (1967) 285 and 291; **46** (1968) 11.
9. MULLIN, J. W. and AMATAVIVADHANA, A., *J. Appl. Chem.* **17** (1967) 151.
10. GARSIDE, J., MULLIN, J. W., and DAS, S. N., *Ind. Eng. Chem. Fundamentals* **13** (1974) 299.
11. PHILLIPS, V. R. and EPSTEIN, N., *AIChEJ* **20** (1974) 678.
12. GARSIDE, J., PHILLIPS, V. R., and SHAH, M. B., *Ind. Eng. Chem. Fundamentals* **15** (1976) 230.
13. GARSIDE, J., JANSEN-VAN ROSMALEN, R., and BENNEMA, P., *J. Cryst. Growth* **29** (1975) 353.
14. JANSE, A. H. and DE JONG, E. J., p. 145 in ref. 3.
15. BENNEMA, P., p. 91 in ref. 3.
16. BUJAC, P. D. B., p. 23 in ref. 3.
17. VAN'T LAND, C. M. and WIENK, B. G., p. 51 in ref. 3.
18. GARSIDE, J. and JANCIC, S. J., *AIChEJ* **22** (1976) 887.
19. ARMSTRONG, A. J., *Chem. Process Engng* **51** (11) (1970) 59.
20. ATKINS, R. S., *Hydrocarbon Processing* Nov. (1970) 127.
21. BARDUHN, A. J., *Chem. Engng Progr.* **71** (11) (1975) 80.
22. USYUKIN, I. P. *et al.*, *Soviet Chem. Ind.* **4** (1973) 267.
23. CERNY, J., *Br. Pat.* 932215 (1963).
24. BAMFORTH, A. W., *Industrial Crystallization*, Leonard Hill, London, 1965.
25. NÝVLT, J., *Industrial Crystallization from Solutions*, Butterworth, London, 1971.
26. SAEMAN, W. C., *AIChEJ* **2** (1956) 107.
27. SAEMAN, W. C., *Ind. Eng. Chem.* **8** (1961) 612.
28. FERNANDEZ-LOZANO, J. A., *Ind. Eng. Chem. Proc. Des. Dev.* **15** (3) (1976) 445.
29. ROBERTS, A. G. and SHAH, K. D., *Chem. Engr. London* Dec. (1975) 748.
30. ZIEF. M. and WILCOX, W. R. (eds.) *Fractional Solidification* Dekker, New York, 1967.
31. MOLINARI, J., chapters 13 and 14 in ref. 30.
32. POWERS, J. E. *et al.*, chapter 11 in ref. 30 and *AIChEJ* **16** (1970) 648 and 1055.
33. PLAYER, M. R., *Ind. Eng. Chem. Proc. Des. Dev.* **8** (1969) 210.
34. MCKAY, D. L., chapter 16 in ref. 30.
35. BRODIE, J. A., *Chem. Engng*, 13 Feb. (1978) 73.
36. ARKENBOUT, G. J., *Chem. Tech.* Sept. (1976) 596.
37. NÝVLT, J. and MULLIN, J. W., *Kristall und Technik* **9** (2) (1974) 141.
38. RANDOLPH, A. D. and LARSON, M. A., *Theory of Particulate Processes*, Academic Press, New York, 1971.
39. RANDOLPH, A. D., *Chem. Engng* 4 May (1970) 80.
40. LARSON, M. A. and GARSIDE, J., *Chem. Engr. London* June (1973) 318.
41. LARSON, M. A., *Chem. Engng* 13 Feb. (1978) 19.

CHAPTER 15

Assessment of Crystalline Perfection

D. B. HOLT

Metallurgy Department, Imperial College, University of London, UK

15.1. Introduction

The determination of the defect structure of crystals is vital from both the consumers' and producers' points of view. Research workers and device producers need to know the degree of purity and perfection of crystals to interpret structure dependent properties and in order to determine whether the material can be successfully employed in the experiments or device production processes in question. Crystal growers need to know the nature and distribution of the imperfections present in the crystals that they grow, whether the crystals are satisfactory or not. Only assessment can determine that the crystals *are* satisfactory. If the crystals are not satisfactory, structural assessment can provide clues from which to deduce how the growth techniques should be modified so that the perfection of the material may be increased. The classical example of the latter type of work is that of Dash (1958) who employed the decoration technique to observe the dislocations grown into silicon crystals and evolve methods for growing large dislocation free crystals of both silicon and germanium. Up to the present only GaAs has been added to this select band of materials which are available in dislocation free bulk form.

Since the discovery of transistor action in 1948 and the emergence of branches of industry based on the use of single crystals, a rapidly growing range of methods for the assessment of the structural perfection of crystals has been developed. This has vastly increased the speed, completeness and precision with which structural defects can be identified. It is neither possible nor desirable in an introductory chapter to attempt to give a complete account of even one of these techniques. References will be given to the standard books and review articles in each case for those who may wish to employ one particular method or another. What will be done here is to attempt to outline all the techniques of importance. In each case the type of structure made visible by the technique will be indicated and the physical principles by which this is done will be described. The advantages and disadvantages of each technique will be given. This account is written with the emphasis mainly directed towards single crystals of relatively high purity and perfection. These are usually crystals of non-metallic materials. Metallographers have evolved numerous techniques for the study of the structure of complex specimens of metals and alloys (Brandon, 1966; Smallman and Ashbee, 1966). These are frequently polycrystalline, polyphase, and of high dislocation densities.

No one technique will serve for all purposes. The ideal is to have them all available and to know which to call upon for particular purposes. Scanning electron microscopy, however, is a whole family of techniques as will be described below. Moreover, the properties of materials which give rise to contrast in the several modes of scanning electron microscopy include those of electronic interest. Thus scanning electron microscopy is the nearest approach to a universal assessment tool for electronic materials and especially for semiconducting materials and devices.

15.2. Volume, Area, Line, and Point Defects

In order to gain some appreciation of the utility of the different assessment techniques it is first necessary to have some idea of the nature of the various types of structural imperfections that must be detected.

Imperfections in crystals are classified in diminishing order of size in terms of the number of dimensions in which they are large compared with interatomic distances in the material.

Volume defects which are many atoms across in all three dimensions are the gross structural imperfections in crystals. They include volumes which differ from the rest of the crystal in one or more of the following characteristics: crystal structure, orientation, or composition.

A volume differing from the matrix only in orientation is called a grain. Certain orientations which are related to the matrix orientation by simple symmetry operations are of particularly low energy and occur frequently. These are called twins or, if very small, microtwins. A number of degrees of deviation from monocrystallinity are distinguished. If the entire volume of the material consists of grains of varying orientation it is a polycrystal. If most of the material has one orientation this is referred to as the matrix and the other volumes are known as included grains. This situation occurs frequently, for example, in epitaxial films. If the variations from the orientation of the matrix are less than about a degree the material is described as a single crystal with a lineage or mosaic structure. Most, if not all, metallic crystals are of this type.

A volume differing only in crystal structure from the remainder of the crystal, referred to as the matrix in this context, is an included grain of second phase material, or a phase transformed region. This type of defect can only occur, of course, in materials which are polymorphic, i.e. that occur in more than one crystal structure. Examples are the II–VI compounds which can have either the cubic sphalerite or the hexagonal wurtzite structures and SiC which can occur in over 150 different stacking sequence structures called polytypes. Domains are volumes differing in some form of alignment property or state parameter of the atoms on the sites of the crystal structure. Examples include ferromagnetic domains within which the spins of the unpaired electrons on the atoms are all aligned in parallel, and anti-phase domains which differ from one another in the occupation by the elements in the compound of the various sublattices of atomic sites in the crystal structure. The development of certain types of domain structures involves crystal structure changes, e.g. in the case of the emergence of ferroelectric domains at the Curie temperature in $BaTiO_3$.

A volume differing in chemical composition from the matrix is called a precipitate. In lightly doped samples of well-developed semiconductors the impurity content is too low for these to occur. However, it is known that silicon oxide precipitates can occur even in high purity silicon and small precipitates are quite common in highly doped silicon, GaAs, and GaP. It must be assumed therefore that precipitates are common in samples of most of the less well-developed semiconducting compounds and in many of the less pure non-metallic materials as well as in many alloys.

Area defects are one or a few interatomic distances in thickness but may be macroscopic in extent. Most area defects are interfaces, many of them being the surfaces of the volume defects just discussed. These can be listed as grain boundaries and twin boundaries, phase interfaces, and domain boundaries including antiphase boundaries.

In addition area defects include such things as stacking faults. Stacking faults are planes across which the stacking order of the crystal structure is altered. It is well known that the f.c.c. (face-centred cubic) structure is built up by stacking close-packed {111} planes of atoms in the order *abc abc* ... that is by cyclic permutation of all three of the geometrically possible positions for close-packed planes atop one another as indicated in Fig. 15.1a and c. The h.c.p. (hexagonal close-packed) structure results from stacking close-packed (0001) planes of atoms in cyclic permutation of any two of the sets of stacking positions, e.g. *abab* ... (Fig. 15.1b). More complex structures can be built up in similar ways. For example, the basic structures of the adamantine semiconducting materials can be built up by analogously stacking double planes of atoms. The atoms in one such plane occupy a close-packed type of array with the atoms of the second plane vertically above them, i.e. in the same type of position. Because the positions of the atoms in the two planes are not crystallographically equivalent it is the convention that one position be referred by a Latin letter as before while the other position vertically above is referred to by means of the Greek letter occupying the same position in the alphabet. Thus α positions are above *a* and *γ* positions above *c* sites. The diamond structure results from cyclically permuting double atom planes in all three positions thus: *aαbβcγaαbβcγ* ... as indicated in Fig. 15.1d. The sphalerite (or zinc-blende) structure has sites in the same array but the Latin letter sites are occupied by atoms of one chemical element and the Greek letter sites by atoms of the other element in the compound. The wurtzite structure is built up by cyclically permuting only two of the types of site in stacking sequences of the form *aαbβaαbβ*. ...

A stacking fault, then, is a plane across which the stacking is not that required to continue the crystal structure. Thus at the sloping line in the following sequence there is a stacking fault *aαbβcγaα/cγaαbβcγ*. ... Stacking faults result in the inclusion of lamellae of material of different structure in the crystal. In the example just given the long-range sequence on either side of the stacking fault plane is of the diamond or sphalerite type permuting all three positions. However, the four double atom layers, two double planes on either side of the fault plane, have the sequence *cγaαcγaα*, which is of the wurtzite type permuting only two of the stacking positions.

The one unavoidable form of imperfection in crystals is the free surface. This is of increasing practical importance, e.g. as electronic devices shrink and the surface to volume ratio consequently rises. It is also, of course, currently a major field of research. The structural features of crystal surfaces that require assessment are their short-range smoothness, long-range flatness, their cleanliness, and their crystal structure. These features are so different from the internal crystalline structural defects with which we are here concerned and the techniques used for the study are so numerous and specialized that it must suffice merely to list the more important of them in passing. Surface topography is normally best assessed by means of optical interferometry (Winterbottom and McLean, 1960; Gifkins, 1970; Tolansky, 1948, 1955), high sensitivity mechanical probes such as the commercial Tallysurf or sometimes, for high resolution, by means of secondary electron,

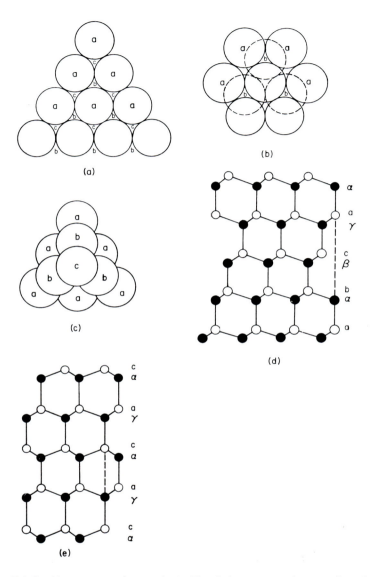

FIG. 15.1. Stacking sequences in crystals. (a) The circles represent the atoms in a close-packed plane. The three sets of sites on which successive planes of atoms can be placed are indicated by the letters *a*, *b*, and *c*. Two structures can be produced by stacking planes like this in two different sequences of positions as follows: (b) The stacking sequence *abab* ... produces the close-packed hexagonal structure. (c) The stacking sequence *abcabc* ... produces the face-centred cubic structure. (d) The diamond and sphalerite (zincblende) structures are built up by stacking pairs of vertically spaced planes of atoms, which are labelled with the corresponding letters of the Latin and Greek alphabets, e.g. *a* and α, in cyclic permutation of all three types of position. The fourth double atom plane then lies vertically above the first. (e) The wurtzite structure involves stacking of double atom planes in cyclic permutation of only two of the three types of position. In this structure, the third double atom plane lies vertically above the first.

emissive mode scanning electron microscopy; see, for example, Gopinath (1974). The chemical composition of the surface of atomically clean crystals can be determined by means of electron spectroscopy, especially Auger spectroscopy (Chang, 1971; Haas *et al.*, 1971; Treitz, 1977), and the existence of ordered structures on crystal surfaces is detected by means of low-energy electron diffraction (LEED) and reflection high-energy electron diffraction in ultra high vacuum (RHEED or HEED) (Farnsworth *et al.*, 1973; Gomer, 1975). Scanning high-energy electron diffraction (SHEED) equipment provides a quantitative, scanning read out of intensities in velocity analysed HEED patterns and can also be used for recording transmission electron diffraction patterns (Grigson, 1969; Tompsett, 1972). Scanning electron microscopes with ultra-high vacuum specimen environments are now commercially available and some of them can be equipped to carry out such surface analysis techniques as scanning Auger microscopy (e.g. Janssen *et al.*, 1977).

Line defects are called dislocations and are a more coherent group of defects than those discussed above. Dislocations are specified by two vectors, the arbitrarily chosen positive direction along the dislocation line **l** and the Burgers vector **b** which is defined once **l** is chosen by a procedure called the Burgers circuit together with a sign convention discussed, for example, by Hirth and Lothe (1968). The two vectors together determine the geometrical form of the dislocation, its energy, possible reactions with other defects, etc. It can be shown that most of the energy of dislocations is in the long-range elastic strain field and that this energy is proportional to the square of the Burgers vector. Consequently energetic considerations dictate that most dislocations have Burgers vectors equal in length to the shortest lattice translation vector. Dislocations with such Burgers vectors are known as unit dislocations.

A number of particular types of dislocations can be distinguished in terms of the angle between **l** and **b**. When the two vectors are parallel the dislocations are screws and when **l** and **b** are at right angles the dislocations are edges. In some crystal structures other types of dislocation are significant. For example, in the diamond and sphalerite structures $60°$ dislocations (for which **l** is inclined at $60°$ to **b**) are important.

Dislocation reactions of the form $\mathbf{b}_1 \to \mathbf{b}_2 + \mathbf{b}_3$ are energetically possible if $b_1^2 > b_2^2 + b_3^2$. This is the case for the unit dislocation Burgers vector dissociating into two partial dislocations connected by a strip of stacking fault as illustrated in Fig. 15.2. The combination of partials and stacking fault is known as an extended dislocation.

Point defects are of atomic dimensions. There are three chief types of point defects. These are impurity atoms, vacant lattice sites in the crystal structure which are known as vacancies, and atoms of the crystal existing in the interstices between sites of the crystal structure which are called interstitials. Individual point defects can just be resolved by three methods: in refractory metals (Brandon, 1966) and in SiC (Smith, 1969) using field ion microscopy, in scanning transmission electron microscopes (STEM) (Crewe, 1974), and in the most recent transmission electron microscopes giving atomic resolution using the direct lattice resolution method (Allpress and Sanders, 1973; Desseaux *et al.*, 1977). These methods are used to study the atomic structure of larger defects such as grain boundaries (e.g. Howell and Ralph, 1974; Krivanek *et al.*, 1977) but have not been systematically used for the examination of point defects. Larger aggregates of point defects are really small volume defects such as pre-precipitates, small voids, etc. There are, of course, a great many techniques of chemical analysis available for determining impurity concentrations and compositions. Some of these, like electron probe microanalysis, will

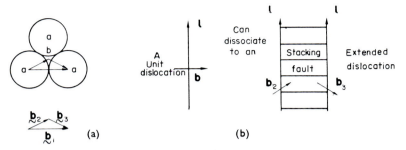

F<small>IG</small>. 15.2. Dislocation dissociation. (a) The Burgers vector $\mathbf{b}_1$ of unit dislocations, in crystals with f.c.c., diamond and sphalerite structures, corresponds to the slipping of the atoms of the crystal from one position to another of the same kind (a to a). The Burgers vectors $\mathbf{b}_2$ and $\mathbf{b}_3$ of the partial dislocations result in the shearing of atoms from positions of one type to those of another type (a to b and vice versa). (b) It is energetically favourable for a unit dislocation of Burgers vector $\mathbf{b}_1$ to dissociate into an extended dislocation consisting of two partial dislocations of Burgers vectors $\mathbf{b}_2$ and $\mathbf{b}_3$ connected by a strip of stacking fault.

also provide information on the spatial distribution of the impurities. Bands of varying concentration known as impurity growth striations are a common problem in crystal growth (Figs. 15.13 and 15.19).

15.3. Threshold Concentrations of Defects in Crystals

As pointed out above, it has been found to be a necessary preliminary to the effective study of the physical properties of a number of materials to produce them as crystals with the concentration of defects reduced below the level at which the defects affect the properties under study. It has also been found equally vital to design production processes to minimize the defect content.

This has not always been accepted. On a number of occasions suggestions have been made for utilizing defects, especially dislocations, to produce desirable characteristics. The reason that these suggestions have not led to practical products has always been that it has proved impossible to introduce desired defect structures with the necessary degree of precision and reproducibility. In contrast it has always proved possible with an adequate development effort to reduce defect concentrations below the threshold for the purpose in question. These thresholds are extremely variable. For example, practical metal components of machines are usually polycrystalline, extremely impure (parts per hundred or more of impurities) and contain 10^8 to 10^{12} dislocations cm^{-2}. Silicon for processing into integrated circuits at the other extreme must be monocrystalline, contain no electrically active impurities in concentrations greater than about one atom in 10^7, and must contain less than 10^4 dislocations cm^{-2}.

Thus the levels of defect concentration that it is necessary to detect are extremely variable. The number of types of defect that may be important is also large. Clearly, therefore, there is no one assessment technique which will do equally well in all cases. Excessive reliance on one technique will inevitably direct attention to those defects that are highly visible by that method and this may well lead to a quite distorted view of the total range of macroscopic and microscopic structural and chemical inhomogeneities in the material. Much emphasis will therefore be given here to the relative advantages of the different techniques and their spheres of application.

15.4. Methods for Detecting Structural Imperfections

15.4.1. OPTICAL METHODS

Optical microscopy was the earliest method of assessment and is still one of the most valuable. It is not particularly fashionable at present but it is relatively inexpensive, quick and easy to use, and fairly simple to interpret. There are a number of more specialized optical microscopy techniques which require rather more knowledge and experience to obtain full value which should not be overlooked. These include phase contrast for observing regions of varying refractive index in transparent materials, interference contrast, e.g. the Nomarski technique (Winterbottom and McLean, 1960; Gifkins, 1970), and multiple beam interferometry discussed by Tolansky (1948, 1955), for observing small-scale surface topographical features such as growth steps or polishing marks and polarized light microscopy for examining inhomogeneities in optically anisotropic materials and stress fields (stress birefringence or photoelasticity) in optically isotropic transparent crystals. For these techniques reference should be made to the literature on optical microscopy and metallography (e.g. Winterbottom and McLean, 1960; Gifkins, 1970.

In some transparent materials precipitates or other volume defects are visible in the as-grown crystals by transmission optical microscopy or by the use of crossed polarizers in transmission. Sometimes these precipitates are arranged in lines along the grown-in dislocations. An important technique produces the effect by deliberately introducing an impurity and heat treating the crystal so that the impurity precipitates at points along the dislocation lines. This is referred to as decorating the dislocations. This technique has been used, principally to study dislocations and grain boundaries in transparent ionic crystals, by Amelinckx and Dekeyser (1959), and to study dislocations in silicon. Copper is used to decorate the dislocations in silicon, and infrared transmission microscopy has to be employed as silicon is not optically transparent. This method was used by Dash (1957) in many classical studies of dislocations, and in the work that led to the development of methods for growing large dislocation-free crystals of both silicon and germanium (Dash, 1958) (Fig. 15.3). Like all the optical methods, the decoration of dislocations in silicon is relatively quick, cheap, and simple to use. It is, of course, destructive in that the introduction of the impurities and the heat treatment in general render the specimen examined of no further value.

The earliest and generally the cheapest and quickest of the methods for assessing structural perfection, however, is etching. This involves immersing the crystal in a specially developed reagent which dissolves the material with most rapid attack at the points at which defects intersect the surface of the crystal. This produces etch pits at the points of emergence of dislocations, grooves along the lines of intersection with the surface of grain boundaries, and stacking faults, etc. Again much classical work on dislocations was done by this means especially using a famous early dislocation etchant for germanium called CP-4 (Fig. 15.4). Etch-pit counts quoted as dislocation densities are the standard method of specifying structural perfection in the case of semiconductors.

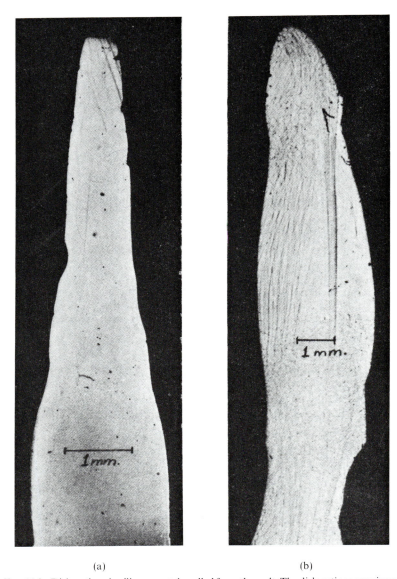

(a)   (b)

FIG. 15.3. Dislocations in silicon crystals pulled from the melt. The dislocations are viewed by transmission infrared microscopy and are decorated with copper precipitates. The micrographs illustrate the effect of different growth conditions. The crystal shown in micrograph (a) was grown under conditions that maintained a small diameter and a high temperature gradient. All the dislocations emanating from the seed crystal at the top grew out in about 5 mm. The crystal of micrograph (b) was pulled much more slowly, the diameter increased rapidly, and dislocations propagated down the crystal and further multiplied under the action of thermal stresses. Micrographs taken from Dash (1958).

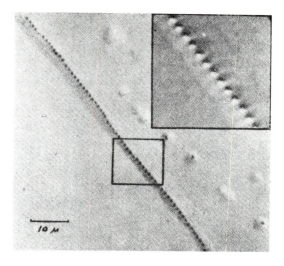

FIG. 15.4. Etch-pits at the dislocations making up a small angle subgrain boundary in germanium. By measuring the pit spacing and determining the angle of misorientation across the boundary the dislocation theory of grain boundaries was confirmed and the one-to-one correspondence of etch-pits and dislocations was checked. (After Vogel *et al.*, 1953.)

There are available many hundreds of defect etchants for all sorts of materials. Tables of these reagents will be found in Johnston (1962), Faust (1962), Warekois *et al.* (1962), Holmes (1962), and Amelinckx (1964). Of these reagents most will not be found to work immediately as described. This is because the action of the reagents depends upon the state of the surface of the crystal, the minor impurities in both the crystal and the reagent, etc. Generally a small amount of trial and error experimentation will result in the discovery of the variations in the proportions of the components in the mixture, the etching temperature, and procedure that are required for success with the material at one's disposal.

There exist theoretical treatments of etching using both kinematic approaches by Irving (1962) and Frank (1958) and electrochemical concepts by Gatos and Lavine (1964).

15.4.2. TRANSMISSION ELECTRON MICROSCOPY

Transmission electron microscopes (TEMs) have a limit of resolution, about 1.5 Å, which gives genuine atomic resolution (Allpress and Sanders, 1973; Desseaux *et al.*, 1977; Iizui *et al.*, 1977). Moreover, a TEM can very easily produce both the micrograph and the electron diffraction pattern for the same area of the specimen under identical conditions. A highly sophisticated theory of diffraction contrast has been developed whereby the information in the diffraction pattern can be used to interpret the detail in the micrograph so that most defects in the specimen can be unambiguously identified. The recently developed weak-beam technique (e.g. Loretto and Smallman, 1975) makes possible the resolution of much additional detail near the cores of defects. This makes transmission electron microscopy the final arbiter for the analysis of defects in crystals.

Of course the TEM has great drawbacks, too, which have meant in fact that it has not been much used for the assessment of crystal perfection. Firstly, TEMs are expensive (about £60,000 for a 100 kV instrument and about £600,000 for a 1 MV instrument) and complicated. Years of experience are required to develop a real mastery of specimen thinning techniques and the sophisticated operating techniques required to obtain the controlled two-beam and weak-beam conditions for optimum applications of diffraction contrast theory. The great resolving power of the TEM is also a limitation. Firstly, it is generally wasted on highly perfect crystals such as silicon. Consequently most TEM work has concentrated on metal and alloy specimens containing high densities of defects such as dislocations and very small precipitates. Secondly, because of the high resolving power, TEMs are normally employed to record micrographs at magnifications of 20,000 × or greater. Because of the small penetration of electrons, specimens must generally be only of the order of 2000 Å thick. In a single micrograph, say 5 cm × 10 cm in area, therefore, the defect content is recorded only for a volume of $5 \times 10 \times (20,000)^{-2} \times 2 \times 10^{-5}$, that is 2.5×10^{-12} cm^3. Thus there is always a serious possibility that the tiny volume examined may be grossly atypical. Nevertheless, it is essential to be aware of the potentialities of this technique for defect analysis.

Due to the limitations on the space available no attempt will be made to explain how electron optical instruments work. Clear accounts of electron optics will be found in many standard texts such as that by Klemperer and Barnett (1971). Here we shall simply accept that coils of wire carrying accurately stabilized direct currents can be made to act as electron lenses and that by varying the current the focal lengths of the lenses can be varied. An electron optical column can be made up of a number of such lenses as shown in Fig. 15.5 which constitutes a transmission electron microscope. The condenser lenses are analogous to the condenser lens of an optical transmission microscope and serve to control the electron "illumination" of the thin foil specimen. The three lenses around and below the specimen act like the objective lens and eyepiece of an optical microscope to produce either the micrograph or the diffraction pattern on the fluorescent screen or photographic emulsion.

The essential facts regarding the formation of the micrograph and diffraction pattern are illustrated in Fig. 15.6. A crystalline specimen thin enough to be transparent to the electron beam is placed inside the objective lens, above the principal plane of the lens. Bragg reflection from various atomic planes in the crystal produces a number of diffracted beams which are referred to by the Miller indices of the reflecting planes *hkl*. These beams are deflected by the lens in a manner which can be schematically represented by deflections at the principal plane of the lens. The beams are so deflected that rays of equal inclination (to the vertical axis of the microscope) intersect in the back focal plane of the lens to form the diffraction pattern. Rays from the same object point, however, intersect in the conjugate plane to form a real image. By varying the current through the intermediate lens to alter its focal length the microscope operator can focus either the diffraction pattern or the image on the viewing screen or photographic emulsion. This is done literally by turning a single knob, and this facility is one of the great advantages of the TEM.

The micrographs are produced with an objective aperture placed in the back focal plane of the objective lens so that generally only one chosen beam is transmitted down the electron optical column. If the single beam that is used to form the diffraction contrast micrograph, as it is called, is the undeflected (centre) beam, a bright-field micrograph is obtained. Imaging a diffracted beam produces a dark-field micrograph. The contrast

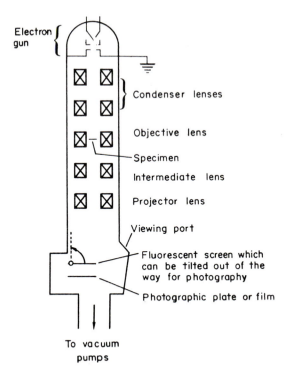

Electron gun

Condenser lenses

Objective lens

Specimen

Intermediate lens

Projector lens

Viewing port

Fluorescent screen which can be tilted out of the way for photography

Photographic plate or film

To vacuum pumps

FIG. 15.5. Vertical cross-section through the electron optical column of a transmission electron microscope. The lenses consist of horizontal coils of wire carrying d.c. currents and containing specially designed pole pieces to produce magnetic fields of the necessary forms.

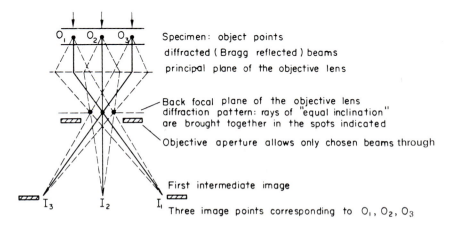

Specimen: object points

diffracted (Bragg reflected) beams

principal plane of the objective lens

Back focal plane of the objective lens diffraction pattern: rays of "equal inclination" are brought together in the spots indicated

Objective aperture allows only chosen beams through

First intermediate image

Three image points corresponding to O_1, O_2, O_3

FIG. 15.6. Formation of the diffraction pattern and image of the specimen by the objective lens in a transmission electron microscope.

(variations in intensity from point to point on the micrographs) is due to variations in the strength of excitation of the diffracted beams at the corresponding points in the specimen. These are due to variation in the orientation or structure of the material. Experimentally it is arranged by observing the diffraction pattern that the specimen is tilted so that only two beams are intense: the undiffracted beam and one reflected beam. This "two-beam" condition simplifies the interpretation and is essential to ensure that diffraction contrast theory can be applied reliably. This extensive and highly developed technique is well treated in Hirsch *et al.* (1965).

Here we will just introduce the physical ideas underlying the treatment as these will also be useful for discussing X-ray topography. The situation in a crystal upon which radiation falls at an angle that satisfies Bragg's law for some set of crystal planes is shown in Fig. 15.7. Both direct (i.e. transmitted) beams and diffracted (i.e. scattered) beams arise. Both are travelling in directions that satisfy Bragg's law and so both continue to be scattered or "Bragg reflected". Initially all the intensity is in the original direct or transmitted beam **K** but gradually more and more of the intensity is scattered into the diffracted beam **K′**. After penetrating a certain depth $\frac{1}{2}\xi_g$ into the crystal all the intensity has been scattered into the diffracted beam **K′**. ξ_g is known as the extinction distance. The intensity is then gradually scattered back into the transmitted beam **K** until after a further depth $\frac{1}{2}\xi_g$, **K** has all the intensity and **K′** has been reduced to zero. Under two-beam conditions this transfer of intensity between the two beams continues while absorption and inelastic scattering effects reduce the total intensity in the beams. The proportion of the total transmitted intensity that is in the transmitted and in the diffracted beam depends on the thickness of the crystal relative to the extinction distance for the particular Bragg planes. The bright-field diffraction contrast micrograph is simply a reproduction of the intensity of the direct beam as it emerges from the various points on the lower surface of the specimen. The dark-field diffraction contrast micrograph is a similar picture of the varying intensity of the diffracted beam as it emerges from the bottom of the crystal. The general level of intensity of both micrographs, therefore, is determined by the thickness of the specimens in relation

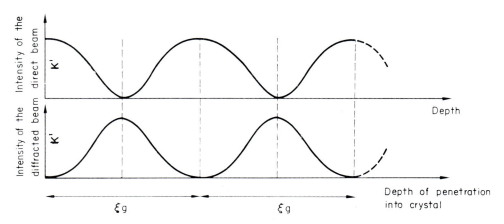

FIG. 15.7. As a result of Bragg scattering, at the exact Bragg condition intensity is scattered back and forth between the direct and the diffracted beams with a period ξ_g known as the extinction distance.

to ξ_g. For a thickness that is an odd half-integral number of extinction distances, for example, all the intensity would occur in the diffracted beam and the general brightness of the dark-field micrograph would be high.

The visibility of defects can be interpreted from this point of view as follows. The elastic stress field round any defect may result in the bending of the Bragg reflecting planes *hkl*. If this happens either the planes will move towards or away from the exact Bragg orientation. Thus the strength of the scattering will increase or decrease in the vicinity of the defect as indicated, for example, in the case of an edge dislocation in Fig. 15.8. The dislocation therefore produces stronger scattering and consequently a shorter ξ_g on one side than on the other and a local change in the intensity of both the diffracted and transmitted beams results. The dislocation therefore appears as a line of changed brightness on both the bright- and dark-field micrographs. This effect only occurs if the strain field of the defect bends the planes that have been tilted to reflect under the chosen two-beam conditions. This is expressed algebraically by means of the important rule that there is no contrast if

$$\mathbf{g} \cdot \mathbf{b} = 0,$$

where $\mathbf{g}$ is the reciprocal lattice vector normal to the reflecting *hkl* planes and $\mathbf{b}$ is the Burgers vector of the dislocation. Thus by tilting to find several reflections $\mathbf{g}$ in which the dislocation is invisible it is possible to determine $\mathbf{b}$. In fact the above rule applies exactly only to screw dislocations and slightly modified rules apply for edge and partial dislocations, stacking faults, etc. However, the precise procedures required for determining the nature of a great variety of types of defect have now been worked out and

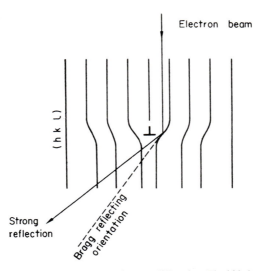

FIG. 15.8. Effect of an edge dislocation on electron diffraction. The *hkl* planes are distorted as shown near the dislocation line. On one side the planes are rotated towards the exact Bragg orientation and strong reflection occurs. On the other side of the dislocation line the planes are rotated even further away from this orientation than are the undistorted planes. Thus, enhanced diffraction takes place on one side of the dislocation line giving rise to the diffraction contrast effects that make the dislocation visible in transmission electron microscopy.

can be routinely applied (Hirsch *et al.*, 1965; Lorretto and Smallman, 1975; Edington, 1974, 1975a, b; Amelinckx *et al.*, 1978).

The extinction distances for electron diffraction are of the order of hundreds of angstroms so that the specimens used in TEMs are many ξ_g thick. The repeated diffraction of electrons back and forth between the direct and diffracted beams therefore is always a characteristic feature of electron diffraction. The "dynamical theory of diffraction", which takes this into account, must thus be used. When scattering is very weak, as in the case of X-ray diffraction so that the thickness of the crystal may be very much less than ξ_g, the direct beam is strong and the diffracted beam is weak. The transfer of intensity back and forth between the beams can then be ignored and the mathematically simpler "kinematic theory of diffraction" can be used.

The penetration of radiation through relatively thick crystals near the conditions for Bragg reflection can also be described in a second way that is more useful mathematically. The two descriptions are in the end equivalent of course. The electrons are actually travelling in a net direction given by the vector sum of **K** and **K**′, that is, parallel to the reflecting planes. The incident electrons can be divided into two groups represented by two types of Bloch waves of different symmetry. In other words there occur two types of wave field $\psi^{(1)}$ and $\psi^{(2)}$. These vary periodically in the direction normal to the reflecting planes as shown in Fig. 15.9. Thus wave field 2 has its maxima at the atomic planes. That is the electrons represented by $\psi^{(2)}$ spend most of their time near the atoms in the reflecting planes and are strongly scattered. Consequently these electrons do not penetrate far

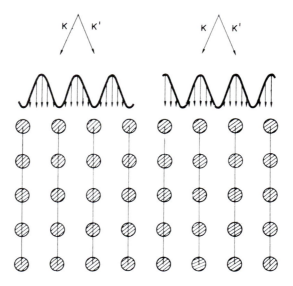

FIG. 15.9. The two forms of wave field that occur for orientations near the exact Bragg reflecting position for the vertical planes of atoms in a simple cubic lattice. Both types of wave field represent electrons with current flow vector parallel to the reflecting planes as indicated by the arrows drawn inside the waves at the top of the diagram. The absorbing regions at the atoms are shaded. The type 2 wave is absorbed more rapidly than the type 1 wave. Diagram taken from Hashimoto *et al.* (1962), and reprinted by permission of the Royal Society.

through the crystal. Conversely wave field 1 has its maxima between the reflecting planes of atoms, that is the electrons spend most of their time between the planes of atoms. Consequently these electrons are less strongly scattered by the atoms and therefore are anomalously (strongly) transmitted. It is this phenomenon of anomalous transmission which occurs near the exact Bragg condition which makes it possible for transmission electron microscopy to be employed at all successfully. Anomalous transmission is essentially the same phenomenon as that known as channelling in the case of irradiation with sub-atomic particles like protons or α-particles. In the case of X-rays it is known as the Borrmann effect.

When absorption effects are taken into account a pair of differential equations known as the Howie–Whelan equations are obtained for the rate of change of the amplitude of the direct and diffracted wave functions with depth. These cannot be solved analytically. Computers are therefore used to obtain numerical solutions for particular defects and operating (diffraction) conditions. These are generally presented as graphs showing the intensity profiles that are predicted on crossing the image of a particular defect in a particular type of micrograph (bright- or dark-field for a particular reflection *khl*). It is by comparing observed contrast effects in TEM micrographs with these predicted profiles for particular types of defects that detailed identifications of defects are carried out. An assumption which is made and which is necessary to make the calculations tractable is the so-called column approximation. This is the assumption that the correct result may be obtained by integrating the Howie–Whelan equations only down narrow cylindrical

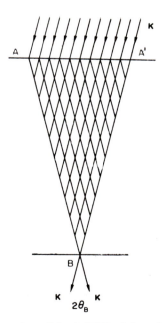

FIG. 15.10. Schematic diagram from Takagi (1962) of the volume of a crystal within which single or multiple Bragg reflections through the angle θ_B can lead to emergence through the point B. Because θ_B is small for electron diffraction the region ABA' is small and the column approximation can be used. In X-ray topography, the Bragg angles θ_B are much larger and ABA' does not approximate to a narrow column.

columns running from top to bottom of the specimen. This is justified for electron diffraction since the Bragg angles for electron diffraction are small. Bragg's law states that $2d \sin \theta_B = n\lambda$ where d is the spacing of the reflecting planes, θ_B the Bragg angle (of incidence of the electron beam on the plane), n is an integer and λ the wavelength. Taking $\lambda = 0.037\,\text{Å}$ for 100 kV electrons, and using the values for first-order ($n = 1$) reflections from the (111) planes of silicon makes $\theta = 20' = \frac{1}{3}°$, which is a typical value. Figure 15.10 shows that because of the small values of θ_B in the electron diffraction case, the volume of the crystal within which Bragg reflections can take place which will contribute to the strength of the beam emerging from a given point on the bottom surface is small. For X-rays, λ is much larger than for the electrons in TEMs. Consequently the Bragg angles for X-ray diffraction are much larger and the volume of crystal contributing to the intensity at a point on the crystal exit surface is large and the column approximation does not apply.

15.4.3. X-RAY TOPOGRAPHY

There are a number of different techniques of X-ray topography which can be classified according to whether the specimen is examined in transmission or reflection and whether the specimen is thick or thin. Only transmission X-ray topography will be considered here as this includes the most widely used current techniques. Complete lists of all the X-ray topography techniques will be found in Lang (1974) and Tanner (1976, 1977).

The second distinction is made on the basis of whether the specimen thickness t is such the $\mu t \lesssim 1$ or $\mu t \gg 1$, say $\gtrsim 10$, where μ is the linear absorption coefficient for the monochromatic X-rays that are used. Two different contrast mechanisms predominate in the two thickness regions. There are two main practical X-ray topographic techniques, one operating in each of these regions, and we shall discuss each in turn.

The Lang technique. Probably the most widely used topographic technique is that shown in Fig. 15.11. A line-by-line map of the intensity in the chosen beam, direct or diffracted, is recorded photographically as the crystal and the film are translated together. A micrograph is thereby recorded on a one-to-one scale. "Magnification" is obtained by photographically enlarging in printing. By using fine-grained emulsions optimum results are obtained and enlargements up to 50 × can be employed. The limit of resolution is a few microns, that is it is similar to those of optical techniques such as etching.

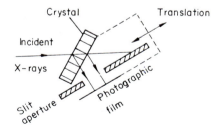

FIG. 15.11. The Lang X-ray topographic technique. The crystal is set at the exact Bragg angle for a particular set of planes *hkl* to reflect. The film is set at right angles to the transmitted or reflected beam, whichever is selected by the aperture (reflected beam or dark-field topography is illustrated here). The film and the crystal are translated together in order to record the whole area line by line.

The specimen thickness and the X-ray wavelength are chosen so that the relation $\mu t \lesssim 1$ is satisfied and the main contrast mechanism is then that known as extinction contrast. This can be discussed in a similar way to that used initially in describing diffraction contrast in transmission electron microscopy. In the present case, as before, those defects that disturb the reflecting *hkl* planes alter the diffraction conditions and therefore change the intensity of both the transmitted and the reflected beams locally. That is, in Lang topography the strain field around imperfections reduces the primary extinction of X-rays which occurs in nearly perfect crystals. Primary extinction is the phenomenon which is responsible for the fact that the integrated intensity of reflection by a perfect crystal is proportional to the structure factor, whereas the integrated intensity reflected by an imperfect crystal is proportional to the square of the structure factor. Thus defects in otherwise perfect crystals cause localized *increases* in the reflected intensity. When the diffracted beam is used to record the topograph, therefore, defects appear as regions of greater intensity than the perfect crystal background under conditions of extinction contrast.

The condition that no contrast arises for a dislocation when $\mathbf{g} \cdot \mathbf{b} = 0$, where $\mathbf{g}$ is the reciprocal lattice vector normal to the *hkl* planes for the operative reflection and $\mathbf{b}$ is the Burgers vector of the dislocation, applies just as for the TEM. The identification of defects by observing the contrast in a number of reflections and finding several non-coplanar $\mathbf{g}$ vectors for which $\mathbf{g} \cdot \mathbf{b} = 0$ is also possible in X-ray topography therefore. However, the interpretive theory is much less complete in the X-ray case. Nevertheless, the fact that the Lang technique is non-destructive, as are the other methods of X-ray topography, makes the method attractive and Lang X-ray topographic equipment is commercially available from several manufacturers. The technique has been used successfully not only to study grown-in arrays of dislocations in crystals, as, for example, in Fig. 15.12, but also to observe impurity growth striations as shown in Fig. 15.13.

FIG. 15.12. X-ray Lang topograph of a natural diamond of width about 4 mm taken from Lang (1974). This configuration of dislocations radiating out from a central nucleus is commonly observed in equiaxed crystals of many materials.

A practical limitation of the Lang technique is that a number of materials have such large values of absorption coefficient for readily available X-ray lines that t must be made small to satisfy the condition $\mu t \lesssim 1$. Consequently in many cases the internal stresses in the large thin crystals or the stresses introduced in mounting the slices cut for examination in the topographic camera suffice to change the orientation of the crystals so that only in a few narrow areas will it be in the Bragg reflecting condition. Hence only a few local areas will be recorded. Schwuttke (1965) therefore developed a scanning oscillation translation technique (SOT). In this method, while the crystal is being scanned by translation, both the film and the crystal are also oscillated simultaneously about the normal to the plane containing the incident and reflected beams, i.e. about the normal to the plane of the diagram of Fig. 15.11. The angle of oscillation is made large enough to cover the whole reflecting range of the crystal. In this way the entire slice can be recorded on one exposure.

Anomalous transmission topography. The phenomenon of anomalous transmission of X-rays occurs in nearly perfect crystals and is known as the Borrmann effect, after its discoverer. The mechanism of anomalous transmission in the X-ray case is basically similar to that in the electron transmission case as outlined in Fig. 15.9. However, the situation is more complicated in the X-ray case in that a total of four wave fields have to be included as both the anomalously transmitted and the strongly absorbed waves occur in two states of polarization. The dynamical theory of X-ray diffraction (James, 1963; Batterman and Cole, 1964; Penning, 1966) has to be used in the case in which anomalous transmission predominates, i.e. when $\mu t \gg 1$ and this theory is complicated. However, the physical situation is relatively simple. When $\mu t \gtrsim 10$, say, the only reason that any intensity penetrates through the specimen is that the anomalously transmitted wave fields are excited and pass through the crystal parallel to the reflecting planes and mainly propagating between the planes of atoms. Defects which set up elastic strains or otherwise

FIG. 15.13. Lang X-ray topograph of a longitudinal section of a silicon crystal grown by the Czochralski technique. The vertical pull direction was [110]. Growth was from a silica crucible and the horizontal striations are due to oxygen segregation. Topograph taken from Schwuttke (1962), and reproduced by permission of the Metallurgical Society of the AIME.

alter the scattering characteristics of the atomic planes locally will disrupt this anomalous transmission of the wave field. Hence such defects will "cast shadows" and appear as regions of locally *reduced* intensity in *both* the topograph recorded with the diffracted beam and that recorded using the direct transmitted beam. This effect is known as anomalous transmission contrast and the images are described as dynamical.

Schwuttke (1962a) developed a very simple apparatus for recording anomalous transmission topographs as shown in Fig. 15.14. This method is particularly useful for materials, such as GaAs, which have large absorption coefficients for X-rays of the readily available wavelengths.

Advantages and disadvantages of X-ray topography. By choice of the appropriate technique as-grown crystals of any material that can be grown with high perfection can be examined non-destructively and *in toto.* The apparatus is relatively inexpensive (thousands of pounds sterling) and a great many types of inhomogeneity such as domain structures and impurity striations can be seen as well as crystallographic structure defects such as dislocations or grain boundaries. These advantages are so great as to outweigh the disadvantages of the method.

The chief drawbacks of X-ray topography are that the theory for the interpretation of details of contrast in the topographs is more complex and less fully developed than in the TEM case, see Authier (1972), and that the limit of resolution (several microns) is much greater than for the TEM. At present, too, exposure times for recording either Lang X-ray topographs or anomalous transmission topographs are long, of the order of several hours, so that only one or two can be taken per day.

However, it has been shown that X-ray sensitive television camera tubes and closed circuit television systems can be used to give an instantaneous television display of X-ray topographs with fair resolution (better than 15μ), by Rozgonyi *et al.* (1970). When this is done electronic methods of magnification, contrast enhancement, etc., can also be employed.

Because of its applicability to the whole bulk crystal as grown, its non-destructiveness and its acceptable levels of resolution and detailed interpretability, X-ray topography is undoubtedly the most valuable single assessment technique for crystal growers. A number of recent reviews of X-ray topography are available (Tanner, 1976, 1977).

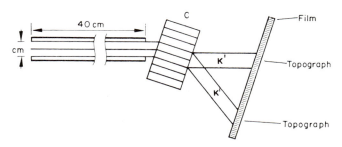

FIG. 15.14. The method of Schwuttke (1962a) for anomalous transmission X-ray topography. A long tube (40 cm long, 1 cm inside diameter) is used as a collimator. The crystal is set at the Bragg angle for one set of planes and both the direct and diffracted beam topographs of the illuminated area are recorded.

15.4.4. SCANNING ELECTRON MICROSCOPY

Scanning electron microscopes provide a whole family of techniques. Because of the physical properties of the specimen which give rise to contrast in the micrographs these are particularly valuable for the study of semiconducting materials in particular. The essential features of scanning electron microscopes are shown in block diagrammatic form in Fig. 15.15. The electron optical column contains an electron gun to produce the necessary electron beam. Two or three demagnifying magnetic lenses act to produce a reduced image of the electron source on the specimen surface. This is an electron probe that can be made as small as 70 Å in diameter in current commercial instruments. Scanning coils are placed in the bore of the objective lens, and these deflect the beam so that it scans line by line a square television type raster over the specimen surface. The energy of the beam is dissipated into several other forms of energy as will be discussed below and any one of these can be detected and turned into an electrical signal by a suitable transducer. The electrical signal so produced is amplified and displayed as video signal on a CRO scanned in synchronism with the scanning of the beam over the specimen.

There is thus a one-to-one correspondence between picture points on the CRO screen and points on the specimen. Variations in brightness on the CRO screen, i.e. contrast on the SEM micrographs, arise from variations in the strength of excitation of the physical signal being detected. The magnification of the micrograph is $M = L/l$, where L is the side of the square area scanned on the CRO display screen and l is the side of the area scanned on the specimen. The latter quantity is varied by tapping a variable fraction of the voltage from the scan generator as indicated in Fig. 15.15. Scanning electron microscopes

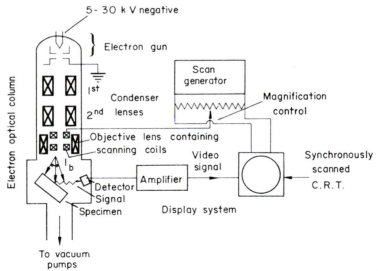

FIG. 15.15. Schematic diagram of the components of scanning electron microscopes. The instrument consists basically of two systems. Firstly, there is an electron column which produces an electron beam of small diameter which is scanned in a square television-type raster over the surface of the specimen. Secondly, there is a display system which in the simplest case as shown here presents the information in the signal from the specimen as a video signal, i.e. as variations in the brightness of the picture points on a television-like micrograph on a cathode-ray tube screen that is scanned in synchronism with the scanning of the beam over the specimen surface.

therefore have magnifications that are continuously variable over a wide range, usually from $20 \times$ to $80,000 \times$.

The five modes of scanning electron microscopy. The forms of energy emitted by solid specimens under electron bombardment and which are used in scanning electron microscopy are shown in Fig. 15.16. These five forms of energy constitute the signals underlying the five modes of scanning electron microscopy. We will outline each of these in the chronological order in which they have been exploited in commercial instruments.

The emitted X-ray photons that are of interest are those belonging to the sharp emission lines known as characteristic X-rays. The wavelength and energy of such rays are characteristic of the chemical element of the emitting atom. By using either X-ray spectrometers or lithium-drifted silicon or germanium detectors together with pulse height analysers, a particular line characteristic of a particular element may be detected. The intensity of such a line is a measure of the concentration of the element in the bombarded volume of the specimen. This mode of operation is known as electron probe microanalysis. The impurity concentrations in as-grown, non-metallic materials are often well below the X-ray detection limit which is one part in 10^4 or more. No further attention will be given to this technique here. There is a large literature on electron probe microanalysis with reference to the use of special purpose scanning electron beam instruments fitted with wavelength dispersive X-ray spectrometers (Tousmis and Marton, 1969; Reed, 1975; Long, 1977). The principles of X-ray microanalysis using scanning electron microscopes are dealt with by Belk (1979), Gedke (1974), Lifshin (1974), and Goldstein and Yakowitz (1975).

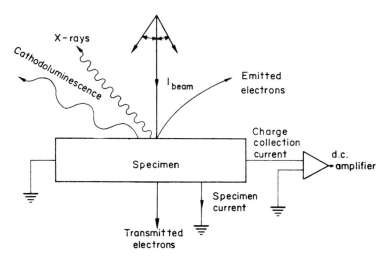

FIG. 15.16. The types of energy dissipation in electron bombarded crystals which are employed as signals in the five modes of scanning electron microscopy. The characteristic X-rays are used in the X-ray mode, i.e. in microanalysis. The emitted electrons are used in the emissive mode. The specimen current or the charge collection currents are used in the conductive mode. Cathodoluminescent light is used in the luminescent mode. Transmitted electrons which occur only if the specimen is thin (less than about 1 μ in thickness) are used in scanning transmission electron microscopy (STEM).

The emitted electrons provide the signals employed in the emissive mode of operation of the SEM. By using velocity analysing detectors various peaks of the electron emission spectrum may be selected and different types of contrast observed, giving information on alternative physical properties.

The electron beam induces two distinct sorts of currents in semiconductors. The first is the leakage current from the beam to earth via the specimen which is known as the specimen current or absorbed electron current (in distinction to the emitted current). The other type of conductive signal is described as a charge collection current in the terminology of semiconductor particle detectors. Again the two sorts of signal contain different types of physical information.

The cathodoluminescence can be spectrally analysed to obtain information regarding the different sorts of activator (radiative recombination centre) impurities present or about the different crystal structures in the specimen. Micrographic display of the signal strength variations with the detector (monochromator or filters and photomultiplier) set for an emission peak characteristic of a particular impurity in the case of strongly emitting materials gives a detection sensitivity far higher than does the X-ray mode in general (Wittry, 1966).

In the case of relatively thin specimens, transmitted electrons can be detected. This type of signal will contain the same sort of information regarding structural defects as is obtained in diffraction contrast transmission electron micrographs. After detection the signal can readily be amplified and signal processing and velocity analysing detectors are readily incorporated in scanning transmission electron microscopy (STEM). Moreover the information obtainable in the STEM mode can immediately be inter-compared with the information obtainable in the other modes of scanning electron microscopy.

The STEM mode has not been used in commercially available "conventional" scanning electron microscopes, that is those providing a high vacuum (10^{-5} torr) in the electron optical column and employing thermionically emitting tungsten filaments in the electron gun. The reason is that the resolution obtainable in this way is much lower than in transmission electron microscopes (50Å as against 1.5Å). To understand this and the method adopted to overcome this limitation we have next to consider the factors that govern the resolution of scanning electron microscopes.

In the case of scanning electron microscopy the resolution is defined as the diameter of the minimum area of the specimen surface from which a signal can be obtained that is strong enough to satisfy the signal/noise ratio requirements of the detector–amplifier–display system of the microscope. The resolution is therefore determined by two factors, the signal strength requirement of the display channel and the electron optical design considerations which determine the minimum beam diameter d_{min} into which can be concentrated the current necessary to give the required strength of signal.

The beam current is reduced when the beam diameter is reduced because to reduce the beam diameter the currents through the lenses have to be increased to increase the demagnification. The lens aberrations, together with diffraction effects, then combine so that a greater fraction of the electrons emitted by the gun are stopped by the apertures in the column, which apertures are themselves largely necessary because of the severe spherical aberrations of magnetic electron lenses. Present electron optical technology makes it possible to get the current necessary for emissive mode signals of acceptable strength, about 10^{-11} A, into a spot of about 50Å diameter in conventional SEMs. This is the basis of manufacturers' quoted resolutions.

Analysis of the electron optical design principles involved shows that the resolution of scanning electron microscopes can be greatly improved only by employing brighter electron sources, that is electron emitters with higher current density outputs. Only one source gives orders of magnitude higher emission current densities than does the thermionic tungsten emitter. This is the field emission tip, which provides emission current densities up to 10^6 times those of the thermionic filament. By using a field emission electron gun Crewe *et al.* (1968a and b) succeeded in reducing the resolution of scanning electron microscopes from the 150 Å or so of current commercial instruments to a few angstroms and they published micrographs in 1970 showing thorium atoms as bright blobs on a dark background. To be able to use a field emission source it is necessary for the electron optical column of the microscope to be pumped down to an ultra-high vacuum pressure of 10^{-9} torr or better. Crewe (1974) has designed a high voltage STEM instrument that is intended to give a resolution of 1 Å.

In order to make use of the smaller electron beam diameter d_{min} that is made possible by the use of a field emission source, it is necessary to work in either the emissive or transmissive modes of scanning electron microscopy. The reason is that due to collisions the electrons of the beam spread out as they penetrate into solid specimens. The energy of the beam is then dissipated in a volume $1-2\,\mu$ in diameter. Bulk signals, that is X-rays, cathodoluminescence and conductive mode signals (electrical currents), can be detected from the whole energy dissipation volume and the resolution in these modes is therefore limited by beam spreading. This is not a great problem in the emissive mode when the low energy (50 eV) secondary electrons are detected because these electrons cannot escape if generated more than about 100 Å below the surface. In the case of the transmissive mode, the specimen must be relatively thin. Therefore, again, beam spreading is limited.

The emissive mode (Gopinath, 1974). The emitted electron spectrum consists of a number of components. The numerous electrons with energies below about 50 eV are called secondaries. The electrons in the peak which occurs at the energy of the incident beam are called primaries or "back-scattered" electrons.

When the secondary electrons are detected and used to form the video signal, micrographs are obtained in which contrast arises mainly or entirely from the specimen surface topography. These micrographs have three advantages as follows. Because of the high resolution of the scanning electron microscope the maximum useful magnification is much greater than in, for example, optical microscopy, and because the magnification is altered by tapping a variable fraction of a voltage the magnification is continuously variable. Secondly, because the angular convergence of the electron beam is extremely small the spot size d_{min} remains constant despite large variations in the height of the specimen surface. In other words, the surface remains in focus and the depth of focus is large. Thus, three-dimensional objects can be examined at high magnification in a way that is impossible in any other instrument. Thirdly, because the dependence of the number of emitted secondary electrons on the angle of incidence of the electron beam on the surface is, fortunately, the same as the dependence of scattered light intensity on the scattering angle, secondary electron SEM micrographs look just like optical views of the object so that interpretation presents no new difficulties. It was because of those advantages of this mode of microscopy that scanning electron microscopes were first developed. Secondary electron emissive mode microscopy is of some value for the examination of crystal growth morphologies as shown in Fig. 15.17.

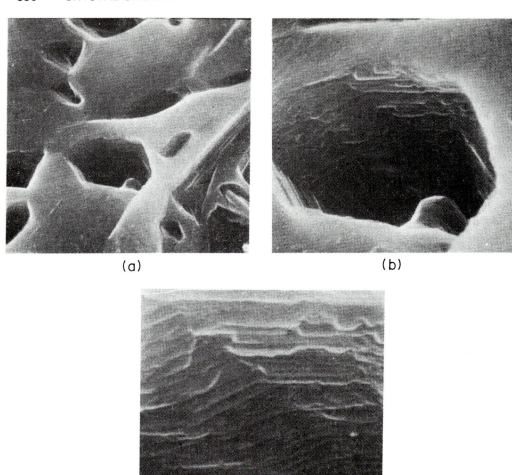

(a)   (b)

(c)

FIG. 15.17. Scanning electron microscope secondary electron micrographs illustrating the exploitation of the great depth of focus available at high magnifications in the study of growth morphology. A hole in a graphite crystal grown in cast iron is shown at three successively larger magnifications. The linear magnification in (c) is approximately 14,000 ×. Micrograph taken from Minkoff and Nixon (1966).

The number of high-energy electrons produced by back-scattering rises monotonically with increasing atomic number. Therefore, a form of contrast arising from variations in the composition of crystals is seen when the primary electrons are detected, amplified, and displayed as video signal. The material in areas which appear bright has a larger average atomic number than the material in areas which appear darker in such primary or back-scattered electron SEM micrographs.

The conductive mode (Holt, 1974a). There are three distinguishable situations included in the conductive mode. Firstly, there is that of the specimen current in which the video signal is the current flowing to earth from the incident electron beam via the specimen. This is sometimes referred to as the absorbed electron current. By Kirchhoff, therefore, $I_{beam} = I_{emitted} + I_{absorbed} =$ constant. As $I_{absorbed} = C - I_{emitted}$, the contrast in the absorbed electron or specimen current micrographs in most cases is simply the reverse of that in the secondary electron micrographs of the same area.

The second situation that can be distinguished within the conductive mode is that in which the video signal is a current flowing in a closed loop between two contacts on either side of the specimen with the beam incident on the specimen between the contacts and with no source of e.m.f. in the circuit external to the specimen. Currents only flow in such a case if electron bombardment generates potential differences between portions of the specimens. The mechanisms whereby this happens are known as electron voltaic effects. They are analogous to the photovoltaic effects whereby light generates potential differences in semiconductors.

The third situation is that in which the video signal is a current flowing in a closed loop including the specimen and an external e.m.f. In addition to any signals arising from possible electron voltaic phenomena that may occur in the specimen, signals due to the alteration of the electrical conductivity of the specimen by electron bombardment will be observed. This is referred to as "β-conductivity". The general term for micrography in the conductive mode using two contacts to the specimen is charge collection. With a single earthed connection, it is referred to as specimen current micrography. No significant conductive mode contrast effects are observed in metals. Conductive mode scanning electron microscopy is principally of interest, of course, in assessing the uniformity of electronic materials.

The barrier electron voltaic effect. The strongest contrast in the conductive mode charge collection signal micrographs is that due to the barrier electron voltaic effect. This is the generation of a potential difference between the portions of material on either side of an electrical barrier by the separation by the field at the barrier of holes and electrons generated by the incident electron beam. The result is that a net negative charge accumulates on one side and a net positive charge on the other side of the barrier. These charges then disappear via the current through the external circuit as illustrated in Fig. 15.18 for the case of a p–n junction. Any other form of barrier such as a Schottky diode rectifying contact or a hetero-junction is similar to a p–n junction in that it is a high resistivity interface across which there is a built-in field. Hence, at any form of electrical barrier hole–electron pairs generated by electron bombardment will be separated and, as a consequence, a current will flow in the external circuit to allow the net negative and positive charges to disappear. This is often called EBIC (electron beam induced current microscopy).

The barrier electron voltaic effect has been employed in two geometries to observe p–n junctions as indicated in Fig. 15.18c and d. Diffusion-induced misfit dislocation networks were seen by Lander *et al.* (1963) and Czaja (1966) in dark contrast on a bright background in plan view observations on silicon p–n junctions. Edge-on viewing of GaAs laser junctions showed the presence of a variety of defects in the material (Holt and Chase, 1968; Donalato, 1979; Fathy *et al.*, 1980; Petroff *et al.*, 1980).

The bulk electron voltaic effect and β-conductivity. The bulk electron voltaic effect is the generation of varying voltages across a semiconducting specimen of non-uniform

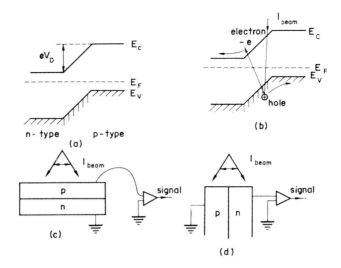

Fig. 15.18. Barrier electron voltaic effect at a p–n junction. (a) Energy band diagram of a p–n junction. E_c is the bottom of the conduction band, E_v the top of the valence band, E_F is the Fermi level. V_D, the diffusion voltage, results in the built-in electric field of the junction. (b) When electron–hole pairs are produced by electron bombardment at or near the junction the hole and electron will be separated by the field of the junction. (c) The flow of current round the external circuit (usually via earth connections as shown) to neutralize the separated positive and negative charges acts as charge collection signal. This is the plan view geometry as originally used by Lander *et al.* (1963) and Czaja (1966). (d) Edge-on geometry is also used, originally by Wittry and Kyser (1965).

resistivity by scanning electron bombardment. Munakata (1967) showed that by integrating the bulk electron voltaic effect signal ΔV the local resistivity ρ can be obtained and resistivity variations observed. The bulk electron voltaic effect produces very small voltages ΔV when the diffusion length for minority carriers is short. Consequently, it is not useful for the observations of non-uniformities in the resistivity, that is in the impurity content, of materials with short minority carrier diffusion lengths.

Munakata (1968a, 1968b) has developed a second method which is usable in the case of short minority carrier diffusion length crystals. This is the technique that he calls "β-conductivity". In this method a d.c. biasing voltage is applied to the specimen. The electron bombardment of the specimen increases the number of free charge carriers and, therefore, increases the specimen conductance in the bombarded region. Therefore, the voltage across the specimen decreases if the bias is supplied by a constant current source, or alternatively an additional current flows if a constant voltage source is used. The voltage decrease or the current increase is then used as the β-conductivity signal. Munakata used a constant current source and measured the voltage change. The β-conductivity is constant but its effect is position dependent because the voltage decrement is dependent on the relative magnitudes of the electron beam induced conductivity and the original conductivity. The contribution of the β-conductivity proper to the voltage decrement has to be separated from the contributions due to the bulk electron voltaic effect and that due to the specimen current. Munakata (1968a, 1968b) derived and experimentally verified formulae for the β-conductivity signal and showed it to be proportional to the square of the local electric field strength in the material, and hence to the square of the local resistivity if the biasing current density is constant. Munakata applied the method to obtain the conductivity variations in as-grown bars of germanium and CdS as shown in Fig. 15.19.

The luminescent mode (Muir and Grant, 1974; Holt, 1974b). Many materials have the property that under electron bombardment they emit light. This is known as cathodoluminescence. This light can be detected by means of a photomultiplier to provide a signal for display to produce the luminescent mode of scanning electron microscopy. Optical filters or monochromators can be used to select a particular line or peak of the emission spectrum for detection as luminescent mode signals.

Cathodoluminescence may be produced by two main types of mechanisms in crystals. The first type of mechanism involves the transfer of energy from the electron beam to "luminescent centres" which are generally impurity atoms or point defect aggregates including impurity atoms. These are raised into an excited state from which they relax by emitting a photon. The second way in which light may be produced is intrinsic to the crystal and involves raising an electron from a lower energy band to a higher one, generally from the filled valence band to the empty conduction band. The electron then drops back into the hole in the lower band emitting a photon. This is known as recombination radiation.

Obviously if the cathodoluminescent peak that is used as luminescent mode radiation arises from impurity atoms the SEM micrographs will be bright at points where the concentration of that impurity is high and dark at points where the impurity concentration is low. Observations of this kind have been made on impurity segregation

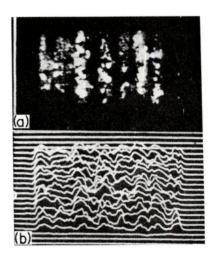

FIG. 15.19. Conductive mode scanning electron miscroscope observations of grown-in resistivity striations in a bar of n-type germanium. (a) is a picture in which the β-conductive signal variations in voltage are used to modulate the brightness of the CRT screen. (b) is a "Y-modulation" micrograph in which the square root of the voltage variation (which is proportional to the field and the resistivity) is used to deflect the electron beam vertically during the line scans. Taken from Munakata (1969) and reproduced by permission of the Institute of Physics.

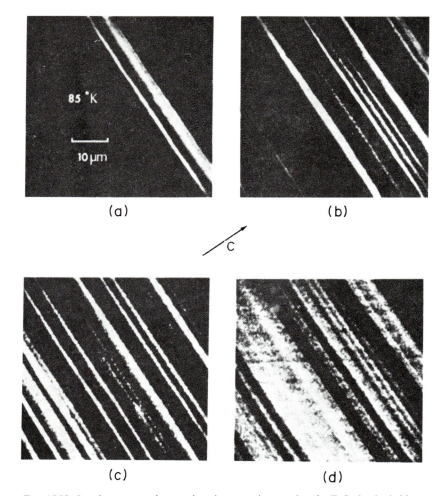

Fɪɢ. 15.20. Luminescent mode scanning electron micrographs of a ZnS platelet held at 80°K. The emission spectrum near 3300 Å contains peaks due to bound exciton transitions in the 2H (hexagonal wurtzite-structure) modification, the 3C (cubic sphalerite-structure) modification, and two polytypes intermediate between 2H and 3C. Scanning micrographs recorded at each of these wavelengths, using a monochromator in the SEM display system, are shown in (a), (b), (c), and (d), respectively. Emission from each of the polytypes is located in faulted bands which are perpendicular to the *c* axis of the crystal. This is from A. D. Yoffe and P. M. Williams (unpublished).

to dislocations in as-grown tellurium-doped GaAs by Kyser and Wittry (1964) and Shaw and Thornton (1968) and on impurity growth striations in tellurium-doped GaAs by Wittry (1966). Casey and Kaiser (1967) showed that several parameters obtained by measurements on the main cathodoluminescent emission peak, including the photon energy of the peak and the half-width of the peak, could be used to measure the impurity concentrations in n-type GaAs.

If the cathodoluminescent emission line or peak that is detected as luminescent mode signal arises from a transition that is characteristic of a particular crystal structure, the

SEM micrographs will be bright only in those regions of the material that have that specific structure. In the case of ZnS, examined by Williams and Yoffe (1969), and of ZnSe, by Williams and Yoffe (1968), it was shown that by selecting ultraviolet emission lines that arose only from sphalerite-structure material or only from wurtzite-structure material it was possible to identify the crystal structures of different regions in striated vapour-phase grown platelet crystals, as shown in Fig. 15.20. Moreover, it was shown that stacking faults and microtwins could be observed by this means. The recent literature on the CL mode is covered by Bröcher and Pfefferkorn (1978) and Holt and Datta (1980).

The transmissive mode (STEM) (Howie, 1974; Crewe, 1974). Transmissive mode scanning electron microscopy is usually referred to as "STEM", an acronym formed of the initial letters of the phrase: scanning transmission electron microscopy. The micrographs so produced closely resemble those produced by transmission electron microscopes. The comparative uses of STEM and TEM are now being explored (e.g. Sparrow and Valdre, 1977). The great advantage of STEM microscopy will then be that it is available together with the other modes of microscopy of the SEM. Intercomparison of the information regarding structural defects in STEM micrographs with the information on electrical and optical property inhomogeneities and on surface topography from the other modes will make the SEM an even more powerful tool for the assessment of the structural perfection and physical and chemical homogeneity of crystals. (See, for example, Kimerling (1978), Pennycook *et al.* (1980), Fathy *et al.* (1980), and Petroff *et al.* (1980).)

Hybrid transmission and scanning electron microscopes. The advantages of combining the information from several types of observation have led to the introduction of hybrid TEM-based instruments. These offer TEM, STEM, X-ray microanalysis, and emissive-mode SEM operation in a combined instrument. They have only recently become available from several different manufacturers. Their obvious combined attractions have resulted in their widespread adoption despite their price of about £120,000.

Suggestions for further reading: Light Microscopy. Southworth (1975) provides a brief introductory account. Gifkins (1970) is a completely non-mathematical account of the physical principles of these techniques. *Transmission Electron Microscopy* (Hirsch *et al.*, 1965) is the standard monograph. Amelinckx *et al.* (1978) contains recent summaries by a number of experts. Loretto and Smallman (1975) is a good brief textbook account of the techniques for new users. Edington (1974, 1975a, 1975b) give full experimental details of the techniques of the TEM. *X-ray Topography* (Tanner, 1976) has at last provided a complete account in book form. *Scanning Electron Microscopy* (Oatley, 1972) covers the design principles of the instrument. The standard monographs are Holt *et al.* (1974), which deals with all modes of operation, Wells (1974), and Goldstein and Yakowitz (1975), which deal mainly with the emissive and X-ray modes. *Electron Probe Microanalysis.* The standard monograph is Reed (1975).

References

ALLPRESS, J. G. and SANDERS, J. V. (1973) The direct observation of the structure of real crystals by lattice imaging, *J. Appl. Cryst.* **6**, 165–190.

AMELINCKX, S. and DEKEYSER, W. (1959) The structure and properties of grain boundaries, *Solid State Phys.* **8**, 325.

AMELINCKX, S., GEVERS, R., and VAN LANDUYT, J. (eds.) (1978) *Diffraction and Imaging Techniques in Materials Science*, Vols. I and II, North-Holland, Amsterdam.

AUTHIER, A. (1972) X-ray topography as a tool in crystal growth studies, *J. Cryst. Growth* **13/14**, 34.

BATTERMAN, B. W. and COLE, H. (1964) Dynamical diffraction of X-rays by perfect crystals, *Revs. Mod. Phys.* **36**, 681.

BELK, J. A. (ed.) (1979) *Electron Microscopy and Microanalysis of Crystalline Materials*, Applied Science Publishers, London.

BRÖCHER, W. and PFEFFERKORN, G. E. (1978) Bibliography on cathodoluminescence, in *SEM 1978*, Vol. I, SEM Inc., A. M. F. O'Hare, IL 60666, U.S.A., pp. 33–351.

CASEY, H. C. and KAISER, R. H. (1967) Analysis of n-type GaAs with electron-beam-excited radiative recombination, *J. Electrochem. Soc.* **114**, 149.

CHADWICK, G. A. and SMITH, D. A. (eds.) (1976) *Grain Boundary Structure and Properties*, Academic Press, London.

CHANG, C. G. (1971) Auger electron spectroscopy, *Surf. Sci.* **25**, 53.

CREWE, A. V. (1974) Scanning microscopes: the approach to 1 Å, in *Quantitative Scanning Electron Microscopy* (D. B. Holt, M. D. Muir, P. R. Grant, and I. M. Boswarva, eds.), Academic Press, London.

CREWE, A. V., EGGENBERGER, D. N., WALL, J., and WELTER, L. M. (1968a) Electron gun using a field emission source, *Rev. Sci. Instrum.* **38**, 576.

CREWE, A. V., WALL, J., and WELTER, L. M. (1968b) A high-resolution scanning transmission electron microscope, *J. Appl. Phys.* **39**, 5861.

CREWE, A. V., WALL, J., and LONGMORE, J. (1970) Visibility of single atoms, *Science* **168**, 1338.

CZAJA, W. (1966) Response of Si and GaP p–n junctions to a 5 to 40 keV electron beam, *J. Appl. Phys.* **37**, 4236.

DASH, W. C. (1957) The observation of dislocations in silicon, in *Dislocations and Mechanical Properties of Crystals* (Fisher, J. C., Johnston, W. G., Thomson, R., and Vreeland, T., eds.), Wiley, New York.

DASH, W. C. (1958) The growth of silicon crystals free from dislocations, in *Growth and Perfection of Crystals* (Doremus, R. H., Roberts, B. W., and Turnbull, D., eds.), Wiley, New York.

DESSEAUX, J., RENAULT, A., and BOURRET, A. (1977) Multi-beam lattice images from germanium oriented in (011), *Phil. Mag.* **35**, 357–372.

DONALATO, C. (1979) Contract formation in SEM charge-collection images of semiconductor defects, in *SEM 1979*, Vol. I, SEM Inc., A. M. F. O'Hare, IL 60666, pp. 257–266 and 274.

EDINGTON, J. W. (1974, 1975a, 1975b) *Practical Electron Microscopy in Materials Science*, Vols. I, II, and III, Macmillan, Philips Tech. Library, London and Basingstoke.

FATHY, D., SPARROW, T. G., and VALDRE, U. (1980) Observation of dislocations and microplasma sites in semiconductors by direct correlations of STEBIC, STEM and ELS, *J. Microscopy* **118**, 263–273.

FAUST, J. W. (1962) Etching of the III–V intermetallic compounds, in *Compound Semiconductors*, **1**, *Preparation of III–V Compounds* (Willardson, R. K. and Goering, H. L., eds.), Reinhold, New York.

FRANK, F. C. (1958) On the kinematic theory of crystal growth and dissolution processes, in *Growth and Perfection of Crystals* (Doremus, R. H., Roberts, B. W., and Turnbull, D., eds.), Wiley, New York.

GATOS, H. C. and LAVINE, M. C. (1964) Chemical behaviour of semiconductors: etching characteristics, *Prog. Semicond.* **9**, 1.

GEDKE, D. A. (1974) The Si(Li) X-ray energy spectrometer for X-ray microanalysis, in *Quantitative Scanning Electron Microscopy* (D. B. Holt, M. D. Muir, P. R. Grant, and I. M. Boswarva, eds.), Academic Press, London.

GIFKINS, R. C. (1970) *Optical Microscopy of Metals*, Pitman, London.

GOLDSTEIN, J. I. and YAKOWITZ, H. (eds.) (1975) *Practical Scanning Electron and Iron Microprobe Analysis*, Plenum Press, New York.

GOMER, R. (ed.) (1975) *Interactions on Metal Surfaces*, Springer-Verlag, Berlin.

GOPINATH, A. (1974) The emissive mode, in *Quantitative Scanning Electron Microscopy* (D. B. Holt, M. D. Muir, P. R. Grant, and I. M. Boswarva, eds.), Academic Press, London.

GRIGSON, C. W. B. (1969) Studies of polycrystalline films by electron beams, in *Adv. Electron. El. Phys.* Suppl. **4** (Marton, L. and El Kareh, A. B., eds.), Academic Press, New York.

HAAS, T. W., DOOLEY, G. J., and GRANT, J. T. (1971) Bibiography of LEED and Auger, *Prog. in Surf. Sci.* **1**, 155.

HASHIMOTO, H., HOWIE, A., and WHELAN, M. J. (1962) Anomalous electron absorption effects in metal foils: theory and comparison with experiment, *Proc. Roy. Soc.* A, **269**, 80.

HEINRICH, K. F. J. (ed.) (1968) *Quantitative Electron Probe Microanalysis*, National Bureau of Standards Special Publication 298.

HIRSCH, P. B., HOWIE, A., NICHOLSON, R. B., PASHLEY, D. W., and WHELAN, M. J. (1965) *Electron Microscopy of Thin Crystals*, Butterworths, London.

HIRTH, J. P. and LOTHE, J. (1968) *Theory of Dislocations*, McGraw-Hill, New York.

HOLMES, P. J. (1962) Practical applications of chemical etching, in *The Electrochemistry of Semiconductors* (Holmes, P. J., ed.), Academic Press, London.

HOLT, D. B. (1974a) Quantitative conductive mode scanning electron microscopy, in *Quantitative Scanning Electron Microscopy* (D. B. Holt, M. D. Muir, P. R. Grant, and I. M. Boswarva, eds.), Academic Press, London.

HOLT, D. B. (1974b) Quantitative scanning electron microscope studies of cathodoluminescence in adamantine semiconductors, in *Quantitative Scanning Electron Microscopy* (D. B. Holt, M. D. Muir, P. R. Grant, and I. M. Boswarva, eds.), Academic Press, London.

HOLT, D. B. and CHASE, B. D. (1968) Scanning-electron-beam-excited charge collection micrography of GaAs lasers. *J. Mater. Sci.* **3**, 178.

HOLT, D. B. and DATTA, S. (1980) The cathodoluminescent mode as an analytical technique: its development and prospects, in *SEM 1980*, SEM Inc., A. M. F. O'Hare, IL 60666, U.S.A. (to be published).

HOLT, D. B., MUIR, M. D., GRANT, P. R., and BOSWARVA, I. M. (eds.) (1974) *Quantitative Scanning Electron Microscopy*, Acdemic Press, London.

HOWELL, P. R. and RALPH, B. (1974) The topographical structure of grain boundaries in tungsten, *J. Microscopy* **102**, 361–370.

HOWIE, A. (1974) Theory of diffraction contrast effects in the scanning electron microscope, in *Quantitative Scanning Electron Microscopy* (D. B. Holt, M. D. Muir, P. R. Grant, and I. M. Boswarva, eds.), Academic Press, London.

IIZUI, K., FURUNO, S., and OTSU, H. (1977) Observations of crystal structure images of silicon, *J. Electron Microscopy*, **26**, 129–132.

IRVING, B. A. (1962) Chemical etching of semiconductors, *The Electrochemistry of Semiconductors* (Holmes, P. J., ed.), Academic Press, London.

JAMES, R. W. (1963) The dynamical theory of X-ray diffraction, *Solid State Phys.* **15**, 53.

JANSSEN, A. P., VENABLES, J. A., HWANG, J. C. M., and BALUFFI, R. W. (1977) Direct observation of grain boundary diffusion by scanning Auger microscopy, *Phil. Mag.* **36**, 1537–1540.

JOHNSTON, W. G. (1962) Dislocation etch pits in non-metallic crystals, with bibliography, *Prog. in Ceram. Sci.* **2**, 1.

KIMERLING, L. C. (1978) *Defect State Microscopy in Physics of Semiconductors 1978*, Conf. Series No. 43, Inst. Phys., Bristol and London, pp. 113–122.

KLEMPERER, O. and BARNETT, M. E. (1971) *Electron Optics*, Cambridge University Press.

KRIVANEK, O. L., ISADA, S., and KOBAYASHI, K. (1977) Lattice imaging of a grain boundary in crystalline germanium, *Phil. Mag.* **36**, 931–940.

KYSER, D. F. and WITTRY, D. B. (1964) Cathodoluminescence in gallium arsenide, *The Electron Microprobe* (McKinley, T. D., Heinrich, K. F. J., and Wittry, D. B., eds.), Wiley, New York.

LANDER, J. J., SCHREIBER, H., BUCK, T. M., and MATTHEWS, J. R. (1963) Microscopy of internal crystal imperfections in Si p–n junction diodes by use of electron beams, *Appl. Phys. Letters* **3**, 206.

LIFSHIN, E. (1974) X-ray generation and detection in the SEM, in *Scanning Electron Microscopy* by O. C. Wells, McGraw-Hill, New York.

LONG, J. V. P. (1977) Electron probe microanalysis, in *Physical Methods in Determination Mineralogy* (J. Zussman, ed.), Academic Press, London.

LORETTO, M. H. and SMALLMAN, R. E. (1975) *Defect Analysis in Electron Microscopy*, Chapman & Hall, London.

MINKOFF, I. and NIXON, W. C. (1966) Scanning electron microscopy of graphite growth in iron and nickel alloys, *J. Appl. Phys.* **37**, 4848.

MUIR, M. D. and GRANT, P. R. (1974) Cathodoluminescence, in *Quantitative Scanning Electron Microscopy* (D. B. Holt, M. D. Muir, P. R. Grant, and I. M. Boswarva, eds.), Academic Press, London.

MUNAKATA, C. (1967) An electron beam method of measuring resistivity distribution in semiconductors, *Japan. J. Appl. Phys.* **6**, 963.

MUNAKATA, C. (1968a) Voltage signal due to electron-beam-induced conductivity in semiconductors, *Japan. J. Appl. Phys.* **7**, 1051.

MUNAKATA, C. (1968b) An application of beta conductivity to the measurement of resistivity distribution, *J. Phys. E. Sci. Instrum.* **1**, 639.

MUNAKATA, C. (1969) Scanning electron micrograph using beta-conductive signal, *J. Phys. E. Sci. Instrum.* **2**, 738.

OATLEY, C. W. (1972) *The Scanning Electron Microscope: Part 1, The Instrument*, Cambridge University Press.

PENNING, P. (1966) Theory of X-ray diffraction in unstrained and lightly strained perfect crystals, *Philips Res. Rpts. Suppl.* No. **5**, 1.

PENNYCOOK, S. J., BROWN, L. M., and CRAVEN, A. J. (1980) Observation of cathodoluminescence at single disolcations by STEM, *Phil. Mag.* **A41**, 589–600.

PETROFF, P. M., LOGAN, R. A., and SAVAGE, A. (1980) Nonradiative recombination at dislocations in III–V compound semiconductors, *J. Microscopy* **118**, 255–261.

REED, S. J. B. (1975) *Electron Microprobe Analysis*, Cambridge University Press, Cambridge.

SCHWUTTKE, G. H. (1962a) Direct observation of imperfections in semiconductor crystals by anomalous transmission of X-rays, *J. Appl. Phys.* **33**, 2760.

SCHWUTTKE, G. H. (1962b) X-ray diffraction microscopy of impurities in silicon single crystals by extinction contrast, in *Direct Observation of Imperfections in Crystals* (NewKirk, J. B. and Wernick, J. H., eds.), Interscience, New York.

SCHWUTTKE, G. H. (1965) New X-ray diffraction microscopy technique for the study of imperfections in semiconductor crystals, *J. Appl. Phys.* **36**, 2712.

SHAW, D. A. and THORNTON, P. R. (1968) Cathodoluminescence studies of laser quality GaAs, *J. Mater. Sci.* **3**, 507.

SOUTHWORTH, H. N. (1975) *Introduction to Modern Microscopy*, Wykeham Publications, London.

SPARROW, T. G. and VALDRE, U. (1977) Application of scanning transmission electron microscopy to semiconductor devices, *Phil. Mag.* **36**, 1517–1528.

SUTFIN, L. V. and OGILVIE, R. E. (1970) A comparison of X-ray analysis techniques available for scanning electron microscopes, in *Scanning Electron Microscopy, Proc. Third Annual SEM Symposium Chicago*, IITRI, Chicago, p. 17.

TAKAGI, S. (1962) Dynamical theory of diffraction applicable to crystals with any kind of small distortion, *Acta Cryst.* **15**, 1311.

TANNER, B. K. (1976) *X-ray Diffraction Topography*, Pergamon Press, Oxford.

TANNER, B. K. (1977) Crystal assessment by X-ray topography using synchrotron radiation, *Prog. Cryst. Growth Characterization* **1**, 23–55.

TOLANSKY, S. (1948) *Multiple Beam Interferometry*, Clarendon Press, Oxford.

TOLANSKY, S. (1955) *An Introduction to Interferometry*, Longmans, Green, London.

TOMPSETT, M. F. (1972) Scanning high energy electron diffraction (SHEED) in materials science, *J. Mater. Sci.* **7**, 1069.

TOUSMIS, A. J. and MARTON, L. (eds.) (1969) *Electron Probe Microanalysis* (Supplement 6 to *Adv. in Electronics and El. Phys.*), Academic Press, New York.

TREITZ, N. (1977) Analysis of solid surface monolayers by mass and energy spectrometry methods, *J. Phys. E. (Sci. Instrum.)* **10**, 573–585.

UEDA, R. and MULLIN, J. B. (eds.) (1975) *Crystal Growth and Characterization*, North-Holland, Amsterdam.

VOGEL, F. L., PFANN, W. G., COREY, H. E., and THOMAS, E. E. (1953) Observations of dislocations in lineage boundaries in germanium, *Phys. Rev.* **90**, 489.

WAREKOIS, E. P., LAVINE, M. C., MARIANO, A. N., and GATOS, H. C. (1962) Crystallographic polarity in the II–VI compounds, *J. Appl. Phys.* **33**, 690.

WELLS, O. C. (1974) *Scanning Electron Microscopy*, McGraw-Hill, New York.

WINTERBOTTOM, A. B. and McLEAN, D. (1960) *The Physical Examination of Metals* (Chalmers, B. and Quarrel, A. G., eds.), Arnold, London.

WITTRY, D. B. (1966) Cathodoluminescence and impurity variations in Te-doped GaAs, *Appl. Phys. Letters*, **8**, 142.

WITTRY, D. B. and KYSER, D. F. (1965) Cathodoluminescence at p–n junctions in GaAs, *J. Appl. Phys.* **36**, 1387.

Subject Index

(Page references in italics refer to important items)

Organic Compounds Index

Inorganic Compounds Index

(Page references in italics refer to important items)

Table of Fundamental Physico-chemical Constants

B. R. Pamplin

Velocity of light in free space	$c = (\varepsilon_0\mu_0)^{-1/2}$		$2.99792458 \times 10^8\,\mathrm{m\,s^{-1}}$
Permittivity of free space	ε_0		$8.85418782 \times 10^{-12}\,\mathrm{F\,m^{-1}}$
Permeability of free space	μ_0	$4\pi \times 10^{-7} =$	$1.25663706 \times 10^{-6}\,\mathrm{H\,m^{-1}}$
Elementary charge	e		$1.6021892 \times 10^{-19}\,\mathrm{C}$
Atomic mass unit	$= 10^{-3}\,\mathrm{kg\,mol^{-1}}\,N^{-1}$		$1.6605655 \times 10^{-27}\,\mathrm{kg}$
Avogadro's constant	N		$6.022045 \times 10^{23}\,\mathrm{mol^{-1}}$
Electron rest mass	m_e		$0.9109534 \times 10^{-30}\,\mathrm{kg}$
Proton rest mass	m_p		$1.6726485 \times 10^{-27}\,\mathrm{kg}$
Neutron rest mass	m_n		$1.6749543 \times 10^{-27}\,\mathrm{kg}$
Specific electronic charge	e/m_e		$1.7588047 \times 10^{11}\,\mathrm{C\,kg^{-1}}$
Faraday's constant	$F = Ne$		$9.648456 \times 10^4\,\mathrm{C\,mol^{-1}}$
Planck's constant	h		$6.626176 \times 10^{-34}\,\mathrm{J\,Hz^{-1}}$
	$\hbar = h/2\pi$		$1.0545887 \times 10^{-34}\,\mathrm{J\,s}$
Bohr radius	a_0		$0.52917706 \times 10^{-10}\,\mathrm{m}$
Molar gas constant	R		$8.31441\,\mathrm{J\,mol^{-1}\,K^{-1}}$
Molar volume (ideal gas at NTP)	$V = \dfrac{RT_0}{P_0}$		$0.02241383\,\mathrm{m^3\,mol^{-1}}$
Boltzmann's constant	$k = R/N$		$1.380662 \times 10^{-23}\,\mathrm{J\,K^{-1}}$
Stefan Boltzmann constant	$\sigma = \dfrac{\pi^2\,k^4}{60\,\hbar^3 c}$		$5.67032\,\mathrm{W\,m^{-2}\,K^{-4}}$
Gravitational constant	G		$6.6720\,\mathrm{N\,m^2\,kg^{-2}}$

Table of Useful Conversions of Units

1 dyne	$= 1\,\mathrm{cm\,g/sec^2}$
1 erg	$= 1\,\mathrm{dyne\,cm}$
	$= 1\,\mathrm{cm^2\,g/sec^2}$
	$= 6.24 \cdot 10^{11}\,\mathrm{eV}$
1 dyne/cm^2	$= 1\,\mathrm{g/cm\,sec^2}$
1 g (gram)	$= 6.24 \times 10^{11}\,\mathrm{eV\,sec^2/cm^2}$
1 kcal/mole	$= 4.33 \times 10^{-2}\,\mathrm{eV}$
1 atm	$= 760\,\mathrm{mmHg}$
	$= 760\,\mathrm{Torr}$
	$= 1.01 \times 10^6\,\mathrm{dyne/cm^2}$
1 bar	$= 10^6\,\mathrm{dyne/cm}$
	$= 0.987\,\mathrm{atm}$
1 cal	$= 4.185 \times 10^7\,\mathrm{erg}$
1 Joule	$= 10^7\,\mathrm{erg}$
	$= 1\,\mathrm{W\,sec}$
	$= 1\,\mathrm{VA\,sec}$
	$= 0.238\,\mathrm{cal}$
1 Coulomb	$= 1\,\mathrm{A\,sec}$
1 eV	$= 1.6 \times 10^{-12}\,\mathrm{erg}$
	$= 1.6 \times 10^{-12}\,\mathrm{dyne\,cm}$
	$= 23.06\,\mathrm{kcal/mole}$
1 Farad (F)	$= 1\,\mathrm{A\,sec/V}$
1 Gauss (G)	$= 10^{-8}\,\mathrm{V\,sec/cm^2}$
1 A	$= 1.036 \times 10^{-5}\,\mathrm{Faraday/sec}$
1 cm	$= 10^4\,\mu\mathrm{m}$
1 cm^2/sec	$= 3.6 \times 10^{11}\,\mu\mathrm{m^2/hr}$
1 μm/min	$= 1.67 \times 10^2\,\mathrm{\AA/sec}$
1 mil	$= 25.4\,\mu\mathrm{m}$
$\log x$	$= 0.4343 \ln x$
	$(\log 10 = 1,\ \ln e = 1;$
	$\ln 10 = 2.30)$

Table of Energy Conversion Factors

		J	eV	cm⁻¹	Hz	°K	amu	kg	Cal	Btu
1 Joule	=	1	6.242 (18)	5.035 (22)	1.509 (33)	7.244 (22)	6.705 (9)	1.113 (−17)	0.2389	9.481 (−4)
1 eV	=	1.602 (−19)	1	8.066 (3)	2.418 (14)	1.161 (4)	1.074 (−9)	1.783 (−36)	3.827 (−20)	1.519 (−22)
1 cm⁻¹	=	1.986 (−23)	1.240 (−4)	1	2.998 (10)	1.439	1.331 (−13)	2.210 (−40)	4.745 (−24)	1.883 (−26)
1 Hz	=	6.626 (−34)	4.136 (−15)	3.336 (−11)	1	4.799 (−11)	4.443 (−24)	7.372 (−51)	1.583 (−34)	6.282 (−37)
1 °K	=	1.381 (−23)	8.617 (−5)	6.950 (−1)	2.084 (10)	1	9.256 (−14)	1.536 (−40)	3.298 (−24)	1.309 (−26)
1 amu	=	1.492 (−10)	9.315 (8)	7.513 (12)	2.251 (23)	1.080 (13)	1	1.660 (−27)	3.564 (−11)	1.415 (−13)
1 kg	=	8.987 (16)	5.610 (35)	4.525 (39)	1.356 (50)	6.510 (39)	6.025 (26)	1	2.147 (16)	8.521 (13)
1 Calorie	=	4.186	2.613 (19)	2.108 (23)	6.317 (33)	3.032 (23)	2.807 (10)	4.659 (−17)	1	3.968 (−3)
1 Btu	=	1.055 (3)	6.585 (21)	5.311 (25)	1.592 (36)	7.645 (25)	7.074 (12)	1.174 (−14)	252	1

Periodic Table (reproduced from the *Concise Dictionary of Physics and Related Subjects*, 2nd Edition by J. Thewlis, Pergamon Press, 1979).

			← Transition Elements — d →														Inert Gases
Group	**IA**	**IIA**	**IIIB**	**IVB**	**VB**	**VIB**	**VIIB**	**VIII**			**IB**	**IIB**	**IIIA**	**IVA**	**VA**	**VIA**	**VIIA**	**0**
Valence shell	s^1	s^2	$d^1s^2f^x$	d^2s^2	$(d^3s^2)^\dagger$	$(d^5s^1)^\dagger$	d^5s^2	$(d^6s^2)^\dagger$	$(d^7s^2)^\dagger$	$(d^8s^2)^\dagger$	s^1d^{10}	s^2	s^2p^1	s^2p^2	s^2p^3	s^2p^4	s^2p^5	s^2p^6
Principal quantum number (Period)																		
$n=1$ $1s$	1 **H** 1.008																	2 **He** 4.003
$n=2$ $2s2p$	3 **Li** 6.94	4 **Be** 9.01											5 **B** 10.81	6 **C** 12.01	7 **N** 14.01	8 **O** 15.999	9 **F** 18.99	10 **Ne** 20.18
$n=3$ $3s3p$	11 **Na** 22.99	12 **Mg** 24.31											13 **Al** 26.98	14 **Si** 28.09	15 **P** 30.97	16 **S** 32.06	17 **Cl** 35.45	18 **Ar** 39.95
$n=4$ $4s3d4p$	19 **K** 39.10	20 **Ca** 40.08	21 **Sc** 44.96	22 **Ti** 47.90	23 **V** 50.94	24 **Cr** 51.996	25 **Mn** 54.94	26 **Fe** 55.85	27 **Co** 58.93	28 **Ni** 58.71	29 **Cu** 63.54	30 **Zn** 65.37	31 **Ga** 69.72	32 **Ge** 72.59	33 **As** 74.92	34 **Se** 78.96	35 **Br** 79.91	36 **Kr** 83.80
$n=5$ $5s4d5p$	37 **Rb** 85.47	38 **Sr** 87.62	39 **Y** 88.91	40 **Zr** 91.22	41 **Nb** 92.91	42 **Mo** 95.94	43 **Tc** 99	44 **Ru** 101.07	45 **Rh** 102.91	46 **Pd** 106.4	47 **Ag** 107.87	48 **Cd** 112.40	49 **In** 114.82	50 **Sn** 118.69	51 **Sb** 121.75	52 **Te** 127.60	53 **I** 126.90	54 **Xe** 131.30
$n=6$ $6s4f5d6p$	55 **Cs** 132.91	56 **Ba** 137.34	57* **La** 138.91	72 **Hf** 178.49	73 **Ta** 180.95	74 **W** 183.85	75 **Re** 186.2	76 **Os** 190.2	77 **Ir** 192.2	78 **Pt** 195.09	79 **Au** 196.97	80 **Hg** 200.59	81 **Tl** 204.37	82 **Pb** 207.19	83 **Bi** 208.98	84 **Po**	85 **At**	86 **Rn**
$n=7$ $7s5f6d7p$	87 **Fr**	88 **Ra** 226	89⊙ **Ac**															

Inner Transition Elements — f

***Lanthanide Series**	58 **Ce** 140.12	59 **Pr** 140.91	60 **Nd** 144.24	61 **Pm** 145	62 **Sm** 150.35	63 **Eu** 151.96	64 **Gd** 157.25	65 **Tb** 158.92	66 **Dy** 162.50	67 **Ho** 164.93	68 **Er** 167.26	69 **Tm** 168.93	70 **Yb** 173.04
													71 **Lu** 174.97
⊙**Actinide Series**	90 **Th** 232.04	91 **Pa**	92 **U** 238.03	93 **Np**	94 **Pu**	95 **Am**	96 **Cm**	97 **Bk**	98 **Cf**	99 **Es**	100 **Fm**	101 **Md**	102 **No**
													103 **Lr**

$\dagger$ Variable valence shells

609